Progress in Mathematical Physics
Volume 26

Editors-in-Chief
Anne Boutet de Monvel, *Université Paris VII Denis Diderot*
Gerald Kaiser, *The Virginia Center for Signals and Waves*

Editorial Board
D. Bao, *University of Houston*
C. Berenstein, *University of Maryland, College Park*
P. Blanchard, *Universität Bielefeld*
A.S. Fokas, *Imperial College of Science, Technology and Medicine*
C. Tracy, *University of California, Davis*
H. van den Berg, *Wageningen University*

Philippe Blanchard
Erwin Brüning

Mathematical Methods in Physics

Distributions, Hilbert Space Operators, and Variational Methods

Birkhäuser
Boston • Basel • Berlin

Philippe Blanchard
University of Bielefeld
Faculty of Physics
Bielefeld, 33615
Germany

Erwin Brüning
University of Durban—Westville
Department of Mathematics
 and Applied Mathematics
Durban, 4000
South Africa

Library of Congress Cataloging-in-Publication Data

Blanchard, Philippe.
  Mathematical methods in physics : distributions, Hilbert space operators, and
variational methods / Philippe Blanchard and Erwin Brüning.
   p. cm.– (Progress in mathematical physics ; v. 26)
  Includes bibliographical references and index.
  ISBN 0-8176-4228-5 (alk. paper) — ISBN 3-7643-4228-5 (alk. paper)
  1. Mathematical physics. I. Brüning, Erwin. II. Title. III. Series.

QC20.7.B545 2002
530.15–dc21
                                                              2002074361
                                                                    CIP

AMS Subject Classifications: Primary: 46-01, 47-01, 49-01. Secondary: 46A03, 46C05, 46Fxx, 46Nxx, 49R50, 49Jxx, 81Q10. Tertiary: 26ER15, 26E02, 34B05, 34B15, 35D05, 35Jxx, 35Qxx

Printed on acid-free paper.
©2003 Birkhäuser Boston   *Birkhäuser*

Based on German edition *Distribitionen und Hilbertraumoperatorem—Mathematische Methoden der Physik*, SV Vienna, 1993.

All rights reserved. This work may not be translated or copied in whole or in part without the written permission of the publisher (Birkhäuser Boston, c/o Springer-Verlag New York, Inc., 175 Fifth Avenue, New York, NY 10010, USA), except for brief excerpts in connection with reviews or scholarly analysis. Use in connection with any form of information storage and retrieval, electronic adaptation, computer software, or by similar or dissimilar methodology now known or hereafter developed is forbidden.
The use of general descriptive names, trade names, trademarks, etc., in this publication, even if the former are not especially identified, is not to be taken as a sign that such names, as understood by the Trade Marks and Merchandise Marks Act, may accordingly be used freely by anyone.

ISBN 0-8176-4228-5   SPIN 10832409
ISBN 3-7643-4228-5

Typeset by the authors.
Printed in the United States of America.

9 8 7 6 5 4 3 2 1

Birkhäuser   Boston • Basel • Berlin
A member of BertelsmannSpringer Science+Business Media GmbH

*Dedicated to the memory of
Yurko Vladimir Glaser and Res Jost,
mentors and friends*

Contents

Preface · xv

Notation · xvii

I Distributions · 1

1 Introduction · 3

2 Spaces of Test Functions · 7
  2.1 Hausdorff locally convex topological vector spaces . . . . · 7
        2.1.1 Examples of HLCTVS . . . . · 14
        2.1.2 Continuity and convergence in a HLCVTVS . . . . · 15
  2.2 Basic test function spaces of distribution theory . . . . · 18
        2.2.1 The test function space $\mathcal{D}(\Omega)$ of $\mathcal{C}^\infty$ functions of compact support . . . . · 19
        2.2.2 The test function space $\mathcal{S}(\Omega)$ of strongly decreasing $\mathcal{C}^\infty$-functions on Ω . . . . · 20
        2.2.3 The test function space $\mathcal{E}(\Omega)$ of all $\mathcal{C}^\infty$-functions on Ω . . · 21
        2.2.4 Relation between the test function spaces $\mathcal{D}(\Omega)$, $\mathcal{S}(\Omega)$, and $\mathcal{E}(\Omega)$ . . . . · 22
  2.3 Exercises . . . . · 22

3 Schwartz Distributions — 27
- 3.1 The topological dual of a HLCTVS 27
- 3.2 Definition of distributions 29
  - 3.2.1 The regular distributions 31
  - 3.2.2 Some standard examples of distributions 33
- 3.3 Convergence of sequences and series of distributions 35
- 3.4 Localization of distributions 40
- 3.5 Tempered distributions and distributions with compact support . . 42
- 3.6 Exercises 44

4 Calculus for Distributions — 47
- 4.1 Differentiation 48
- 4.2 Multiplication 51
- 4.3 Transformation of variables 54
- 4.4 Some applications 56
  - 4.4.1 Distributions with support in a point 56
  - 4.4.2 Renormalization of $(\frac{1}{x})_+ = \frac{\theta(x)}{x}$ 58
- 4.5 Exercises 60

5 Distributions as Derivatives of Functions — 63
- 5.1 Weak derivatives 63
- 5.2 Structure theorem for distributions 65
- 5.3 Radon measures 67
- 5.4 The case of tempered and compactly supported distributions . . . 68
- 5.5 Exercises 70

6 Tensor Products — 71
- 6.1 Tensor product for test function spaces 71
- 6.2 Tensor product for distributions 75
- 6.3 Exercises 81

7 Convolution Products — 83
- 7.1 Convolution of functions 83
- 7.2 Regularization of distributions 87
- 7.3 Convolution of distributions 90
- 7.4 Exercises 96

8 Applications of Convolution — 99
- 8.1 Symbolic Calculus – ordinary linear differential equations 100
- 8.2 Integral equation of Volterra 104
- 8.3 Linear partial differential equations with constant coefficients . . . 105
- 8.4 Elementary solutions of partial differential operators 108
  - 8.4.1 The Laplace operator $\Delta_n = \sum_{i=1}^{n} \frac{\partial^2}{\partial x_i^2}$ in $\mathbb{R}^n$ 108
  - 8.4.2 The PDE operator $\frac{\partial}{\partial t} - \Delta_n$ of the heat equation in $\mathbb{R}^{n+1}$. 110
  - 8.4.3 The wave operator $\Box_4 = \partial_0^2 - \Delta_3$ in $\mathbb{R}^4$ 111
- 8.5 Exercises 113

9 Holomorphic Functions — 115
- 9.1 Hypo-ellipticity of $\bar{\partial}$ 115
- 9.2 Cauchy theory 118
- 9.3 Some properties of holomorphic functions 121
- 9.4 Exercises 126

10 Fourier Transformation — 127
- 10.1 Fourier transformation for integrable functions 128
- 10.2 Fourier transformation on $\mathcal{S}(\mathbb{R}^n)$ 134
- 10.3 Fourier transformation for tempered distributions 137
- 10.4 Some applications 143
  - 10.4.1 Examples of tempered elementary solutions 145
  - 10.4.2 Summary of properties of the Fourier transformation 148
- 10.5 Exercises 149

11 Distributions and Analytic Functions — 153
- 11.1 Distributions as boundary values of analytic functions 153
- 11.2 Exercises 157

12 Other Spaces of Generalized Functions — 159
- 12.1 Generalized functions of Gelfand type $\mathcal{S}$ 160
- 12.2 Hyperfunctions and Fourier hyperfunctions 164
- 12.3 Ultradistributions 167

II Hilbert Space Operators — 171

13 Hilbert Spaces: A Brief Historical Introduction — 173
- 13.1 Survey: Hilbert spaces 173
- 13.2 Some historical remarks 179
- 13.3 Hilbert spaces and Physics 181

14 Inner Product Spaces and Hilbert Spaces — 185
- 14.1 Inner product spaces 185
  - 14.1.1 Basic definitions and results 186
  - 14.1.2 Basic topological concepts 190
  - 14.1.3 On the relation between normed spaces and inner product spaces 192
  - 14.1.4 Examples of Hilbert spaces 193
- 14.2 Exercises 196

15 Geometry of Hilbert Spaces — 199
- 15.1 Orthogonal complements and projections 199
- 15.2 Gram determinants 203
- 15.3 The dual of a Hilbert space 205
- 15.4 Exercises 209

16 Separable Hilbert Spaces — 211
16.1 Basic facts 211
16.2 Weight functions and orthogonal polynomials 217
16.3 Examples of complete orthonormal systems for $L^2(I, \rho dx)$ 221
16.4 Exercises 223

17 Direct Sums and Tensor Products — 227
17.1 Direct sums of Hilbert spaces 227
17.2 Tensor products 229
17.3 Some applications of tensor products and direct sums 232
    17.3.1 State space of particles with spin 232
    17.3.2 State space of multi-particle systems 233
17.4 Exercises 234

18 Topological Aspects — 235
18.1 Compactness 235
18.2 The weak topology 237
18.3 Exercises 245

19 Linear Operators — 247
19.1 Basic facts 247
19.2 Adjoints, closed and closable operators 250
19.3 Symmetric and self-adjoint operators 256
19.4 Examples 259
    19.4.1 Operator of multiplication 259
    19.4.2 Momentum operator 260
    19.4.3 Free Hamilton operator 261
19.5 Exercises 262

20 Quadratic Forms — 265
20.1 Basic concepts. Examples 265
20.2 Representation of quadratic forms 268
20.3 Some applications 271
20.4 Exercises 274

21 Bounded Linear Operators — 275
21.1 Preliminaries 275
21.2 Examples 277
21.3 The space $\mathfrak{L}(\mathcal{H}, \mathcal{K})$ of bounded linear operators 281
21.4 The C^*-algebra $\mathfrak{B}(\mathcal{H})$ 283
21.5 Calculus in the C^*-algebra $\mathfrak{B}(\mathcal{H})$ 286
    21.5.1 Preliminaries 286
    21.5.2 Polar decomposition of operators 288
21.6 Exercises 289

22 Special Classes of Bounded Operators — 293
- 22.1 Projection operators 293
- 22.2 Unitary operators 297
    - 22.2.1 Isometries 297
    - 22.2.2 Unitary operators 297
    - 22.2.3 Examples of unitary operators 300
- 22.3 Compact operators 300
- 22.4 Trace class operators 304
- 22.5 Some applications in Quantum Mechanics 308
- 22.6 Exercises 311

23 Self-adjoint Hamilton Operators — 313
- 23.1 Kato perturbations 314
- 23.2 Kato perturbations of the free Hamiltonian 315
- 23.3 Exercises 316

24 Elements of Spectral Theory — 317
- 24.1 Basic concepts and results 318
- 24.2 The spectrum of special operators 322
- 24.3 Comments on spectral properties of linear operators 324
- 24.4 Exercises 325

25 Spectral Theory of Compact Operators — 327
- 25.1 The results of Riesz and Schauder 327
- 25.2 The Fredholm alternative 329
- 25.3 Exercises 331

26 The Spectral Theorem — 333
- 26.1 Geometric characterization of self-adjointness 334
    - 26.1.1 Preliminaries 334
    - 26.1.2 Subspaces of controlled growth 335
- 26.2 Spectral families and their integrals 340
    - 26.2.1 Spectral families 341
    - 26.2.2 Integration with respect to a spectral family 342
- 26.3 The spectral theorem 347
- 26.4 Some applications 351
- 26.5 Exercises 353

27 Some Applications of the Spectral Representation — 355
- 27.1 Functional calculus 355
- 27.2 Decomposition of the spectrum – Spectral subspaces 357
- 27.3 Interpretation of the spectrum of a self-adjoint Hamiltonian 364
- 27.4 Exercises 369

III Variational Methods   371

28 Introduction   373
28.1 Roads to Calculus of Variations 374
28.2 Classical approach versus direct methods 375
28.3 The objectives of the following chapters 378

29 Direct Methods in the Calculus of Variations   379
29.1 General existence results 379
29.2 Minimization in Banach spaces 381
29.3 Minimization of special classes of functionals 383
29.4 Exercises 384

30 Differential Calculus on Banach Spaces and Extrema of Functions   387
30.1 The Fréchet derivative 388
30.2 Extrema of differentiable functions 393
30.3 Convexity and monotonicity 395
30.4 Gâteaux derivatives and variations 397
30.5 Exercises 401

31 Constrained Minimization Problems (Method of Lagrange Multipliers)   403
31.1 Geometrical interpretation of constrained minimization 404
31.2 Tangent spaces of level surfaces 405
31.3 Existence of Lagrange multipliers 407
      31.3.1 Comments on Dido's problem 409
31.4 Exercises 410

32 Boundary and Eigenvalue Problems   413
32.1 Minimization in Hilbert spaces 413
32.2 The Dirichlet–Laplace operator and other elliptic differential operators 416
32.3 Nonlinear convex problems 420
32.4 Exercises 426

33 Density Functional Theory of Atoms and Molecules   429
33.1 Introduction 429
33.2 Semi-classical theories of density functionals 431
33.3 Hohenberg–Kohn theory 432
      33.3.1 Hohenberg–Kohn variational principle 435
      33.3.2 The Kohn–Sham equations 437
33.4 Exercises 438

IV Appendix   **439**

A Completion of Metric Spaces   **441**

B Metrizable Locally Convex Topological Vector Spaces   **445**

C The Theorem of Baire   **447**
   C.1 The uniform boundedness principle 449
   C.2 The open mapping theorem 452

D Bilinear Functionals   **455**

   References   **457**

   Index   **465**

Preface

Courses in modern theoretical physics have to assume some basic knowledge of the theory of generalized functions (in particular distributions) and of the theory of linear operators in Hilbert spaces. Accordingly the Faculty of Physics of the University of Bielefeld offered a compulsory course *Mathematische Methoden der Physik* for students in the second semester of the second year which now has been given for many years. This course has been offered by the authors over a period of about ten years. The main goal of this course is to provide basic mathematical knowledge and skills as they are needed for modern courses in quantum mechanics, relativistic quantum field theory and related areas. The regular repetitions of the course allowed, on the one hand, testing of a number of variations of the material and on the other hand the form of the presentation. From this course the book *Distributionen und Hilbertraumoperatoren. Mathematische Methoden der Physik. Springer-Verlag Wien, 1993* emerged. The present book is a translated, considerably revised and extended version of this book. It contains much more than this course since we added many detailed proofs, many examples and exercises as well as hints linking the mathematical concepts or results to the relevant physical concepts or theories.

This book addresses students of physics who are interested in a conceptually and mathematically clear and precise understanding of physical problems, and it addresses students of mathematics who want to learn about physics as a source and as an area of application of mathematical theories, i.e., all those students with interest in the fascinating interaction between physics and mathematics.

It is assumed that the reader has a solid background in analysis and linear algebra (in Bielefeld this means three semesters of analysis and two of linear algebra). On this basis the book starts in Part A with an introduction to basic linear functional analysis as needed for the Schwartz theory of distributions and continues in Part B with the particularities of Hilbert spaces and the core aspects of the theory of linear operators in Hilbert spaces. Part C develops the basic mathematical foundations for modern computations of the ground state energies and charge densities in atoms and molecules, i.e., basic aspects of the direct methods of the calculus of variations including constrained minimization. A powerful strategy for solving linear and nonlinear boundary and eigenvalue problems, which covers the Dirichlet problem and its nonlinear generalizations, is presented as well. An appendix gives detailed proofs of the fundamental principles and results of functional analysis to the extent they are needed in our context.

With great pleasure we would like to thank all those colleagues and friends who have contributed to this book through their advice and comments, in particular G. Bolz, J. Loviscach, G. Roepstorff and J. Stubbe. Last but not least we thank the editorial team of Birkhäuser – Boston for their professional work.

Bielefeld and Durban *Ph. Blanchard*
June 2002 *E. Brüning*

Notation

$\mathbb{N}$	the natural numbers
$\mathbb{R}$	field of real numbers
$\mathbb{C}$	field of complex numbers
$\mathbb{K}$	field of real or of complex numbers
$\mathbb{R}_+$	the set of nonnegative real numbers
$\mathbb{K}^n$	$\mathbb{K}$ vector space of n-tuples of numbers in $\mathbb{K}$
$A \pm B$	$\{a \pm b;\ a \in A;\ b \in B\}$ for subsets A and B of a vector space V
ΛM	$\{\lambda \cdot u;\ \lambda \in \Lambda,\ u \in M\}$ for a subset $\Lambda \subset \mathbb{K}$ and a subset M of a vector space V over $\mathbb{K}$
$A \backslash B$	the set of all points in a set A which do not belong to the subset B of A
$\mathcal{C}(\Omega) = \mathcal{C}(\Omega; \mathbb{K})$	vector space of all continuous functions $f : \Omega \to \mathbb{K}$, for an open set $\Omega \subset \mathbb{K}^n$
supp f	support of the function f
$\mathcal{C}_0(\Omega)$	vector space of all continuous functions $f : \Omega \to \mathbb{K}$ with compact support in Ω
$\mathcal{C}^k(\Omega)$	vector space of all functions which have continuous derivatives up to order k, for $k = 0, 1, 2, \ldots$

xviii   Notation

$D^\alpha = \frac{\partial^{|\alpha|}}{\partial x_1^{\alpha_1} \cdots \partial x_n^{\alpha_n}}$	derivative monomial of order $|\alpha| = \alpha_1 + \cdots + \alpha_n$, defined on spaces $\mathcal{C}^k(\Omega)$, for open sets $\Omega \subset \mathbb{R}^n$ and $k \geq |\alpha|$
$\mathcal{D}_K(\Omega)$	vector space of all functions $f : \Omega \to \mathbb{K}$ which have continuous derivatives of any order and which have a compact support $\operatorname{supp} f$ contained in the compact subset $\mathbb{K}$ of $\Omega \subset \mathbb{R}^n$, equipped with the topology of uniform convergence of all derivatives
$\mathcal{D}(\Omega)$	inductive limit of the spaces $\mathcal{D}_K(\Omega)$ with respect to all subsets $K \subset \Omega$, K compact; test function space of all $\mathcal{C}^\infty$-functions $f : \Omega \to \mathbb{K}$ which have a compact support in the open set $\Omega \subset \mathbb{R}^n$
$|x|$	Euclidean norm $\sqrt{x_1^2 + \cdots + x_n^2}$ of the vector $x = (x_1, \ldots, x_n) \in \mathbb{R}^n$
$\mathcal{S}(\Omega)$	test function space of all $\mathcal{C}^\infty$-functions $f : \Omega \to \mathbb{K}$ which, together with all their derivatives decrease faster than $\operatorname{const}.(1 + |x|)^{-k}$ for $k = 0, 1, 2, \ldots$, for some constant and $x \in \Omega$
$\mathcal{E}(\Omega)$	test function space of all $\mathcal{C}^\infty$-functions $f : \Omega \to \mathbb{K}$, equipped with the topology of uniform convergence of all derivatives $f^\alpha = D^{(\alpha)} f$ on all compact subsets K of Ω
lctvs	locally convex topological vector space
hlctvs	Hausdorff locally convex topological vector space
X^*	algebraic dual of a vector space X
X'	topological dual of a topological vector space X
$\mathcal{D}'(\Omega) \equiv \mathcal{D}(\Omega)'$	space of all distributions on the open set $\Omega \subseteq \mathbb{R}^n$
$\mathcal{S}'(\Omega) \equiv \mathcal{S}(\Omega)'$	space of all tempered (i.e., slowly growing) distributions on $\Omega \subseteq \mathbb{R}^n$
$\mathcal{E}'(\Omega) \equiv \mathcal{E}(\Omega)'$	space of all distributions on $\Omega \subseteq \mathbb{R}^n$ with compact support
I_f	the regular distribution defined by the locally integrable function f
$\mathcal{D}'_{\operatorname{reg}}(\Omega)$	the space of all regular distributions on the open set $\Omega \subseteq \mathbb{R}^n$
$\mathcal{D}'_+(\mathbb{R})$	space of all distributions on $\mathbb{R}$ with support in $\mathbb{R}_+$
$L^p(\Omega)$	space of equivalence classes of Lebesgue measurable functions on $\Omega \subseteq \mathbb{R}^n$ for which $|f|^p$ is Lebesgue integrable over Ω; $1 \leq p < \infty$, Ω Lebesgue measurable

$L^\infty(\Omega)$	space of all equivalence classes of Lebesgue measurable functions on Ω which are essentially bounded; $\Omega \subseteq \mathbb{R}^n$ Lebesgue measurable
$p_{K,m}$	for $m = 0, 1, 2, \ldots$, $K \subset \Omega$, K compact, $\Omega \subseteq \mathbb{R}^n$ open, the semi-norm on $\mathcal{D}_K(\Omega)$ defined by $$p_{K,m}(f) = \sup_{|\alpha| \leq m,\, x \in K} |D^\alpha f(x)|$$
$q_{K,m}$	the semi-norm on $\mathcal{D}_K(\Omega)$ defined by $$q_{K,m}(f) = \left(\sum_{|\alpha| \leq m} \int_K |D^\alpha f(x)|^2 dx \right)^{1/2}$$ K, m, Ω as above
$p_{m,k}$	the norm on $\mathcal{S}(\mathbb{R}^n)$ defined by $$p_{m,k}(f) = \sup_{x \in \mathbb{R}^n,\, |\alpha| \leq k} (1+x^2)^{\frac{m}{2}} |D^\alpha f(x)|$$ for $m, k = 0, 1, 2, \ldots$
$B_{p,r}(x_0)$	open ball of radius $r > 0$ and centre x_0, with respect to the semi-norm p
δ_a	Dirac's delta distribution centered at $x = a \in \mathbb{R}^n$; for $a = 0$ we write δ instead of δ_0
θ	Heaviside function
$vp\frac{1}{x}$	Cauchy's principal value
$\frac{1}{x \pm io}$	$\lim_{\epsilon \searrow 0} \frac{1}{x \pm i\epsilon}$ in $\mathcal{D}'(\mathbb{R})$
supp T	support of a distribution T
supp sing T	singular support of a distribution T
$f \otimes g$	tensor product of two functions f and g
$T \otimes S$	tensor product of two distributions T and S
$\mathcal{D}(\mathbb{R}^n) \otimes \mathcal{D}(\mathbb{R}^m)$	algebraic tensor product of the test function spaces $\mathcal{D}(\mathbb{R}^n)$ and $\mathcal{D}(\mathbb{R}^m)$
$\mathcal{D}(\mathbb{R}^n) \otimes_\pi \mathcal{D}(\mathbb{R}^m)$	the space $\mathcal{D}(\mathbb{R}^n) \otimes \mathcal{D}(\mathbb{R}^m)$ equipped with the projective tensor product topology
$\mathcal{D}(\mathbb{R}^n) \tilde{\otimes}_\pi \mathcal{D}(\mathbb{R}^m)$	completion of the space $\mathcal{D}(\mathbb{R}^n) \otimes_\pi \mathcal{D}(\mathbb{R}^m)$
$u * v$	convolution of two functions u and v
$T * u$	the convolution of a distribution $T \in \mathcal{D}'(\Omega)$ with a test function $u \in \mathcal{D}(\Omega)$; regularization of T

$T * S$	convolution of two distributions T and S, if defined
$\bar{\partial}$	the differential operator $\frac{1}{2}(\frac{\partial}{\partial x} + i\frac{\partial}{\partial y})$ on $\mathcal{D}'(\mathbb{R}^2)$
$\mathcal{F}$	operator of Fourier transform, on $L^1(\mathbb{R}^n)$ or $\mathcal{S}(\mathbb{R}^n)$
$\mathcal{F}'$	Fourier transform on $\mathcal{S}'(\mathbb{R}^n)$
$<\cdot,\cdot>$	inner product on a vector space
$\|\cdot\|$	norm on a vector space
$l^2(\mathbb{K})$	Hilbert space of square summable sequences of numbers in $\mathbb{K}$
$M^\perp$	orthogonal complement of a set M in a Hilbert space
lin M	the linear span of the set M in a vector space
$[M]$	the closure of lin M in a topological vector space, i.e., the smallest closed subspace which contains M
dim V	dimension of a vector space V
$D(A)$	domain (of definition) of the (linear) operator A
ker $A = N(A)$	the kernel or null-space of a linear operator A
ran A	the range or set of values of a linear operator A
$\Gamma(A)$	graph of a linear operator A
A^*	the adjoint of the densely defined linear operator A
A_F	Friedrichs extension of the densely defined non-negative linear operator A
$A + B$	form sum of the linear operators A and B
$\mathcal{L}(X, Y)$	space of continuous linear operators $X \to Y$, X and Y topological vector space over the field $\mathbb{K}$
$\mathcal{B}(\mathcal{H}) = \mathcal{L}(\mathcal{H}, \mathcal{H})$	space of bounded linear operators on a Hilbert space $\mathcal{H}$
$\hat{A} = (D, A)$	linear operator with domain D and rule of assignment A
$\mathcal{K}(\mathcal{H})$	space of compact operators on a Hilbert space $\mathcal{H}$
$\mathcal{P}(\mathcal{H})$	space of all orthogonal projections on a Hilbert space $\mathcal{H}$
$\mathcal{S}(\mathcal{H})$	space of all trace class operators on a Hilbert space $\mathcal{H}$
$\mathcal{U}(\mathcal{H})$	space of all unitary operators on a Hilbert space $\mathcal{H}$
$\rho(A)$	resolvent set of a linear operator A

Notation   xxi

$R_A(z)$	resolvent operator at the point $z \in \rho(A)$ for the linear operator A
$\sigma(A)$	$= \mathbb{C}\setminus\rho(A)$, spectrum of the linear operator A
$\sigma_p(A)$	point spectrum of A
$\sigma_c(A)$	$= \sigma(A)\setminus\sigma_p(A)$, continuous spectrum of A
$\sigma_d(A)$	discrete spectrum of A
$\sigma_{ac}(A)$	absolutely continuous spectrum of A
$\sigma_{sc}(A)$	singular continuous spectrum of A
$\mathcal{H}_p(A)$	discontinuous subspace of A
$\mathcal{H}_c(A)$	continuous subspace of A
$\mathcal{H}_{sc}(A)$	singular continuous subspace of a self-adjoint operator A
$\mathcal{H}_{ac}(A)$	$= \mathcal{H}_c(A) \cap \mathcal{H}_{sc}(A)^{\perp}$, absolute continuous subspace of a self-adjoint operator A
$\mathcal{H}_s(A)$	$= \mathcal{H}_p(A) \oplus \mathcal{H}_{sc}(A)$, singular subspace of a self-adjoint operator A
$\mathcal{M}_b(H)$	subspace of bounded states of a self-adjoint Schrödinger operator H
$\mathcal{M}_\infty(H)$	subspace of scattering states of H, H as above
proj_M	orthogonal projection operator onto the closed subspace M of a Hilbert space
$[f \leq r]$	for a function $f : M \to \mathbb{R}$ and $r \in \mathbb{R}$ the sub-level set $\{x \in M : f(x) \leq r\}$
$proj_K$	projection onto the closed convex subset K of a Hilbert space $\mathcal{H}$
$[f = c]$	for a function $f : M \to \mathbb{R}$ and $c \in \mathbb{R}$ the level set $\{x \in M : f(x) = c\}$
$f'(x) = D_x f = Df(x)$	the Fréchet derivative of a function $f : U \to F$ at a point $x \in U$, for $U \subset E$ open, E, F Banach spaces
$\mathcal{B}(E^{\times n}, F)$	the Banach space of all continuous n-linear operators $E^{\times n} = E \times \cdots \times E \to F$, for Banach spaces E, F
$\delta f(x_0, h)$	Gâteaux differential of a function $f : U \to F$ at a point $x_0 \in U$ in the direction $h \in E$, $U \subset E$ open, E, F Banach spaces
$\delta_{x_0} f(h)$	Gâteaux derivative of f at $x_0 \in U$, applied to $h \in E$

$\Delta^n f(x_0, h)$ $= \frac{d^n}{dt^n} f(x_0 + th)|_{t=0}$, nth variation of a function f at the point x_0 in the direction h

$T_x M$ tangent space of the differential manifold M at the point $x \in M$

Mathematical Methods in Physics

Distributions, Hilbert Space Operators, and Variational Methods

Part I
Distributions

1
Introduction

One of the earliest and most famous examples of a generalized function or distribution is "Dirac's delta function". It was originally defined by Dirac (1926–1927) as a function

$$\mathbb{R} \ni x \to \delta_{x_0}(x) \in \bar{\mathbb{R}} \equiv \mathbb{R} \cup \{\infty\}$$

with the following properties (x_0 is a given real number):

(a)
$$\delta_{x_0}(x) = \begin{cases} 0 : & x \in \mathbb{R}, \ x \neq x_0, \\ +\infty : & x = x_0. \end{cases}$$

(b) $\int_\mathbb{R} f(x)\delta_{x_0}(x)dx = f(x_0)$ for all sufficiently smooth functions $f : \mathbb{R} \to \mathbb{R}$.

However, elementary results from integration theory show that the conditions (a) and (b) contradict each other. Indeed, by (a), $f(x)\delta_{x_0}(x) = 0$ for almost all $x \in \mathbb{R}$ (with respect to the Lebesgue measure on $\mathbb{R}$), and thus the Lebesgue integral of $f(x)\delta_{x_0}(x)$ vanishes:

$$\int_\mathbb{R} f(x)\delta_{x_0}(x)dx = 0$$

and this contradicts (b) for all f with $f(x_0) \neq 0$. An appropriate reading of condition (b) is to interpret $f(x)\delta_{x_0}(x)dx$ as a measure of total mass 1 which is concentrated in $x = x_0$. But this is in conflict with condition (a).

Nevertheless, physicists continued to work with this contradictory object quite successfully, in the sense of formal calculations. This showed that this mathematical object was useful in principle. In addition numerous other examples hinted at the usefulness of mathematical objects similar to Dirac's distribution. These objects, respectively concepts, were introduced initially in an often rather vague way in order to deal with concrete problems. The concepts we have in mind here were mainly those which later in the theory of generalized functions found their natural formulation as weak derivative, generalized solution, Green's function etc. This is to say that distribution theory should be considered as the natural result, through a process of synthesis and simplification, of several attempts to extend classical analysis which arose from various concrete problems. With the formulation of distribution theory one had an analogous situation to the invention of differential and integral calculus by Leibniz and Newton. In both cases many, mainly ad-hoc methods, were known for the solutions of many concrete problems which then found their "synthesis and simplification" in a comprehensive theory.

The main contributions to the development of distribution theory came from S. Bochner, J. Leray, K. Friedrichs, S. Sobolev, I. M. Gelfand and, in particular, Laurent Schwartz (1945–1949). New general ideas and methods from topology and functional analysis were used, mainly by L. Schwartz, in order to solve many, often old, problems and to extract their common general mathematical framework. Distribution theory, as created through this process, allows us to consider well defined mathematical objects with the conditions (a) and (b) from above by giving these conditions a new interpretation. In a first step, condition (b) becomes the definition of an object δ_{x_0} which generalizes the concept of the Lebesgue integral in the original formulation, i.e., our preliminary definition for δ_{x_0} reads:

$$\delta_{x_0}: \{f : \mathbb{R} \to \mathbb{C},\ f \text{ sufficiently smooth}\} \to \mathbb{C} \text{ defined by}$$

$$\delta_{x_0}(f) = f(x_0).$$

According to this δ_{x_0} assigns numbers to sufficiently smooth functions f in a linear way, just as ordinary integrals

$$I_g(f) = \int_{\mathbb{R}} g(x) f(x) dx$$

if they do exist (here g is a given function). Property (a) then becomes a 'support property' of this newly defined object on a vector space of sufficiently smooth functions:

$$\delta_{x_0}(f) = 0 \quad \text{whenever} \quad f(x_0) = 0.$$

In this sense one can also consider functions as 'linear functions' or 'functionals' on a suitable vector space of functions ϕ. The idea is quite simple: Consider the vector space $\mathcal{C}_0(\mathbb{R}^n)$ of continuous functions $\phi : \mathbb{R}^n \to \mathbb{C}$ with compact support $\text{supp}\,\phi$. Recall: The support of a function is by definition the closure of the set of those points where the function does not vanish, i.e.,

$$\text{supp}\,\phi = \overline{\{x \in \mathbb{R}^n : \phi(x) \neq 0\}}.$$

Then every continuous function g on $\mathbb{R}^n$ can be considered, in a natural way, as a linear functional I_g on the vector space $\mathcal{C}_0(\mathbb{R}^n)$ by defining

$$I_g(\phi) \equiv \int_{\mathbb{R}^n} g(x)\phi(x)dx. \tag{1.1}$$

When we think about the fact that the values of measurements of physical quantities are obtained by an averaging process, then the interpretation appears reasonable that many physical quantities can be described mathematically only by objects of the type (1.1). Later, when we have progressed with the precise formulation, we will call objects of the type (1.1) *regular distributions*. Distributions are a special class of *generalized functions* which indeed generalize functions along the lines indicated in (1.1). This will be discussed in more detail later. The theory of generalized functions has been developed to overcome various difficulties in classical analysis, in particular the following problems:

(i) the existence of continuous but not differentiable functions (B. Riemann 1861, K. Weierstraß 1872), e.g., $f(x) = \sum_{n=0}^{\infty} \frac{\sin 3^n x}{2^n}$;

(ii) the problem of interchangeability of limit operations.

A brief illustration of the kind of problems we have in mind in (ii) is the existence of sequences of C^∞-functions f_n which converge uniformly to a limit function which is of class C^∞ too, but the sequence of derivatives does not converge (in the sense of classical analysis). A simple example is the sequence $f_n(x) = \frac{1}{n}\sin nx$ which converges to 0 uniformly on $\mathbb{R}$, but the sequence of derivatives $f'_n(x) = \cos nx$ does not converge, not even point-wise.

Our focus will be the distribution theory as developed mainly by L. Schwartz. The final section discusses some other important classes of generalized functions.

Distribution theory addresses the problem of generalizing the classical concept of a function in such a way that the difficulties related to this classical concept are resolved in the new theory. In concrete terms, this envisaged generalization of the classical concept of functions should satisfy the following four conditions:

1. Every (locally integrable) function is a distribution.

2. Every distribution is differentiable, and the derivative is again a distribution.

3. As far as possible, the rules of calculation of classical analysis remain valid.

4. In distribution theory the interchangeability of the main limit operations is guaranteed 'automatically'.

As mentioned above, the realization of this program leads to a synthesis and a simplification. Nevertheless, we do not get mathematically well-defined objects with the very convenient properties (1), (2), (3), (4) for free. The mathematical work has to be done at the level of definition of these objects. At this point distribution theory might appear to be difficult. However, in reality it is quite simple, and for practical applications only a rather limited amount of mathematical knowledge is required.

There are different ways to define distributions; we mention the main three. One can define distributions as:

D_1 continuous linear functions on suitable spaces of smooth functions ('test functions');

D_2 certain equivalence classes of suitable Cauchy sequences of (smooth) functions;

D_3 'weak' derivatives of continuous functions (locally).

We consider the first way as the most convenient and most powerful since many results from functional analysis can be used directly. Accordingly we define distributions according to D_1 and derive D_2 and D_3 as important characterizations of distributions.

Remark 1.0.1 *Many details about the historical development of distribution theory can be found in the book by J. Lutzen 'The Prehistory of the Theory of Distributions,' Springer-Verlag 1982. Here we mention only two important aspects very briefly: It was not in order to give the Dirac function a mathematical meaning that L. Schwartz was interested in what later became the theory of distributions, but in order to solve a relatively abstract problem formulated by Choquet and Deny (1944). But without hesitation L. Schwartz addressed also practical problems in his new theory. As early as 1946 he gave a talk entitled "Generalization of the concepts of functions and derivatives" addressing an audience of electrical engineers.*

2
Spaces of Test Functions

The spaces of test functions we are going to use are vector spaces of smooth (i.e., sufficiently often continuously differentiable) functions on open nonempty subsets $\Omega \subseteq \mathbb{R}^n$ equipped with a 'natural' topology. Accordingly we start with a general method to equip a vector space V with a topology such that the vector space operations of addition and scalar multiplication become continuous, i.e., such that

$$A : V \times V \to V, \quad A(x, y) = x + y, \quad x, y \in V,$$
$$M : \mathbb{K} \times V \to V, \quad M(\lambda, x) = \lambda x, \quad \lambda \in \mathbb{K}, \ x \in V$$

become continuous functions for this topology. This can be done in several different but equivalent ways. The way we describe has the advantage of being the most natural one for the spaces of test functions we want to construct. A vector space V which is equipped with a topology $\mathcal{T}$ such that the functions A and M are continuous is called a *topological vector space*, usually abbreviated as *TVS*. The test function spaces used in distribution theory are concrete examples of topological vector spaces where, however, the topology has the additional property that every point has a neighborhood basis consisting of (absolutely) convex sets. These are called *locally convex topological vector spaces*, abbreviated as *LCVTVS*.

2.1 Hausdorff locally convex topological vector spaces

To begin we recall the concept of a topology. To define a *topology* on a set X means to define a system $\mathcal{T}$ of subsets of X which has the following properties:

T_1 $X, \emptyset \in \mathcal{T}$ ($\emptyset$ denotes the empty set);

T2 $W_i \in \mathcal{T}, i \in I \Rightarrow \bigcup_{i \in I} W_i \in \mathcal{T}$ (I any index set);

T3 $W_1, \ldots, W_N \in \mathcal{T}, N \in \mathbb{N} \Rightarrow \bigcap_{j=1}^{N} W_j \in \mathcal{T}$.

The elements of $\mathcal{T}$ are called *open* and their complements *closed* sets of the topological space $(X, \mathcal{T})$.

Example 2.1.1 1. Define $\mathcal{T}_t = \{\emptyset, X\}$. $\mathcal{T}_t$ is called the trivial topology on X.

2. Define $\mathcal{T}_d$ to be the system of all subsets of X including X and $\emptyset$. $\mathcal{T}_d$ is called the discrete topology on X.

3. The usual topology on the real line $\mathbb{R}$ has as open sets all unions of open intervals $]a, b[= \{x \in \mathbb{R} : a < x < b\}$.

Note that according to T3 only finite intersections are allowed. If one would take here the intersection of infinitely many sets, the resulting concept of a topology would not be very useful. For instance, every point $a \in \mathbb{R}$ is the intersection of infinitely many open intervals $I_n =]a - \frac{1}{n}, a + \frac{1}{n}[$, $a = \bigcap_{n \in \mathbb{N}}$. Hence, if in T3 infinite intersections were allowed, all points would be open, thus every subset would be open (see discrete topology), a property which in most cases is not very useful.

If we put any topology on a vector space, it is not assured that the basic vector space operations of addition and scalar multiplication will be continuous. A fairly concrete method to define a topology $\mathcal{T}$ on a vector space V so that the resulting topological space $(V, \mathcal{T})$ is actually a topological vector space is described in the following paragraphs. The starting point is the concept of a *semi-norm* on a vector space as a real valued, sub-additive, positive homogeneous and symmetric function.

Definition 2.1.1 *Let V be a vector space over $\mathbb{K}$. Any function $q : V \to \mathbb{R}$ with the properties*

(i) $q(x + y) \leq q(x) + q(y) \ \forall x, y \in V$ *(sub-additive),*

(ii) $q(\lambda x) = |\lambda| q(x), \ \forall \lambda \in \mathbb{K}, \forall x \in V$ *(symmetric and positive homogeneous),*

is called a **semi-norm** *on V. If a semi-norm q has the additional property*

(iii) $q(x) = 0 \Rightarrow x = 0$,

then it is called a **norm**.

There are some immediate consequences which are used very often:

Lemma 2.1.1 *For every semi-norm q on a vector space V one has*

1. $q(0) = 0$;

2. $|q(x) - q(y)| \leq q(x - y) \ \forall x, y \in V$;

3. $0 \leq q(x) \ \forall x \in V$.

Proof. The second condition in the definition of a semi-norm gives for $\lambda = 0$ that $q(0x) = 0$. But for any $x \in V$ one has $0x = 0 \equiv$ the neutral element 0 in V and the first part follows. Apply subadditivity of q to $x = y + (x - y)$ to get $q(x) = q(y + (x - y)) \leq q(y) + q(x - y)$. Similarly one gets for $y = x + (y - x)$ that $q(y) \leq q(x) + q(y - x)$. The symmetry condition ii) of a semi-norm says in particular $q(-x) = q(x)$, hence $q(x - y) = q(y - x)$, and thus the above two estimates together say $\pm(q(x) - q(y)) \leq q(x - y)$ and this proves the second part. For $y = 0$ the second part says $|q(x) - q(0)| \leq q(x)$, hence by observing $q(0) = 0$ we get $|q(x)| \leq q(x)$ and therefore a semi-norm takes only nonnegative values and we conclude. □

Example 2.1.2   1. It is easy to show that the functions $q_i : \mathbb{R}^n \to \mathbb{R}$ defined by $q_i(x) = |x_i|$ for $x = (x_1, \ldots, x_n) \in \mathbb{R}^n$ are semi-norms on the real vector space $\mathbb{R}^n$ but not norms if $n > 1$. And it is well known that the system $\mathcal{P} = \{q_1, \ldots, q_n\}$ can be used to define the usual Euclidean topology on $\mathbb{R}^n$.

2. More generally, consider any vector space V over the field $\mathbb{K}$ and its **algebraic dual space** $V^* = L(V; \mathbb{K})$ defined as the set of all linear functions $T : V \to \mathbb{K}$, i.e., those functions which satisfy

$$T(\alpha x + \beta y) = \alpha T(x) + \beta T(y) \quad \forall x, y \in V, \quad \forall \alpha, \beta \in \mathbb{K}.$$

Each such $T \in V^*$ defines a semi-norm q_T on V by

$$q_T(x) = |T(x)| \quad \forall x \in V.$$

3. For an open nonempty set $\Omega \subset \mathbb{R}^n$, the set $\mathcal{C}^k(\Omega)$ of all functions $f : \Omega \to \mathbb{K}$ which have continuous derivatives up to order k is actually a vector space over $\mathbb{K}$ and on it the following functions $p_{K,m}$ and $q_{K,m}$ are indeed semi-norms. Here $K \subset \Omega$ is any compact subset and $k \in \mathbb{N}$ is any non-negative integer. For $0 \leq m \leq k$ and $\phi \in \mathcal{C}^k(\Omega)$ define

$$p_{K,m}(\phi) = \sup_{x \in K, |\alpha| \leq m} |D^\alpha \phi(x)|, \tag{2.1}$$

$$q_{K,m}(\phi) = \left(\sum_{|\alpha| \leq m} \int_K |D^\alpha \phi(x)|^2 dx \right)^{1/2}. \tag{2.2}$$

The notation is as follows. For a multi-index $\alpha = (\alpha_1, \ldots, \alpha_n) \in \mathbb{N}^n$ we denote by $D^\alpha = \frac{\partial^{|\alpha|}}{\partial x_1^{\alpha_1} \cdots x_n^{\alpha_n}}$ the derivative monomial of order $|\alpha| = \alpha_1 + \cdots + \alpha_n$, i.e., $D^\alpha \phi(x) = \frac{\partial^{|\alpha|} \phi}{\partial x_1^{\alpha_1} \cdots x_n^{\alpha_n}}(x)$, $x = (x_1, \ldots, x_n)$. Thus, for example for $f \in \mathcal{C}^3(\mathbb{R}^3)$, one has in this notation: If $\alpha = (1, 0, 0)$, then $|\alpha| = 1$ and $D^\alpha f = \frac{\partial f}{\partial x_1}$; if $\alpha = (1, 1, 0)$, then $|\alpha| = 2$ and $D^\alpha f = \frac{\partial^2 f}{\partial x_1 \partial x_2}$; if $\alpha = (0, 0, 2)$, then $|\alpha| = 2$ and $D^\alpha f = \frac{\partial^2 f}{\partial^2 x_3}$; if $\alpha = (1, 1, 1)$ then $|\alpha| = 3$ and $D^\alpha f = \frac{\partial^3 f}{\partial x_1 \partial x_2 \partial x_3}$.

A few comments on these examples are in order. The semi-norms given in the second example play an important role in general functional analysis, those of the third will be used later in the definition of the topology on the test function spaces used in distribution theory.

Recall that in a Euclidean space $\mathbb{R}^n$ the open ball $B_r(x)$ with radius $r > 0$ and centre x is defined by

$$B_r(x) = \{y \in \mathbb{R}^n : |y - x| < r\}$$

where $|y - x| = \sqrt{\sum_{i=1}^n (y_i - x_i)^2}$ is the Euclidean distance between the points $y = (y_1, \ldots, y_n)$ and $x = (x_1, \ldots, x_n)$. Similarly one proceeds in a vector space V on which a semi-norm p is given: The *open p-ball* in V with centre x and radius $r > 0$ is defined by

$$B_{p,r}(x) = \{y \in V : p(y - x) < r\}.$$

In this definition the Euclidean distance is replaced by the semi-distance $d_p(y, x) = p(y - x)$ between the points $y, x \in V$. Note: If p is not a norm, then one can have $d_p(y, x) = 0$ for $y \neq x$. In this case the open p-ball $B_{p,r}(0)$ contains the nontrivial subspace $N(p) = \{y \in V : p(y) = 0\}$. Nevertheless these p-balls share all essential properties with balls in Euclidean space.

1. $B_{p,r}(x) = x + B_{p,r}$, i.e., every point $y \in B_{p,r}(x)$ has the unique representation $y = x + z$ with $z \in B_{p,r} \equiv B_{p,r}(0)$;

2. $B_{p,r}$ is circular, i.e., $y \in B_{p,r}$, $\alpha \in \mathbb{K}$, $|\alpha| \leq 1$ implies $\alpha x \in B_{p,r}$;

3. $B_{p,r}$ is convex, i.e., $x, y \in B_{p,r}$ and $0 \leq \lambda \leq 1$ implies $\lambda x + (1-\lambda)y \in B_{p,r}$;

4. $B_{p,r}$ absorbs the points of V, i.e., for every $x \in V$ there is a $\lambda > 0$ such that $\lambda x \in B_{p,r}$;

5. The nonempty intersection $B_{p_1,r_1}(x_1) \cap B_{p_2,r_2}(x_2)$ of two open p-balls contains an open p-ball: $B_{p,r}(x) \subset B_{p_1,r_1}(x_1) \cap B_{p_2,r_2}(x_2)$.

For the proof of these statements see the Exercises.

In a finite dimensional vector space all norms are equivalent, i.e., they define the same topology. However, this statement does not hold in an infinite dimensional vector space (see Exercises). As the above examples indicate, in an infinite dimensional vector space there are many different semi-norms. This raises naturally two questions: How do we compare semi-norms? When do two systems of semi-norms define the same topology? A natural way to compare two semi-norms is to compare their values in all points. Accordingly one has:

Definition 2.1.2 *For two semi-norms p and q on a vector space V one says*

*a) p is **smaller than** q, in symbols $p \leq q$ if, and only if, $p(x) \leq q(x)$ $\forall x \in V$;*

b) p and q are **comparable** if, and only if, either $p \leq q$ or $q \leq p$.

The semi-norms q_i in our first example above are not comparable. Among the semi-norms $q_{K,m}$ and $p_{K,m}$ from the third example there are many which are comparable. Suppose two compact subsets K_1 and K_2 satisfy $K_1 \subset K_2$ and the nonnegative integers m_1 is smaller than or equal to the nonnegative integer m_2, then obviously

$$p_{K_1,m_1} \leq p_{K_2,m_2} \quad \text{and} \quad q_{K_1,m_1} \leq q_{K_2,m_2}.$$

In the Exercises we show the following simple facts about semi-norms: If p is a semi-norm on a vector space V and r a positive real number, then rp defined by $(rp)(x) = rp(x)$ for all $x \in V$ is again a semi-norm on V. The maximum $p = \max\{p_1, \ldots, p_n\}$ of finitely many semi-norms $p_1, \ldots, p_n$ on V, which is defined by $p(x) = \max\{p_1(x), \ldots, p_n(x)\}$ for all $x \in V$, is a semi-norm on V such that $p_i \leq p$ for $i = 1, \ldots, n$. This prepares us for a discussion of systems of semi-norms on a vector space.

Definition 2.1.3 *A system $\mathcal{P}$ of semi-norms on a vector space V is called **filtering** if, and only if, for any two semi-norms $p_1, p_2 \in \mathcal{P}$ there is a semi-norm $q \in \mathcal{P}$ and there are positive numbers $r_1, r_2 \in \mathbb{R}_+$ such that $r_1 p_1 \leq q$ and $r_2 p_2 \leq q$ hold.*

Certainly, not all systems of semi-norms are filtering (see our first finite-dimensional example). However it is straightforward to construct a filtering system which contains a given system: Given a system $\mathcal{P}_0$ on a vector space V one defines the system $\mathcal{P} = \mathcal{P}(\mathcal{P}_0)$ generated by $\mathcal{P}_0$ as follows:

$$q \in \mathcal{P} \Leftrightarrow \exists\, p_1, \ldots, p_n \in \mathcal{P}_0\, \exists\, r_1, \ldots, r_n \in \mathbb{R}_+ : q = \max\{r_1 p_1, \ldots, r_n p_n\}.$$

One can show that $\mathcal{P}(\mathcal{P}_0)$ is the minimal filtering system of semi-norms on V that contains $\mathcal{P}_0$. In our third example above we considered the following two systems of semi-norms on $V = \mathcal{C}^k(\Omega)$:

$$\mathcal{P}_k(\Omega) = \{p_{K,m} : K \subset \Omega,\ K \text{ compact},\ 0 \leq m \leq k\},$$
$$\mathcal{Q}_k(\Omega) = \{q_{K,m} : K \subset \Omega,\ K \text{ compact},\ 0 \leq m \leq k\}.$$

In the Exercises it is shown that both are filtering.

Our first use of the open p-balls is to define a topology.

Theorem 2.1.1 *Suppose that $\mathcal{P}$ is a filtering system of semi-norms on a vector space V. Define a system $\mathcal{T}_\mathcal{P}$ of subsets of V as follows: A subset $U \subset V$ belongs to $\mathcal{T}_\mathcal{P}$ if, and only if, either $U = \emptyset$ or*

$$\forall x \in U\ \exists\, p \in \mathcal{P},\ \exists\, r > 0 :\ B_{p,r}(x) \subset U.$$

Then $\mathcal{T}_\mathcal{P}$ is a topology on V in which every point $x \in V$ has a neighborhood basis $\mathcal{V}_x$ consisting of open p-balls, $\mathcal{V}_x = \{B_{p,r}(x) : p \in \mathcal{P},\ r > 0\}$.

Proof. Suppose we are given $U_i \in \mathcal{T}_\mathcal{P}, i \in I$. We are going to show that $U = \cup_{i \in I} U_i \in \mathcal{T}_\mathcal{P}$. Take any $x \in U$, then $x \in U_i$ for some $i \in I$. Thus $U_i \in \mathcal{T}_\mathcal{P}$ implies: There are $p \in \mathcal{P}$ and $r > 0$ such that $B_{p,r}(x) \subset U_i$. It follows that $B_{p,r}(x) \subset U$, hence $U \in \mathcal{T}_\mathcal{P}$. Next assume that $U_1, \ldots, U_n \in \mathcal{T}_\mathcal{P}$ are given. Denote $U = \cap_{i=1}^n U_i$ and consider $x \in U \subset U_i, i = 1, \ldots, n$. Therefore, for $i = 1, \ldots, n$, there are $p_i \in \mathcal{P}$ and $r_i > 0$ such that $B_{p_i, r_i}(x) \subset U_i$. Since the system $\mathcal{P}$ is filtering, there is a $p \in \mathcal{P}$ and there are $\rho_i > 0$ such that $\rho_i p_i \leq p$ for $i = 1, \ldots, n$. Define $r = \min\{\rho_1 r_1, \ldots, \rho_n r_n\}$. It follows that $B_{p,r}(x) \subset B_{p_i, r_i}(x)$ for $i = 1, \ldots, n$ and therefore $B_{p,r}(x) \subset \cap_{i=1}^n U_i = U$. Hence the system $\mathcal{T}_\mathcal{P}$ satisfies the three axioms of a topology. By definition $\mathcal{T}_\mathcal{P}$ is the topology defined by the system V_x of open p-balls as a neighborhood basis of a point $x \in V$. □

This result shows that there is a unique way to construct a topology on a vector space as soon as one is given a filtering system of semi-norms. Suppose now that two filtering systems $\mathcal{P}$ and $\mathcal{Q}$ of semi-norms are given on a vector space V. Then we get two topologies $\mathcal{T}_\mathcal{P}$ and $\mathcal{T}_\mathcal{Q}$ on V and naturally one would like to know how these topologies compare, in particular when they are equal. This question is answered in the following proposition.

Proposition 2.1.2 *Given two filtering systems $\mathcal{P}$ and $\mathcal{Q}$ on a vector space V, construct the topologies $\mathcal{T}_\mathcal{P}$ and $\mathcal{T}_\mathcal{Q}$ on V according to Theorem (2.1.1). Then the following two statements are equivalent:*

(i) $\mathcal{T}_\mathcal{P} = \mathcal{T}_\mathcal{Q}$.

(ii) $\forall p \in \mathcal{P} \, \exists q \in \mathcal{Q} \, \exists \lambda > 0 : p \leq \lambda q$ and $\forall q \in \mathcal{Q} \, \exists p \in \mathcal{P} \, \exists \lambda > 0 : q \leq \lambda p$.

Two systems $\mathcal{P}$ and $\mathcal{Q}$ of semi-norms on a vector space V are called **equivalent** *if, and only if, any of these equivalent conditions holds.*

The main technical element of the proof of this proposition is the following elementary but widely used lemma about the relation of open p-balls and their defining semi-norms. Its proof is left as an exercise.

Lemma 2.1.2 *Suppose that p and q are two semi-norms on a vector space V. Then, for any $r > 0$ and $R > 0$, the following holds:*

$$p \leq \frac{r}{R} q \quad \Leftrightarrow \quad \text{for any } x \in V : B_{q,R}(x) \subseteq B_{p,r}(x). \tag{2.3}$$

Proof of 2.1.2. Assume condition i). Then every open p-ball $B_{p,r}(x)$ is open for the topology $\mathcal{T}_\mathcal{Q}$, hence there is an open q-ball $B_{q,R}(x) \subset B_{p,r}(x)$. By the lemma we conclude that $p \leq \frac{r}{R} q$. Condition (i) also implies that every open q-ball is open for the topology $\mathcal{T}_\mathcal{P}$, hence we deduce $p \leq \lambda q$ for some $0 < \lambda$. Therefore condition (ii) holds.

Conversely, suppose that condition (ii) holds. Then, using again the lemma one deduces: For every open p-ball $B_{p,r}(x)$ there is an open q-ball $B_{q,R}(x) \subset B_{p,r}(x)$ and for every open q-ball $B_{q,R}(x)$ there is an open p-ball $B_{p,r}(x) \subset B_{q,R}(x)$. This then implies that the two topologies $\mathcal{T}_\mathcal{P}$ and $\mathcal{T}_\mathcal{Q}$ coincide. □

Recall that a topological space is called *Hausdorff* if any two distinct points can be separated by disjoint neighborhoods. There is a convenient way to decide when the topology $\mathcal{T}_\mathcal{P}$ defined by a filtering system of semi-norms is Hausdorff.

2.1 Hausdorff locally convex topological vector spaces

Proposition 2.1.3 *Suppose $\mathcal{P}$ is a filtering system of semi-norms on a vector space V. Then the topology $\mathcal{T}_\mathcal{P}$ is Hausdorff if, and only if, for every $x \in V$, $x \neq 0$, there is a semi-norm $p \in \mathcal{P}$ such that $p(x) > 0$.*

Proof. Suppose that the topological space $(V, \mathcal{T}_\mathcal{P})$ is Hausdorff and $x \in V$ is given, $x \neq 0$. Then there are two open balls $B_{p,r}(0)$ and $B_{q,R}(x)$ which do not intersect. By definition of these balls it follows that $p(x) \geq r > 0$ and the condition of the proposition holds. Conversely assume that the condition holds and two points $x, y \in V$, $x - y \neq 0$ are given. There is a $p \in \mathcal{P}$ such that $0 < 2r = p(x - y)$. Then the open balls $B_{p,r}(x)$ and $B_{p,r}(y)$ do not intersect. (If $z \in V$ were a point belonging to both balls, then we would have $p(z - x) < r$ and $p(z - y) < r$ and therefore $2r = p(x - y) = p(x - z + z - y) \leq p(x - z) + p(z - y) < r + r = 2r$, a contradiction). Hence the topology $\mathcal{T}_\mathcal{P}$ is Hausdorff. □

Finally we discuss the continuity of the basic vector space operations of addition and scalar multiplication with respect to the topology $\mathcal{T}_\mathcal{P}$ defined by a filtering system $\mathcal{P}$ of semi-norms on a vector space V. Recall that a function $f : E \to F$ from a topological space E into a topological space F is continuous at a point $x \in E$ if, and only if, the following condition is satisfied: For every neighborhood U of the point $y = f(x)$ in F there is a neighborhood V of x in E such that $f(V) \subset U$, and it is enough to consider instead of general neighborhoods U and V only elements of a neighborhood basis of $f(x)$, respectively x.

Proposition 2.1.4 *Let $\mathcal{P}$ be a filtering system of semi-norms on a vector space V. Then addition (A) and scalar multiplication (M) of the vector space V are continuous with respect to the topology $\mathcal{T}_\mathcal{P}$, hence $(V, \mathcal{T}_\mathcal{P})$ is a topological vector space. This topological vector space is usually denoted by*

$$(V, \mathcal{P}) \quad \text{or} \quad V[\mathcal{P}].$$

Proof. We show that the addition $A : V \times V \to V$ is continuous at any point $(x, y) \in V \times V$. Naturally, the product space $V \times V$ is equipped with the product topology of $\mathcal{T}_\mathcal{P}$. Given any open p-ball $B_{p,2r}(x + y)$ for some $r > 0$, then $A(B_{p,r}(x) \times B_{p,r}(y)) \subset B_{p,2r}(x + y)$ since for all $(x', y') \in B_{p,r}(x) \times B_{p,r}(y)$ we have $p(A(x', y') - A(x, y)) = p((x' + y') - (x + y)) = p(x' - x + y - y') \leq p(x' - x) + p(y' - y) < r + r = 2r$. Continuity of scalar multiplication M is proved in a similar way. □

We summarize our results in the following theorem.

Theorem 2.1.5 *Let $\mathcal{P}$ be a filtering system of semi-norms on a vector space V. Equip V with the induced topology $\mathcal{T}_\mathcal{P}$. Then $(V, \mathcal{T}_\mathcal{P}) = V[\mathcal{T}_\mathcal{P}]$ is a locally convex topological vector space. It is Hausdorff or a **HLCVTVS** if, and only if, for every $x \in V$, $x \neq 0$, there is a $p \in \mathcal{P}$ such that $p(x) > 0$.*

Proof. By Theorem 2.1.1 every point $x \in V$ has a neighborhood basis V_x consisting of open p-balls. These balls are absolutely convex (i.e., $y, z \in B_{p,r}(x)$, $\alpha, \beta \in \mathbb{K}$, $\alpha + \beta = 1$, $|\alpha| + |\beta| \leq 1$ implies $\alpha y + \beta z \in B_{p,r}(x)$) by the properties of p-balls listed earlier. Hence by Proposition 2.1.4 $V[\mathcal{T}_\mathcal{P}]$ is a LCTVS. Finally by Proposition 2.1.3 we conclude. □

2.1.1 Examples of HLCTVS

The examples of HLCTVS which we are going to discuss serve a dual purpose. Naturally they are considered in order to illustrate the concepts and results introduced above. Then later they will be used as building blocks of the test function spaces used in distribution theory.

1. Recall the filtering systems of semi-norms $\mathcal{P}_k(\Omega)$ and $\mathcal{Q}_k(\Omega)$ introduced earlier on the vector space $\mathcal{C}^k(\Omega)$ of k times continuously differentiable functions on an open nonempty subset $\Omega \subseteq \mathbb{R}^n$. With the help of Theorem 2.1.5 it is easy to show that both $(\mathcal{C}^k(\Omega), \mathcal{P}_k(\Omega))$ and $(\mathcal{C}^k(\Omega), \mathcal{Q}_k(\Omega))$ are Hausdorff locally convex topological vector spaces.

2. Fix a compact subset K of some open nonempty set $\Omega \subseteq \mathbb{R}^n$ and consider the space $\mathcal{C}_K^\infty(\Omega)$ of all functions $\phi : \Omega \to \mathbb{K}$ which are infinitely often differentiable on Ω and which have their support in K, i.e., supp $f \subseteq K$. On $\mathcal{C}_K^\infty(\Omega)$ consider the systems of semi-norms

$$\mathcal{P}_K(\Omega) = \{p_{K,m} : m = 0, 1, 2, \ldots\} \quad \mathcal{Q}_K(\Omega) = \{q_{K,m} : m = 0, 1, 2, \ldots\}$$

introduced in equation (2.1), respectively in equation (2.2). Both systems are obviously filtering, and both $p_{K,m}$ and $q_{K,m}$ are norms on $\mathcal{C}_K^\infty(\Omega)$. In the Exercises it is shown that both systems are equivalent and thus we get that

$$\mathcal{D}_K(\Omega) = (\mathcal{C}_K^\infty(\Omega), \mathcal{P}_K(\Omega)) = (\mathcal{C}_K^\infty(\Omega), \mathcal{Q}_K(\Omega)) \quad (2.4)$$

is a Hausdorff locally convex topological vector space.

3. Now let $\Omega \subseteq \mathbb{R}^n$ be an open nonempty subset which may be unbounded. Consider the vector space $\mathcal{C}^k(\Omega)$ of functions $\phi : \Omega \to \mathbb{K}$ which have continuous derivatives up to order k. Introduce two families of symmetric and sub-additive functions $\mathcal{C}^k(\Omega) \to [0, +\infty]$ by defining, for $l = 0, 1, 2, \ldots, k$ and $m = 0, 1, 2, \ldots$,

$$p_{m,l}(\phi) = \sup_{x \in \Omega, |\alpha| \leq l} (1 + x^2)^{m/2} |D^\alpha \phi(x)|,$$
$$q_{m,l}(\phi) = \left(\sum_{|\alpha| \leq l} \int_\Omega (1 + x^2)^{m/2} |D^\alpha \phi(x)|^2 dx\right)^{1/2}.$$

For $x = (x_1, \ldots, x_n) \in \mathbb{R}^n$ we use the notation $x^2 = x_1^2 + \cdots + x_n^2$ and $|x| = \sqrt{x^2}$. Define the following subspace of $\mathcal{C}^k(\Omega)$:

$$\mathcal{C}_m^k(\Omega) = \left\{\phi \in \mathcal{C}^k(\Omega) : p_{m,l}(\phi) < \infty, \, l = 0, 1, \ldots, k\right\}.$$

Then the system of norms $\{p_{m,l} : 0 \leq l \leq k\}$ is filtering on this subspace and thus $(\mathcal{C}_m^k(\Omega), \{p_{m,l} : 0 \leq l \leq k\})$ is a HLCTVS. $\mathcal{C}_m^k(\Omega)$ is the space of continuously differentiable functions which decay at infinity (if Ω is unbounded), with all derivatives of order $\leq k$, at least as $|x|^{-m}$. Similarly one can build a HLCTVS space by using the system of norms $q_{m,l}, 0 \leq l \leq k$.

4. In this example we use some basic facts from Lebesgue integration theory [GF68]. Let $\Omega \subset \mathbb{R}^n$ be a nonempty measurable set. On the vector space $L^1_{loc}(\Omega)$ of all measurable functions $f : \Omega \to \mathbb{K}$ which are *locally integrable*, i.e., for which

$$\|f\|_K = \int_K |f(x)| dx$$

is finite for every compact subset $K \subset \Omega$, consider the system of semi-norms $\mathcal{P} = \{\|\cdot\|_K : K \subset \Omega,\ K \text{ compact}\}$. Since the finite union of compact sets is compact, it follows easily that this system is filtering. If $f \in L^1_{loc}(\Omega)$ is given and if $f \neq 0$, then there is a compact set K such that $\|f\|_K > 0$, since $f \neq 0$ means that f is different from zero on a set of positive Lebesgue measure. Therefore, by Theorem 2.1.5, the space

$$(L^1_{loc}(\Omega), \{\|\cdot\|_K : K \subset \Omega,\ K \text{ compact}\})$$

is a HLCTVS.

2.1.2 Continuity and convergence in a HLCVTVS

Since the topology of a LCTVS $V[\mathcal{P}]$ is defined in terms of a filtering system $\mathcal{P}$ of semi-norms it is, in most cases, much more convenient to have a characterization of the basic concepts of convergence, of a Cauchy sequence, and of continuity in terms of the semi-norms directly instead of having to rely on the general topological definitions. Such characterizations will be given in this subsection.

Recall: A sequence $(x^i)_{i \in \mathbb{N}}$ of points $x^i = (x^i_1, \ldots, x^i_n) \in \mathbb{R}^n$ is said to converge if, and only if, there is a point $x \in \mathbb{R}^n$ such that for every open Euclidean ball $B_r(x) = \{y \in \mathbb{R}^n : |y - x| < r\}$ only a finite number of elements of the sequence are not contained in this ball, i.e., there is an index i_0, depending on $r > 0$, such that $x^i \in B_r(x)$ for all $i \geq i_0$, or expressed directly in terms of the Euclidean norm, $|x^i - x| < r$ for all $i \geq i_0$.

Similarly one proceeds in a general HLCTVS $V[\mathcal{P}]$ where now however instead of the Euclidean norm $|\cdot|$ all the semi-norms $p \in \mathcal{P}$ have to be taken into account.

Definition 2.1.4 *Let $V[\mathcal{P}]$ be a HLCTVS and $(x_i)_{i \in \mathbb{N}}$ a sequence in $V[\mathcal{P}]$. Then one says:*

1. *The sequence $(x_i)_{i \in \mathbb{N}}$ **converges** (in $V[\mathcal{P}]$) if, and only if, there is an $x \in V$ (called a **limit point** of the sequence) such that for every $p \in \mathcal{P}$ and for every $r > 0$ there is an index $i_0 = i_0(p, r)$ depending on p and r such that $p(x - x_i) < r$ for all $i \geq i_0$.*

2. *The sequence $(x_i)_{i \in \mathbb{N}}$ is a **Cauchy sequence** if, and only if, for every $p \in \mathcal{P}$ and every $r > 0$ there is an index $i_0 = i_0(p, r)$ such that $p(x_i - x_j) < r$ for all $i, j \geq i_0$.*

The following immediate results are well known in $\mathbb{R}^n$.

16   2. Spaces of Test Functions

Theorem 2.1.6 *(a) Every convergent sequence in a LCTVS $V[\mathcal{P}]$ is a Cauchy sequence.*

(b) In a HLCTVS $V[\mathcal{P}]$ the limit point of a convergent sequence is unique.

Proof. Suppose a sequence $(x_i)_{i \in \mathbb{N}}$ converges in $V[\mathcal{P}]$ to $x \in V$. Then, for any $p \in \mathcal{P}$ and any $r > 0$, there is an $i_0 \in \mathbb{N}$ such that $p(x - x_i) < r/2$ for all $i \geq i_0$. Therefore, for all $i, j \geq i_0$, one has $p(x_i - x_j) = p((x - x_j) + (x_i - x)) \leq p(x - x_j) + p(x_i - x) < \frac{r}{2} + \frac{r}{2} = r$, hence $(x_i)_{i \in \mathbb{N}}$ is a Cauchy sequence and part (a) follows.

Suppose $V[\mathcal{P}]$ is a HLCTVS and $(x_i)_{i \in \mathbb{N}}$ is a convergent sequence in $V[\mathcal{P}]$. Assume that for $x, y \in V$ the condition in the definition of convergence holds, i.e., for every $p \in \mathcal{P}$ and every $r > 0$ there is an i_1 such that $p(x - x_i) < r$ for all $i \geq i_1$ and there is an i_2 such that $p(y - x_i) < r$ for all $i \geq i_2$. Then, for all $i \geq \max\{i_1, i_2\}$, $p(x - y) = p(x - x_i + x_i - y) \leq p(x - x_i) + p(x_i - y) < r + r = 2r$, and since $r > 0$ is arbitrary, it follows that $p(x - y) = 0$. Since this holds for every $p \in \mathcal{P}$ and $V[\mathcal{P}]$ is Hausdorff, we conclude (see Proposition 2.1.3) that $x = y$ and thus part (b) follows. □

Part a) of Theorem 2.1.6 raises naturally the question whether the converse holds too, i.e., whether every Cauchy sequence converges. In general, this is not the case. Spaces in which this statement holds are distinguished according to the following definition.

Definition 2.1.5 *A HLCTVS in which every Cauchy sequence converges is called* **sequentially complete**.

Example 2.1.3 *1. Per construction, the field $\mathbb{R}$ of real numbers equipped with the absolute value $|\cdot|$ as a norm is a sequentially complete HLCTVS.*

2. *The Euclidean spaces $(\mathbb{R}^n, |\cdot|)$, $n=1,2,...$ are HLCTVS. Here $|\cdot|$ denotes the Euclidean norm.*

3. *For any $\Omega \subset \mathbb{R}^n$, Ω open and nonempty, and $k=0,1,2,...$, the space*

$$\mathcal{C}^k(\Omega)[\mathcal{P}_k(\Omega)]$$

is a sequentially complete HLCTVS. This is shown in the Exercises. Recall the definition

$$\mathcal{P}_k(\Omega) = \{p_{K,m} : K \subset \Omega, \ K \text{ compact}, \ 0 \leq m \leq k\}.$$

Note that $\mathcal{C}^k(\Omega)[\mathcal{P}_k(\Omega)]$ is equipped with the **topology of uniform convergence of all derivatives of order $\leq k$ on all compact subsets of Ω**.

Compared to a general topological vector space one has a fairly explicit description of the topology in a locally convex topological vector space. Here, as we have learned, each point has a neighborhood basis consisting of open balls, and thus formulating the definition of continuity one can completely rely on these open balls. This then has an immediate translation into conditions involving only the systems of semi-norms which define the topology. Suppose that $X[\mathcal{P}]$ and $Y[\mathcal{Q}]$ are two LCTVS. Then a function $f : X \to Y$ is said to be continuous at $x_0 \in X$

if, and only if, for every open q-ball $B_{q,R}(f(x_0))$ in $Y[\mathcal{Q}]$ there is an open p-ball $B_{p,r}(x)$ in $X[\mathcal{P}]$ which is mapped by f into $B_{q,R}(f(x_0))$. This can also be expressed as follows:

Definition 2.1.6 *Assume that $X[\mathcal{P}]$ and $Y[\mathcal{Q}]$ are two LCTVS. A function $f : X \to Y$ is said to be* **continuous at** $x_0 \in X$ *if, and only if, for every semi-norm $q \in \mathcal{Q}$ and every $R > 0$ there are $p \in \mathcal{P}$ and $r > 0$ such that for all $x \in X$ the condition $p(x - x_0) < r$ implies $q(f(x) - f(x_0)) < R$. f is called* **continuous on** X *if, and only if, f is continuous at every point $x_0 \in X$.*

Our main interest however are linear functions from one locally convex topological vector space to another. For them one can give a characterization of continuity which in most cases, in particular in concrete examples, is much easier to verify. This characterization is prepared by the following definition.

Definition 2.1.7 *Assume that $X[\mathcal{P}]$ and $Y[\mathcal{Q}]$ are two LCTVS. A linear function $f : X \to Y$ is said to be* **bounded** *if, and only if, for every semi-norm $q \in \mathcal{Q}$ there are $p \in \mathcal{P}$ and $\lambda \geq 0$ such that for all $x \in X$ one has*

$$q(f(x)) \leq \lambda p(x). \tag{2.5}$$

The announced characterization of continuity now has a simple formulation.

Theorem 2.1.7 *Let $X[\mathcal{P}]$ and $Y[\mathcal{Q}]$ be two LCTVS and $f : X \to Y$ a linear function. Then f is continuous if, and only if, it is bounded.*

Proof. Suppose that f is bounded, i.e., given $q \in \mathcal{Q}$ there are $p \in \mathcal{P}$ and $\lambda \geq 0$ such that $q \circ f \leq \lambda p$. It follows for any $x, y \in X$: $q(f(y) - f(x)) = q(f(x - y)) \leq \lambda p(y - x)$. Continuity of f at x is now evident: Given $q \in \mathcal{Q}$ and $R > 0$, take $r = \frac{R}{\lambda}$ and the semi-norm $p \in \mathcal{P}$ from the boundedness condition.

Conversely assume that f is continuous. Then f is continuous at $0 \in X$. Hence, given $q \in \mathcal{Q}$ and $R > 0$ there are $p \in \mathcal{P}$ and $r > 0$ such that $p(x) < r$ implies $q(f(x)) < R$ (we use here that $f(0) = 0$ for a linear function). This shows: $B_{p,r}(0) \subseteq B_{q \circ f, R}(0)$ and therefore by Lemma 2.1.2 we conclude that $q \circ f \leq \frac{R}{r} p$, i.e., f is bounded. $\square$

The proof of this theorem shows actually some further details about continuity of linear functions on LCTVS. We summarize them as a corollary.

Corollary 2.1.1 *Let $X[\mathcal{P}]$ and $Y[\mathcal{Q}]$ be two LCTVS and $f : X \to Y$ a linear function. Then the following statements are equivalent.*

1. *f is continuous at the origin $x = 0$.*
2. *f is continuous at some point $x \in X$.*
3. *f is continuous.*
4. *f is bounded.*
5. *f is bounded on some open ball $B_{p,r}(0)$ in $X[\mathcal{P}]$.*

Definition 2.1.8 *The* **topological dual** $X'[\mathcal{P}]$ *of a Hausdorff toplogical vector space* $X[\mathcal{P}]$ *over the field* $\mathbb{K}$ *is by definition the space of all continuous linear functions* $X[\mathcal{P}] \to \mathbb{K}$.

We conclude this subsection with a discussion of an important special case of a HLCTVS. Suppose that $X[\mathcal{P}]$ is a HLCTVS and that the filtering system of semi-norms $\mathcal{P}$ is countable, i.e., $\mathcal{P} = \{p_i : i \in \mathbb{N}\}$ with $p_i \leq p_{i+1}$ for all $i = 0, 1, 2, \ldots$. Then the topology $\mathcal{T}_\mathcal{P}$ of $X[\mathcal{P}]$ can be defined in terms of a *metric d*, i.e., a function $d: X \times X \to \mathbb{R}$ with the following properties:

1. $d(x, y) \geq 0$ for all $x, y \in X$;
2. $d(x, y) = d(y, x)$ for all $x, y \in X$;
3. $d(x, y) \leq d(x, z) + d(z, y)$ for all $x, y, z \in X$;
4. $d(x, y) = 0 \Leftrightarrow x = y$.

In terms of the given system of semi-norms, the metric can be expressed as:

$$d(x, y) = \sum_{i=0}^{\infty} \frac{1}{2^i} \frac{p_i(x-y)}{1 + p_i(x-y)}. \tag{2.6}$$

In the Exercises we show that this function is indeed a metric on X which defines the given topology by using as open balls with centre x and radius $r > 0$ the sets $B_{d,r}(x) = \{y \in X : d(y, x) < r\}$. A HLCTVS $X[\mathcal{P}]$ is called *metrizable* if, and only if, its topology $\mathcal{T}_\mathcal{P}$ can be defined in terms of a metric. Some other special cases are addressed in the Exercises as well.

We conclude this section with an example of a complete metrizable HLCTVS which will play an important role in the definition of the basic test function spaces.

Proposition 2.1.8 *Let* $\Omega \subset \mathbb{R}^n$ *be any nonempty open set and* $K \subset \Omega$ *any compact subset. Then the space* $\mathcal{D}_K(\Omega)$ *introduced in (2.4) is a complete metrizable HLCTVS.*

Proof. That this space is metrizable is clear from the definition. The proof of completeness is left as an exercise. □

2.2 Basic test function spaces of distribution theory

The previous sections provide nearly all concepts and results which are needed for the definition of the standard test function spaces and the study of their basic properties. The important items that are missing are the concepts of inductive and projective limits of TVS. Here we take a practical approach by defining these concepts not abstractly but only in the context where they are used. We discuss now the underlying test function spaces of general (Schwartz) distributions, of tempered distributions, and of distributions with compact support.

2.2.1 The test function space $\mathcal{D}(\Omega)$ of C^∞ functions of compact support

For a nonempty open subset $\Omega \subset \mathbb{R}^n$ recall the spaces $\mathcal{D}_K(\Omega)$, $K \subset \Omega$ compact, as introduced in equation (2.4) and note the following:

$$K_1 \subset K_2 \subset \Omega, \quad K_1, K_2 \text{ compact} \Rightarrow \mathcal{D}_{K_1}(\Omega) \subset \mathcal{D}_{K_2}(\Omega).$$

The statement "$\mathcal{D}_{K_1}(\Omega) \subset \mathcal{D}_{K_2}(\Omega)$" actually means two things:

1. The vector space $C^\infty_{K_1}(\Omega)$ is a subspace of the vector space $C^\infty_{K_2}(\Omega)$.

2. The restriction of the topology of $\mathcal{D}_{K_2}(\Omega)$ to the subspace $\mathcal{D}_{K_1}(\Omega)$ equals the original topology of $\mathcal{D}_{K_1}(\Omega)$ as defined in equation (2.4).

Now denote by $\mathcal{K} = \mathcal{K}(\Omega)$ the set of all compact subsets of Ω and define

$$\mathcal{D}(\Omega) = \bigcup_{K \in \mathcal{K}} \mathcal{D}_K(\Omega). \tag{2.7}$$

Then $\mathcal{D}(\Omega)$ is the set of functions $\phi : \Omega \to \mathbb{K}$ of class C^∞ which have a compact support in Ω. It is easy to show that this set is actually a vector space over $\mathbb{K}$. In order to define a topology on $\mathcal{D}(\Omega)$ denote, for $K \subset \Omega$, K compact, by $i_K : \mathcal{D}_K(\Omega) \to \mathcal{D}(\Omega)$ the identical embedding of $\mathcal{D}_K(\Omega)$ into $\mathcal{D}(\Omega)$. Define on $\mathcal{D}(\Omega)$ the strongest locally convex topology such that all these embeddings i_K, $K \subset \Omega$ compact, are continuous. Thus $\mathcal{D}(\Omega)$ becomes a HLCTVS (see Exercises). In this way the test function space $\mathcal{D}(\Omega)$ of C^∞-functions of compact support is defined as the *inductive limit* of the spaces $\mathcal{D}_K(\Omega)$, $K \subset \Omega$ compact. According to this definition a function $\phi \in C^\infty(\Omega)$ belongs to $\mathcal{D}(\Omega)$ if, and only if, it vanishes in some neighborhood of the boundary $\partial\Omega$ of Ω.

In the Exercises it is shown that given $\Omega \subset \mathbb{R}^n$, Ω open and nonempty, there is a sequence of compact sets K_i, $i \in \mathbb{N}$, with nonempty interior such that

$$K_i \Subset K_{i+1} \subset \Omega \quad \forall i \in \mathbb{N}, \quad \cup_{i=1}^\infty K_i = \Omega.$$

It follows that, for all $i \in \mathbb{N}$,

$$\mathcal{D}_{K_i}(\Omega) \Subset \mathcal{D}_{K_{i+1}}(\Omega) \tag{2.8}$$

with the understanding that $\mathcal{D}_{K_i}(\Omega)$ is a proper subspace of $\mathcal{D}_{K_{i+1}}(\Omega)$ and that the restriction of the topology of $\mathcal{D}_{K_{i+1}}(\Omega)$ to $\mathcal{D}_{K_i}(\Omega)$ is just the original topology of $\mathcal{D}_{K_i}(\Omega)$.

One deduces that $\mathcal{D}(\Omega)$ is actually the strict (because of (2.8)) inductive limit of the sequence of complete metrizable spaces $\mathcal{D}_{K_i}(\Omega)$, $i \in \mathbb{N}$:

$$\mathcal{D}(\Omega) = \cup_{i=1}^\infty \mathcal{D}_{K_i}(\Omega). \tag{2.9}$$

We collect some basic properties of the test function space $\mathcal{D}(\Omega)$.

Theorem 2.2.1 *The following statements hold for the test function space $\mathcal{D}(\Omega)$ of compactly supported C^∞-functions on $\Omega \subset \mathbb{R}^n$, Ω open and not empty:*

1. *$\mathcal{D}(\Omega)$ is the strict inductive limit of a sequence of complete metrizable Hausdorff locally convex topological vector spaces $\mathcal{D}_{K_i}(\Omega)$.*

2. *$\mathcal{D}(\Omega)$ is a HLCTVS.*

3. *A subset $U \subset \mathcal{D}(\Omega)$ is a neighborhood of zero if, and only if, $U \cap \mathcal{D}_K(\Omega)$ is a neighborhood of zero in $\mathcal{D}_K(\Omega)$, for every compact subset $K \subset \Omega$.*

4. *$\mathcal{D}(\Omega)$ is sequentially complete.*

5. *$\mathcal{D}(\Omega)$ is not metrizable.*

Proof. The first statement has been established above. After further preparation the remaining statements are shown in the Appendix. □

For many practical purposes it is important to have a concrete description of the notion of convergence in $\mathcal{D}(\Omega)$. The following characterization results from basic properties of inductive limits and is addressed in the Appendix.

Proposition 2.2.2 *Let $\Omega \subset \mathbb{R}^n$ be a nonempty open set. Then a sequence $(\phi_i)_{i \in \mathbb{N}}$ converges in the test function space $\mathcal{D}(\Omega)$ if, and only if, there is a compact subset $K \subset \Omega$ such that $\phi_i \in \mathcal{D}_K(\Omega)$ for all $i \in \mathbb{N}$ and this sequence converges in the space $\mathcal{D}_K(\Omega)$.*

According to the definition given earlier, a sequence $(\phi_i)_{i \in \mathbb{N}}$ converges in $\mathcal{D}_K(\Omega)$ to $\phi \in \mathcal{D}_K(\Omega) \Leftrightarrow \forall_{r>0} \forall_{m \in \mathbb{N}} \exists_{i_0} \forall_{i \geq i_0} \; p_{K,m}(\phi - \phi_i) < r$.

Proposition 2.2.3 *Let $Y[\mathcal{Q}]$ be a locally convex topological vector space and $f : \mathcal{D}(\Omega) \to Y[\mathcal{Q}]$ a linear function. Then f is continuous if, and only if, for every compact set $K \subset \Omega$ the map $f \circ i_K : \mathcal{D}_K(\Omega) \to Y[\mathcal{Q}]$ is continuous.*

Proof. By definition the test function space carries the strongest locally convex topology such that all the embeddings $i_K : \mathcal{D}_K(\Omega) \to \mathcal{D}(\Omega)$, $K \subset \Omega$ compact, are continuous. Thus, if f is continuous, all maps $f \circ i_K$ are continuous as compositions of continuous maps. Conversely assume that all maps $f \circ i_K$ are continuous; then given any neighborhood of zero U in $Y[\mathcal{Q}]$, we know that $(f \circ i_K)^{-1}(U) = f^{-1}(U) \cap \mathcal{D}_K(\Omega)$ is a neighborhood of zero in $\mathcal{D}_K(\Omega)$. Since this holds for every compact subset K it follows, by part 3 of Theorem 2.2.1, that $f^{-1}(U) \subset \mathcal{D}(\Omega)$ is a neighborhood of zero, hence f is continuous. □

2.2.2 The test function space $\mathcal{S}(\Omega)$ of strongly decreasing C^∞-functions on Ω

Again, Ω is an open nonempty subset of $\mathbb{R}^n$, often $\Omega = \mathbb{R}^n$. A function $\phi \in C^\infty(\Omega)$ is called **strongly decreasing** if, and only if, it and all its derivatives decrease faster than $C(1+x^2)^{-k}$, for any $k \in \mathbb{N}$, i.e., if, and only if, the following condition holds:

$$\forall_{\alpha \in \mathbb{N}^n} \forall_{m \in \mathbb{N}_0} \exists_C \forall_{x \in \Omega} \quad |D^\alpha \phi(x)| \leq \frac{C}{(1+x^2)^{\frac{m}{2}}}. \qquad (2.10)$$

Certainly, in this estimate the constant C depends in general on the function ϕ, the order α of the derivative, and the exponent m of decay. Introduce

$$\mathcal{S}_0(\Omega) = \{\phi \in \mathcal{C}^\infty(\Omega) : \phi \text{ is strongly decreasing}\}.$$

It is straightforward to show that $\mathcal{S}_0(\Omega)$ is a vector space. The norms

$$p_{m,l}(\phi) = \sup_{x \in \Omega, |\alpha| \leq l} (1+x^2)^{m/2} |D^\alpha \phi(x)|$$

are naturally defined on it for all $m, l = 0, 1, 2, \ldots$. Equip this space with the topology defined by the filtering system $\mathcal{P}(\Omega) = \{p_{m,l} : m, l = 0, 1, 2, \ldots\}$ and introduce the *test function space of strongly decreasing $\mathcal{C}^\infty$-functions* as the Hausdorff locally convex topological vector space

$$\mathcal{S}(\Omega) = (\mathcal{S}_0(\Omega), \mathcal{P}(\Omega)). \tag{2.11}$$

Note that $\mathcal{S}_0(\Omega)$ can be expressed in terms of the function spaces $\mathcal{C}_m^k(\Omega)$ introduced earlier as:

$$\mathcal{S}_0(\Omega) = \cap_{k,m=0}^\infty \mathcal{C}_m^k(\Omega).$$

Elementary facts about $\mathcal{S}(\Omega)$ are collected in the following theorem.

Theorem 2.2.4 *The test function space $\mathcal{S}(\Omega)$ of strongly decreasing $\mathcal{C}^\infty$-functions, for any open and nonempty subset $\Omega \subseteq \mathbb{R}^n$, is a complete metrizable HLCTVS.*

Proof. Since the filtering system of norms of this space is countable, $\mathcal{S}(\Omega)$ is a metrizable HLCTVS. Completeness of this space is shown in the Exercises. Further properties will be presented in the Appendix. □

2.2.3 The test function space $\mathcal{E}(\Omega)$ of all $\mathcal{C}^\infty$-functions on Ω

On the vector space $\mathcal{C}^\infty(\Omega)$ we use the filtering system of semi-norms $\mathcal{P}_\infty(\Omega) = \{p_{K,m} : K \subset \Omega \text{ compact}, m = 0, 1, 2, \ldots\}$ and then introduce

$$\mathcal{E}(\Omega) = (\mathcal{C}^\infty(\Omega), \mathcal{P}_\infty(\Omega)) \tag{2.12}$$

as the *test function space of all $\mathcal{C}^\infty$-functions with uniform convergence for all derivatives on all compact subsets*.

Note that in contrast to elements in $\mathcal{S}(\Omega)$ or $\mathcal{D}(\Omega)$, elements in $\mathcal{E}(\Omega)$ are not restricted in their growth near the boundary of Ω. Again we give the basic facts about this test function space.

Theorem 2.2.5 *The test function space $\mathcal{E}(\Omega)$ is a complete metrizable HLCTVS.*

Proof. By taking an increasing sequence of compact subsets K_i which exhaust Ω (compare problem 14 of the Exercises) one shows that the topology can be defined in terms of a countable set of seminorms; hence this space is metrizable. Completeness of the spaces $\mathcal{C}^k(\Omega)[\mathcal{P}_k(\Omega)]$ for all $k = 0, 1, 2, \ldots$ easily implies completeness of $\mathcal{E}(\Omega)$. □

2.2.4 Relation between the test function spaces $\mathcal{D}(\Omega)$, $\mathcal{S}(\Omega)$, and $\mathcal{E}(\Omega)$

It is fairly obvious from their definitions that as sets one has

$$\mathcal{D}(\Omega) \subset \mathcal{S}(\Omega) \subset \mathcal{E}(\Omega). \tag{2.13}$$

The following result shows that this relation also holds for the topological structures as well.

Theorem 2.2.6 *Let $\Omega \subset \mathbb{R}^n$ be a nonempty open subset. Then for the three test function spaces introduced in the previous subsections the following holds: $\mathcal{D}(\Omega)$ is continuously embedded into $\mathcal{S}(\Omega)$ and $\mathcal{S}(\Omega)$ is continuously embedded into $\mathcal{E}(\Omega)$.*

Proof. Denote $i : \mathcal{D}(\Omega) \to \mathcal{S}(\Omega)$ and $j : \mathcal{S}(\Omega) \to \mathcal{E}(\Omega)$ the identical embeddings. We have to show that both are continuous. According to Proposition 2.2.3 the embedding i is continuous if, and only if, the embeddings $i \circ i_K : \mathcal{D}_K(\Omega) \to \mathcal{S}(\Omega)$ are continuous, for every compact subset $K \subset \Omega$. By Theorem 2.1.7 it suffices to show that these linear maps are bounded. Given any semi-norm $p_{m,l} \in \mathcal{P}(\Omega)$ we estimate, for all $\phi \in \mathcal{D}_K(\Omega)$, as follows:

$$p_{m,l}(i \circ i_K(\phi)) = \sup_{\substack{x \in \Omega \\ |\alpha| \le l}} (1+x^2)^{m/2} |D^\alpha \phi(x)| = \sup_{\substack{x \in K \\ |\alpha| \le l}} (1+x^2)^{m/2} |D^\alpha \phi(x)|.$$

We deduce that, for all $\phi \in \mathcal{D}_K(\Omega)$, all $K \subset \Omega$ compact, and all $m, l = 0, 1, 2, \ldots$,

$$p_{m,l}(i \circ i_K(\phi)) \le C p_{K,l}(\phi)$$

where $C = \sup_{x \in K}(1+x^2)^{m/2} < \infty$. Hence the map $i \circ i_K$ is bounded and we conclude continuity of the embedding i.

Similarly we proceed for the embedding j. Take any semi-norm $p_{K,L} \in \mathcal{P}_\infty(\Omega)$ and estimate, for all $\phi \in \mathcal{S}(\Omega)$,

$$p_{K,l}(j(\phi)) = \sup_{\substack{x \in K \\ |\alpha| \le l}} |D^\alpha \phi(x)| \le \sup_{\substack{x \in \Omega \\ |\alpha| \le l}} (1+x^2)^{m/2} |D^\alpha \phi(x)|,$$

i.e., $p_{K,l}(j(\phi)) \le p_{m,l}(\phi)$ for all $\phi \in \mathcal{S}(\Omega)$, for all $K \subset \Omega$ compact and all $m, l = 0, 1, 2, \ldots$. Hence the embedding j is bounded and thus continuous. □

2.3 Exercises

1. Let p be a semi-norm on a vector space V. Show: The null space $N(p) = \{x \in V : p(x) = 0\}$ is a linear subspace of V. $N(p)$ is trivial if, and only if, p is a norm on V.

2. Show: If p is a semi-norm on a vector space V and $r > 0$, then rp, defined by $(rp)(x) = rp(x)$ for all $x \in V$, is again a semi-norm on V. If $p_1, \ldots, p_n$ are semi-norms on V, then their maximum $p = \max\{p_1, \ldots, p_n\}$, defined by $p(x) = \max\{p_1(x), \ldots, p_n(x)\}$ for all $x \in V$, is a semi-norm such that $p_i \le p$ for $i = 1, \ldots, n$.

2.3 Exercises

3. Prove the five properties of open p-balls stated in the text.

4. Let p and q be two norms on $\mathbb{R}^n$. Show: There are positive numbers $r > 0$ and $R > 0$ such that $rq \leq p \leq Rq$. Thus on a finite dimensional space all norms are equivalent.

5. Prove: The systems of semi-norms $\mathcal{P}_k(\Omega)$ and $\mathcal{Q}_k(\Omega)$ on $\mathcal{C}^k(\Omega)$ are filtering.

6. Let $\mathcal{P}$ be a filtering system of semi-norms on a vector space V. Define the p-balls $B_{p,r}(x)$ for $p \in \mathcal{P}$ and $r > 0$ and the topology $\mathcal{T}_\mathcal{P}$ as in Theorem 2.1.1. Show: $B_{p,r}(x) \in \mathcal{T}_\mathcal{P}$, i.e., the balls $B_{p,r}(x)$ are open with respect to the topology $\mathcal{T}_\mathcal{P}$ and thus it is consistent to call them open p-balls.

7. Prove Lemma 2.1.2.

    Hints: Observe that $B_{q,R}(x) \subseteq B_{p,r}(x)$ implies: Whenever $z \in V$ satisfies $q(z) < R$, then it follows that $p(z) < r$. Now fix any $y \in V$ and define, for any $\sigma > 0$, $z = \frac{R}{q(y)+\sigma} y$; it follows that $q(z) = \frac{R}{q(y)+\sigma} q(y) < R$, hence $p(z) = \frac{R}{q(y)+\sigma} p(y) < r$ or $p(y) < \frac{r}{R}(q(y) + \sigma)$. Since $\sigma > 0$ is arbitrary, we conclude that $p(y) \leq \frac{r}{R} q(y)$ and since this holds for any $y \in V$ we conclude that $p \leq \frac{r}{R} q$. The converse direction is straightforward.

8. On the vector space $V = \mathbb{K}^n$, define the following functions:

    (a) $q(x) = \sqrt{\sum_{i=1}^n x_i^2}, \, x = (x_1, \ldots, x_n) \in \mathbb{K}^n$;

    (b) $p(x) = \max\{|x_1|, \ldots, |x_n|\}$;

    (c) $r(x) = |x_1| + \cdots + |x_n|$.

    Show that these functions are actually norms on $\mathbb{K}^n$ and all define the same topology.

9. Show that the two systems of semi-norms $\mathcal{P}_k(\Omega)$ and $\mathcal{Q}_k(\Omega)$ on $\mathcal{C}_K^\infty(\Omega)$ (see section "Examples of HLCVTVS") are equivalent.

    Hints: It is a straightforward estimate to get $q_{K,l}(\phi) \leq C_{K,l} p_{K,l}(\phi)$ for some constant $C_{K,l}$ depending on l and $|K| = \int_K dx$. The converse estimate is particularly simple for $n = 1$. There we use for $\phi \in \mathcal{C}_K^\infty(\Omega)$ and $\alpha = 0, 1, 2, \ldots$ the representation $\phi^{(\alpha)}(x) = \int_{-\infty}^x \phi^{(\alpha+1)}(y) dy$ to estimate $|\phi^{(\alpha)}(x)| \leq |K|^{1/2} (\int_K |\phi^{(\alpha+1)}(y)|^2 dy)^{1/2}$ and therefore $p_{K,l}(\phi) \leq |K|^{1/2} q_{K,l+1}(\phi)$. The general case uses the same idea.

10. Using the fact that $(\mathbb{R}, |\cdot|)$ is a sequentially complete HLCTVS, show that the Euclidean spaces $(\mathbb{R}^n, |\cdot|)$ are sequentially complete HLCTVS too, for any $n \in \mathbb{N}$.

11. Show that $\mathcal{C}^k(\Omega)[\mathcal{P}_k(\Omega)]$ is sequentially complete for $\Omega \subset \mathbb{R}^n$, Ω open and nonempty, $k = 0, 1, 2, \ldots$.

24   2. Spaces of Test Functions

Hints: The underlying ideas of the proof can best be explained for the case $\Omega \subset \mathbb{R}$ and $k = 1$. Given a Cauchy sequence $(f_i)_{i \in \mathbb{N}}$ in $\mathcal{C}^1(\Omega)[\mathcal{P}_1(\Omega)]$ and any compact set $K \subset \Omega$ and any $r > 0$, there is $i_0 \in \mathbb{N}$ such that $p_{K,1}(f_i - f_j) < r$ for all $i, j \geq i_0$. Observe, for $m = 0$ and $m = 1$ and every $x \in K$: $|f_i^{(m)}(x) - f_j^{(m)}(x)| \leq p_{K,1}(f_i - f_j)$. It follows, for $m \in \{0, 1\}$ and all $x \in K$, that $(f_i^{(m)}(x))_{i \in \mathbb{N}}$ is a Cauchy sequence in $\mathbb{K}$ which is known to be complete. Hence each of these Cauchy sequences converges to some number which we call $f_{(m)}(x)$, i.e., $f_{(m)}(x) = \lim_{i \to \infty} f_i^{(m)}(x)$. Thus we get two functions $f_{(m)} : \Omega \to \mathbb{K}$. From the assumed uniform convergence on all compact subsets we deduce that both functions are continuous. Apply uniform convergence again to show for any $x, y \in \Omega$ the following chain of identities: $f_{(0)}(x) - f_{(0)}(y) = \lim_{i \to \infty}(f_i(x) - f_i(y)) = \lim_{i \to \infty} \int_y^x f_i^{(1)}(z)dz = \int_y^x f_{(1)}(z)dz$. Deduce that $f_{(0)}$ is continuously differentiable with derivative $f_{(1)}$ and that the given sequence converges to $f_{(0)}$ in $\mathcal{C}^1(\Omega)[\mathcal{P}_1(\Omega)]$.

12. Using the results of the previous problem show that the spaces $\mathcal{D}_K(\Omega)$ defined in (2.4) are complete.

13. Consider the spaces $\mathcal{D}_K(\Omega)$ and $\mathcal{D}(\Omega)$ as introduced in (2.4), respectively (2.7) and denote by $i_K : \mathcal{D}_K(\Omega) \to \mathcal{D}(\Omega)$ the identical embedding for $K \subset \Omega$ compact. Show: There is a strongest locally convex topology $\mathcal{T}$ on $\mathcal{D}(\Omega)$ such that all embeddings i_K are continuous. This topology is Hausdorff.

14. Prove: For any open nonempty subset $\Omega \subseteq \mathbb{R}^n$ there is a sequence of compact sets $K_i \subset \Omega$ with the following properties: Each set K_i has a nonempty interior. K_i is properly contained in K_{i+1}. $\bigcup_{i=1}^\infty K_i = \Omega$.

    Hints: For $i \in \mathbb{N}$ define $\Omega_i = \{x \in \Omega : \text{dist}(x, \partial\Omega) \geq \frac{1}{i}\}$ and $B_i = \{x \in \mathbb{R}^n : |x| \leq i\}$. Here $\text{dist}(x, \partial\Omega)$ denotes the Euclidean distance of the point $x \in \Omega$ from the boundary of Ω. Then show that the sets $K_i = B_i \cap \Omega_i$, for i sufficiently large, have the properties as claimed.

15. Let $\Omega \subset \mathbb{R}^n$ be an open nonempty set. Show: For every closed ball $K_r(x) = \{y \in \mathbb{R}^n : |y - x| \leq r\} \subset \Omega$ with centre $x \in \Omega$ and radius $r > 0$ there is a $\phi \in \mathcal{D}(\Omega), \phi \neq 0$, with support $\text{supp}\,\phi \subseteq K_r(x)$. Thus, in particular, $\mathcal{D}(\Omega)$ is not empty.

    Hints: Define a function $\rho : \mathbb{R}^n \to \mathbb{R}$ by

    $$\rho(x) = \begin{cases} 0 & : \text{ for } |x| \geq 1, \\ \exp\frac{-1}{1-x^2} & : \text{ for } |x| < 1, \end{cases} \quad (2.14)$$

    and show that $\rho \in \mathcal{C}^\infty(\mathbb{R}^n)$. Then define $\phi_r(y) = \rho(\frac{y-x}{r})$ and deduce that $\phi_r \in \mathcal{D}(\Omega)$ has the desired support properties.

16. Prove: The space $\mathcal{S}(\Omega)$ is complete.

    Hints: One can use the fact that the spaces $\mathcal{C}^k(\Omega)[\mathcal{P}_k(\Omega)]$ are complete, for any $k \in \mathbb{N}$. The decay properties need some additional considerations.

3
Schwartz Distributions

As we had mentioned in the introduction the Schwartz approach to distribution theory defines distributions as continuous linear functions on a test function space. The various classes of distributions are distinguished by the underlying test function spaces. Before we come to the definition of the main classes of Schwartz distribution we collect some basic facts about continuous linear functions or functionals on a HLCTVS and about spaces of such functionals. Then the definition of the three main spaces of Schwartz distributions is straightforward. Numerous examples explain this definition.

The remainder of this chapter introduces convergence of sequences and series of distributions, discusses localization, in particular support and singular support of distributions.

3.1 The topological dual of a HLCTVS

Suppose that X is a vector space over the field $\mathbb{K}$ on which a filtering system $\mathcal{P}$ of semi-norms is given such that $X[\mathcal{P}]$ is a HLCTVS. The *algebraic dual* X^* of X has been defined as the set of all linear functions or *functionals* $f : X \to \mathbb{K}$. The *topological dual* is defined as the subset of those linear functions which are continuous, i.e.,

$$X' \equiv X[\mathcal{P}]' = \{f \in X^* : f \text{ continuous}\} \tag{3.1}$$

In a natural way, both X^* and X' are vector spaces over $\mathbb{K}$. As a special case of Theorem 2.1.7 the following result is a convenient characterization of the elements of the topological dual of a HLCTVS.

3. Schwartz Distributions

Proposition 3.1.1 *Suppose that $X[\mathcal{P}]$ is a HLCTVS and $f : X \to \mathbb{K}$ a linear function. Then the following statements are equivalent.*

(a) *f is continuous, i.e., $f \in X'$.*

(b) *There is a semi-norm $p \in \mathcal{P}$ and a nonnegative number λ such that $|f(x)| \leq \lambda p(x)$ for all $x \in X$.*

(c) *There is a semi-norm $p \in \mathcal{P}$ such that f is bounded on the p-ball $B_{p,1}(0)$.*

Proof. The equivalence of statements (a) and (b) is just the special case $Y[\mathcal{Q}] = \mathbb{K}[\{|\cdot|\}]$ of Theorem 2.1.7.

The equivalence of (b) and (c) follows easily from Lemma 2.1.2 if we introduce the semi-norm $q(x) = |f(x)|$ on X and if we observe that then (b) says $q \leq \lambda p$ while (c) translates into $B_{p,1}(0) \subseteq B_{q,\lambda}(0)$. □

The *geometrical interpretation* of linear functionals is often helpful, in particular in infinite dimensional spaces. We give a brief review. Recall: A *hyperplane through the origin* is a maximal proper subspace of a vector space X. If such a hyperplane is given there is a point $a \in X \setminus H$ such that the vector space X over the field $\mathbb{K}$ has the representation

$$X = H + \mathbb{K}a,$$

i.e., every point $x \in X$ has the unique representation $x = h + \alpha a$ with $h \in H$ and $\alpha \in \mathbb{K}$. The announced geometrical characterization now is

Proposition 3.1.2 *Let $X[\mathcal{P}]$ be a HLCTVS over the field $\mathbb{K}$.*

(a) *A linear functional $f \in X^*$, $f \neq 0$, is characterized by*

   (i) *a hyperplane $H \subset X$ through the origin and*

   (ii) *the value in a point $x_0 \in X \setminus H$.*

   The connection between the functional f and the hyperplane is given by

   $$H = \ker f = \{x \in X : f(x) = 0\}.$$

(b) *A linear functional f on X is continuous if, and only if, in the geometric characterization a) the hyperplane H is closed.*

Proof. Given $f \in X^*$ the kernel or null space $\ker f$ is easily seen to be a linear subspace of X. Since $f \neq 0$ there is a point in X at which f does not vanish. By re-scaling this point we get a point $a \in X \setminus \ker f$ with $f(a) = 1$. We claim that $H = \ker f$ is a hyperplane. Given any point $x \in X$ observe $x = x - f(x)a + f(x)a$ where $h = x - f(x)a \in \ker f$ since $f(h) = f(x) - f(x)f(a) = 0$ and $f(x)a \in \mathbb{K}a$. The representation $x = h + \alpha a$ with $h \in \ker f$ and $\alpha \in \mathbb{K}$ is unique: If one has, for some $x \in X$, $x = h_1 + \alpha_1 a = h_2 + \alpha_2 a$ with $h_i \ker f$ then $h_1 - h_2 = (\alpha_1 - \alpha_2)$ and thus $0 = f(h_1 - h_2) = (\alpha_1 - \alpha_2)f(a) = \alpha_1 - \alpha_2$, hence $\alpha_1 = \alpha_2$ and $h_1 = h_2$.

Conversely assume that H is a hyperplane through the origin and $a \in X \setminus H$. Then every point $x \in X$ has the unique representation $x = h + \alpha a$ with $h \in H$ and $\alpha \in \mathbb{K}$. Now define $f_H : X \to \mathbb{K}$

by $f_H(x) = f_H(h + \alpha a) = \alpha$. It is an elementary calculation to show that f_H is a well defined linear function. Certainly one has $\ker f_H = H$. This proves part (a).

In order to prove part (b) we have to show that $H = \ker f$ is closed if, and only if, the linear functional f is continuous. When f is continuous then $\ker f$ is closed as the inverse image of the closed set $\{0\}$. Conversely assume that $H = \ker f$ is closed. Then its complement $X \setminus H$ is open and there is some open p-ball $B_{p,r}(a) \subset X \setminus H$ around the point a, $f(a) = 1$. In order to prove continuity of f it suffices, according to Proposition 3.1.1, to show that f is bounded on the open ball $B_{p,r}(0)$. This is done indirectly. If there were some $x \in B_{p,r}(0)$ with $|f(x)| \geq 1$ then $y = a - \frac{x}{f(x)} \in B_{p,r}(a)$ and $f(y) = f(a) - \frac{f(x)}{f(x)} = 1 - 1 = 0$, i.e., $y \in H$, a contradiction. Therefore f is bounded on $B_{p,r}(0)$ by 1 and we conclude. $\square$

3.2 Definition of distributions

For an open nonempty subset $\Omega \subset \mathbb{R}^n$ we have introduced the test function spaces $\mathcal{D}(\Omega)$, $\mathcal{S}(\Omega)$, and $\mathcal{E}(\Omega)$ as Hausdorff locally convex topological vector spaces. Furthermore the relation

$$\mathcal{D}(\Omega) \subset \mathcal{S}(\Omega) \subset \mathcal{E}(\Omega)$$

with continuous embeddings in both cases has been established (see Theorem 2.2.6). This section gives the basic definitions of the three basic classes of distributions as elements of the topological dual space of these test function spaces. Elements of the topological dual $\mathcal{D}'(\Omega)$ of $\mathcal{D}(\Omega)$ are called *distributions on Ω*. Elements of the topological dual $\mathcal{S}'(\Omega)$ of $\mathcal{S}(\Omega)$ are called *tempered distributions* and elements of topological $\mathcal{E}'(\Omega)$ of $\mathcal{E}(\Omega)$ are called *distributions of compact support*. Later, after further preparation, the names for the latter two classes of distributions will be apparent. The continuous embeddings mentioned above imply the following relation between these three classes of distributions and it justifies calling elements in $\mathcal{S}'(\Omega)$, respectively in $\mathcal{E}'(\Omega)$, distributions:

$$\mathcal{E}'(\Omega) \subset \mathcal{S}'(\Omega) \subset \mathcal{D}'(\Omega). \tag{3.2}$$

We proceed with a more explicit discussion of distributions.

Definition 3.2.1 *A* **distribution** *T on an open nonempty subset $\Omega \subset \mathbb{R}^n$ is a continuous linear functional on the test function space $\mathcal{D}(\Omega)$ of C^∞-functions of compact support. The* **set of all distributions on Ω** *equals the topological dual $\mathcal{D}'(\Omega)$ of $\mathcal{D}(\Omega)$.*

Another way to define a distribution on a nonempty open subset $\Omega \subset \mathbb{R}^n$ is to recall Proposition 2.2.3 and to define: A linear functional T on $\mathcal{D}(\Omega)$ is a distribution on Ω if, and only if, its restriction to the spaces $\mathcal{D}_K(\Omega)$ is continuous for every compact subset $K \subset \Omega$. Taking Theorem 2.1.7 into account one arrives at the following characterization of distributions.

Theorem 3.2.1 *A linear functional $T : \mathcal{D}(\Omega) \to \mathbb{K}$ is a distribution on the open nonempty set $\Omega \subset \mathbb{R}^n$ if, and only if, for every compact subset $K \subset \Omega$ there exist* a number $C \in \mathbb{R}_+$ and a natural number $m \in \mathbb{N}$, both depending in general on K and T, such that for all $\phi \in \mathcal{D}_K(\Omega)$ the estimate

$$|T(\phi)| \leq C p_{K,m}(\phi) \tag{3.3}$$

holds.

An equivalent way to express this is the following:

Corollary 3.2.1 *A linear function $T : \mathcal{D}(\Omega) \to \mathbb{K}$ is a distribution on Ω if, and only if, for every compact subset $K \subset \Omega$ there is an integer m such that*

$$p'_{K,m}(T) = \sup\{|T(\phi)| : \phi \in \mathcal{D}_K(\Omega),\ p_{K,m}(\phi) \leq 1\} \tag{3.4}$$

is finite and then

$$|T(\phi)| \leq p'_{K,m}(T) p_{K,m}(\phi) \quad \forall \phi \in \mathcal{D}_K(\Omega).$$

The proof of the corollary is left as an exercise. This characterization leads to the important concept of the *order* of a distribution.

Definition 3.2.2 *Let T be a distribution on $\Omega \subset \mathbb{R}^n$, Ω open and nonempty, and let $K \subset \Omega$ be a compact subset. Then the **local order** $O(T, K)$ of T on K is defined as the minimum of all natural numbers m for which 3.3 holds. The **order** $O(T)$ of T is the supremum over all local orders.*

In terms of the concept of order, Theorem 3.2.1 says: Locally every distribution is of finite order, i.e., a finite number of derivatives of the test functions ϕ are used in the estimate (3.3) (recall the definition of the semi-norms $p_{K,m}$ in equation (2.1)).

Remark 3.2.1   1. *As the topological dual of the HLCTVS $\mathcal{D}(\Omega)$, the set of all distributions on an open set $\Omega \subset \mathbb{R}^n$ forms naturally a vector space over the field $\mathbb{K}$. Addition and scalar multiplication are explicitly given as follows: For all $T, T_i \in \mathcal{D}'(\Omega)$ and all $\lambda \in \mathbb{K}$,*

$$\forall_{\phi \in \mathcal{D}(\Omega)} \quad (T_1 + T_2)(\phi) = T_1(\phi) + T_2(\phi), \qquad (\lambda T)(\phi) = \lambda T(\phi).$$

Thus $(T, \phi) \mapsto T(\phi)$ is a bilinear function $\mathcal{D}' \times \mathcal{D} \to \mathbb{K}$.

2. *According to their definition, distributions assign real or complex numbers $T(\phi)$ to a test function $\phi \in \mathcal{D}(\Omega)$. A frequently used other notation for the value $T(\phi)$ of the function T is*

$$T(\phi) = \langle T, \phi \rangle = \langle T(x), \phi(x) \rangle.$$

3. *In physics textbooks one often finds the notation $\int_\Omega T(x) \phi(x) dx$ for the value $T(\phi)$ of the distribution T at the test function ϕ. This suggestive notation is rather formal since when one wants to make sense out of this expression the integral sign used has little to do with the standard integrals (further details are provided in the section on representation of distributions as 'generalized' derivatives of continuous functions).*

4. The axiom of choice allows us to show that there are linear functionals on $\mathcal{D}_K(\Omega)$ which are not continuous. But nobody has succeeded in giving an explicit example of such a noncontinuous functional. Thus in practice one does not encounter these exceptional functionals.

5. One may wonder why we spoke about $\mathcal{D}(\Omega)$ as **the** test function space of distribution theory. Naturally, $\mathcal{D}(\Omega)$ is not given à priori. One has to make a choice. The use of $\mathcal{D}(\Omega)$ is justified à posteriori by many successful applications. Nevertheless there are some guiding principles for the choice of test function spaces (compare the introductory remarks on the goals of distribution theory).

   a) The choice of test function spaces as subspaces of the space of $\mathcal{C}^\infty$-functions on which all derivative monomials D^α act linearly and continuously ensure that all distributions will be infinitely often differentiable too.

   b) Further restrictions on the subspace of $\mathcal{C}^\infty$-functions as a test function space depends on the intended use of the resulting space of generalized functions. For instance, the choice of $\mathcal{C}^\infty$-functions on Ω with compact support ensures that the resulting distributions on Ω are not restricted in their behavior at the boundary of the set Ω. Later we will see that the test function space of $\mathcal{C}^\infty$-functions which are strongly decreasing ensures that the resulting space of generalized functions admits the Fourier transformation as an isomorphism, which has many important consequences.

A number of concrete **Examples** will help to explain how the above definition operates in concrete cases. The first class of examples show furthermore how distributions generalize functions so that it is appropriate to speak about distributions as special classes of generalized functions. Later we will give an overview of some other classes of generalized functions.

3.2.1 The regular distributions

Suppose that $f : \Omega \to \mathbb{K}$ is a continuous function on the open nonempty set $\Omega \subset \mathbb{R}^n$. Then, for every compact subset $K \subset \Omega$ the (Riemann) integral $\int_K |f(x)|dx = C$ is known to exist. Hence for all $\phi \in \mathcal{D}_K(\Omega)$ one has

$$\left| \int_K f(x)\phi(x)dx \right| \leq \int_K |f(x)\phi(x)|dx \leq \sup_{x \in K} |\phi(x)| \int_K |f(x)|dx.$$

It follows that $I_f : \mathcal{D}(\Omega) \to \mathbb{K}$ is well defined by

$$\langle I_f, \phi \rangle = \int f(x)\phi(x)dx \qquad \forall \phi \in \mathcal{D}(\Omega)$$

and that for all $\phi \in \mathcal{D}_K(\Omega)$ one has the estimate

$$|\langle I_f, \phi \rangle| \leq C p_{K,0}(\phi).$$

Elementary properties of the Riemann integral imply that I_f is a linear functional on $\mathcal{D}(\Omega)$. Since we could establish the estimate 3.3 in Theorem 3.2.1 it follows that I_f is continuous and thus a distribution on Ω. In addition this estimate shows that the local order and the order of the distribution I_f is 0.

Obviously these considerations apply to any $f \in C(\Omega)$. Therefore $f \mapsto I_f$ defines a map $I : C(\Omega) \to \mathcal{D}'(\Omega)$ which is easily seen to be linear. In the Exercises it is shown that I is injective and thus provides an embedding of the space of all continuous functions into the space of distributions.

Note that the decisive property we used for the embedding of continuous functions into the space of distributions was that, for $f \in C(\Omega)$ and every compact subset, the Riemann integral $C = \int_K |f(x)|dx$ is finite. Therefore the same ideas allows us to consider a much larger space of functions on Ω as distributions, namely the space $L^1_{loc}(\Omega)$ of all *locally integrable functions on* Ω. $L^1_{loc}(\Omega)$ is the space of all (equivalence classes of) Lebesgues measurable functions on Ω for which the Lebesgue integral

$$\|f\|_{1,K} = \int_K |f(x)|dx \qquad (3.5)$$

is finite for every compact subset $K \subset \Omega$. Thus the map I can be extended to a map $I : L^1_{loc}(\Omega) \to \mathcal{D}'(\Omega)$ by the same formula: For every $f \in L^1_{loc}(\Omega)$ define $I_f : \mathcal{D}(\Omega) \to \mathbb{K}$ by

$$I_f(\phi) = \int f(x)\phi(x)dx \qquad \forall \phi \in \mathcal{D}(\Omega).$$

The bound $|I_f(\phi)| \leq \|f\|_{1,K}\, p_{K,0}(\phi)$ for all $\phi \in \mathcal{D}_K(\Omega)$ proves as above that $I_f \in \mathcal{D}'(\Omega)$ for all $f \in L^1_{loc}(\Omega)$. A simple argument implies that I is a linear map and in the Exercises we prove that I is injective, i.e., $I_f = 0$ in $\mathcal{D}'(\Omega)$ if, and only if, $f = 0$ in $L^1_{loc}(\Omega)$. Therefore I is an embedding of $L^1_{loc}(\Omega)$ into $\mathcal{D}'(\Omega)$. The space $L^1_{loc}(\Omega)$ is a HLCTVS when it is equipped with the filtering system of semi-norms $\{\|\cdot\|_{1,K} : K \subset \Omega, \text{ compact}\}$. With respect to this topology the embedding I is continuous in the following sense. If $(f_j)_{j \in \mathbb{N}}$ is a sequence which converges to zero in $L^1_{loc}(\Omega)$, then, for every $\phi \in \mathcal{D}(\Omega)$, one has $\lim_{j \to \infty} I_{f_j}(\phi) = 0$ which follows easily from the bound given above. We summarize our discussion as the so-called *embedding theorem*.

Theorem 3.2.2 *The space $L^1_{loc}(\Omega)$ of locally integrable functions on an open nonempty set $\Omega \subset \mathbb{R}^n$ is embedded into the space $\mathcal{D}'(\Omega)$ of distributions on Ω by the linear and continuous injection I. The image of $L^1_{loc}(\Omega)$ under I is called the space of* **regular distributions** *on Ω:*

$$\mathcal{D}'_{\text{reg}}(\Omega) = I(L^1_{loc}(\Omega)) \subset \mathcal{D}'(\Omega). \qquad (3.6)$$

Note that under the identification of f and I_f we have established the following chain of relations:

$$\mathcal{C}(\Omega) \subset L^r_{loc}(\Omega) \subset L^1_{loc}(\Omega) \subset \mathcal{D}'(\Omega)$$

for any $r \geq 1$, since for $r > 1$ the space of measurable functions f on Ω for which $|f|^r$ is locally integrable is known to be contained in $L^1_{loc}(\Omega)$.

3.2.2 Some standard examples of distributions

Dirac's delta distribution.

For any point $a \in \Omega \subset \mathbb{R}^n$ define a functional $\delta_a : \mathcal{D}(\Omega) \to \mathbb{K}$ by

$$\delta_a(\phi) = \phi(a) \qquad \forall \phi \in \mathcal{D}(\Omega).$$

Obviously δ_a is linear. For any compact subset $K \subset \Omega$ one has the following estimate:

$$|\delta_a(\phi)| \leq C(a, K) p_{K,0}(\phi) \qquad \forall \phi \in \mathcal{D}_K(\Omega)$$

where the constant $C(a, K)$ equals 1 if $a \in K$ and $C(a, K) = 0$ otherwise. Therefore the linear functional δ_a is continuous on $\mathcal{D}(\Omega)$ and thus a distribution. Its order obviously is zero. In the Exercises it is shown that δ_a is not a regular distribution, i.e., there is no $f \in L^1_{loc}(\Omega)$ such that $\delta_a(\phi) = \int f(x)\phi(x)dx$ for all $\phi \in \mathcal{D}(\Omega)$.

Cauchy's principal value.

It is easy to see that $x \mapsto \frac{1}{x}$ is not a locally integrable function on the real line $\mathbb{R}$, hence $I_{\frac{1}{x}}$ does not define a regular distribution. Nevertheless one can define a distribution on $\mathbb{R}$ which agrees with $I_{\frac{1}{x}}$ on $\mathbb{R} \setminus \{0\}$. This distribution is called *Cauchy's principal value* and is defined by

$$\langle \mathrm{vp}\frac{1}{x}, \phi \rangle = \lim_{r \to 0} \int_{|x| \geq r} \frac{\phi(x)}{x} dx. \tag{3.7}$$

We have to show that this limit exists and that it defines a continuous linear functional on $\mathcal{D}(\mathbb{R})$. For $a > 0$ consider the compact interval $K = [-a, a]$. Take $0 < r < a$ and calculate, for all $\phi \in \mathcal{D}_K(\mathbb{R})$,

$$\int_{|x| \geq r} \frac{\phi(x)}{x} dx = \int_r^a \frac{\phi(x) - \phi(-x)}{x} dx.$$

If we observe that $\phi(x) - \phi(-x) = x \int_{-1}^{+1} \phi'(xt) dt$, we get the estimate

$$\left| \frac{\phi(x) - \phi(-x)}{x} \right| \leq 2 \sup_{y \in K} |\phi'(y)| \leq 2 p_{K,1}(\phi),$$

and thus $|\int_r^a \frac{\phi(x)-\phi(-x)}{x}dx| \leq 2ap_{K,1}(\phi)$ uniformly in $0 < r < a$, for all $\phi \in \mathcal{D}_K(\mathbb{R})$. It follows that this limit exists and that it has the value:

$$\lim_{r \to 0} \int_{|x| \geq r} \frac{\phi(x)}{x} dx = \int_0^\infty \frac{\phi(x) - \phi(-x)}{x} dx.$$

Furthermore the continuity bound

$$|\langle \mathrm{vp} \frac{1}{x}, \phi \rangle| \leq |K| p_{K,1}(\phi)$$

for all $\phi \in \mathcal{D}_K(\mathbb{R})$ follows. Therefore $\mathrm{vp}\frac{1}{x}$ is a well defined distribution on $\mathbb{R}$ according to Theorem 3.2.1. Its order obviously is 1.

The above proof gives the following convenient formula for Cauchy's principal value:

$$\langle \mathrm{vp} \frac{1}{x}, \phi \rangle = \int_0^\infty \frac{\phi(x) - \phi(-x)}{x} dx. \tag{3.8}$$

Test functions in $\mathcal{D}(\mathbb{R}\setminus\{0\})$ have the property that they vanish in some neighborhood of the origin (depending on the function). Hence for these test function the singular point $x = 0$ of $\frac{1}{x}$ is avoided and thus it follows that

$$\lim_{r \to 0} \int_{|x| \geq r} \frac{\phi(x)}{x} dx = \int_\mathbb{R} \frac{\phi(x)}{x} dx = \langle I_{\frac{1}{x}}, \phi \rangle \qquad \forall \phi \in \mathcal{D}(\mathbb{R}\setminus\{0\}).$$

Sometimes one also finds the notation $\mathrm{vp} \int_\mathbb{R} \frac{\phi(x)}{x} dx$ for $\langle \mathrm{vp} \frac{1}{x}, \phi \rangle$. The letters 'vp' in the notation for Cauchy's principal value stand for the original French name 'valeur principale'.

Hadamard's principal values.

Closely related to Cauchy's principal value is a family of distributions on $\mathbb{R}$ which can be traced back to Hadamard. Certainly, for $1 < \beta < 2$ the function $\frac{1}{x^\beta}$ is not locally integrable on $\mathbb{R}_+$. We are going to define a distribution T on $\mathbb{R}_+$ which agrees on $\mathbb{R}_+\setminus\{0\} = (0, \infty)$ with the regular distribution $I_{x^{-\beta}}$. For all $\phi \in \mathcal{D}(\mathbb{R})$ define

$$\langle T, \phi \rangle = \int_0^\infty \frac{\phi(x) - \phi(0)}{x^\beta} dx.$$

Since again $\phi(x) - \phi(0) = x \int_0^1 \phi'(xt) dt$ we can estimate

$$|\frac{\phi(x) - \phi(0)}{x^\beta}| \leq |x|^{1-\beta} p_{K,1}(\phi)$$

if $\phi \in \mathcal{D}_K(\mathbb{R})$. Since now the exponent $\gamma = 1 - \beta$ is larger than -1, the integral exists over compact subsets. Hence T is well defined on $\mathcal{D}(\mathbb{R})$. Elementary properties of integrals imply that T is linear and the above estimate implies, as in the previous example, the continuity bound. Therefore T is a distribution on $\mathbb{R}$.

If $\phi \in \mathcal{D}(\mathbb{R}\setminus\{0\})$, then in particular $\phi(x) = 0$ for all $x \in \mathbb{R}$, $|x| \le r$ for some $r > 0$, and we get $\langle T, \phi \rangle = \int_0^\infty \frac{\phi(x)}{x^\beta} dx = I_{\frac{1}{x^\beta}}(\phi)$. Hence on $\mathbb{R}\setminus\{0\}$ the distribution T is regular.

Distributions like Cauchy's and Hadamard's principal values are also called *pseudo functions* since away from the origin $x = 0$ they coincide with the corresponding regular distributions. Thus we can consider the pseudo functions as extensions of the regular distributions to the point $x = 0$.

3.3 Convergence of sequences and series of distributions

Often the need arises to approximate given distributions by 'simpler' distributions, for instance functions. For this one obviously needs a topology on the space $\mathcal{D}'(\Omega)$ of all distributions on a nonempty open set $\Omega \subset \mathbb{R}^n$. A topology which suffices for our purposes is the so-called *weak topology* which is defined on $\mathcal{D}'(\Omega)$ by the system of semi-norms $\mathcal{P}_\sigma = \{\rho_\phi : \phi \in \mathcal{D}(\Omega)\}$. Here ρ_ϕ is defined by

$$\rho_\phi(T) = |\langle T, \phi \rangle| = |T(\phi)| \quad \forall T \in \mathcal{D}'(\Omega).$$

This topology is usually denoted by $\sigma \equiv \sigma(\mathcal{D}', \mathcal{D})$.

If not stated explicitly otherwise we consider $\mathcal{D}'(\Omega)$ always equipped with this topology σ. Then, from our earlier discussions on HLCTVS, we know in principle what convergence in $\mathcal{D}'$ means or what a Cauchy sequence of distributions is. For clarity we write down these definitions explicitly.

Definition 3.3.1 *Let $\Omega \subset \mathbb{R}^n$ be open and nonempty and let $(T_j)_{j \in \mathbb{N}}$ be a sequence of distributions on Ω, i.e., a sequence in $\mathcal{D}'(\Omega)$. One says:*

1. *$(T_j)_{j \in \mathbb{N}}$ **converges in** $\mathcal{D}'(\Omega)$ if, and only if, there is a $T \in \mathcal{D}'(\Omega)$ such that for every $\phi \in \mathcal{D}(\Omega)$ the numerical sequence $(T_j(\phi))_{j \in \mathbb{N}}$ converges in $\mathbb{K}$ to $T(\phi)$.*

2. *$(T_j)_{j \in \mathbb{N}}$ is a **Cauchy sequence** in $\mathcal{D}'(\Omega)$ if, and only if, for every $\phi \in \mathcal{D}(\Omega)$ the numerical sequence $(T_j(\phi))_{j \in \mathbb{N}}$ is a Cauchy sequence in $\mathbb{K}$.*

Several simple examples will illustrate these definitions and how these concepts are applied to concrete problems. All sequences we consider here are sequences of regular distributions defined by sequences of functions which have no limit in the sense of functions.

Example 3.3.1 1. *The sequence of $\mathcal{C}^\infty$-functions $f_j(x) = \sin jx$ on $\mathbb{R}$ certainly has no limit in the sense of functions. We claim that the sequence of regular distributions $T_j = I_{f_j}$ defined by these functions converges in $\mathcal{D}'(\mathbb{R})$ to zero. For the proof take any $\phi \in \mathcal{D}(\mathbb{R})$. A partial integration shows that*

36   3. Schwartz Distributions

$$\langle T_j, \phi \rangle = \int \sin(jx)\phi(x)dx = \frac{1}{j}\int \cos(jx)\phi'(x)dx$$

and we conclude that $\lim_{j\to\infty}\langle T_j, \phi\rangle = 0$.

2. **Delta sequences:** δ-sequences are sequences of functions which converge in $\mathcal{D}'$ to Dirac's delta distribution. We present three examples of such sequences.

   a) Consider the sequence of continuous functions $t_j(x) = \frac{\sin(jx)}{x}$ and denote $T_j = I_{t_j}$. Then

$$\lim_{j\to\infty} T_j = \pi\delta \quad \text{in } \mathcal{D}'(\mathbb{R}).$$

   For the proof take any $\phi \in \mathcal{D}(\mathbb{R})$. Then the support of ϕ is contained in $[-a, a]$ for some $a > 0$. It follows that

$$\langle T_j, \phi\rangle = \int_{-a}^{+a} \frac{\sin(jx)}{x}\phi(x)dx$$
$$= \int_{-a}^{+a} \frac{\sin(jx)}{x}[\phi(x) - \phi(0)]dx + \int_{-a}^{+a} \frac{\sin(jx)}{x}\phi(0)dx.$$

   As in the first example one shows that

$$\int_{-a}^{+a} \frac{\sin(jx)}{x}[\phi(x)-\phi(0)]dx = \frac{1}{j}\int_{-a}^{+a}\cos(jx)\frac{d}{dx}(\frac{\phi(x)-\phi(0)}{x})dx$$

   converges to zero for $j \to \infty$. Then recall the integral:

$$\int_{-a}^{+a} \frac{\sin(jx)}{x}dx = \int_{-ja}^{+ja}\frac{\sin y}{y}dy \to_{j\to\infty} \int_{-\infty}^{+\infty}\frac{\sin y}{y}dy = \pi.$$

   We conclude that $\lim_{j\to\infty}\langle T_j, \phi\rangle = \pi\phi(0)$ for every $\phi \in \mathcal{D}(\mathbb{R})$ which proves the statement.

   b) Take any nonnegative function $f \in L^1(\mathbb{R}^n)$ with $\int_{\mathbb{R}^n} f(x)dx = 1$. Introduce the sequence of functions $f_j(x) = j^n f(jx)$ and the associated sequence of regular distributions $T_j = I_{f_j}$. We claim:

$$\lim_{j\to\infty} T_j = \delta \quad \text{in } \mathcal{D}'(\mathbb{R}^n).$$

   The proof is simple. Take any $\phi \in \mathcal{D}(\mathbb{R}^n)$ and calculate as above,

$$\langle T_j, \phi\rangle = \int_{\mathbb{R}^n} f_j(x)\phi(x)dx$$
$$= \int_{\mathbb{R}^n} f_j(x)[\phi(x) - \phi(0)]dx + \int_{\mathbb{R}^n} f_j(x)\phi(0)dx.$$

   To the first term

$$\int_{\mathbb{R}^n} f_j(x)[\phi(x) - \phi(0)]dx = \int_{\mathbb{R}^n} j^n f(jx)[\phi(x) - \phi(0)]dx$$
$$= \int_{\mathbb{R}^n} f(y)[\phi(\frac{y}{j}) - \phi(0)]dy$$

we apply Lebesgue's dominated convergence theorem to conclude that the limit $j \to \infty$ of this term vanishes. For the second term note that $\int_{\mathbb{R}^n} f_j(x)dx = \int_{\mathbb{R}^n} f(y)dy = 1$ for all $j \in \mathbb{N}$ and we conclude.

As a special case of this result we mention that we can take in particular $f \in \mathcal{D}(\mathbb{R}^n)$. This then shows that Dirac's delta distribution is the limit in $\mathcal{D}'$ of a sequence of C^∞-functions of compact support.

c) For the last example of a delta sequence we start with the Gauss function on $\mathbb{R}^n$: $g(x) = (\pi)^{-\frac{n}{2}} \exp{-x^2}$. Certainly $0 \leq g \in L^1(\mathbb{R}^n)$ and thus we can proceed as in the previous example. The sequence of scaled Gauss functions $g_j(x) = j^n g(jx)$ converges in the sense of distributions to Dirac's delta distribution, i.e., for every $\phi \in \mathcal{D}(\mathbb{R}^n)$:

$$\lim_{j \to \infty} \langle I_{g_j}, \phi \rangle = \phi(0) = \langle \delta, \phi \rangle.$$

This example shows that Dirac's delta can also be approximated by a sequence of strongly decreasing C^∞-functions.

3. Now we prove the **Breit–Wigner formula**. For each $\epsilon > 0$ define a function $f_\epsilon \to \mathbb{R}$ by

$$f_\epsilon(x) = \frac{\epsilon}{x^2 + \epsilon^2} = \operatorname{Im} \frac{1}{x - i\epsilon} = \frac{i}{2}\left[\frac{1}{x + i\epsilon} - \frac{1}{x - i\epsilon}\right].$$

We claim that

$$\lim_{\epsilon \to 0} I_{f_\epsilon} = \pi \delta \quad \text{in } \mathcal{D}'(\mathbb{R}). \tag{3.9}$$

Often this is written as

$$\lim_{\epsilon \to 0} \frac{\epsilon}{x^2 + \epsilon^2} = \pi \delta$$

(Breit–Wigner formula).

This is actually a special case of a delta sequence: The function $h(x) = \frac{1}{1+x^2}$ satisfies $0 \leq h \in L^1(\mathbb{R})$ and $\int_{\mathbb{R}} h(x)dx = \pi$. Thus one can take $h_j(x) = jh(jx) = f_\epsilon(x)$ for $\epsilon = \frac{1}{j}$ and apply the second result on delta sequences..

4. Closely related to the Breit–Wigner formula is the **Sokhotski–Plemelji formula**. It reads

$$\lim_{\epsilon \to 0} \frac{1}{x \pm i\epsilon} = \mp i\pi\delta + \operatorname{vp}\frac{1}{x} \quad \text{in } \mathcal{D}'(\mathbb{R}). \tag{3.10}$$

Both formulas are used quite often in quantum mechanics.

For any $\epsilon > 0$ we have

$$\frac{1}{x \pm i\epsilon} = \operatorname{Re}\frac{1}{x \pm i\epsilon} + i\operatorname{Im}\frac{1}{x \pm i\epsilon}$$

where
$$\operatorname{Re}\frac{1}{x\pm i\epsilon} = \frac{x}{x^2+\epsilon^2} \equiv g_\epsilon(x),$$
$$\operatorname{Im}\frac{1}{x\pm i\epsilon} = \mp\frac{\epsilon}{x^2+\epsilon^2} \equiv \mp f_\epsilon(x).$$

The limit of f_ϵ for $\epsilon \to 0$ has been determined for the Breit–Wigner formula. To find the same limit for the functions g_ϵ note first that g_ϵ is not integrable on $\mathbb{R}$. It is only locally integrable. Take any $\phi \in \mathcal{D}(\mathbb{R})$ and observe that the functions g_ϵ are odd. Thus we get

$$\langle I_{g_\epsilon}, \phi \rangle = \int_\mathbb{R} g_\epsilon(x)\phi(x)dx = \int_0^\infty g_\epsilon(x)[\phi(x) - \phi(-x)]dx.$$

Rewrite the integrand as

$$g_\epsilon(x)[\phi(x) - \phi(-x)] = xg_\epsilon(x)\frac{\phi(x) - \phi(-x)}{x}$$

and observe that the function $\frac{\phi(x)-\phi(-x)}{x}$ belongs to $L^1(\mathbb{R})$ while the functions $xg_\epsilon(x)$ are bounded on $\mathbb{R}$ by 1 and converge, for $x \neq 0$, pointwise to 1 as $\epsilon \to 0$. Lebesgue's dominated convergence theorem thus implies that

$$\lim_{\epsilon\to 0}\int_\mathbb{R} g_\epsilon(x)\phi(x)dx = \int_0^\infty \frac{\phi(x)-\phi(-x)}{x}dx,$$

or

$$\lim_{\epsilon\to 0}\frac{x}{x^2+\epsilon^2} = \operatorname{vp}\frac{1}{x} \quad \text{in } \mathcal{D}'(\mathbb{R}) \tag{3.11}$$

where we have taken equation (3.8) into account. Equation (3.11) and the Breit–Wigner formula together imply easily the Sokhotski–Plemelj formula.

These concrete examples illustrate various practical aspects which have to be addressed in the proof of convergence of sequences of distributions. Now we formulate a fairly general and powerful result which simplifies the convergence proofs for sequences of distributions in an essential way: It says that for the convergence of a sequence of distributions, it suffices to show that this sequence is a Cauchy sequence, i.e., the space of distributions equipped with the weak topology is sequentially complete. Because of the great importance of this result we present a detailed proof.

Theorem 3.3.1 *Equip the space of distributions $\mathcal{D}'(\Omega)$ on an open nonempty set $\Omega \subset \mathbb{R}^n$ with the weak topology $\sigma = \sigma(\mathcal{D}'(\Omega), \mathcal{D}(\Omega))$. Then $\mathcal{D}'(\Omega)$ is a sequentially complete Hausdorff locally convex topological vector space.*

In particular, for any sequence $(T_i)_{i\in\mathbb{N}} \subset \mathcal{D}'(\Omega)$ such that for each $\phi \in \mathcal{D}(\Omega)$ the numerical sequence $(T_i(\phi))_{i\in\mathbb{N}}$ converges, there are, for each compact subset $K \subset \Omega$, a constant C and an integer $m \in \mathbb{N}$ such that

$$|T_i(\phi)| \leq Cp_{K,m}(\phi) \quad \forall \phi \in \mathcal{D}_K(\Omega), \forall i \in \mathbb{N}; \tag{3.12}$$

3.3 Convergence of sequences and series of distributions

i.e., the sequence $(T_i)_{i \in \mathbb{N}}$ is equi-continuous on $\mathcal{D}_K(\Omega)$ for each compact set $K \subset \Omega$.

Proof. Since its topology is defined in terms of a system of semi-norms, the space of all distributions on Ω is certainly a locally convex topological vector space. Now given $T \in \mathcal{D}'(\Omega)$, $T \neq 0$, there is a $\phi \in \mathcal{D}(\Omega)$ such that $T(\phi) \neq 0$, thus $p_\phi(T) = |T(\phi)| > 0$ and Proposition 2.1.3 implies that the weak topology is Hausdorff, hence $\mathcal{D}'(\Omega)$ is a HLCTVS.

In order to prove sequential completeness we take any Cauchy sequence $(T_i)_{i \in \mathbb{N}}$ in $\mathcal{D}'(\Omega)$ and construct an element $T \in \mathcal{D}'(\Omega)$ to which this sequence converges.

For any $\phi \in \mathcal{D}(\Omega)$ we know (by definition of a Cauchy sequence) $(T_i(\phi))_{i \in \mathbb{N}}$ to be a Cauchy sequence in the field $\mathbb{K}$ which is complete. Hence this Cauchy sequence of numbers converges to some number which we call $T(\phi)$. Since this argument applies to any $\phi \in \mathcal{D}(\Omega)$, we can define a function $T : \mathcal{D}(\Omega) \to \mathbb{K}$ by

$$T(\phi) = \lim_{i \to \infty} T_i(\phi) \qquad \forall \phi \in \mathcal{D}(\Omega).$$

Since each T_i is linear, basic rules of calculation for limits of convergent sequences of numbers imply that the limit function T is linear too.

In order to show continuity of this linear functional T it suffices, according to Theorem 3.2.1, to show that $T_K = T|\mathcal{D}_K(\Omega)$ is continuous on $\mathcal{D}_K(\Omega)$ for every compact subset $K \subset \Omega$. This is done by constructing a neighborhood U of zero in $\mathcal{D}_K(\Omega)$ on which T is bounded and by using Corollary 2.1.1 to deduce continuity.

Since T_i is continuous on $\mathcal{D}_K(\Omega)$, we know that

$$U_i = \{\phi \in \mathcal{D}_K(\Omega) : |T_i(\phi)| \leq 1\}$$

is a closed absolutely convex neighborhood of zero in $\mathcal{D}_K(\Omega)$ (see also the Exercises). Now define

$$U = \cap_{i=1}^\infty U_i$$

and observe that U is a closed absolutely convex set on which the functional T is bounded by 1. Hence in order to deduce continuity of T one has to show that U is actually a neighborhood of zero in $\mathcal{D}_K(\Omega)$. This part is indeed the core of the proof which relies on some fundamental properties of the space $\mathcal{D}_K(\Omega)$ which are proven in the Appendix.

Take any $\phi \in \mathcal{D}_K(\Omega)$; since the sequence $(T_i(\phi))_{i \in \mathbb{N}}$ converges, it is bounded and there is an $n = n(\phi) \in \mathbb{N}$ such that $|T_i(\phi)| \leq n$ for all $i \in \mathbb{N}$. It follows that $|T(\phi)| = \lim_{i \to \infty} |T_i(\phi)| \leq n$ and thus $\phi = n \cdot \frac{1}{n}\phi \in nU$. Since ϕ was arbitrary in $\mathcal{D}_K(\Omega)$, this proves

$$\mathcal{D}_K(\Omega) = \cup_{n=1}^\infty nU.$$

In Proposition 2.1.8 it is shown that $\mathcal{D}_K(\Omega)$ is a complete metrizable HLCTVS. Hence the theorem of Baire (see Appendix, Theorem C.0.5) applies to this space, and it follows that one of the sets nU and hence U itself must have a nonempty interior. This means that some open ball $B = \phi_0 + B_{p,r} \equiv \phi_0 + \{\phi \in \mathcal{D}_K(\Omega) : p(\phi) < r\}$ is contained in the set U. Here ϕ_0 is some element in U, r some positive number and $p = p_{K,m}$ is some continuous semi-norm of the space $\mathcal{D}_K(\Omega)$. Since T is bounded on U by 1 it is bounded on the neighborhood of zero $B_{p,r}$ by $1 + |T(\phi_0)|$ and thus T is continuous.

All elements of T_i and the limit element T are bounded on this neighborhood U by 1. From the above it follows that there are a constant C and some integer $m \in \mathbb{N}$ such that

$$|T_i(\phi)| \leq C p_{K,m}(\phi) \qquad \forall \phi \in \mathcal{D}_K(\Omega), \forall i \in \mathbb{N};$$

i.e., the sequence $(T_i)_{i \in \mathbb{N}}$ is equi-continuous on $\mathcal{D}_K(\Omega)$ for each compact set $K \subset \Omega$ and we conclude. □

The convergence of a series of distributions is defined in the usual way through convergence of the corresponding sequence of partial sums. This can easily be translated into the following concrete formulation.

Definition 3.3.2 *Given a sequence* $(T_i)_{i \in \mathbb{N}}$ *of distributions on a nonempty open set* $\Omega \subset \mathbb{R}^n$ *one says that the* **series** $\sum_{i \in \mathbb{N}} T_i$ **converges** *if, and only if, there is a* $T \in \mathcal{D}'(\Omega)$ *such that for every* $\phi \in \mathcal{D}(\Omega)$ *the numerical series* $\sum_{i \in \mathbb{N}} T_i(\phi)$ *converges to the number* $T(\phi)$.

As a first important application of Theorem 3.3.1, one has a rather convenient characterization of the convergence of a series of distributions.

Corollary 3.3.1 *A series* $\sum_{i \in \mathbb{N}} T_i$ *of distributions* $T_i \in \mathcal{D}'(\Omega)$ *converges if, and only if, for every* $\phi \in \mathcal{D}(\Omega)$ *the numerical series* $\sum_{i \in \mathbb{N}} T_i(\phi)$ *converges.*

As a simple *example* consider the distributions $T_i = c_i \delta_{ia}$ for some $a > 0$ and any sequence of numbers c_i. Then the series

$$\sum_{i \in \mathbb{N}} c_i \delta_{ia}$$

converges in $\mathcal{D}'(\mathbb{R})$. The proof is simple. For every $\phi \in \mathcal{D}(\mathbb{R})$ one has

$$\sum_{i \in \mathbb{N}} T_i(\phi) = \sum_{i \in \mathbb{N}} c_i \phi(ia) = \sum_{i=1}^{m} c_i \phi(ia)$$

for some $m \in \mathbb{N}$ depending on the support of the test function ϕ (for $ia > m$ the point ia is not contained in $\operatorname{supp} \phi$).

3.4 Localization of distributions

Distributions on a nonempty open set $\Omega \subset \mathbb{R}^n$ have been defined as continuous linear functionals on the test function space $\mathcal{D}(\Omega)$ over Ω but not directly in points of Ω. Nevertheless we consider these distributions to be localized. In this section we explain in which sense this localization is understood.

Suppose $\Omega_1 \subset \Omega_2 \subset \mathbb{R}^n$. Then every test function $\phi \in \mathcal{D}(\Omega_1)$ vanishes in a neighborhood of the boundary of Ω_1 and thus can be continued by 0 to Ω_2 to give a compactly supported test function $i_{\Omega_2, \Omega_1}(\phi)$ on Ω_2. This defines a mapping $i_{\Omega_2, \Omega_1} : \mathcal{D}(\Omega_1) \to \mathcal{D}(\Omega_2)$ which is evidently linear and continuous. Thus we can consider $\mathcal{D}(\Omega_1)$ to be embedded into $\mathcal{D}(\Omega_2)$ as $i_{\Omega_2, \Omega_1}(\mathcal{D}(\Omega_1))$, i.e.,

$$i_{\Omega_2, \Omega_1}(\mathcal{D}(\Omega_1)) \subset \mathcal{D}(\Omega_2).$$

Hence every continuous linear functional T on $\mathcal{D}(\Omega_2)$ defines also a continuous linear functional $T \circ i_{\Omega_2, \Omega_1} \equiv \rho_{\Omega_1, \Omega_2}(T)$ on $\mathcal{D}(\Omega_1)$. Therefore every distribution T on Ω_2 can be restricted to any open nonempty subset Ω_1 by

$$T | \Omega_1 = \rho_{\Omega_1, \Omega_2}(T). \tag{3.13}$$

In particular this allows us to express the fact that a distribution T on Ω_2 *vanishes on an open subset* Ω_1: $\rho_{\Omega_1, \Omega_2}(T) = 0$, or in concrete terms

$$T \circ i_{\Omega_2, \Omega_1}(\phi) = 0 \quad \forall \phi \in \mathcal{D}(\Omega_1).$$

3.4 Localization of distributions

For convenience of notation the trivial extension map i_{Ω_2,Ω_1} is usually omitted and one writes

$$T(\phi) = 0 \quad \forall \phi \in \mathcal{D}(\Omega_1)$$

to express the fact that a distribution T on Ω_2 vanishes on the open subset Ω_1. As a slight extension we state: Two distributions T_1 and T_2 on Ω_2 *agree on an open subset* Ω_1 if, and only if,

$$\rho_{\Omega_1,\Omega_2}(T_1) = \rho_{\Omega_1,\Omega_2}(T_2)$$

or in more convenient notation if, and only if,

$$T_1(\phi) = T_2(\phi) \quad \forall \phi \in \mathcal{D}(\Omega_1).$$

The *support of a function* $f : \Omega \to \mathbb{K}$ is defined as the closure of the set of those points in which the function does not vanish, or equivalently as the complement of the largest open subset of Ω on which f vanishes. The above preparations thus allow us to define the *support of a distribution* T *on* Ω as the complement of the largest open subset $\Omega_1 \subset \Omega$ on which T vanishes. The support of T is denoted by supp T. It is characterized by the formula

$$\operatorname{supp} T = \bigcap_{A \in C_T} A \tag{3.14}$$

where C_T denotes the set of all closed subsets of Ω such that T vanishes on $\Omega \setminus A$. Accordingly a point $x \in \Omega$ belongs to the support of the distribution T on Ω if, and only if, T does not vanish in every open neighborhood U of x, i.e., for every open neighborhood U of x there is a $\phi \in \mathcal{D}(U)$ such that $T(\phi) \neq 0$.

In the Exercises one shows that this concept of support of distributions is compatible with the embedding of functions and the support defined for functions, i.e., one shows

$$\operatorname{supp} I_f = \operatorname{supp} f \quad \forall f \in L^1_{loc}(\Omega).$$

A simple example shows that distributions can have a support consisting of one point: The support of the distribution T on Ω defined by

$$T(\phi) = \sum_{|\alpha| \leq m} c_\alpha D^\alpha \phi(x_0) \tag{3.15}$$

is the point $x_0 \in \Omega$, for any choice of the constants c_α and any $m \in \mathbb{N}$. If a distribution is of the form (3.15) then certainly $T(\phi) = 0$ for all $\phi \in \mathcal{D}(\mathbb{R}^n \setminus \{x_0\})$ since such test functions vanish in a neighborhood of x_0 and thus all derivatives vanish there. And, if not all coefficients c_α vanish, there are, in any neighborhood U of the point x_0, test functions $\phi \in \mathcal{D}(U)$ such that $T(\phi) \neq 0$. This claim is addressed in the Exercises.

Furthermore, this formula actually gives the general form of a distribution whose support is the point x_0. We show this later in Proposition 4.4.3.

Since we have learned above when two distributions on Ω agree on an open subset, we know in particular when a distribution is equal to a $\mathcal{C}^\infty$-function, or more precisely when a distribution is equal to the regular distribution defined by a $\mathcal{C}^\infty$-function, on some open subset. This is used in the definition of the singular support of a distribution, which seems somewhat ad hoc but which has proved itself to be quite useful in the analysis of constant coefficient partial differential operators.

Definition 3.4.1 *Let T be a distribution on a nonempty open set $\Omega \subset \mathbb{R}^n$. The* **singular support of** T*, denoted* **sing supp** T*, is the smallest closed subset of Ω in the complement of which T is equal to a $\mathcal{C}^\infty$-function.*

We mention a simple one dimensional example, Cauchy's principal value $\mathrm{vp}\,\frac{1}{x}$. In the discussion following formula (3.8) we saw that $\mathrm{vp}\,\frac{1}{x} = I_{\frac{1}{x}}$ on $\mathbb{R}\setminus\{0\}$. Since $\frac{1}{x}$ is a $\mathcal{C}^\infty$-function on $\mathbb{R}\setminus\{0\}$, sing supp $\mathrm{vp}\,\frac{1}{x} \subseteq \{0\}$. And since $\{0\}$ is obviously the smallest closed subset of $\mathbb{R}$ outside which the Cauchy principal value is equal to a $\mathcal{C}^\infty$-function, it follows that

$$\text{sing supp vp}\,\frac{1}{x} = \{0\}\,.$$

3.5 Tempered distributions and distributions with compact support

Tempered distributions are distributions which admit the Fourier transform as an isomorphism of topological vector spaces and accordingly we will devote later a separate chapter to Fourier transformation and tempered distributions. This section just gives the basic definitions and properties of tempered distributions and distributions with compact support.

Recall the beginning of the section on the definition of distributions. What has been done there for general distributions will be done here for the subclasses of tempered and compactly supported distributions.

Definition 3.5.1 *A* **tempered distribution** *T on an open nonempty subset $\Omega \subset \mathbb{R}^n$ is a continuous linear functional on the test function space $\mathcal{S}(\Omega)$ of strongly decreasing $\mathcal{C}^\infty$-functions on Ω. The* **set of all tempered distributions on** Ω *equals the topological dual $\mathcal{S}'(\Omega)$ of $\mathcal{S}(\Omega)$.*

In analogy with Thereom 3.2.1 we have the following explicit characterization of tempered distributions.

Theorem 3.5.1 *A linear functional $T : \mathcal{S}(\Omega) \to \mathbb{K}$ is a tempered distribution on the open nonempty set $\Omega \subset \mathbb{R}^n$ if, and only if, there exist a number $C \in \mathbb{R}_+$ and natural numbers $m, k \in \mathbb{N}$, depending on T, such that for all $\phi \in \mathcal{S}(\Omega)$ the* *estimate*

$$|T(\phi)| \leq C p_{m,k}(\phi) \tag{3.16}$$

holds.

Proof. Recall the definition of the filtering system of norms of the space $\mathcal{S}(\Omega)$ and the condition of boundedness for a linear function $T : \mathcal{S}(\Omega) \to \mathbb{K}$. Then it is clear that the above estimate characterizes T as being bounded on $\mathcal{S}(\Omega)$. Thus, by Theorem 2.1.7, this estimate characterizes continuity and we conclude. □

According to relation (3.2) we know that every tempered distribution is a distribution and therefore all results established for distributions apply to tempered distributions. Also the basic definitions of convergence and of a Cauchy sequence are formally the same as soon as we replace the test function space $\mathcal{D}(\Omega)$ by the smaller test function space $\mathcal{S}(\Omega)$ and the topological dual $\mathcal{D}'(\Omega)$ of $\mathcal{D}(\Omega)$ by the topological dual $\mathcal{S}'(\Omega)$ of $\mathcal{S}(\Omega)$. Hence we do not repeat these definitions, but we formulate the important counterpart of Theorem 3.3.1 explicitly.

Theorem 3.5.2 *Equip the space of distributions $\mathcal{S}'(\Omega)$ of tempered distributions on an open nonempty set $\Omega \subseteq \mathbb{R}^n$ with the weak topology $\sigma = \sigma(\mathcal{S}'(\Omega), \mathcal{S}(\Omega))$. Then $\mathcal{S}'(\Omega)$ is a sequentially complete Hausdorff locally convex topological vector space.*

Proof. As in the proof of Theorem 3.3.1 one sees that $\mathcal{S}'(\Omega)$ is a Hausdorff locally convex topological vector space. By this theorem one also knows that a Cauchy sequence in $\mathcal{S}'(\Omega)$ converges to some distribution T on Ω. In order to show that T is actually tempered, one proves that T is bounded on some open ball in $\mathcal{S}(\Omega)$. Since $\mathcal{S}(\Omega)$ is a complete metrizable space this can be done as in the proof of Theorem 3.3.1. Thus we conclude. □

Finally we discuss briefly the space of *distributions of compact support*. Recall that a distribution $T \in \mathcal{D}'(\Omega)$ is said to have a compact support if there is a compact set $K \subset \Omega$ such that $T(\phi) = 0$ for all $\phi \in \mathcal{D}(\Omega \setminus K)$. The smallest of the compact subsets K for which this condition holds is called the *support* of T, denoted by $\operatorname{supp} T$. As we are going to explain now, distributions of compact support can be characterized topologically as elements of the topological dual of the test function space $\mathcal{E}(\Omega)$. According to (2.12) the space $\mathcal{E}(\Omega)$ is the space $\mathcal{C}^\infty(\Omega)$ equipped with the filtering system of semi-norms $\mathcal{P}_\infty(\Omega) = \{p_{K,m} : K \subset \Omega \text{ compact}, m = 0, 1, 2, \ldots\}$. Hence a linear function $T : \mathcal{E}(\Omega) \to \mathbb{K}$ is continuous if, and only if, there are a compact set $K \subset \Omega$, a constant $C \in \mathbb{R}_+$ and an integer m such that

$$|T(\phi)| \leq C p_{K,m}(\phi) \quad \forall \phi \in \mathcal{E}(\Omega). \tag{3.17}$$

Now suppose $T \in \mathcal{E}'(\Omega)$ is given. Then T satisfies condition (3.17) and by relation (3.2) we know that T is a distribution on Ω. Take any $\phi \in \mathcal{D}(\Omega \setminus K)$. Then ϕ vanishes in some open neighborhood U of K and thus $D^\alpha \phi(x) = 0$ for all $x \in K$ and all $\alpha \in \mathbb{N}^n$. It follows that $p_{K,m}(\phi) = 0$ and thus $T(\phi) = 0$ for all $\phi \in \mathcal{D}(\Omega \setminus K)$, hence $\operatorname{supp} T \subseteq K$. This shows that elements in $\mathcal{E}'(\Omega)$ are distributions with compact support.

Conversely suppose that $T \in \mathcal{D}'(\Omega)$ has a support contained in a compact set $K \subset \Omega$. There are functions $u \in \mathcal{D}(\Omega)$ which are equal to 1 in an open neighborhood of K and which have their support in a slightly larger compact set K' (see Exercises). It follows that $(1 - u) \cdot \phi \in \mathcal{D}(\Omega \setminus K)$ and therefore $T((1 - u) \cdot \phi) = 0$ or $T(\phi) = T(u \cdot \phi)$ for all $\phi \in \mathcal{D}(\Omega)$. For any $\psi \in \mathcal{E}(\Omega)$ one knows $u \cdot \psi \in \mathcal{D}_{K'}(\Omega)$ and thus $T_0(\psi) = T(u \cdot \psi)$ is a well defined linear function $\mathcal{E}(\Omega) \to \mathbb{K}$. (If $v \in \mathcal{D}(\Omega)$ is another function which is equal to 1 in some open neighborhood of K, then $u \cdot \psi - v \cdot \psi \in \mathcal{D}(\Omega \setminus K)$ and therefore $T(u \cdot \psi - v \cdot \psi) = 0$). Since T is a distribution there are a constant $C \in \mathbb{R}_+$ and $m \in \mathbb{N}$ such that $|T(\phi)| \le c p_{K',m}(\phi)$ for all $\phi \in \mathcal{D}_{K'}(\Omega)$. For all $\psi \in \mathcal{E}(\Omega)$ we thus get

$$|T_0(\psi)| = |T(u \cdot \psi)| \le C p_{K',m}(u \cdot \psi) \le C p_{K',m}(u) p_{K',m}(\psi).$$

This shows that T_0 is continuous on $\mathcal{E}(\Omega)$, i.e., $T_0 \in \mathcal{E}'(\Omega)$. On $\mathcal{D}(\Omega)$ the functionals T_0 and T agree: $T_0(\phi) = T(u \cdot \phi) = T(\phi)$ for all $\phi \in \mathcal{D}(\Omega)$ as we have seen above and therefore we can formulate the following result.

Theorem 3.5.3 *The topological dual $\mathcal{E}'(\Omega)$ of the test function space $\mathcal{E}(\Omega)$ equals the space of distributions on Ω which have a compact support. Equipped with the weak topology $\sigma = \sigma(\mathcal{E}'(\Omega), \mathcal{E}(\Omega))$ the space $\mathcal{E}'(\Omega)$ of distributions with compact support is a sequentially complete Hausdorff locally convex topological vector space.*

Proof. The proof that $\mathcal{E}'(\Omega)$ is a sequentially complete HLCTVS is left as an exercise. The other statements have been proven above. □

3.6 Exercises

1. Let $f : \Omega \to \mathbb{R}$ be a continuous function on an open nonempty set $\Omega \subset \mathbb{R}^n$. Show: If $\int f(x)\phi(x)dx = 0$ for all $\phi \in \mathcal{D}(\Omega)$, then $f = 0$, i.e., the map $I : \mathcal{C}(\Omega) \to \mathcal{D}'(\Omega)$ of Theorem 3.2.2 is injective. Deduce that I is injective on all of $L^1_{loc}(\Omega)$.

2. Prove: There is no $f \in L^1_{loc}(\Omega)$ such that $\delta_a(\phi) = \int f(x)\phi(x)dx$ for all $\phi \in \mathcal{D}(\Omega)$.

   Hints: It suffices to consider the case $a = 0$. Then take the function $\rho : \mathbb{R}^n \to \mathbb{R}$ by

   $$\rho(x) = \begin{cases} 0 & : \text{ for } |x| \ge 1, \\ \exp\frac{-1}{1-x^2} & : \text{ for } |x| < 1, \end{cases}$$

   and define $\rho_r(x) = \rho(\frac{x}{r})$ for $r > 0$. Recall that $\rho_r \in \mathcal{D}(\Omega)$ and $\rho_r(x) = 0$ for all $x \in \mathbb{R}^n$ with $|x| > r$. Finally observe that for $f \in L^1_{loc}(\Omega)$ one has

   $$\lim_{r \to 0} \int_{\{x : |x| \le r\}} |f(x)| dx = 0.$$

3. Consider the hyperplane $H = \{x = (x_1, \ldots, x_n) \in \mathbb{R}^n : x_1 = 0\}$. Define a function $\delta_H : \mathcal{D}(\mathbb{R}^n) \to \mathbb{K}$ by

$$\langle \delta_H, \phi \rangle = \int_{\mathbb{R}^{n-1}} \phi(0, x_2, \ldots, x_n) dx_2 \cdots dx_n \qquad \forall \phi \in \mathcal{D}(\mathbb{R}^n).$$

Show that δ_H is a distribution on $\mathbb{R}^n$. It is called *Dirac's delta distribution on the hyperplane H*.

4. For any point $a \in \Omega \subset \mathbb{R}^n$, Ω open and not empty, define a functional $T : \mathcal{D}(\Omega) \to \mathbb{K}$ by

$$\langle T, \phi \rangle = \sum_{i=1}^{n} \frac{\partial^2 \phi}{\partial x_i^2}(a) \equiv \Delta \phi(a) \qquad \forall \phi \in \mathcal{D}(\mathbb{R}^n).$$

Prove: T is a distribution on Ω of order 2. On $\Omega \setminus \{a\}$ this distribution is equal to the regular distribution I_0 defined by the zero function.

5. Let $S_{n-1} = \{x \in \mathbb{R}^n : \sum_{i=1}^n x_i^2 = 1\}$ be the unit sphere in $\mathbb{R}^n$ and denote by $d\sigma$ the uniform measure on S_{n-1}. The derivative in the direction of the outer normal of S_{n-1} is denoted by $\frac{\partial}{\partial n}$. Now define a function $T : \mathcal{D}(\mathbb{R}^n) \to \mathbb{K}$ by

$$\langle T, \phi \rangle = \int_{S_{n-1}} \frac{\partial \phi}{\partial n} d\sigma \qquad \forall \phi \in \mathcal{D}(\Omega)$$

and show that T is a distribution on $\mathbb{R}^n$ of order 1 which is equal to the regular distribution I_0 on $\mathbb{R}^n \setminus S_{n-1}$.

6. Given a Cauchy sequence $(T_i)_{i \in \mathbb{N}}$ of distributions on a nonempty open set $\Omega \subset \mathbb{R}^n$, prove in detail that the (pointwise or weak) limit T is a linear function $\mathcal{D}(\Omega) \to \mathbb{K}$.

7. Let $X[\mathcal{P}]$ be a HLCTVS, $T \in X'[\mathcal{P}]$ and $r > 0$. Show:

$$U = \{x \in X : |T(x)| \leq r\}$$

is a closed absolutely convex neighborhood of zero.

4
Calculus for Distributions

This chapter deals with the basic parts of calculus, i.e., with differentiation of distributions, multiplication of distributions with smooth functions and with other distributions, and change of variables for distributions. There are other parts which will be addressed in separate chapters since they play a prominent role in distribution theory, viz., Fourier transform for a distinguished subclass of distributions and convolution of distributions with functions and with other distributions.

Certainly, when we define differentiation, multiplication and variable transformations for distributions, we insist that these definitions be consistent with these operations on functions and the embedding of functions into the space of distributions.

As preparation we mention a small but important observation. Let $\Omega \subset \mathbb{R}^n$ be nonempty and open and $A : \mathcal{D}(\Omega) \to \mathcal{D}(\Omega)$ a continuous linear function of the test function space on Ω into itself. Such a map induces a map on the space of distributions on Ω: $A' : \mathcal{D}'(\Omega) \to \mathcal{D}'(\Omega)$ according to the formula

$$A'(T) = T \circ A \qquad \forall T \in \mathcal{D}'(\Omega). \tag{4.1}$$

As a composition of two linear and continuous functions, $A'(T)$ is a continuous linear function $\mathcal{D}(\Omega) \to \mathbb{K}$ and thus a distribution. Therefore A' is well defined and is called the *adjoint of A*. Obviously $A' : \mathcal{D}'(\Omega) \to \mathcal{D}'(\Omega)$ is linear, but it is also continuous, since for every $\phi \in \mathcal{D}(\Omega)$ we have, for all $T \in \mathcal{D}'(\Omega)$, $p_\phi(A'(T)) = p_{A(\phi)}(T)$ so that Definition 2.1.7 and Theorem 2.1.7 imply continuity.

The adjoint itself (or a slight modification thereof in order to ensure consistency with the embedding of functions) will be used to define differentiation of distributions, their multiplication and change of variables.

4.1 Differentiation

Let D^α be a derivative monomial of order $\alpha = (\alpha_1, \ldots, \alpha_n) \in \mathbb{N}^n$. It is certainly a linear map $\mathcal{D}(\Omega) \to \mathcal{D}(\Omega)$ for any open nonempty set $\Omega \subset \mathbb{R}^n$. Continuity of $D^\alpha : \mathcal{D}(\Omega) \to \mathcal{D}(\Omega)$ follows easily from the estimate

$$p_{K,m}(D^\alpha \phi) \leq p_{K,m+|\alpha|}(\phi) \qquad \forall \phi \in \mathcal{D}_K(\Omega)$$

in conjunction with Definition 2.1.7 and Theorem 2.1.7. Therefore the adjoint of the derivative monomial D^α is a continuous linear map $\mathcal{D}'(\Omega) \to \mathcal{D}'(\Omega)$ and thus appears to be a suitable candidate for the definition of the derivative of distributions. However, since we insist on consistency of the definition with the embedding of differentiable functions into the space of distributions, a slight adjustment has to be made. To determine this adjustment take any $f \in \mathcal{C}^1(\mathbb{R}^n)$ and calculate for every $\phi \in \mathcal{D}(\mathbb{R}^n)$,

$$\begin{aligned}\langle I_{\partial_1 f}, \phi \rangle &= \int_{\mathbb{R}^n} \tfrac{\partial f}{\partial x_1}(x)\phi(x)dx = \int \cdots \int (\int_{\mathbb{R}} \tfrac{\partial f}{\partial x_1}(x)\phi(x)dx_1)dx_2 \cdots dx_n \\ &= -\int \cdots \int (\int_{\mathbb{R}} f(x)\tfrac{\partial \phi}{\partial x_1}(x)dx_1)dx_2 \ldots dx_n = -\langle I_f, \partial_1 \phi \rangle.\end{aligned}$$

Here we use the abbreviation $\partial_1 \equiv \tfrac{\partial}{\partial x_1}$. Similarly, by repeated partial integration, one obtains (see Exercises) for $f \in \mathcal{C}^k(\mathbb{R}^n)$,

$$\langle I_{D^\alpha f}, \phi \rangle = (-1)^{|\alpha|} \langle I_f, D^\alpha \phi \rangle \quad \forall \phi \in \mathcal{D}(\mathbb{R}^n), \quad \forall \alpha \in \mathbb{N}^n, \ |\alpha| \leq k.$$

Denoting the derivative of order α on $\mathcal{D}'(\Omega)$ with the same symbol as for functions, the condition of consistency with the embedding reads

$$D^\alpha I_f = I_{D^\alpha f} \quad \forall \alpha \in \mathbb{N}^n, \ |\alpha| \leq k, \quad \forall f \in \mathcal{C}^k(\Omega).$$

Accordingly one takes as the derivative monomial of order α on distributions the following modification of the adjoint of the derivative monomial on functions.

Definition 4.1.1 *The* **derivative of order** $\alpha = (\alpha_1, \ldots, \alpha_n) \in \mathbb{N}^n$ *is defined on* $\mathcal{D}'(\Omega)$ *by the formula*

$$D^\alpha T = (-1)^{|\alpha|} T \circ D^\alpha \qquad \forall T \in \mathcal{D}'(\Omega),$$

i.e., for each $T \in \mathcal{D}'(\Omega)$ *one has*

$$\langle D^\alpha T, \phi \rangle = (-1)^{|\alpha|} \langle T, D^\alpha \phi \rangle \qquad \forall \phi \in \mathcal{D}(\Omega).$$

There are a number of immediate powerful consequences of this definition. The proof of these results is straightforward.

Theorem 4.1.1 *Differentiation on the space of distributions $\mathcal{D}'(\Omega)$ on a nonempty open set $\Omega \subset \mathbb{R}^n$ as defined in Definition 4.1.1 has the following properties:*

1. Every distribution has derivatives of all orders and the order in which derivatives are calculated does not matter, i.e.,

$$D^\alpha(D^\beta T) = D^\beta(D^\alpha T) = D^{\alpha+\beta}T \qquad \forall T \in \mathcal{D}'(\Omega), \quad \forall \alpha, \beta \in \mathbb{N}^n.$$

2. The local order of a distribution increases by the order of differentiation.

3. Differentiation on $\mathcal{D}'(\Omega)$ is consistent with the embedding of $\mathcal{C}^\infty(\Omega)$ into $\mathcal{D}'(\Omega)$.

4. The derivative monomials $D^\alpha : \mathcal{D}'(\Omega) \to \mathcal{D}'(\Omega)$ are linear and continuous, hence in particular

   a) If $T = \lim_{i \to \infty} T_i$ in $\mathcal{D}'(\Omega)$, then

   $$D^\alpha(\lim_{i \to \infty} T_i) = \lim_{i \to \infty} D^\alpha T_i.$$

   b) If a series $\sum_{i \in \mathbb{N}} T_i$ converges in $\mathcal{D}'(\Omega)$, then

   $$D^\alpha \sum_{i \in \mathbb{N}} T_i = \sum_{i \in \mathbb{N}} D^\alpha T_i.$$

Proof. The first part has been shown in the definition of the derivative for distributions. The order of differentiation does not matter since on $\mathcal{C}^\infty(\Omega)$ the order of differentiation can be interchanged.

If for some compact set $K \subset \Omega$ we have $|T(\phi)| \leq C p_{K,m}(\phi)$ for all $\phi \in \mathcal{D}_K(\Omega)$, we get $|D^\alpha T(\phi)| = |T(D^\alpha \phi)| \leq C p_{K,m}(D^\alpha \phi) \leq C p_{K,m+|\alpha|}(\phi)$ and the second part follows easily from the definition of the local order (Definition 3.2.2).

The consistency of the derivative for distributions with the embedding of differentiable functions has been built into the definition.

Since the derivative D^α on $\mathcal{D}'(\Omega)$ equals the adjoint of the derivative on functions multiplied by $(-1)^{|\alpha|}$, the continuity of the derivative follows immediately from that of the adjoint of the linear continuous map on $\mathcal{D}(\Omega)$ as discussed after equation (4.1). □

Remark 4.1.1 1. *Obviously, the fact that every distribution has derivatives of all orders comes from the definition of the test function space as a subspace of the space of all $\mathcal{C}^\infty$-functions and the definition of a topology on this subspace which ensures that all derivative monomials are continuous.*

2. *In the sense of distributions, every locally integrable function has derivatives of all orders. But certainly, in general the result will be a distribution and not a function. We mention a famous example. Consider the* **Heaviside function** θ *on the real line $\mathbb{R}$ defined by*

$$\theta(x) = \begin{cases} 0 & : \text{ for } x < 0, \\ 1 & : \text{ for } x \geq 0. \end{cases}$$

θ is locally integrable and thus has a derivative in the sense of distributions which we calculate now. For all $\phi \in \mathcal{D}(\mathbb{R})$ one has

$$\begin{aligned}\langle DI_\theta, \phi\rangle &= -\langle I_\theta, D\phi\rangle = -\int_\mathbb{R} \theta(x)\phi'(x)dx \\ &= -\int_0^\infty \phi'(x)dx = \phi(0) = \langle \delta, \phi\rangle.\end{aligned}$$

This shows that $DI_\theta = \delta$, which is often written as

$$\theta'(x) = \delta(x),$$

i.e., the derivative (in the sense of distributions) of Heaviside's function equals Dirac's delta function.

Some other examples of derivatives are given in the Exercises.

3. Part 4 of Theorem 4.1.1 represents a remarkable contrast to classical analysis. Recall the example of the sequence of $\mathcal{C}^\infty$-functions $f_j(x) = \frac{1}{j}\sin(jx)$ on $\mathbb{R}$ which converges uniformly on $\mathbb{R}$ to the $\mathcal{C}^\infty$-function 0, but for which the sequence of derivatives $f_j'(x) = \cos(jx)$ does not converge (not even point-wise). In the sense of distributions the sequence of derivatives also converges to 0: For all $\phi \in \mathcal{D}(\mathbb{R})$ we have

$$\langle DI_{f_j}, \phi\rangle = -\langle I_{f_j}, D\phi\rangle = -\frac{1}{j}\int_\mathbb{R} \sin(jx)\phi'(x)dx \to 0$$

as $j \to \infty$.

One of the major goals in the development of distribution theory was to get a suitable framework for solving linear partial differential equations with constant coefficients. This goal has been achieved (see [Hör83a, Hör83b]). Here we mention only a few elementary aspects. Knowing the derivative monomials on $\mathcal{D}'(\Omega)$ we can consider *linear constant coefficient partial differential operators* on this space, i.e., operators of the form

$$P(D) = \sum_{|\alpha|\le k} a_\alpha D^\alpha \tag{4.2}$$

with certain coefficients $a_\alpha \in \mathbb{K}$ and $k = 1, 2, \ldots$. Now given $f \in \mathcal{C}(\Omega)$ we can consider the equation

$$P(D)u = f \tag{4.3}$$

in two ways: A *classical or strong solution* is a function $u \in \mathcal{C}^k(\Omega)$ such that this equation holds in the sense of functions. A distribution $T \in \mathcal{D}'(\Omega)$ for which $P(D)T = I_f$ holds in $\mathcal{D}'(\Omega)$ is called a *distributional or weak solution*.

Since the space of distributions $\mathcal{D}'(\Omega)$ is much larger than the space $\mathcal{C}^k(\Omega)$ of continuously differentiable functions, one expects that it is easier to find a solution in this larger space. This expectation has been proven to be correct in many important classes of problems. However in most cases, in particular in those arising from physics, one does not look for a weak but for a classical solution. So it is very important to have a theory which ensures that for special classes of partial differential equations the weak solutions are actually classical ones. The so-called elliptic regularity theory provides these results also for 'elliptic' partial differential equations (see Part III). Here we discuss a very simple class of examples of this type.

Proposition 4.1.2 *Suppose $T \in \mathcal{D}'(\mathbb{R})$ satisfies the constant coefficient ordinary differential equation*

$$D^n T = 0 \quad \text{in} \quad \mathcal{D}'(\mathbb{R}).$$

Then T is a polynomial P_{n-1} of degree $\leq n-1$, i.e., $T = I_{P_{n-1}}$. Hence the sets of classical and of distributional solutions of this differential equation coincide.

Proof. The proof is by induction on the order n of this differential equation. Hence in a first step we show: If a distribution $T \in \mathcal{D}(\mathbb{R})$ satisfies $DT = T' = 0$, then T is a constant, i.e., of the form $T = I_c$ for some constant c.

Choose some test function $\psi \in \mathcal{D}(\mathbb{R})$ which is normalized by the condition $I(\psi) \equiv \int \psi(x)dx = 1$. Next consider any test function $\phi \in \mathcal{D}(\mathbb{R})$. Associate with it the auxiliary test function $\chi = \phi - I(\phi)\psi$ which has the property $I(\chi) = 0$. Hence χ is the derivative of a test function ρ defined by $\rho(x) = \int_{-\infty}^{x} \chi(y)dy$, $\rho' = \chi$ (see Exercises). $T' = 0$ in $\mathcal{D}'(\mathbb{R})$ implies that

$$T(\chi) = T(\rho') = -T'(\rho) = 0$$

and therefore

$$T(\phi) = T(\psi)I(\phi) = I_c(\phi)$$

with the constant $c = T(\psi)$.

Now suppose that the conclusion of the proposition holds for some $n \geq 1$. We are going to show that then this conclusion also holds for $n + 1$.

Assume $D^{n+1}T = 0$ in $\mathcal{D}'(\mathbb{R})$. It follows that $D(D^n T) = 0$ in $\mathcal{D}'(\mathbb{R})$ and hence $D^n T = I_c$ for some constant c. In the Exercises we show the identity $I_c = D^n I_{P_n}$ where P_n is a polynomial of degree n of the form $P_n(x) = \frac{c}{n!}x^n + P_{n-1}(x)$. Here P_{n-1} is any polynomial of degree $\leq n-1$. Therefore $D^n T = D^n I_{P_n}$ or $D^n(T - I_{P_n}) = 0$ in $\mathcal{D}'(\mathbb{R})$. The induction hypothesis implies that

$$T - I_{P_n} = I_{Q_{n-1}}$$

for some polynomial Q_{n-1} of degree $\leq n-1$ and we conclude that T is a polynomial of degree $\leq n$. In the Exercises we will also show that any classical solution is also a distributional solution. □

4.2 Multiplication

As is well known from classical analysis, the (pointwise) product of two continuous functions $f, g \in \mathcal{C}(\Omega)$, defined by $(f \cdot g)(x) = f(x)g(x)$ for all $x \in \Omega$, is again a continuous function on Ω. Similarly, the product of two continuously differentiable functions $f, g \in \mathcal{C}^1(\Omega)$ is again a continuously differentiable function, due to the product rule of differentiation.

However, the product of two locally integrable functions $f, g \in L^1_{loc}(\Omega)$ is in general not a locally integrable function. As a typical case we mention: $f \cdot g$ is not locally integrable when both functions have a sufficiently strong singularity at the same point. A simple example is the function

$$f(x) = \begin{cases} +\infty & : \text{ for } x = 0, \\ \frac{1}{\sqrt{|x|}} & : \text{ for } x \neq 0. \end{cases}$$

Obviously $f \in L^1_{loc}(\mathbb{R})$, but $f \cdot f = f^2$ is not locally integrable. Nevertheless the product of two locally integrable functions which have a singularity at the same point will be locally integrable if these singularities are sufficiently weak; for example take the function

$$g(x) = \begin{cases} +\infty & : \text{ for } x = 0, \\ \frac{1}{|x|^s} & : \text{ for } x \neq 0. \end{cases}$$

for some exponent $s > 0$. If $2s < 1$, then g^2 is locally integrable on $\mathbb{R}$.

On the other hand there are many subspaces of $L^1_{loc}(\Omega)$ with the property that any element in this subspace can multiply any element in $L^1_{loc}(\Omega)$ such that the product is again in $L^1_{loc}(\Omega)$ ($\Omega \subset \mathbb{R}^n$ open and nonempty), for instance the subspace $\mathcal{C}(\Omega)$ of continuous functions on Ω or the bigger subspace $L^\infty_{loc}(\Omega)$ of those functions which are essentially bounded on every compact subset $K \subset \Omega$.

These few examples show that in spaces of functions whose elements can have singularities the multiplication cannot be done in general. Accordingly we cannot expect to have unrestricted multiplication in the space $\mathcal{D}'(\Omega)$ of distributions and therefore only some special but important cases of multiplication for distributions are discussed.

Proposition 4.2.1 *In the space $\mathcal{D}'(\Omega)$ of distributions on a nonempty open set $\Omega \subset \mathbb{R}^n$, multiplication with $\mathcal{C}^\infty$-functions is well defined by*

$$(u \cdot T)(\phi) = T(u \cdot \phi) \quad \forall \phi \in \mathcal{D}(\Omega)$$

for every $T \in \mathcal{D}'(\Omega)$ and every $u \in \mathcal{C}^\infty(\Omega)$. This product has the following properties:

1. *For fixed $u \in \mathcal{C}^\infty(\Omega)$ the map $T \mapsto u \cdot T$ is linear and continuous on $\mathcal{D}'(\Omega)$.*

2. *The **product rule of differentiation** holds:*

$$\frac{\partial}{\partial x_j}(u \cdot T) = \frac{\partial u}{\partial x_j} \cdot T + u \cdot \frac{\partial T}{\partial x_j} \tag{4.4}$$

   for all $j = 1, \ldots, n$, all $u \in \mathcal{C}^\infty(\Omega)$, and all $T \in \mathcal{D}'(\Omega)$.

3. *This multiplication is compatible with the embedding of functions, i.e.,*

$$u \cdot I_f = I_{uf}$$

   for all $u \in \mathcal{C}^\infty(\Omega)$ and all $f \in L^1_{loc}(\Omega)$.

Proof. For each $u \in \mathcal{C}^\infty(\Omega)$ introduce the mapping $M_u : \mathcal{D}(\Omega) \to \mathcal{D}(\Omega)$ which multiplies a test function ϕ with the function u: $M_u(\phi) = u \cdot \phi$ (pointwise product). Obviously we have $\operatorname{supp} u \cdot \phi \subseteq \operatorname{supp} \phi$. Hence $M_u(\phi)$ has a compact support. The product rule of differentiation for functions shows that $M_u(\phi) \in \mathcal{C}^\infty(\Omega)$, and therefore $M_u(\phi) \in \mathcal{D}(\Omega)$ and M_u is well defined. Clearly, M_u is a linear map $\mathcal{D}(\Omega) \to \mathcal{D}(\Omega)$.

In order to prove continuity recall first the *Leibniz formula*

$$D^\alpha(f \cdot g) = \sum_{\beta+\gamma=\alpha} \frac{\alpha!}{\beta!\gamma!} D^\beta \cdot D^\gamma g \quad \forall f, g \in \mathcal{C}^{|\alpha|}(\Omega). \tag{4.5}$$

Here we use the multi-index notation: For $\alpha = (\alpha_1, \ldots, \alpha_n) \in \mathbb{N}^n$ one defines $\alpha! = \alpha_1! \cdots \alpha_n!$ and addition of multi-indices is as usual component-wise.

Now given any compact set $K \subset \Omega$ and $m \in \mathbb{N}$ we estimate as follows, for all $\phi \in \mathcal{D}_K(\Omega)$:

$$p_{K,m}(M_u(\phi)) \leq C p_{K,m}(u) p_{K,m}(\phi).$$

Here C is a constant depending only on m and n. The details of this estimate are left as an Exercise. Since $u \in \mathcal{C}^\infty(\Omega)$ we know that $p_{K,m}(u)$ is finite for every compact set K and every $m \in \mathbb{N}$. Proposition 2.2.3 thus implies continuity of M_u. Therefore its adjoint M'_u is a continuous linear map $\mathcal{D}'(\Omega) \to \mathcal{D}'(\Omega)$ (see the arguments following equation (4.1)). Hence the multiplication with $\mathcal{C}^\infty$-functions u,

$$u \cdot T = T \circ M_u = M'_u(T)$$

acts continuously on $\mathcal{D}'(\Omega)$.

The proof of the product rule for differentiation is a straightforward calculation. Take any $T \in \mathcal{D}'(\Omega)$ and any $u \in \mathcal{C}^\infty(\Omega)$. Using the abbreviation $\partial_j = \frac{\partial}{\partial x_j}$ we have for all $\phi \in \mathcal{D}(\Omega)$,

$$\begin{aligned}\langle \partial_j(u \cdot T), \phi \rangle &= -\langle (u \cdot T), \partial_j \phi \rangle = -\langle T, u\partial_j\phi \rangle = -\langle T, \partial_j(u\phi) - \phi\partial_j u \rangle \\ &= -\langle T, \partial_j(u\phi)\rangle + \langle T, \phi\partial_j u\rangle = \langle \partial_j T, u\phi\rangle + \langle T, \phi\partial_j u\rangle \\ &= \langle u \cdot \partial_j T, \phi\rangle + \langle \partial_j u \cdot T, \phi\rangle = \langle u \cdot \partial_j T + \partial_j u \cdot T, \phi\rangle,\end{aligned}$$

and the product rule follows.

Finally we prove compatibility of the multiplication for distribution with the multiplication for $L^1_{loc}(\Omega)$ under the embedding I. As we have seen earlier, $u \cdot f \in L^1_{loc}(\Omega)$ for all $u \in \mathcal{C}^\infty(\Omega)$ and all $f \in L^1_{loc}(\Omega)$. Thus, given $f \in L^1_{loc}(\Omega)$ and $u \in \mathcal{C}^\infty(\Omega)$, we calculate, for all $\phi \in \mathcal{D}(\Omega)$,

$$\langle u \cdot I_f, \phi\rangle = \langle I_f, u\phi\rangle = \int f(x) u(x) \phi(x) dx = \langle I_{uf}, \phi\rangle,$$

and we conclude. $\square$

This proposition shows that the *multiplicator space* for distributions on Ω is all of $\mathcal{C}^\infty(\Omega)$, i.e., every $T \in \mathcal{D}'(\Omega)$ can be multiplied by every $u \in \mathcal{C}^\infty(\Omega)$ to give a distribution $u \cdot T$ on Ω. In the case of tempered distributions one has to take growth restrictions into account and accordingly the multiplicator space for tempered distributions on Ω is considerably smaller as the following proposition shows:

Proposition 4.2.2 *Denote by $\mathcal{O}_m(\mathbb{R}^n)$ the space of all $\mathcal{C}^\infty$-functions u on $\mathbb{R}^n$ such that for every $\alpha \in \mathbb{N}^n$ there are a constant C and an integer m such that*

$$|D^\alpha u(x)| \leq C(1+x^2)^{\frac{m}{2}} \quad \forall x \in \mathbb{R}^n.$$

Then every $T \in \mathcal{S}'(\mathbb{R}^n)$ can be multiplied by every $u \in \mathcal{O}_m(\mathbb{R}^n)$ and $u \cdot T \in \mathcal{S}'(\mathbb{R}^n)$.

Proof. We have to show that multiplication by $u \in \mathcal{O}_m(\mathbb{R}^n)$, $\phi \mapsto u \cdot \phi$ is a continuous linear map $\mathcal{S}(\mathbb{R}^n) \to \mathcal{S}(\mathbb{R}^n)$. Using Leibniz' formula, this is a straightforward calculation. For the details we refer to the Exercises. $\square$

4.3 Transformation of variables

As in classical analysis it is often helpful to be able to work with distributions in different coordinate systems. This amounts to a change of variables in which the distributions are considered. Since in general distributions are defined through their action on test functions, these changes of variables have to take place first on the level of test functions and then by taking adjoints, on the level of distributions. This requires that admissible transformations of variables have to take test function spaces into test functions and in this way they are considerably more restricted than in classical analysis.

Let $\Omega_x \subset \mathbb{R}^n_x$ be a nonempty open set and $\sigma : \Omega_x \to \Omega_y$, $\Omega_y \subset \mathbb{R}^n_y$, a differentiable bijective mapping from Ω_x onto the open set $\Omega_y = \sigma(\Omega_x)$. Then the determinant of the derivative of this mapping does not vanish: $\det \frac{\partial \sigma}{\partial x} \neq 0$ on Ω_x. We assume $\sigma \in C^\infty(\Omega_x)$. It follows that the inverse transformation σ^{-1} is a C^∞-transformation from Ω_y onto Ω_x and compact subsets $K \subset \Omega_x$ are transformed onto compact subsets $\sigma(K)$ in Ω_y. The chain rule for functions implies that $\phi \circ \sigma^{-1}$ is of class C^∞ on Ω_y for every $\phi \in \mathcal{D}(\Omega_x)$. Hence

$$\phi \mapsto \phi \circ \sigma^{-1}$$

is a well defined mapping $\mathcal{D}(\Omega_x) \to \mathcal{D}(\Omega_y)$. In the Exercises we show that this mapping is actually continuous. In the Exercises we also prove that

$$|\det \frac{\partial \sigma^{-1}}{\partial y}| \in C^\infty(\Omega_y).$$

The wellknown formula for the change of variables in integrals will guide us to a definition of the change of variables for distributions, which is compatible with the embedding of functions into the space of distributions. Take any $f \in L^1_{loc}(\Omega_y)$ and calculate for all $\phi \in \mathcal{D}(\Omega_x)$,

$$\int_{\Omega_x} f(\sigma(x))\phi(x)dx = \int_{\Omega_y} f(y)\phi(\sigma^{-1}(y))|\det \frac{\partial \sigma^{-1}}{\partial y}|dy,$$

i.e., $\langle I_{f \circ \sigma}, \phi \rangle = \langle |\det \frac{\partial \sigma^{-1}}{\partial y}| \cdot I_f, \phi \circ \sigma^{-1} \rangle$. Accordingly one defines the change of variables for distributions.

Definition 4.3.1 *Let $\sigma : \Omega_x \to \Omega_y$ be a bijective C^∞-transformation from a nonempty open set $\Omega_x \subset \mathbb{R}^n$ onto a (nonempty open) set Ω_y. To every distribution T on $\Omega_y = \sigma(\Omega_x)$ one assigns a distribution $T \circ \sigma$ of new variables on Ω_x which is defined in the following* **formula for the transformation of variables:**

$$\langle T \circ \sigma, \phi \rangle = \langle |\det \frac{\partial \sigma^{-1}}{\partial y}| \cdot T, \phi \circ \sigma^{-1} \rangle \qquad \forall \phi \in \mathcal{D}(\Omega_x). \qquad (4.6)$$

Proposition 4.3.1 *For the transformation of variables as defined above, the* **chain rule** *holds, i.e., if $T \in \mathcal{D}'(\Omega_y)$ and $\sigma = (\sigma_1, \ldots, \sigma_n)$ is a bijective C^∞-transformation, then one has for $j = 1, \ldots, n$,*

$$D_j(T \circ \sigma) = \sum_{i=1}^{n}(\partial_i \sigma) \cdot (D_i T) \circ \sigma.$$

Proof. Since we will not use this rule in an essential way we refer for a proof to the literature [Zem87]. □

In applications in physics, typically, rather special cases of this general formula are used, mainly to formulate symmetry or invariance properties of the system. Usually these symmetry properties are defined through transformations of the co-ordinate space, such as translations, rotations, and Galileo or Lorentz transformations. We give a simple concrete example.

Let A be a constant $n \times n$ matrix with nonvanishing determinant and $a \in \mathbb{R}^n$ some vector. Define a transformation $\sigma : \mathbb{R}^n \to \mathbb{R}^n$ by $y = \sigma(x) = Ax + a$ for all $x \in \mathbb{R}^n$. This transformation certainly satisfies all our assumptions. Its inverse is $x = \sigma^{-1}(y) = A^{-1}(y - a)$ for all $y \in \mathbb{R}^n$ and thus $\frac{\partial \sigma^{-1}}{\partial y} = A^{-1}$. Given $T \in \mathcal{D}'(\mathbb{R}_y^n)$ we want to determine its transform under σ. According to equation (4.6) it is given by $\langle T \circ \sigma, \phi \rangle = |\det A^{-1}|\langle T, \phi \circ \sigma^{-1}\rangle$ for all $\phi \in \mathcal{D}(\mathbb{R}_x^n)$. The situation becomes more transparent when we write the different variables explicitly as arguments of the distribution and the test function:

$$\langle (T \circ \sigma)(x), \phi(x) \rangle = \langle T(Ax + a), \phi(x) \rangle = \frac{1}{|\det A|} \langle T(y), \phi(A^{-1}(y - a)) \rangle.$$

In particular for $A = 1_n$ (1_n is the identity matrix in dimension n) this formula describes *translations* by $a \in \mathbb{R}^n$. With the abbreviations $T_a(x) = T(x + a)$ and $\phi_a(y) = \phi(y - a)$ we have

$$\langle T_a, \phi \rangle = \langle T_a(x), \phi(x) \rangle = \langle T(y), \phi_a(y) \rangle.$$

Knowing what the translation of a distribution by $a \in \mathbb{R}^n$ is one can easily formulate periodicity of distributions: A distribution $T \in \mathcal{D}'(\mathbb{R}^n)$ is said to be *periodic with period* $a \in \mathbb{R}^n$ if, and only if, $T_a = T$.

Another interesting application of the translation of distributions is to define the derivative as the limit of difference quotients as is done for functions. One would expect that this definition agrees with the definition of the derivative for distributions given earlier. This is indeed the case as the following corollary shows.

Corollary 4.3.1 *Let T be a distribution on a nonempty open set $\Omega \subset \mathbb{R}^n$ and $a \in \mathbb{R}^n$ some vector. Denote by T_a the translated distribution as introduced above. Then*

$$\lim_{t \to 0} \frac{T_{ta} - T}{t} = a \cdot DT \quad \text{in } \mathcal{D}'(\Omega). \tag{4.7}$$

Here $DT = (\partial_1 T, \ldots, \partial_n T)$ denotes the distributional derivative as given in Definition 4.1.1.

56   4. Calculus for Distributions

Proof. Given $a \in \mathbb{R}^n$ and $T \in \mathcal{D}'(\Omega)$ choose any $\phi \in \mathcal{D}(\Omega)$. Then, for $t \in \mathbb{R}, t \neq 0$ and sufficiently small, we know that $\phi_{ta} \in \mathcal{D}(\Omega)$ too. For these numbers t we have

$$\langle \frac{T_{ta} - T}{t}, \phi \rangle = \langle T, \frac{\phi_{ta} - \phi}{t} \rangle.$$

In the Exercises we show that

$$\lim_{t \to 0} \frac{\phi_{ta} - \phi}{t} = -a \cdot D\phi \quad \text{in } \mathcal{D}(\Omega).$$

Here our notation is $a \cdot D\phi = a_1 \partial_1 \phi + \cdots + a_n \partial_n \phi$. Using continuity of T on $\mathcal{D}(\Omega)$ in the first step and Definition 4.1.1 in the last step, it follows that

$$\lim_{t \to 0} \langle \frac{T_{ta} - T}{t}, \phi \rangle = -\langle T, a \cdot D\phi \rangle = \langle a \cdot DT, \phi \rangle,$$

and thus we conclude. □

4.4 Some applications

4.4.1 Distributions with support in a point

Thus far we have developed elementary calculus for distributions and we have learned about the localization of distributions. This subsection discusses some related results. A first proposition states that the differentiation of distributions and the multiplication of distributions with $\mathcal{C}^\infty$-functions are *local operations* on distributions since under these operations the support is 'conserved'.

Proposition 4.4.1 *Suppose $\Omega \subset \mathbb{R}^n$ is open and nonempty. Then:*

1. $\operatorname{supp}(D^\alpha T) \subseteq \operatorname{supp} T$ *for every* $T \in \mathcal{D}'(\Omega)$ *and every* $\alpha \in \mathbb{N}^n$.

2. $\operatorname{supp}(u \cdot T) \subseteq \operatorname{supp} T$ *for every* $T \in \mathcal{D}'(\Omega)$ *and every* $u \in \mathcal{C}^\infty(\Omega)$.

The proof of these two simple statements is suggested as an exercise. Here we want to point out that in both statements the relation $\subseteq$ cannot be replaced by $=$. This can be seen by looking at some simple examples, for instance take $T = I_c$ on $\mathbb{R}$ for some constant $c \neq 0$. Then $\operatorname{supp} T = \mathbb{R}$ but for $\alpha \geq 1$ we have $D^\alpha T = I_{D^\alpha c} = I_0 = 0$. And for $u(x) = x$, $u \in \mathcal{C}^\infty(\mathbb{R})$, and $T = \delta \in \mathcal{D}'(\mathbb{R})$ one has $u \cdot \delta = 0$ while $\operatorname{supp} \delta = \{0\}$.

It is also instructive to observe that $\phi(x) = 0$ for all $x \in \operatorname{supp} T$ does not imply $T(\phi) = 0$, in contrast to the situation for measures. Take for example $T = D^\alpha \delta$ with $|\alpha| \geq 1$ and a test function ϕ with $\phi(0) = 0$ and $\phi^\alpha(0) \neq 0$.

Recall Proposition 4.1.2 where we showed that the simple ordinary differential equation $D^n T = 0$ has also in $\mathcal{D}'(\mathbb{R})$ only the classical solutions. For ordinary differential equations whose coefficients are not constant the situation can be very different. We look at the simplest case, the equation $x^{n+1} \cdot T = 0$ on $\mathbb{R}$. In $L^1_{loc}(\mathbb{R})$ we only have the trivial solution, but not in $\mathcal{D}'(\mathbb{R})$ as the following proposition shows.

4.4 Some applications

Proposition 4.4.2 $T \in \mathcal{D}'(\mathbb{R})$ *solves the equation*

$$x^{n+1} \cdot T = 0 \quad \text{in} \quad \mathcal{D}'(\mathbb{R}) \tag{4.8}$$

if, and only if, T is of the form

$$T = \sum_{i=0}^{n} c_i D^i \delta \tag{4.9}$$

with certain constants c_i.

Proof. If T is of the form (4.9) then, for all $\phi \in \mathcal{D}(\mathbb{R})$, we have $\langle x^{n+1} \cdot T, \phi \rangle = \langle T, x^{n+1}\phi \rangle = \sum_{i=0}^{n} c_i \langle D^i \delta, x^{n+1}\phi \rangle = \sum_{i=0}^{n} c_i(-1)^i (x^{n+1}\phi)^{(i)}(0) = 0$, since $(x^{n+1}\phi)^{(i)}(0) = 0$ for all $i \leq n$. hence T solves equation (4.8).

Now assume conversely that T is a solution of equation (4.8). In a first step we show indirectly that T has a support contained in $\{0\}$. Suppose $x_0 \in \text{supp}\, T$ and $x_0 \neq 0$. Then there is a neighborhood U of x_0 which does not contain the point $x = 0$ and there is a test function $\psi \in \mathcal{D}(U)$ such that $T(\psi) \neq 0$. It follows that $\phi = x^{-(n+1)}\psi \in \mathcal{D}(U)$. Since $x^{n+1} \cdot T = 0$ we get $0 = (x^{n+1} \cdot T)(\phi) = T(x^{n+1}\phi) = T(\psi)$, a contradiction. Therefore $\text{supp}\, T \subseteq \{0\}$.

Now choose some test function $\rho_r \in \mathcal{D}(\mathbb{R})$ with $\rho_r(x) = 1$ for all $x \in (-s, s)$ for some $s > 0$, as constructed in the Exercises. Then, for any $\phi \in \mathcal{D}(\mathbb{R})$, we know that $\psi = (1 - \rho_r)\phi$ has its support in $\mathbb{R} \setminus \{0\}$ and hence $0 = T(\psi) = T(\phi) - T(\rho_r \phi)$. Using Taylor's Theorem one can write

$$\phi(x) = \sum_{i=0}^{n} \frac{x^i}{i!} \phi^{(i)}(0) + x^{n+1} \phi_1(x)$$

with

$$\phi_1(x) = \int_0^1 \frac{(1-t)^n}{n!} \phi^{(n+1)}(tx)\, dt \quad \in C^\infty(\mathbb{R}).$$

This allows us to approximate the test function ϕ near $x = 0$ by a polynomial, and the resulting approximation in $\mathcal{D}(\mathbb{R})$ is

$$\rho_r \phi = \sum_{i=0}^{n} \frac{\phi^{(i)}(0)}{i!} x^i \rho_r x^{n+1} + \phi_2$$

with $\phi_2 = \rho_r \phi_1 \in \mathcal{D}(\mathbb{R})$. Thus

$$T(\phi) = T(\rho_r \phi) = \sum_{i=0}^{n} \frac{\phi^{(i)}(0)}{i!} T(x^i \rho_r) + T(x^{n+1}\phi_2) = \sum_{i=0}^{n} \frac{T(x^i \rho_r)}{i!} (-1)^i \delta^{(i)}(\phi),$$

since $T(x^{n+1}\phi_2) = (x^{n+1}T)(\phi_2) = 0$. And we conclude that (4.9) holds with $c_i = \frac{(-1)^i}{i!} T(x^i \rho_r)$. □

There is a multi-dimensional version of this result which will be addressed in the Exercises. Though its proof relies on the same principle it is technically more involved.

Proposition 4.4.3 *A distribution $T \in \mathcal{D}'(\mathbb{R}^n)$ has its support in the point $x_0 \in \mathbb{R}^n$ if, and only if, T is of the form (3.15) for some $m \in \mathbb{N}$ and some coefficients $c_\alpha \in \mathbb{K}$, i.e.,*

$$T = \sum_{|\alpha| \leq m} c_\alpha D^\alpha \delta_{x_0}.$$

Proof. The proof that any distribution $T \in \mathcal{D}'(\mathbb{R}^n)$ which has its support in a point $x_0 \in \mathbb{R}^n$ is necessarily of the form (3.15) is given here explicitly only for the case $n = 1$ and $x_0 = 0$. The general case is left as an exercise.

Thus we assume that $T \in \mathcal{D}'(\mathbb{R})$ has its support in $\{0\}$. And we will show that then T solves the equation $x^{m+1} \cdot T = 0$ for some $m \in \mathbb{N}$, and we conclude by Proposition 4.4.2.

As in the proof of this proposition we choose some test function $\rho \in \mathcal{D}(\mathbb{R})$ with $\rho(x) = 1$ for all $x \in (-s, s)$ for some $0 < s < 1$ and support in $K = [-1, 1]$, as constructed in the Exercises, and define for $0 < r < 1$ the function $\rho_r(x) = \rho(\frac{x}{r})$. This function belongs to $\mathcal{D}(\mathbb{R})$, has its support in $[-r, r]$ and is equal to 1 in $(-rs, rs)$. Then, for any $\psi \in \mathcal{D}(\mathbb{R})$ the function $\phi = (1 - \rho_r)\psi$ belongs to $\mathcal{D}(\mathbb{R} \setminus \{0\})$ and thus $T(\phi) = 0$ since supp $T \subseteq \{0\}$, or by linearity of T, $T(\psi) = T(\rho_r \psi)$.

Since T is continuous on $\mathcal{D}_K(\mathbb{R})$ there are a constant $C \in \mathbb{R}_+$ and $m \in \mathbb{N}$ such that $|T(\phi)| \leq C p_{K,m}(\phi)$ for all $\phi \in \mathcal{D}_K(\mathbb{R})$. Apply this estimate to $\psi = x^{m+1} \rho_r \phi$ for all $\phi \in \mathcal{D}_K(\mathbb{R})$ to get

$$|T(x^{m+1} \rho_r \phi)| \leq C p_{K,m}(x^{m+1} \rho_r \phi).$$

In the proof that multiplication by $\mathcal{C}^\infty$-functions is continuous on $\mathcal{D}(\Omega)$, we have shown the estimate $p_{K,m}(u\phi) \leq c p_{K,m}(u) p_{K,m}(\phi)$ for all $u \in \mathcal{C}^\infty(\Omega)$ and all $\phi \in \mathcal{D}_K(\Omega)$ with some constant $c \in \mathbb{R}_+$ depending only on m and the dimension n. We apply this here for $u = x^{m+1} \rho_r$ and get

$$|T(x^{m+1} \rho_r \phi)| \leq C p_{K,m}(x^{m+1} \rho_r) p_{K,m}(\phi) \quad \forall \phi \in \mathcal{D}_K(\mathbb{R}).$$

The first factor we estimate as follows, using Leibniz' formula and the identities $D^\beta x^{m+1} = \frac{(m+1)!}{\beta!} x^{m+1-\beta}$ and $D^\gamma \rho_r(x) = r^{-\gamma} (D^\gamma \rho)(\frac{x}{r})$:

$$p_{K,m}(x^{m+1} \rho_r)$$
$$\leq \sup_{\alpha \leq m} \sup_{x \in K} \sum_{\beta + \gamma = \alpha} \frac{\alpha!}{\beta! \gamma!} \left| \frac{(m+1)!}{\beta!} x^{m+1-\beta} \right| r^{-\gamma} |D^\gamma \rho(\tfrac{x}{r})|$$
$$\leq \sup_{\alpha \leq m} \sup_{x \in K} \sum_{\beta + \gamma = \alpha} \frac{\alpha!}{\beta! \gamma!} \left| \frac{(m+1)!}{\beta!} \right| r^{m+1-\alpha} |(\tfrac{x}{r})^{m+1-\beta} (D^\gamma \rho)(\tfrac{x}{r})|$$
$$\leq r C p_{K,m}(\rho).$$

Now collect all estimates to get, for each $\phi \in \mathcal{D}_K(\mathbb{R})$ and all $0 < r < 1$,

$$|T(x^{m+1} \phi)| = |T(x^{m+1} \rho_r \phi)| \leq C r p_{K,m}(\rho) p_{K,m}(\phi).$$

Taking the limit $r \to 0$, it follows that $T(x^{m+1} \phi) = 0$ for every $\phi \in \mathcal{D}_K(\mathbb{R})$ and hence

$$x^{m+1} \cdot T = 0 \quad \text{in} \quad \mathcal{D}'(\mathbb{R})$$

and we conclude. $\square$

4.4.2 Renormalization of $(\frac{1}{x})_+ = \frac{\theta(x)}{x}$

As an application of Proposition 4.4.3 we discuss a problem which plays a fundamental role in relativistic quantum field theory. Renormalization is about giving formal integrals which do not exist in the Lebesgue sense a mathematically consistent meaning. Here the perspective given by distribution theory is very helpful.

Denote by θ as usual Heaviside's function. As we have seen earlier, in the context of introducing Cauchy's principal value, the function $(\frac{1}{x})_+ = \frac{1}{x}\theta(x)$ is not locally integrable on $\mathbb{R}$ and thus $(\frac{1}{x})_+$ cannot be used directly to define a regular distribution. Consider the subspace $\mathcal{D}_0(\mathbb{R}) = \{\phi \in \mathcal{D}(\mathbb{R}) : \phi(0) = 0\}$ of the test function space $\mathcal{D}(\mathbb{R})$. Every $\phi \in \mathcal{D}_0(\mathbb{R})$ has the representation $\phi(x) = x\psi(x)$ with $\psi(x) = \int_0^1 \phi'(tx)dt \in \mathcal{D}(\mathbb{R})$. Thus we get a definition of $(\frac{1}{x})_+$ as a continuous linear function $\mathcal{D}_0(\mathbb{R}) \to \mathbb{K}$ which agrees on $(0, \infty)$, i.e., on the test function space $\mathcal{D}(\mathbb{R}\setminus\{0\})$, with the function $\frac{1}{x}\theta x$, by the formula

$$\langle (\frac{1}{x})_+, \phi \rangle = \int_0^\infty \frac{\phi(x)}{x} dx. \quad (4.10)$$

If K is any compact subset of $\mathbb{R}$ we get, for all $\phi \in \mathcal{D}_K(\mathbb{R}) \cap \mathcal{D}_0(\mathbb{R})$, the estimate ($|K|$ denotes the measure of the set K)

$$|\langle (\frac{1}{x})_+, \phi \rangle| \leq |K| p_{K,1}(\phi)$$

which shows that equation (4.10) defines $(\frac{1}{x})_+ = T_0$ as a continuous linear functional of order 1 on $\mathcal{D}_0(\mathbb{R})$. By the Hahn–Banach Theorem (see for instance [Rud73, RS80]) the functional T_0 has many continuous linear extensions T to all of $\mathcal{D}(\mathbb{R})$, of the same order 1 as T_0. This means the following: T is a continuous linear functional $\mathcal{D}(\mathbb{R}) \to \mathbb{K}$ such that $|T(\phi)| \leq |K| p_{K,1}(\phi)$ for all $\phi \in \mathcal{D}_K(\mathbb{R})$ and $T|\mathcal{D}_0(\mathbb{R}) = T_0$. Such extensions T of T_0 are called *renormalizations* of $(\frac{1}{x})_+$. How many renormalizations of $(\frac{1}{x})_+$ do we get? This can be decided with the help of Proposition 4.4.3. Since $T|\mathcal{D}_0(\mathbb{R}) = T_0$ we find that T can differ from T_0 only by a distribution with support in $\{0\}$, and since we know the orders of T and T_0 it follows from Proposition 4.4.3 that

$$T = T_0 + c_0 \delta + c_1 \delta'$$

with some constants $c_i \in \mathbb{K}$.

In physics a special 1-parameter family of renormalizations is considered. This choice is motivated by the physical context in which the renormalization problem occurs. For any $0 < M < \infty$ define for all $\phi \in \mathcal{D}(\mathbb{R})$,

$$\langle (\frac{1}{x})_{+,M}, \phi \rangle = \int_0^M \frac{\phi(x) - \phi(0)}{x} dx + \int_M^\infty \frac{\phi(x)}{x} dx.$$

It follows easily that $(\frac{1}{x})_{+,M}$ is a distribution on $\mathbb{R}$ and $(\frac{1}{x})_{+,M}|\mathcal{D}_0(\mathbb{R}) = (\frac{1}{x})_+$. Thus $(\frac{1}{x})_{+,M}$ is a renormalization of $(\frac{1}{x})_+$. If $(\frac{1}{x})_{+,M'}$ is another renormalization of this family, a straightforward calculation shows that

$$(\frac{1}{x})_{+,M} - (\frac{1}{x})_{+,M'} = -\ln(\frac{M}{M'})\delta.$$

Therefore $(\frac{1}{x})_{+,M}$, $0 < M < \infty$, is a 1-parameter family of renormalizations of $(\frac{1}{x})_+$. Now compare $(\frac{1}{x})_{+,M}$ with any other renormalization T of $(\frac{1}{x})_+$. Since both renormalizations are equal to T_0 on $\mathcal{D}_0(\mathbb{R})$, and since we know $T - T_0 = c_0\delta + c_1\delta'$, we get $0 = c_0\phi(0) + c_1\phi'(0)$ for all $\phi \in \mathcal{D}_0(\mathbb{R})$. But $\phi(0) = 0$ for functions in $\mathcal{D}_0(\mathbb{R})$, hence $c_1 = 0$. We conclude: Any renormalization of $(\frac{1}{x})_+$ differs from the renormalization $(\frac{1}{x})_{+,M}$ only by $c_0\delta$. Thus in this renormalization procedure only one free constant appears.

Similar to the term $\ln\left(\frac{M}{M'}\right)$ above, in the renormalization theory of relativistic quantum field theory free constants occur (as renormalized mass or renormalized charge for instance). In this way our simple example reflects the basic ideas of the renormalization theory of relativistic quantum field theory as developed by N. Bogoliubov, O. S. Parasiuk, K. Hepp and later H. Epstein and V. Glaser ([Hep69, EG73, BLOT90]).

4.5 Exercises

1. For $f \in C^k(\mathbb{R}^n)$ show that
$$\langle I_{D^\alpha f}, \phi \rangle = (-1)^{|\alpha|} \langle I_f, D^\alpha \phi \rangle \qquad \forall \phi \in \mathcal{D}(\mathbb{R}^n).$$

2. Prove the following equation in the sense of distributions on $\mathbb{R}$:
$$\frac{d}{dx} \log |x| = \text{vp} \frac{1}{x}.$$
   Hints: Since $\log |x| \in L^1_{loc}(\mathbb{R})$ one has, for any $\phi \in \mathcal{D}(\mathbb{R})$,
$$\int_\mathbb{R} \log(|x|)\phi(x)dx = \lim_{\epsilon \to 0} \int_{|x|\geq \epsilon} \log(|x|)\phi(x)dx.$$
   Recall in addition: $\lim_{\epsilon \to 0} \epsilon \log \epsilon = 0$.

3. Using the relation $\log(x+iy) = \log|x+iy| + i\arg(x+iy)$ for $x, y \in \mathbb{R}$ prove that the following equation holds in the sense of distributions on $\mathbb{R}$:
$$\frac{d}{dx} \log(x+io) = \text{vp}\frac{1}{x} - i\pi \delta(x).$$

4. Show: A test function $\phi \in \mathcal{D}(\mathbb{R})$ is the derivative of some other test function $\psi \in \mathcal{D}(\mathbb{R})$, $\phi = \psi'$ if, and only if, $I(\phi) = \int_\mathbb{R} \phi(x)dx = 0$.

5. In calculus we certainly have the identity $c = D^n P_n$ with $P_n(x) = \frac{c}{n!}x^n + P_{n-1}(x)$ for any polynomial P_{n-1} of degree $\leq n-1$. Show that this identity also holds in $\mathcal{D}'(\mathbb{R})$, i.e., show the identity $I_c = D^n I_{P_n}$.

6. Let $u \in C^k(\Omega)$ be a classical solution of the constant coefficient partial differential equation (4.3). Prove: $P(D)I_u = I_f$ in $\mathcal{D}'(\Omega)$, hence u solves this partial differential equation in the sense of distributions.

7. For $u \in C^\infty(\Omega)$, $\phi \in \mathcal{D}(\Omega)$, $K \subset \Omega$ compact, and $m \in \mathbb{N}$, show that
$$p_{K,m}(M_u(\phi)) \leq C p_{K,m}(u) p_{K,m}(\phi)$$
for some constant C which depends only on m and n.

Hints: For all $x \in K$ and $|\alpha| \leq m$ one can estimate as follows:

$$\left| \sum_{\beta+\gamma=\alpha} \frac{\alpha!}{\beta!\gamma!} D^\beta u(x) D^\gamma \phi(x) \right| \leq \sum_{\beta+\gamma=\alpha} \frac{\alpha!}{\beta!\gamma!} \sup_K |D^\beta u(x)| \sup_K |D^\gamma \phi(x)|.$$

8. Show: If $u \in \mathcal{O}_m(\mathbb{R}^n)$ and $\phi \in \mathcal{S}(\mathbb{R}^n)$, then $M_u(\phi) = u \cdot \phi \in \mathcal{S}(\mathbb{R}^n)$ and $M_u : \mathcal{S}(\mathbb{R}^n) \to \mathcal{S}(\mathbb{R}^n)$ is linear and continuous.

9. Let $\sigma : \Omega_x \to \Omega_y$ be a bijective $\mathcal{C}^\infty$-transformation from a nonempty open set $\Omega_x \subset \mathbb{R}^n$ onto a (nonempty open) set Ω_y. Show:

   a) $\phi \mapsto \phi \circ \sigma^{-1}$ is a continuous linear mapping $\mathcal{D}(\Omega_x) \to \mathcal{D}(\Omega_y)$.

   b) $|\det \frac{\partial \sigma^{-1}}{\partial y}| \in \mathcal{C}^\infty(\Omega_y)$.

10. Given any $\phi \in \mathcal{D}(\Omega)$ and $a \in \mathbb{R}^n$ prove that

$$\lim_{t \to 0} \frac{\phi_{ta} - \phi}{t} = -a \cdot D\phi \quad \text{in } \mathcal{D}(\Omega).$$

Hints: Show first:

$$(a \cdot D\phi)(x) + \frac{\phi(x - ta) - \phi(x)}{t} = \int_0^1 a \cdot [D\phi(x) - D\phi(x - sta)] ds$$

and then estimate the relevant semi-norms for $t \to 0$.

11. Given a closed interval $[a, b] \subset \mathbb{R}$ and $\epsilon > 0$, construct a function $\phi \in \mathcal{D}(\mathbb{R})$ such that supp $\phi \subseteq [a - \epsilon, b + \epsilon]$ and $\phi(x) = 1$ for all $x \in (a + \epsilon, b - \epsilon)$. (We assume $\epsilon \ll b - a$.)

    Hints: Normalize the function ρ in (2.14) such that $\int \rho(x) dx = 1$ and define, for $0 < r < 1$, $\rho_r(x) = \frac{1}{r}\rho(\frac{x}{r})$. Then define a function u_r on $\mathbb{R}$ by $u_r(x) = \int_{-1}^1 \rho_r(x - y) dy$. Show: $u_r \in \mathcal{D}(\mathbb{R})$, supp $u_r \subseteq [-1 - r, 1 + r]$, and $u_r(x) = 1$ for all $x \in (-1 + r, 1 - r)$. Finally translation and rescaling produces a function with the required properties.

12. Given a closed ball $B_r(x_0) = \{x \in \mathbb{R}^n : |x - x_0| \leq r\}$ and $0 < \epsilon < r$, construct a function $\phi \in \mathcal{D}(\mathbb{R}^n)$ such that $\phi(x) = 1$ for all $x \in \mathbb{R}^n$ with $|x - x_0| < r - \epsilon$ and supp $\phi \subseteq K_{r+\epsilon}(x_0)$.

    Hints: The strategy of the one-dimensional case applies.

13. Prove: For every $u \in \mathcal{O}_m(\mathbb{R}^n)$ (see Proposition 4.2.2) the multiplication by u, $\phi \mapsto u \cdot \phi$, is a continuous linear map from $\mathcal{S}(\mathbb{R}^n)$ into $\mathcal{S}(\mathbb{R}^n)$.

5
Distributions as Derivatives of Functions

The general form of a distribution on a nonempty open set can be determined in a relatively simple way as soon as the topological dual of a certain function space is known. As we are going to learn in the second part the dual of a Hilbert space is easily determined. Thus we use the freedom to define the topology on the test function space through various equivalent systems of norms so that we can use the simple duality theory for Hilbert spaces.

This chapter gives the general form of a distribution. Among other things the results of this chapter show that the space of distributions $\mathcal{D}'(\Omega)$ on a nonempty open set $\Omega \subset \mathbb{R}^n$ is the smallest extension of the space $\mathcal{C}(\Omega)$ of continuous functions on Ω in which one can differentiate without restrictions in the order of differentiation, naturally in the *weak or distributional sense*. Thus we begin with a discussion of weak differentiation and mention a few examples. The following section provides a result which gives the general form of a distribution on a nonempty open set $\Omega \subset \mathbb{R}^n$. How measures and distributions are related and in which way they differ is explained in a section on Radon measures. The final section presents tempered distributions and those which have a compact support as weak derivatives of functions.

5.1 Weak derivatives

In general, a locally integrable function f on a nonempty open set $\Omega \subset \mathbb{R}^n$ cannot be differentiated. But we have learned how to interpret such functions as (regular) distributions I_f, and we have learned to differentiate distributions. Thus in this way we know how to differentiate locally integrable functions.

Definition 5.1.1 *The* **weak or distributional derivative** $D^\alpha f$ *of order* $\alpha \in \mathbb{N}^n$ *of a function* $f \in L^1_{loc}(\Omega)$ *is a distribution on* Ω *defined by the equation*

$$\langle D^\alpha f, \phi \rangle = (-1)^{|\alpha|} \int f(x) D^\alpha \phi(x) dx \qquad \forall \phi \in \mathcal{D}(\Omega). \tag{5.1}$$

From the section on the derivatives of distributions we recall that on the subspace $C^{|\alpha|}(\Omega)$ of $L^1_{loc}(\Omega)$ the weak and the classical derivative agree.

For $m = 0, 1, 2, \ldots$ introduce the space of all weak derivatives of order $|\alpha| \le m$ of all functions in $L^1_{loc}(\Omega)$, i.e.,

$$\mathcal{D}'_{reg,m}(\Omega) = \left\{ D^\alpha f : f \in L^1_{loc}(\Omega), \; |\alpha| \le m \right\} \tag{5.2}$$

and then

$$\mathcal{D}'_{reg,\infty}(\Omega) = \bigcup_{m=0}^{\infty} \mathcal{D}'_{reg,m}(\Omega). \tag{5.3}$$

Certainly, the space $\mathcal{D}'_{reg,\infty}(\Omega)$ is a subspace of space $\mathcal{D}'(\Omega)$ of all distributions on Ω. In the following section we show that both spaces are almost equal, more precisely, we are going to show that locally every distribution T is a weak derivative of functions in $L^1_{loc}(\Omega)$. And this statement is still true if we replace $L^1_{loc}(\Omega)$ by the much smaller space $\mathcal{C}(\Omega)$ of continuous functions on Ω. The term 'locally' in this statement means that the restriction T_K of T to $\mathcal{D}_K(\Omega)$ is a weak derivative of continuous functions.

Now let us look at some concrete examples. Suppose we are given $m \in \mathbb{N}$ and a set $\{f_\alpha : |\alpha| \le m\}$ of continuous functions on Ω. Define a function $T : \mathcal{D}(\Omega) \to \mathbb{K}$ by

$$T(\phi) = \sum_{|\alpha| \le m} \int f_\alpha(x) D^\alpha \phi(x) dx \qquad \forall \phi \in \mathcal{D}(\Omega).$$

Elementary properties of Riemann integrals ensure that T is linear. On each subspace $\mathcal{D}_K(\Omega)$ one has the estimate $|T(\phi)| \le C p_{K,m}(\phi)$ with the constant $C = \sum_{|\alpha| \le m} \int_K |f_\alpha(x)| dx$. Thus T is a distribution of constant local order m and therefore the order of T is finite and equal to m.

Next we consider a class of concrete examples of distributions for which the local order is not constant. To this end recall the representation of $\mathcal{D}(\Omega)$ as the strict inductive limit of the sequence of complete metrizable spaces $\mathcal{D}_{K_i}(\Omega)$, $i \in \mathbb{N}$:

$$\mathcal{D}(\Omega) = \bigcup_{i=1}^{\infty} \mathcal{D}_{K_i}(\Omega). \tag{5.4}$$

Here K_i is a strictly increasing sequence of compact sets which exhaust Ω. Take a strictly increasing sequence of integers m_i and choose functions $f_\alpha \in L^1_{loc}(\Omega)$ with the following specifications: supp $f_\alpha \subset K_0$ for $|\alpha| \leq m_0$ and supp $f_\alpha \subset K_{i+1}\setminus K_i$ for $m_i < |\alpha| \leq m_{i+1}, i = 1, 2, \ldots$. Then define linear functions $T_i : \mathcal{D}_{K_i}(\Omega) \to \mathbb{K}$ by

$$T_i(\phi) = \sum_{|\alpha| \leq m_i} \int f_\alpha(x) D^\alpha \phi(x) dx \qquad \forall \phi \in \mathcal{D}_{K_i}(\Omega).$$

As above one sees that T_i is continuous on $\mathcal{D}_{K_i}(\Omega)$ with the bound $|T_i(\phi)| \leq C_i p_{K_i, m_i}(\phi)$ where $C_i = \sum_{|\alpha| \leq m_i} \int_{K_i} |f_\alpha(x)| dx$. For all $\phi \in \mathcal{D}_{K_i}(\Omega)$ we find

$$T_{i+1}(\phi) = \sum_{|\alpha| \leq m_i} \int f_\alpha(x) D^\alpha \phi(x) dx = T_i(\phi)$$

since

$$\sum_{m_i < |\alpha| \leq m_{i+1}} \int f_\alpha(x) D^\alpha \phi(x) dx = 0$$

because of the support properties of the functions f_α. Hence we get a well-defined continuous linear function $T : \mathcal{D}(\Omega) \to \mathbb{K}$ by defining $T|\mathcal{D}_{K_i}(\Omega) = T_i$ for all i. This distribution T is not of finite order on Ω.

5.2 Structure theorem for distributions

Again it is convenient to start with the representation of the test function space $\mathcal{D}(\Omega)$ as the strict inductive limit of the sequence of complete metrizable spaces $\mathcal{D}_{K_i}(\Omega)$ for a strictly increasing and exhaustive sequence of compact K_i. Then we can say that T is a distribution on Ω if, and only if, $T_i = T|\mathcal{D}_{K_i}(\Omega) \in \mathcal{D}'_{K_i}(\Omega)$ for all $i \in \mathbb{N}$. This leads to the first step in analyzing the structure of distributions.

Proposition 5.2.1 *Let $\Omega \subset \mathbb{R}^n$ be a nonempty open set. Represent the test functions space $\mathcal{D}(\Omega)$ as the strict inductive limit of the complete metrizable spaces $\mathcal{D}_{K_i}(\Omega)$ for a strictly increasing and exhaustive sequence of compact sets K_i (see equation (5.4)). Then the following characterization of distribution holds.*

1. *A distribution $T \in \mathcal{D}'(\Omega)$ determines in a unique way a sequence of functionals $T_i \in \mathcal{D}'_{K_i}(\Omega)$ which satisfies the compatibility condition*

$$T_{i+1}|\mathcal{D}_{K_i}(\Omega) = T_i \qquad \forall i \in \mathbb{N}. \tag{5.5}$$

2. *Conversely, any sequence of functionals $T_i \in \mathcal{D}'_{K_i}(\Omega)$ which satisfies the compatibility condition (5.5) determines in a unique way a distribution T on Ω by defining*

$$T|\mathcal{D}_{K_i}(\Omega) = T_i \qquad \forall i \in \mathbb{N}. \tag{5.6}$$

66   5. Distributions as Derivatives of Functions

Proof. Since we know $\mathcal{D}_{K_i}(\Omega) \subseteq \mathcal{D}_{K_{i+1}}(\Omega)$ the proof of the first part is obvious. For the proof of the second part note that the compatibility condition (5.5) ensures that a linear function $T : \mathcal{D}(\Omega) \to \mathbb{K}$ is well defined by equation (5.6). Continuity of T follows from the definition of the inductive topology on $\mathcal{D}(\Omega)$ and the continuity of the T_i. □

According to this result the general form of a distribution is known as soon as we know the general form of continuous linear functionals on the spaces $\mathcal{D}_K(\Omega)$. And this can be achieved in a fairly easy way on the basis of a fundamental result from the theory of Hilbert spaces which determines the general form of continuous linear functionals on a Hilbert space. According to the Riesz–Fréchet Theorem of Part II (Theorem 15.3.1) every continuous linear functional on a Hilbert space $\mathcal{H}$ is of the form $h \mapsto \langle u, h \rangle, \forall h \in \mathcal{H}$, where $\langle \cdot, \cdot \rangle$ is the inner product of the Hilbert space and the element $u \in \mathcal{H}$ is determined uniquely by the functional.

As a second input we use the fact that the topology of the space $\mathcal{D}_K(\Omega)$ can be defined in terms of the filtering system of semi-norms $q_{K,m}(\phi) = \sqrt{\langle \phi, \phi \rangle_{K,m}}$ where

$$\langle \phi, \psi \rangle_{K,m} = \sum_{|\alpha| \leq m} \int_K D^\alpha \phi(x) D^\alpha \psi(x) dx \qquad \forall \phi, \psi \in \mathcal{D}_K(\Omega),$$

is a scalar product on $\mathcal{D}_K(\Omega)$. (See the subsection 2.1.1). The completion of the space $\mathcal{D}_K(\Omega)$ with respect to this scalar product produces a Hilbert space $\mathcal{H}_{K,m}$ whose scalar product is denoted in the same way.

Proposition 5.2.2 *Let T be a continuous linear functional on the space $\mathcal{D}_K(\Omega)$, $\Omega \subset \mathbb{R}^n$ open and nonempty, $K \subset \Omega$ compact. Then there is an $m \in \mathbb{N}$ and there are elements u_α in the Hilbert space $L^2(K)$ of square integrable functions on K such that*

$$T(\phi) = \sum_{|\alpha| \leq m} \int_K u_\alpha(x) D^\alpha \phi(x) dx \qquad \forall \phi \in \mathcal{D}_K(\Omega),$$

i.e., T is a sum of weak derivatives of square integrable functions on K:

$$T = \sum_{|\alpha| \leq m} (-1)^{|\alpha|} D^\alpha I_{u_\alpha}.$$

Proof. By definition of the topology on $\mathcal{D}_K(\Omega)$, given $T \in \mathcal{D}'_K(\Omega)$, there are a constant C and there is $m \in \mathbb{N}$ such that $|T| \leq C q_{K,m}$. Then T has a continuous and linear extension T_K to the Hilbert space $\mathcal{H}_{K,m}$ which is obtained as the completion of $\mathcal{D}_K(\Omega)$ with respect to the norm $q_{K,m}$. As mentioned above, continuous linear functions on the Hilbert space $\mathcal{H}_{K,m}$ are defined in terms of the scalar product $\langle \cdot, \cdot \rangle_{K,m}$ and some element $u \in \mathcal{H}_{K,m}$. Therefore we have $T_K(v) = \langle u, v \rangle_{K,m}$ for all $v \in \mathcal{H}_{K,m}$. Taking the specific form of the scalar product into account we thus get for all $\phi \in \mathcal{D}_K(\Omega) \subset \mathcal{H}_{K,m}$, since T_K is an extension of T

$$T(\phi) = T_K(\phi) = \langle u, \phi \rangle_{K,m} = \sum_{|\alpha| \leq m} \int_K \overline{D^\alpha u(x)} D^\alpha \phi(x) dx.$$

Introducing the functions $u_\alpha = \overline{D^\alpha u}$ the formula for T follows. □

Propositions 5.2.1 and 5.2.2 together determine the general form of distributions. In terms of the results in Proposition 5.2.2 the compatibility condition of Proposition 5.2.1 could be evaluated more explicitly but we omit this since it is not used later.

Consider for a moment the case $n = 1$. If we integrate $u_\alpha \in L^2(K)$ we get a continuous function $v_\alpha(x) = \int_a^x u_\alpha(y)dy$ where $a \in K$ is arbitrary such that $Dv_\alpha = u_\alpha$. Thus in the representation formula for $T \in \mathcal{D}'_K(\Omega)$ in Proposition 5.2.2 we can use continuous functions instead of square integrable functions by increasing the order of differentiation correspondingly. In particular, this representation is not unique. Though formally more involved these statements hold for the general case too.

Collecting the results from above we arrive at the structure theorem for distributions.

Theorem 5.2.3 *Let $\Omega \subset \mathbb{R}^n$ be a nonempty open set and K_i be a strictly increasing sequence of compact sets which exhaust Ω. T is a distribution on Ω if, and only if, there is a sequence of nonnegative integers m_i and for each $i \in \mathbb{N}$ there are elements $u_{i,\alpha} \in L^2(K_i)$, $|\alpha| \leq m_i$ such that for $i = 0, 1, 2, \ldots$,*

$$T(\phi) = T_i(\phi) \equiv \sum_{|\alpha| \leq m_i} \int_{K_i} u_{i,\alpha}(x) D^\alpha \phi(x) dx \quad \forall \phi \in \mathcal{D}_{K_i}(\Omega) \quad (5.7)$$

and, for all $\phi \in \mathcal{D}_{K_i}(\Omega)$,

$$\sum_{|\alpha| \leq m_{i+1}} \int_{K_{i+1}} u_{i+1,\alpha}(x) D^\alpha \phi(x) dx = \sum_{|\alpha| \leq m_i} \int_{K_i} u_{i,\alpha}(x) D^\alpha \phi(x) dx. \quad (5.8)$$

Proof. Note that equation (5.8) is just the compatibility condition for the sequence of functionals $T_i \in \mathcal{D}'_{K_i}(\Omega)$ defined in equation (5.7) according to Proposition 5.2.2. Thus by Proposition 5.2.1 and Proposition 5.2.2 we conclude. □

5.3 Radon measures

As previously Ω denotes a nonempty open subset of $\mathbb{R}^n$. Introduce the space $\mathcal{C}_0(\Omega)$ of all continuous functions $f : \Omega \to \mathbb{R}$ which have a compact support in Ω. For a compact subset K of Ω denote by $\mathcal{C}_K(\Omega)$ the subspace of all functions in $\mathcal{C}_0(\Omega)$ which have a support contained in K. On the spaces $\mathcal{C}_K(\Omega)$, $K \subset \Omega$ compact, we use the norms $p_{K,0}$ introduced in Chapter 2. Equip the space $\mathcal{C}_0(\Omega)$ with the inductive limit topology of the spaces $(\mathcal{C}_K(\Omega), p_{K,0})$. A continuous linear functional on this space $\mathcal{C}_0(\Omega)$ is called a *real Radon measure* on Ω. In more concrete terms one has the following characterization.

Corollary 5.3.1 *A linear functional $\mu : \mathcal{C}_0(\Omega) \to \mathbb{R}$ is a real **Radon measure** on Ω if, and only if, for every compact subset $K \subset \Omega$ there is a constant C such that*

$$|\mu(f)| \leq C p_{K,0}(f) \quad \forall f \in \mathcal{C}_K(\Omega).$$

Obviously one has $\mathcal{D}_K(\Omega) \subset \mathcal{C}_K(\Omega)$ and $\mathcal{D}(\Omega) \subset \mathcal{C}_0(\Omega)$ and the natural embeddings are continuous. Hence, every real Radon measure is a distribution.

Now we discuss some order theoretic properties of the test function space $\mathcal{C}_0(\Omega)$ for Radon measures which the test function space $\mathcal{D}(\Omega)$ for distributions does not have. Denote by $\mathcal{C}_{0,+}(\Omega)$ the set of all nonnegative functions in $\mathcal{C}_0(\Omega)$. Given $f \in \mathcal{C}_0(\Omega)$, define $f_\pm(x) = \max\{\pm f(x), 0\}$. It follows that $f_\pm \in \mathcal{C}_{0,+}(\Omega)$ and $f = f_+ - f_-$. This shows

$$\mathcal{C}_0(\Omega) = \mathcal{C}_{0,+}(\Omega) - \mathcal{C}_{0,+}(\Omega).$$

We deduce that every real Radon measure μ is the difference of two positive Radon measures μ_+ and μ_-: $\mu = \mu_+ - \mu_-$. Such decompositions do not hold in distribution theory, neither on the level of test functions nor on the level of distributions. In general, a continuously differentiable real valued function cannot be written as the difference of two nonnegative differentiable functions (take for instance the example of the sine function). Nevertheless there is an interesting order theoretic implication for distributions.

Theorem 5.3.1 *Every nonnegative linear form $T : \mathcal{D}(\Omega) \to \mathbb{R}$, i.e., $T(\phi) \geq 0$ for all nonnegative $\phi \in \mathcal{D}(\Omega)$, is the restriction of a positive Radon measure μ to $\mathcal{D}(\Omega)$, and thus in particular is continuous.*

Proof. Suppose that T is a nonnegative linear function $\mathcal{D}(\Omega) \to \mathbb{R}$. Introduce the restrictions T_K of T to $\mathcal{D}_K(\Omega)$. Clearly, T_K is a nonnegative linear function on $\mathcal{D}_K(\Omega)$ and the net T_K, $K \subset \Omega$ compact, satisfies the compatibility condition $T_{K_2}|\mathcal{D}_{K_1} = T_{K_1}$ for all compact sets $K_1 \subset K_2 \subset \Omega$.

Given a compact set $K \subset \Omega$ there are a compact set $K' \subset \Omega$ such that $K \Subset K'$ and a nonnegative function $\psi \in \mathcal{D}_{K'}(\Omega)$ which is equal to 1 on K. Therefore the estimate

$$-\psi(x) p_{K,0}(\phi) \leq \phi(x) \leq p_{K,0}(\phi) \psi(x)$$

holds for all $x \in \Omega$ and all $\phi \in \mathcal{D}_K(\Omega)$. Since $T_{K'}$ is nonnegative it preserves this estimate:

$$-T_{K'}(\psi) p_{K,0}(\phi) \leq T_{K'}(\phi) \leq T_{K'}(\psi) p_{K,0}(\phi)$$

and thus, since $T_K(\phi) = T_{K'}(\phi)$ for all $\phi \in \mathcal{D}_K(\Omega)$,

$$|T_K(\phi)| = |T_{K'}(\phi)| \leq T_{K'}(\psi) p_{K,0}(\phi)$$

for all $\phi \in \mathcal{D}_K(\Omega)$. This shows continuity of T_K for every compact subset $K \subset \Omega$. Hence T is a distribution on Ω, of order 0.

This continuity estimate for T_K allows us to extend T_K to a nonnegative linear function $\mu_K : \mathcal{C}_K(\Omega) \to \mathbb{R}$ with the same bound. This extension process preserves the compatibility condition of the net T_K, $K \subset \Omega$ compact. Thus we can define a continuous linear function $\mu : \mathcal{C}_0(\Omega) \to \mathbb{R}$ by setting $\mu|\mathcal{C}_K(\Omega) = \mu_K$ for $K \subset \Omega$) compact. Since each μ_K is nonnegative, μ is a nonnegative Radon measure on Ω, and by construction one has $\mu|\mathcal{D}(\Omega) = T$.

Further details of the proof are given in [Don69, Sch57]. □

5.4 The case of tempered and compactly supported distributions

The results on the structure of distributions show that locally every distribution is a weak derivative of functions. In the case of tempered distributions and those

5.4 The case of tempered and compactly supported distributions

distributions which have a compact support this result holds globally, as we are going to prove.

For the case of distributions with compact support this is fairly obvious. We have learned that a distribution T on a nonempty open set Ω with compact support is a continuous linear functional on the test function space $\mathcal{E}(\Omega)$, i.e., there is a compact subset $K \subset \Omega$ and there are a constant $C \in \mathbb{R}_+$ and $m \in \mathbb{N}$ such that $|T(\phi)| \leq Cq_{K,m}(\phi)$ for all $\phi \in \mathcal{E}(\Omega)$. Here we have used again the fact that the two filtering systems of semi-norms $\{p_{K,m} : m = 0, 1, 2, \ldots\}$ and $\{q_{K,m} : m = 0, 1, 2, \ldots\}$ are equivalent. Now we can proceed as in Proposition 5.2.2 and conclude. Note however that the distribution and the functions representing this distribution through a process of taking weak derivatives need not have the same support. As an example consider Dirac's delta function δ which has its support in the point $x = 0$. And we have learned that δ can be represented as the weak derivative θ' of the Heaviside function θ which has its support in $\mathbb{R}_+$.

By definition, a tempered distribution T on $\Omega \subseteq \mathbb{R}^n$ is a linear functional $T : \mathcal{S}(\Omega) \to \mathbb{K}$ for which there are a constant $C \in \mathbb{R}_+$ and $m, k \in \mathbb{N}$ such that

$$|T(\phi)| \leq Cp_{m,k}(\phi) \qquad \forall \phi \in \mathcal{S}(\Omega).$$

Again the filtering system of norms $\{p_{m,k} : m, k = 0, 1, 2, \ldots\}$ is equivalent to the filtering system $\{q_{m,k} : m, k = 0, 1, 2, \ldots\}$ of norms $q_{m,k}(\phi) = \sqrt{\langle \phi, \phi \rangle_{m,k}}$ defined by the scalar product

$$\langle \phi, \psi \rangle_{m,k} = \sum_{|\alpha| \leq k} \int_\Omega \overline{D^\alpha \phi(x)} D^\alpha \psi(x)(1 + x^2)^m dx \qquad \forall \phi, \psi \in \mathcal{S}(\Omega). \quad (5.9)$$

Thus we can assume $|T| \leq Cq_{m,k}$ for some constant C and some nonnegative integers m and k. This allows us to proceed as in the proof of Proposition 5.2.2. Thus there is an element u in the Hilbert space $\mathcal{H}_{m,k}$ defined as the completion of $\mathcal{S}(\Omega)$ with respect to the norm $q_{m,k}$ such that

$$T(\phi) = \langle u, \phi \rangle_{m,k} = \sum_{|\alpha| \leq k} \int_\Omega \overline{D^\alpha u(x)} D^\alpha \phi(x)(1 + x^2)^m dx \qquad \forall \phi \in \mathcal{S}(\Omega).$$

Introduce the functions $u_\alpha(x) = (1 + x^2)^m \overline{D^\alpha u(x)}$ on Ω. Since $u \in \mathcal{H}_{m,k}$ we know, for all $|\alpha| \leq k$, that

$$\int_\Omega |u_\alpha(x)(1 + x^2)^{-\frac{m}{2}}|^2 dx \qquad (5.10)$$

is finite and thus we formulate the structure theorem for tempered distributions.

Theorem 5.4.1 *Let $\Omega \subseteq \mathbb{R}^n$ be an open nonempty set. T is a tempered distribution on Ω if, and only if, there are nonnegative integers m, k and there are measurable functions u_α on Ω, $|\alpha| \leq k$, for which the integrals (5.10) are finite such that*

$$T(\phi) = \sum_{|\alpha| \leq k} \int_\Omega u_\alpha(x) D^\alpha \phi(x) dx \qquad \forall \phi \in \mathcal{S}(\Omega), \quad (5.11)$$

i.e.,
$$T = \sum_{|\alpha|\le k}(-1)^{|\alpha|}D^\alpha I_{u_\alpha}.$$

Proof. In the Exercises we show that equation (5.11) indeed defines a tempered distribution on Ω. That conversely every tempered distribution is of this form we have shown above. Thus we conclude. □

Note that this theorem says that tempered distributions are globally of finite order.

5.5 Exercises

1. Prove: The two filtering systems of norms $\mathcal{P} = \{p_{m,k} : m, k = 0, 1, 2, \ldots\}$ and $\mathcal{Q} = \{q_{m,k} : m, k = 0, 1, 2, \ldots\}$ on $\mathcal{S}(\Omega)$ are equivalent.

2. Show that equation (5.11) defines a tempered distribution.

3. Find an example of a distribution which is not a tempered distribution.

   Hints: Try regular distributions.

4. Show: Every continuous polynomially bounded function on $\mathbb{R}^n$ defines a distribution in $\mathcal{S}'(\mathbb{R}^n) \cap \mathcal{D}'_{reg}(\mathbb{R}^n)$, but not every continuous function on $\mathbb{R}^n$ which defines a distribution in $\mathcal{S}'(\mathbb{R}^n) \cap \mathcal{D}'_{reg}(\mathbb{R}^n)$ is polynomially bounded.

   Hints: Try the function $f(x) = e^x \sin e^x = -\frac{d}{dx}(\cos e^x)$ on $\mathbb{R}$.

6
Tensor Products

The tensor product of distributions is a very important tool in the analysis of distributions. We will use it mainly in the definition of the convolution for distributions which in turn has many important applications, some of which we will discuss in later chapters (approximation of distributions by smooth functions, analysis of partial differential operators with constant coefficients). The tensor product for distributions is naturally based on the tensor product of the underlying test function spaces and their completions. Accordingly we start by developing the theory of tensor products of test function spaces to the extent which is needed later. The following section gives the definition and the main properties of the tensor product for distributions. We assume that the reader is familiar with the definition of the algebraic tensor product of general vector spaces. A short reminder is given in Section 17.2.

6.1 Tensor product for test function spaces

In the chapter on (elementary aspects of) calculus for distributions we discussed among other things a product between functions, between distributions and certain classes of functions, and between distributions if the distributions involved satisfied certain restrictions. This point-wise product assigns to two functions (or distributions) on a set $\Omega \subset \mathbb{R}^n$ a new function (distribution) on the same set Ω.

On the other side, the tensor product assigns to two functions (distributions) f_i on (in general) two different open sets $\Omega_i, i = 1, 2$, a new function (distribution) on the product set $\Omega_1 \times \Omega_2$. To be more specific, assume that $\Omega_1 \subset \mathbb{R}^{n_1}$ and $\Omega_2 \subset \mathbb{R}^{n_2}$ are two nonempty open sets and $\phi_i \in \mathcal{D}(\Omega_i)$ are two test functions on Ω_1 respectively Ω_2. The *tensor product* of ϕ_1 and ϕ_2 is the function $\phi_1 \otimes \phi_2 : \Omega_1 \times \Omega_2 \to \mathbb{K}$ defined by

$$\phi_1 \otimes \phi_2(x_1, x_2) = \phi_1(x_1)\phi_2(x_2) \qquad \forall\, (x_1, x_2) \in \Omega_1 \times \Omega_2. \tag{6.1}$$

Certainly, the tensor product $\phi_1 \otimes \phi_2$ is a C^∞-function on $\Omega_1 \times \Omega_2$ which has a compact support; thus $\phi_1 \otimes \phi_2 \in \mathcal{D}(\Omega_1 \times \Omega_2)$ for all $\phi_i \in \mathcal{D}(\Omega_i)$. The vector space spanned by all these tensor products $\phi_1 \otimes \phi_2$ is denoted by $\mathcal{D}(\Omega_1) \otimes \mathcal{D}(\Omega_2)$. A general element in $\mathcal{D}(\Omega_1) \otimes \mathcal{D}(\Omega_2)$ is of the form

$$\sum_{i=1}^{N} \phi_i \otimes \psi_i, \quad \phi_i \in \mathcal{D}(\Omega_1), \quad \psi_i \in \mathcal{D}(\Omega_2), \quad i = 1, 2, \ldots, N; \tag{6.2}$$

and it follows that the algebraic tensor product $\mathcal{D}(\Omega_1) \otimes \mathcal{D}(\Omega_2)$ of the test function spaces $\mathcal{D}(\Omega_1)$ and $\mathcal{D}(\Omega_2)$ is contained in the test function space over the product set $\Omega_1 \times \Omega_2$:

$$\mathcal{D}(\Omega_1) \otimes \mathcal{D}(\Omega_2) \subset \mathcal{D}(\Omega_1 \times \Omega_2).$$

As a subspace of the test function space $\mathcal{D}(\Omega_1 \times \Omega_2)$ the tensor product space carries naturally the relative topology of $\mathcal{D}(\Omega_1 \times \Omega_2)$. A first important observation is that this tensor product space is dense in the test function space $\mathcal{D}(\Omega_1 \times \Omega_2)$.

Proposition 6.1.1 *Suppose that $\Omega_i \subseteq \mathbb{R}^{n_i}$, $i = 1, 2$, are nonempty open sets. Then the tensor product space $\mathcal{D}(\Omega_1) \otimes \mathcal{D}(\Omega_2)$ of the test function spaces $\mathcal{D}(\Omega_i)$ is sequentially dense in the test function space $\mathcal{D}(\Omega_1 \times \Omega_2)$ over the product set $\Omega_1 \times \Omega_2$, i.e.,*

$$\overline{\mathcal{D}(\Omega_1) \otimes \mathcal{D}(\Omega_2)}^{\tau} = \mathcal{D}(\Omega_1 \times \Omega_2), \tag{6.3}$$

where τ indicates that the closure is taken with respect to the topology of the space $\mathcal{D}(\Omega_1 \times \Omega_2)$.

Proof. We have to show that any given $\psi \in \mathcal{D}(\Omega_1 \times \Omega_2)$ is the limit of a sequence of elements in $\mathcal{D}(\Omega_1) \otimes \mathcal{D}(\Omega_2)$, in the sense of uniform convergence for all derivatives on every compact subset $K \subset \Omega_1 \times \Omega_2$. This is done in several steps.

Given $\psi \in \mathcal{D}(\Omega_1 \times \Omega_2)$ we introduce in a first step the sequence of auxiliary functions ψ_k defined by

$$\psi_k(z) = \int_{\mathbb{R}^n} e_k(z - \xi)\psi(\xi)d\xi = \int_{\mathbb{R}^n} e_k(\xi)\psi(z - \xi)d\xi.$$

Here the following notation is used: $n = n_1 + n_2$, $z = (x_1, x_2) \in \Omega_1 \times \Omega_2$ and

$$e_k(z) = \left(\frac{k}{\sqrt{2\pi}}\right)^n e^{-k^2 z^2}.$$

Observe that $e_k \in C^\infty(\mathbb{R}^n)$ and $\int_{\mathbb{R}^n} e_k(z)dz = 1$ for all $k \in \mathbb{N}$. Without giving the details of the straightforward proof we state, for all $\alpha \in \mathbb{N}^n$, for the derivatives,

$$D^\alpha \psi_k(z) = \int_{\mathbb{R}^n} e_k(\xi)(D^\alpha \psi)(z - \xi)d\xi$$

for all $k \in \mathbb{N}$. Since all derivatives $D^\alpha \psi$ of ψ are uniformly continuous on $\mathbb{R}^n$, given $\epsilon > 0$ there is a $\delta > 0$ such that $|(D^\alpha \psi)(z) - (D^\alpha \psi)(z - \xi)| < \epsilon$ for all $z \in \mathbb{R}^n$ and all $\xi \in \mathbb{R}^n$ with $|\xi| < \delta$. The

6.1 Tensor product for test function spaces 73

normalization of e_k allows us to write $(D^\alpha \psi)(z) - (D^\alpha \psi)_k(z)$ as the integral $\int e_k(\xi)[(D^\alpha \psi)(z) - (D^\alpha \psi)(z - \xi)]d\xi$ which can be estimated, in absolute value, by

$$\int_{|\xi| \leq \delta} e_k(\xi) |(D^\alpha \psi)(z) - (D^\alpha \psi)(z - \xi)| d\xi + \int_{|\xi| > \delta} e_k(\xi) |(D^\alpha \psi)(x) - (D^\alpha \psi)(z - \xi)| d\xi.$$

By choice of δ, using the notation $\|D^\alpha \psi\|_\infty = \sup_{z \in \mathbb{R}^n} |(D^\alpha \psi)(z)|$, this estimate can be continued by

$$\leq \epsilon \int_{|\xi| \leq \delta} e_k(\xi) d\xi + 2\|D^\alpha \psi\|_\infty \int_{|\xi| > \delta} e_k(\xi) d\xi.$$

The first integral is obviously bounded by 1 while for the second integral we find

$$\int_{|\xi| > \delta} e_k(\xi) d\xi = \pi^{-\frac{n}{2}} \int_{|z| > k\delta} e^{-z^2} dz \leq \epsilon$$

for all $k \geq k_0$ for some sufficiently large $k_0 \in \mathbb{N}$. Therefore, uniformly in $z \in \mathbb{R}^n$, for all $k \geq k_0$,

$$|(D^\alpha \psi)(z) - (D^\alpha \psi_k)(z)| \leq 2\epsilon.$$

We deduce: Every derivative $D^\alpha \psi$ of ψ is the uniform limit of the sequence of corresponding derivatives $D^\alpha \psi_k$ of the sequence ψ_k.

In a second step, by using special properties of the exponential function, we prepare the approximation of the elements of the sequence ψ_k by functions in the tensor product space $\mathcal{D}(\Omega_1) \otimes \mathcal{D}(\Omega_2)$. To this end we use the power series representation of the exponential function and introduce, for each $k \in \mathbb{N}$, the sequence of functions defined by the formula

$$\psi_{k,N}(z) = (\frac{k}{\sqrt{\pi}})^n \sum_{i=0}^{N} \frac{1}{i!} \int_{\mathbb{R}^n} [-k^2(z-\xi)^2]^i \psi(\xi) d\xi.$$

As in the first step the derivative of these functions can easily be calculated. One finds

$$D^\alpha \psi_{k,N}(z) = \sum_{i=0}^{N} \frac{1}{i!} (\frac{k}{\sqrt{\pi}})^n \int_{\mathbb{R}^n} [-k^2(z-\xi)^2]^i (D^\alpha \psi)(\xi) d\xi$$

and therefore we estimate, for all $z \in \mathbb{R}^n$ such that $|z - \xi| \leq R$ for all ξ in the support of ψ for some finite R, as follows:

$$|D^\alpha \psi_{k,N}(z) - D^\alpha \psi_k(z)| \leq (\frac{k}{\sqrt{\pi}})^n \sum_{i=N+1}^{\infty} \frac{1}{i!} |k^2(z-\xi)^2|^i |(D^\alpha \psi)(\xi)| d\xi$$
$$\leq I_N(k, R) \int_{\mathbb{R}^n} |(D^\alpha \psi)(\xi)| d\xi$$

where

$$I_N(k, R) = (\frac{k}{\sqrt{\pi}})^n \sum_{i=N+1}^{\infty} \frac{(kR)^{2i}}{i!} \to 0 \quad \text{as } N \to \infty.$$

Using the binomial formula to expand $(z-\xi)^{2i}$ and evaluating the resulting integrals, we see that the functions $\psi_{k,N}$ are actually polynomials in z of degree $\leq 2N$, and recalling that z stands for the pair of variables $(x_1, x_2) \in \mathbb{R}^{n_1} \times \mathbb{R}^{n_2}$, we see that these functions are of the form

$$\psi_{k,N}(x_1, x_2) = \sum_{|\alpha|, |\beta| \leq 2N} C_{\alpha, \beta} x_1^\alpha x_2^\beta.$$

Since $\psi \in \mathcal{D}(\Omega_1 \times \Omega_2)$, there are compact subsets $K_j \subset \Omega_j$ such that supp $\psi \subseteq K_1 \times K_2$. Now choose test functions $\chi_j \in \mathcal{D}(\Omega_j)$ which are equal to 1 on K_j, $j = 1, 2$. It follows that $(\chi_1 \otimes \chi_2) \cdot \psi = \psi$ and

$$\phi_{k,N} = (\chi_1 \otimes \chi_2) \cdot \psi_{k,N} \in \mathcal{D}(\Omega_1) \otimes \mathcal{D}(\Omega_2) \quad \forall k, N \in \mathbb{N},$$

since the $\psi_{k,N}$ are polynomials.

For any compact set $K \subset \mathbb{R}^n$ there is a positive real number R such that $|z - \xi| \leq R$ for all $z \in K$ and all $\xi \in K_1 \times K_2$. From the estimates of the second step, for all $\alpha \in \mathbb{N}^n$, we know that $D^\alpha \psi_{k,N}(z)$ converges uniformly in $z \in K$ to $D^\alpha \psi_k(z)$. Using Leibniz' rule we deduce

$$\lim_{N \to \infty} \phi_{k,N} = (\chi_1 \otimes \chi_2) \cdot \psi_k \equiv \phi_k \quad \text{in } \mathcal{D}(\Omega_1 \times \Omega_2).$$

Again using Leibniz' rule we deduce from the estimates of the first step that

$$\lim_{k \to \infty} \phi_k = (\chi_1 \otimes \chi_2) \cdot \psi = \psi \quad \text{in } \mathcal{D}(\Omega_1 \times \Omega_2)$$

and thus we conclude. □

On the algebraic tensor product $E \otimes F$ of two Hausdorff locally convex topological vector space E and F over the same field, several interesting locally convex topologies can be defined. We discuss here briefly the *projective tensor product topology* which plays an important role in the definition and study of tensor products for distributions. Let $\mathcal{P}$ (respectively $\mathcal{Q}$) be the filtering system of seminorms defining the topology of the space E (respectively of F). Recall that the general element χ in $E \otimes F$ is of the form

$$\chi = \sum_{i=1}^m e_i \otimes f_i \quad \text{with } e_i \in E \text{ and } f_i \in F, \ i = 1, \ldots, m \text{ any } m \in \mathbb{N}. \quad (6.4)$$

Note that this representation of the element χ in terms of factors $e_i \in E$ and $f_i \in F$ is not unique. In the following definition of a semi-norm on $E \otimes F$, this is taken into account by taking the infimum over all such representations of χ. Now given two semi-norms $p \in \mathcal{P}$ and $q \in \mathcal{Q}$, the projective tensor product $p \otimes_\pi q$ of p and q is defined by

$$p \otimes_\pi q(\chi) = \inf \left\{ \sum_{i=1}^m p(e_i) q_i(f_i) : \chi = \sum_{i=1}^m e_i \otimes f_i \right\}. \quad (6.5)$$

In the Exercises we show that this formula defines indeed a semi-norm on the tensor product $E \otimes F$. It follows immediately that

$$p \otimes_\pi q(e \otimes f) = p(e) q(f) \quad \forall e \in E, \ \forall f \in F. \quad (6.6)$$

From the definition it is evident that $p \otimes_\pi q \leq p' \otimes q$ and $p \otimes_\pi q \leq p \otimes q'$ whenever $p, p' \in \mathcal{P}$ satisfy $p \leq p'$ and $q, q' \in \mathcal{Q}$ satisfy $q \leq q'$. Therefore the system

$$\mathcal{P} \otimes_\pi \mathcal{Q} = \{p \otimes_\pi q : p \in \mathcal{P}, q \in \mathcal{Q}\} \quad (6.7)$$

of semi-norms on $E \otimes F$ is filtering and thus defines a locally convex topology on $E \otimes F$, called the *projective tensor product topology*. The vector space $E \otimes F$ equipped with this topology is denoted by

$$E \otimes_\pi F$$

and is called the *projective tensor product of the spaces E and F*.

This definition applies in particular to the test function spaces $E = \mathcal{D}(\Omega_1)$, $\Omega_1 \subseteq \mathbb{R}^{n_1}$, and $F = \mathcal{D}(\Omega_2)$, $\Omega_2 \subseteq \mathbb{R}^{n_2}$. Thus we arrive at the projective tensor product $\mathcal{D}(\Omega_1) \otimes_\pi \mathcal{D}(\Omega_2)$ of these test function spaces. The following theorem identifies the completion of this space which plays an important role in the definition of tensor products for distributions. The general construction of the completion is given in the Appendix A.

Theorem 6.1.2 *Assume $\Omega_j \subseteq \mathbb{R}^{n_j}$ are nonempty open sets. The completion of the projective tensor product $\mathcal{D}(\Omega_1) \otimes_\pi \mathcal{D}(\Omega_2)$ of the test function spaces over Ω_j is equal to the test function space $\mathcal{D}(\Omega_1 \times \Omega_2)$ over the product $\Omega_1 \times \Omega_2$ of the sets Ω_j:*

$$\mathcal{D}(\Omega_1) \tilde{\otimes}_\pi \mathcal{D}(\Omega_2) = \mathcal{D}(\Omega_1 \times \Omega_2). \tag{6.8}$$

6.2 Tensor product for distributions

Knowing the tensor product of two functions $f, g \in L^1_{loc}(\Omega_i)$, we are going to define the tensor product for distributions in such a way that it is compatible with the embedding of functions into the space of distributions and the tensor product for functions. Traditionally the same symbol $\otimes$ is used to denote the tensor product for distributions and for functions. Thus our compatibility condition means $I_f \otimes I_g = I_{f \otimes g}$ for all $f, g \in L^1_{loc}(\Omega)$. Since we know how to evaluate $I_{f \otimes g}$ we get

$$\begin{aligned}\langle I_f \otimes I_g, \phi \otimes \psi \rangle &= \langle I_{f \otimes g}, \phi \otimes \psi \rangle \\ &= \int_{\Omega_1 \times \Omega_2} (f \otimes g)(x, y)(\phi \otimes \psi)(x, y) dx dy \\ &= \int_{\Omega_1 \times \Omega_2} f(x) g(y) \phi(x) \psi(y) dx dy = \langle I_f, \phi \rangle \langle I_g, \psi \rangle\end{aligned}$$

for all $\phi \in \mathcal{D}(\Omega_1)$ and all $\psi \in \mathcal{D}(\Omega_2)$. Hence the compatibility with the embedding is assured as soon as the tensor product for distributions is required to satisfy the following identity, for all $T \in \mathcal{D}'(\Omega_1)$, all $S \in \mathcal{D}'(\Omega_2)$, all $\phi \in \mathcal{D}(\Omega_1)$, and all $\psi \in \mathcal{D}(\Omega_2)$,

$$\langle T \otimes S, \phi \otimes \psi \rangle = \langle T, \phi \rangle \langle S, \psi \rangle. \tag{6.9}$$

Since the tensor product is to be defined in such a way that it is a continuous linear functional on the test function space over the product set, this identity determines the tensor product of two distributions immediately on the tensor product $\mathcal{D}(\Omega_1) \otimes \mathcal{D}(\Omega_2)$ of the test function spaces by linearity:

$$\langle T \otimes S, \chi \rangle = \sum_{i=1}^N \langle T, \phi_i \rangle \langle S, \psi_i \rangle \quad \forall \chi = \sum_{i=1}^N \phi_i \otimes \psi_i \in \mathcal{D}(\Omega_1) \otimes \mathcal{D}(\Omega_2). \tag{6.10}$$

Thus we know the tensor product on the dense subspace $\mathcal{D}(\Omega_1) \otimes \mathcal{D}(\Omega_2)$ of $\mathcal{D}(\Omega_1 \times \Omega_2)$ and this identity allows us to read off the natural continuity requirement for $T \otimes S$. Suppose $K_i \subset \Omega_i$ are compact subsets. Then there are constants $C_i \in \mathbb{R}_+$

and integers m_i such that $|\langle T, \phi \rangle| \leq C_1 p_{K_1, m_1}(\phi)$ for all $\phi \in \mathcal{D}_{K_1}(\Omega_1)$ and $|\langle S, \psi \rangle| \leq C_2 p_{K_2, m_2}(\psi)$ for all $\psi \in \mathcal{D}_{K_2}(\Omega_2)$ and thus, using the abbreviations $p_i = p_{K_i, m_i}$,

$$|\langle T \otimes S, \chi \rangle| \leq \sum_{i=1}^{N} |\langle T, \phi_i \rangle| |\langle S, \psi_i \rangle| \leq \sum_{i=1}^{N} C_1 C_2 p_1(\phi_i) p_2(\psi_i)$$

for all representations of $\chi = \sum_{i=1}^{N} \phi_i \otimes \psi_i$, and it follows that

$$|\langle T \otimes S, \chi \rangle| \leq C_1 C_2 \inf \left\{ \sum_{i=1}^{N} p_1(\phi_i) p_2(\psi_i) : \chi = \sum_{i=1}^{N} \phi_i \otimes \psi_i \right\},$$

i.e.,

$$|\langle T \otimes S, \chi \rangle| \leq C_1 C_2 (p_1 \otimes_\pi p_2)(\chi). \tag{6.11}$$

Hence the tensor product $T \otimes S$ of the distributions T on Ω_1 and S on Ω_2 is a continuous linear function $\mathcal{D}(\Omega_1) \otimes_\pi \mathcal{D}(\Omega_2) \to \mathbb{K}$ which can be extended by continuity to the completion of this space. In Theorem 6.1.2 this completion has been identified as $\mathcal{D}(\Omega_1 \times \Omega_2)$.

We prepare our further study of the tensor product for distributions by some technical results. These results are also used for the study of the convolution for distributions in the next chapter.

Lemma 6.2.1 *Suppose $\Omega_i \subseteq \mathbb{R}^{n_i}$ are nonempty open sets and $\phi : \Omega_1 \times \Omega_2 \to \mathbb{K}$ is a function with the following properties:*

a) *For every $y \in \Omega_2$ define $\phi_y(x) = \phi(x, y)$ for all $x \in \Omega_1$. Then $\phi_y \in \mathcal{D}(\Omega_1)$ for all $y \in \Omega_2$.*

b) *For all $\alpha \in \mathbb{N}^{n_1}$ the function $D_x^\alpha \phi(x, y)$ is continuous on $\Omega_1 \times \Omega_2$.*

c) *For every $y_0 \in \Omega_2$ there is a neighborhood V of y_0 in Ω_2 and a compact set $K \subset \Omega_1$ such that for all $y \in V$ the functions ϕ_y have their support in K.*

Then, for every distribution $T \in \mathcal{D}'(\Omega_1)$ on Ω_1, the function $y \mapsto f(y) = \langle T, \phi_y \rangle$ is continuous on Ω_2.

Proof. Suppose $y_0 \in \Omega_2$ and $r > 0$ are given. Choose a neighborhood V of y_0 and a compact set $K \subset \Omega_1$ according to hypothesis c). Since T is a distribution on Ω_1 there are a constant C and an integer m such that $|\langle T, \phi \rangle| \leq C p_{K,m}(\phi)$ for all $\phi \in \mathcal{D}_K(\Omega_1)$. By hypothesis b) the derivatives $D_x^\alpha \phi(x, y)$ are continuous on $K \times \Omega_2$. It follows (see the Exercises) that there is a neighborhood W of y_0 in Ω_2 such that for all $y \in W$,

$$p_{K,m}(\phi_y - \phi_{y_0}) \leq \frac{r}{C}.$$

Since for all $y \in V \cap W$ the functions ϕ_y belong to $\mathcal{D}_K(\Omega_1)$ we get the estimate

$$|f(y) - f(y_0)| = |\langle T, \phi_y \rangle - \langle T, \phi_{y_0} \rangle| = |\langle T, \phi_y - \phi_{y_0} \rangle| \leq C p_{K,m}(\phi_y - \phi_{y_0}) \leq r.$$

Therefore f is continuous at y_0 and since y_0 was arbitrary in Ω_2, continuity of f on Ω_2 follows. $\square$

6.2 Tensor product for distributions

Corollary 6.2.1 *Under the hypotheses of Lemma 6.2.1 with hypothesis b) replaced by the assumption $\phi \in C^\infty(\Omega_1 \times \Omega_2)$, the function $y \mapsto f(y) = \langle T, \phi_y \rangle$ is of class C^∞ on Ω_2 for every distribution $T \in \mathcal{D}'(\Omega_1)$, and one has*

$$D_y^\beta \langle T, \phi_y \rangle = \langle T, D_y^\beta \phi_y \rangle.$$

Proof. Differentiation is known to be a local operation in the sense that it preserves support properties. Thus we have

1. $D_y^\beta \phi_y \in \mathcal{D}(\Omega_1)$ for all $y \in \Omega_2$;
2. $D_x^\alpha D_y^\beta \phi(x, y)$ is continuous on $\Omega_1 \times \Omega_2$ for all $\alpha \in \mathbb{N}^{n_1}$ and all $\beta \in \mathbb{N}^{n_2}$;
3. For every $\beta \in \mathbb{N}^{n_2}$ and every $y_0 \in \Omega_2$ there are a neighborhood V of y_0 in Ω_2 and a compact set $K \subset \Omega_1$ such that supp $D_y^\beta \phi_y \subseteq K$ for all $y \in V$.

By Lemma 6.2.1 it follows that, for each $T \in \mathcal{D}'(\Omega_1)$ and each $\beta \in \mathbb{N}^{n_2}$, the functions $y \mapsto \langle T, D_y^\beta \phi_y \rangle$ are continuous on Ω_2. In order to conclude we have to show that the functions $\langle T, D_y^\beta \phi_y \rangle$ are just the derivatives of order β of the function $\langle T, \phi_y \rangle$. This is quite a tedious step. We present this step explicitly for $|\beta| = 1$.

Take any $y_0 \in \Omega_2$ and choose a neighborhood V of y_0 and the compact set $K \subset \Omega_1$ according to the third property above. Take any $T \in \mathcal{D}'(\Omega_1)$. For this compact set K and this distribution there are a constant C and an integer m such that $|\langle T, \psi \rangle| \leq C p_{K,m}(\psi)$ for all $\psi \in \mathcal{D}_K(\Omega_1)$. The neighborhood V contains an open ball $y_0 + B_r(0)$ around y_0, for some $r > 0$. Abbreviate $\partial_i = \frac{\partial}{\partial y_i}$ and calculate for $h \in B_r(0)$, as an identity for C^∞-functions of compact support in $K \subset \Omega_1$,

$$\phi_{y_0+h} - \phi_{y_0} = \int_0^1 \frac{d}{dt} \phi_{y_0+th} dt =$$
$$= \sum_{i=1}^{n_2} (\partial_i \phi)_{y_0} h_i + \sum_{i=1}^{n_2} \int_0^1 [(\partial_i \phi)_{y_0+th} - (\partial_i \phi)_{y_0}] h_i dt.$$

Applying the distribution T to this identity gives

$$\langle T, \phi_{y_0+h} - \phi_{y_0} \rangle = \sum_{i=1}^{n_2} \langle T, (\partial_i \phi)_{y_0} \rangle h_i$$
$$+ \sum_{i=1}^{n_2} \langle T, \int_0^1 [(\partial_i \phi)_{y_0+th} - (\partial_i \phi)_{y_0}] \rangle h_i.$$

For all $|\alpha| \leq m$ and $i = 1, 2, \ldots, n_2$ the functions $D_x^\alpha (\partial_i \phi)(x, y)$ are continuous on $\Omega_1 \times \Omega_2$ and have a compact support in the compact set K for all $y \in V$. Thus, as in the proof of Lemma 6.2.1, given $\epsilon > 0$ there is $\delta > 0$ such that for all $i = 1, \ldots, n_2$ and all $|y - y_0| < \delta$ one has $p_{K,m}((\partial_i \phi)_y - (\partial_i \phi)_{y_0}) \leq \frac{\epsilon}{C}$; and we can assume $\delta \leq r$. It follows that

$$p_{K,m}\left(\int_0^1 [(\partial_i \phi)_{y_0+th} - (\partial_i \phi)_{y_0}] dt\right) \leq \int_0^1 p_{K,m}([(\partial_i \phi)_{y_0+th} - (\partial_i \phi)_{y_0}]) dt \leq \frac{\epsilon}{C}$$

and thus

$$|\langle T, \int_0^1 [(\partial_i \phi)_{y_0+th} - (\partial_i \phi)_{y_0}] dt \rangle| \leq C p_{K,m}\left(\int_0^1 [(\partial_i \phi)_{y_0+th} - (\partial_i \phi)_{y_0}] dt\right) \leq \epsilon.$$

We deduce that

$$\langle T, \phi_{y_0+h} - \phi_{y_0} \rangle = \sum_{i=1}^{n_2} \langle T, (\partial_i \phi)_{y_0} \rangle h_i + o(h).$$

Therefore the function $f(y) = \langle T, \phi_y \rangle$ is differentiable at the point y_0 and the derivative is given by

$$\partial_i \langle T, \phi_{y_0} \rangle = \langle T, (\partial_i \phi)_{y_0} \rangle, \quad i = 1, \ldots, n_2.$$

The functions $\partial_i \phi$ satisfy the hypotheses of Lemma 6.2.1, hence the functions $y \mapsto \langle T, (\partial_i \phi)_y \rangle$ are continuous and thus the function $f(y) = \langle T, \phi_y \rangle$ has continuous first order derivatives.

Since with a function ϕ all the functions $(x, y) \mapsto D_y^\beta \phi(x, y)$, $\beta \in \mathbb{N}^{n_2}$, satisfy the hypothesis of the corollary, the above arguments can be iterated and thus we conclude. $\square$

6. Tensor Products

The hypotheses of the above corollary are satisfied in particular for test functions on $\Omega_1 \times \Omega_2$. This case will be used for establishing an important property of the tensor product for distributions.

Theorem 6.2.1 *Suppose that $\Omega_i \subset \mathbb{R}^{n_i}$, $i = 1, 2$, are nonempty open sets.*

a) *For $\phi \in \mathcal{D}(\Omega_1 \times \Omega_2)$ and $T \in \mathcal{D}'(\Omega_1)$ define a function ψ on Ω_2 by*

$$\psi(y) = \langle T, \phi_y \rangle.$$

Then ψ is a test function on Ω_2: $\psi \in \mathcal{D}(\Omega_2)$.

b) *Given compact subsets $K_i \subset \Omega_i$ and an integer m_2, there is an integer m_1 depending on K_1 and the distribution T such that*

$$p_{K_2,m_2}(\psi) \leq p'_{K_1,m_1}(T) p_{K_1 \times K_2, m_1 + m_2}(\phi) \tag{6.12}$$

for all $\phi \in \mathcal{D}_{K_1 \times K_2}(\Omega_1 \times \Omega_2)$.

c) *The assignment $(T, \phi) \mapsto \psi$ defined in part a) defines a bi-linear map $F : \mathcal{D}'(\Omega_1) \times \mathcal{D}(\Omega_1 \times \Omega_2) \to \mathcal{D}(\Omega_2)$ by $F(T, \phi) \equiv \psi$.*

d) *The map $F : \mathcal{D}'(\Omega_1) \times \mathcal{D}(\Omega_1 \times \Omega_2) \to \mathcal{D}(\Omega_2)$ has the following continuity property: F is continuous in $\phi \in \mathcal{D}(\Omega_1 \times \Omega_2)$, uniformly in $T \in B$, B a weakly bounded subset of $\mathcal{D}'(\Omega_1)$.*

Proof. It is straightforward to check that a test function $\phi \in \mathcal{D}(\Omega_1, \Omega_2)$ satisfies the hypotheses of Corollary 6.2.1. Hence this corollary implies that $\psi \in C^\infty(\Omega_2)$. There are compact subsets $K_i \subset \Omega_i$ such that $\text{supp } \phi \subseteq K_1 \times K_2$. Thus the functions ϕ_y are the zero function on Ω_1 for all $y \in \Omega_2 \setminus K_2$ and therefore $\text{supp } \psi \subseteq K_2$. This proves the first part.

For $\phi \in \mathcal{D}_{K_1 \times K_2}(\Omega_1 \times \Omega_2)$ one knows that all the functions $(D_y^\beta \phi)_y$, $y \in K_2$, $\beta \in \mathbb{N}^{n_2}$ belong to $\mathcal{D}_{K_1}(\Omega_1)$. Since $T \in \mathcal{D}'(\Omega_1)$, there is an $m_1 \in \mathbb{N}$ such that $p'_{K_1,m_1}(T)$ is finite and

$$|\langle T, (D_y^\beta \phi)_y \rangle| \leq p'_{K_1,m_1}(T) p_{K_1,m_1}((D_y^\beta \phi)_y)$$

for all $y \in K_2$ and all β. By Corollary 6.2.1 we know that

$$D^\beta \psi(y) = \langle T, (D_y^\beta \phi)_y \rangle,$$

therefore

$$|D^\beta \psi(y)| \leq p'_{K_1,m_1}(T) p_{K_1,m_1}((D_y^\beta \phi)_y) = p'_{K_1,m_1}(T) \sup_{x \in K_1, |\alpha| \leq m_1} |D_x^\alpha D_y^\beta \phi(x, y)|$$

and we conclude that

$$p_{K_2,m_2}(\psi) \leq p'_{K_1,m_1}(T) p_{K_1 \times K_2, m_1 + m_2}(\phi).$$

Thus the second part follows.

Since $F(T, \phi) = \langle T, \phi. \rangle$, F is certainly linear in $T \in \mathcal{D}'(\Omega_1)$. It is easy to see that for every fixed $y \in \Omega_2$ the map $\phi \mapsto \phi_y$ is a linear map $\mathcal{D}(\Omega_1 \times \Omega_2) \to \mathcal{D}(\Omega_1)$. Hence F is linear in ϕ too and Part c) is proven.

For Part d) observe that by the uniform boundedness principle a (weakly) bounded set $B \subset \mathcal{D}'(\Omega_1)$ is equi-continuous on $\mathcal{D}_{K_1}(\Omega_1)$ for every compact subset $K_1 \subset \Omega_1$. This means that we can find some $m_1 \in \mathbb{N}$ such that

$$\sup_{T \in B} p'_{K_1,m_1}(T) < \infty$$

and thus by estimate (6.12) we conclude. □

Theorem 6.2.2 (Tensor product for distributions) *Suppose that $\Omega_i \subseteq \mathbb{R}^{n_i}$, $i = 1, 2$, are nonempty open sets.*

a) *Given $T_i \in \mathcal{D}'(\Omega_i)$ there is exactly one distribution $T \in \mathcal{D}'(\Omega_1 \times \Omega_2)$ on $\Omega_1 \times \Omega_2$ such that*

$$\langle T, \phi_1 \otimes \phi_2 \rangle = \langle T_1, \phi_1 \rangle \langle T_2, \phi_2 \rangle \qquad \forall \phi_i \in \mathcal{D}(\Omega_i), \ i = 1, 2.$$

*T is called the **tensor product** of T_1 and T_2, denoted by $T_1 \otimes T_2$.*

b) *The tensor product satisfies **Fubini's Theorem** (for distributions), i.e., for every $T_i \in \mathcal{D}'(\Omega_i)$, $i = 1, 2$, and for every $\chi \in \mathcal{D}(\Omega_1 \times \Omega_2)$ one has*

$$\langle T_1 \otimes T_2, \chi \rangle = \langle (T_1 \otimes T_2)(x, y), \chi(x, y) \rangle$$
$$= \langle T_1(x), \langle T_2(y), \chi(x, y) \rangle \rangle = \langle T_2(y), \langle T_1(x), \chi(x, y) \rangle \rangle.$$

c) *Given compact subsets $K_i \subset \Omega_i$ there are integers $m_i \in \mathbb{N}$ such that $p'_{K_i, m_i}(T_i)$ are finite for $i = 1, 2$ and for all $\chi \in \mathcal{D}_{K_1 \times K_2}(\Omega_1 \times \Omega_2)$,*

$$|\langle T_1 \otimes T_2, \chi \rangle| \leq p'_{K_1, m_1}(T_1) p'_{K_2, m_2}(T_2) p_{K_1 \times K_2, m_1 + m_2}(\chi). \tag{6.13}$$

Proof. Given $T_i \in \mathcal{D}'(\Omega_i)$ and $\chi \in \mathcal{D}(\Omega_1 \times \Omega_2)$ we know by Theorem 6.2.1 that $F(T_1, \chi) \in \mathcal{D}(\Omega_2)$. Thus

$$\langle T, \chi \rangle = \langle T_2, F(T_1, \chi) \rangle \tag{6.14}$$

is well defined for all $\chi \in \mathcal{D}(\Omega_1 \times \Omega_2)$. Since F is linear in χ, linearity of T_2 implies linearity of T. In order to show that T is a distribution on $\Omega_1 \times \Omega_2$, it suffices to show that T is continuous on $\mathcal{D}_{K_1 \times K_2}(\Omega_1 \times \Omega_2)$ for arbitrary compact sets $K_i \subset \Omega_i$. For any $\chi \in \mathcal{D}_{K_1 \times K_2}(\Omega_1 \times \Omega_2)$ we know by Theorem 6.2.1 that $F(T_1, \chi) \in \mathcal{D}_{K_2}(\Omega_2)$. Since $T_2 \in \mathcal{D}'(\Omega_2)$ there is $m_2 \in \mathbb{N}$ such that $p'_{K_2, m_2}(T_2)$ is finite, and we have the estimate

$$|\langle T, \chi \rangle| = |\langle T_2, F(T_1, \chi) \rangle| \leq p'_{K_2, m_2}(T_2) p_{K_2, m_2}(F(T_1, \chi)).$$

Similarly, since $T_1 \in \mathcal{D}'(\Omega_1)$, there is an $m_1 \in \mathbb{N}$ such that $p'_{K_1, m_1}(T_1)$ is finite so that the estimate (6.12) applies. Combining these two estimates yields

$$|\langle T, \chi \rangle| \leq p'_{K_1, m_1}(T_1) p'_{K_2, m_2}(T_2) p_{K_1 \times K_2, m_1 + m_2}(\chi)$$

for all $\chi \in \mathcal{D}_{K_1 \times K_2}(\Omega_1 \times \Omega_2)$ with integers m_i depending on T_i and K_i. Thus continuity of T follows.

For $\chi = \phi_1 \otimes \phi_2$, $\phi_i \in \mathcal{D}(\Omega_i)$, we have $F(T_1, \chi) = \langle T_1, \phi_1 \rangle \phi_2$ and therefore the distribution T factorizes as claimed:

$$\langle T, \phi_1 \otimes \phi_2 \rangle = \langle T_1, \phi_1 \rangle \langle T_2, \phi_2 \rangle.$$

By linearity this property determines T uniquely on the tensor product space $\mathcal{D}(\Omega_1) \otimes \mathcal{D}(\Omega_2)$ which is known to be dense in $\mathcal{D}(\Omega_1 \times \Omega_2)$ by Proposition 6.1.1. Now continuity of T on $\mathcal{D}(\Omega_1 \times \Omega_2)$ implies that T is uniquely determined by T_1 and T_2. This proves part a).

Above we defined $T = T_1 \otimes T_2$ by the formula $\langle T, \chi \rangle = \langle T_2(y), \langle T_1(x), \chi(x, y) \rangle \rangle$ for all $\chi \in \mathcal{D}(\Omega_1 \times \Omega_2)$. With minor changes in the argument one can show that there is a distribution S on $\Omega_1 \times \Omega_2$, well defined by the formula

$$\langle S, \chi \rangle = \langle T_1(x), \langle T_2(y), \chi(x, y) \rangle \rangle$$

for all $\chi \in \mathcal{D}(\Omega_1 \times \Omega_2)$. Clearly, on the dense subspace $\mathcal{D}(\Omega_1) \otimes \mathcal{D}(\Omega_2)$ the continuous functionals S and T agree. Hence they agree on $\mathcal{D}(\Omega_1 \times \Omega_2)$ and this proves Fubini's theorem for distributions.

The estimate given in Part c) has been shown in the proof of continuity of $T = T_1 \otimes T_2$. □

6. Tensor Products

The following corollary collects some basic properties of the tensor product for distributions.

Corollary 6.2.2 *Suppose that T_i are distributions on nonempty open sets $\Omega_i \subset \mathbb{R}^{n_i}$. Then the following holds:*

a) $\operatorname{supp}(T_1 \otimes T_2) = \operatorname{supp} T_1 \otimes \operatorname{supp} T_2$.

b) $D_x^\alpha (T_1 \otimes T_2) = (D_x^\alpha T_1) \otimes T_2$. *Here x refers to the variable of T_1.*

Proof. The straightforward proof is done as an exercise. □

Proposition 6.2.3 *The tensor product for distributions is jointly continuous in both factors, i.e., if $T = \lim_{j\to\infty} T_j$ in $\mathcal{D}'(\Omega_1)$ and $S = \lim_{j\to\infty} S_j$ in $\mathcal{D}'(\Omega_2)$, then*

$$T \otimes S = \lim_{j\to\infty} T_j \otimes S_j \quad \text{in } \mathcal{D}'(\Omega_1 \times \Omega_2).$$

Proof. Recall that we consider spaces of distributions equipped with the weak topology σ (compare Theorem 3.3.1). Thus, for every $\chi \in \mathcal{D}(\Omega_1 \times \Omega_2)$, we have to show that

$$\langle T \otimes S, \chi \rangle = \lim_{j\to\infty} \langle T_j \otimes S_j, \chi \rangle.$$

By Proposition 6.1.1 and its proof we know: Given $\chi \in \mathcal{D}_K(\Omega_1 \times \Omega_2)$ there are compact sets $K_i \subset \Omega_i$, $K \subset K_1 \times K_2$, such that χ is the limit in $\mathcal{D}_{K_1 \times K_2}(\Omega_1 \times \Omega_2)$ of a sequence in $\mathcal{D}_{K_1}(\Omega_1) \otimes \mathcal{D}_{K_2}(\Omega_2)$. Since $T = \lim_{j\to\infty} T_j$, equation (3.12) of Theorem 3.3.1 implies that there is an $m_1 \in \mathbb{N}$ such that

$$p'_{K_1,m_1}(T_j) \leq M_1 \quad \forall j \in \mathbb{N}$$

and similarly there is an $m_2 \in \mathbb{N}$ such that

$$p'_{K_2,m_2}(S_j) \leq M_2 \quad \forall j \in \mathbb{N}.$$

These bounds also apply to the limits T, respectively S.

Now, given $\epsilon > 0$, there is a $\chi_\epsilon \in \mathcal{D}_{K_1}(\Omega_1) \otimes \mathcal{D}_{K_2}(\Omega_2)$ such that

$$p_{K_1 \times K_2, m_1 + m_2}(\chi - \chi_\epsilon) < \frac{\epsilon}{4 M_1 M_2}.$$

By Part c) of Theorem 6.2.2 this implies the following estimate:

$$|(T_j \otimes S_j)(\chi - \chi_\epsilon)| \leq M_1 M_2 p_{K_1 \times K_2, m_1 + m_2}(\chi - \chi_\epsilon) \leq \epsilon/4 \quad \forall j \in \mathbb{N}.$$

And the same bound results for $T \otimes S$.

Finally we put all information together and get, for all $j \in \mathbb{N}$,

$$|(T \otimes S - T_j \times S_j)(\chi)| \leq |(T \otimes S - T_j \times S_j)(\chi - \chi_\epsilon)| + |(T \otimes S - T_j \times S_j)(\chi_\epsilon)|$$
$$\leq 2 M_1 M_2 p_{K_1 \times K_2, m_1 + m_2}(\chi - \chi_\epsilon) + |(T \otimes S - T_j \times S_j)(\chi_\epsilon)|$$
$$\leq \epsilon/2 + |(T \otimes S - T_j \times S_j)(\chi_\epsilon)|.$$

On $\mathcal{D}(\Omega_1) \otimes \mathcal{D}(\Omega_2)$ the sequence $(T_j \otimes S_j)_{j \in \mathbb{N}}$ certainly converges to $T \otimes S$ (see Exercises). Hence there is $j_0 \in \mathbb{N}$ such that $|(T \otimes S - T_j \times S_j)(\chi_\epsilon)| < \epsilon/2$ for all $j \geq j_0$. It follows that

$$|(T \otimes S - T_j \times S_j)(\chi)| < \epsilon \quad \forall j \geq j_0.$$

This concludes the proof. □

6.3 Exercises

1. Prove: Formula 6.5 for the projective tensor product of two semi-norms p, q on E respectively on F defines indeed a semi-norm on the tensor product $E \otimes F$.

2. Prove Theorem 6.1.2!

   Hints: Consult the book [Trè67].

3. Complete the proof of Lemma 6.2.1.

4. Prove the following: Assume that a sequence $(\phi_j)_{j \in \mathbb{N}}$ converges in $\mathcal{D}(\Omega)$ to $\phi \in \mathcal{D}(\Omega)$ and the sequence of distributions $(T_j)_{j \in \mathbb{N}} \subset \mathcal{D}'(\Omega)$ converges weakly to $T \in \mathcal{D}'(\Omega)$. Then the sequence of numbers $(T_j(\phi_j))_{j \in \mathbb{N}}$ converges to the number $T(\phi)$, i.e.,

$$\lim_{j \to \infty} T_j(\phi_j) = T(\phi).$$

   Hints: In the Appendix C.1 it is shown that a weakly bounded set in $\mathcal{D}'(\Omega)$ is equi-continuous.

5. Prove Corollary 6.2.2.

6. Assume $T = \lim_{j \to \infty} T_j$ in $\mathcal{D}'(\Omega_1)$ and $S = \lim_{j \to \infty} S_j$ in $\mathcal{D}'(\Omega_2)$. Prove: For every $\chi \in \mathcal{D}(\Omega_1) \otimes \mathcal{D}(\Omega_2)$ one has

$$\lim_{j \to \infty} T_j \otimes S_j(\chi) = T \otimes S(\chi).$$

7
Convolution Products

Our goal is to introduce and to study the convolution product for distributions. In order to explain the difficulties which will arise there we discuss first the convolution product for functions. Also for functions the convolution product is only defined under certain restrictions. Thus we start with the class $C_0(\mathbb{R}^n)$ of continuous functions on $\mathbb{R}^n$ which have a compact support.

7.1 Convolution of functions

Suppose $u, v \in C_0(\mathbb{R}^n)$; then for each $x \in \mathbb{R}^n$ we know that $y \mapsto u(x-y)v(y)$ is a continuous function of compact support and therefore the integral of this function over $\mathbb{R}^n$ is well defined. This integral then defines the convolution product $u * v$ of u and v at the point x:

$$u * v(x) = \int_{\mathbb{R}^n} u(x-y)v(y)dy \qquad \forall x \in \mathbb{R}^n. \tag{7.1}$$

The following proposition presents elementary properties of the convolution product on $C_0(\mathbb{R}^n)$.

Proposition 7.1.1 *The convolution (i.e., the convolution product) is a well-defined map $C_0(\mathbb{R}^n) \times C_0(\mathbb{R}^n) \to C_0(\mathbb{R}^n)$. For $u, v \in C_0(\mathbb{R}^n)$ one has*

i) $u * v = v * u$,

ii) $\mathrm{supp}\,(u * v) \subseteq \mathrm{supp}\,u + \mathrm{supp}\,v$.

Proof. We saw above that $u * v$ is a well-defined function on $\mathbb{R}^n$. Note that $u(x-y)v(y) = 0$ whenever $y \notin \operatorname{supp} v$ or $x - y \notin \operatorname{supp} u$. It follows that the integral $\int_{\mathbb{R}^n} u(x-y)v(y)dy$ vanishes whenever $x \in \mathbb{R}^n$ cannot be represented as the sum of a point in $\operatorname{supp} u$ and a point in $\operatorname{supp} v$. This implies that $\operatorname{supp}(u * v) \subseteq \overline{\operatorname{supp} u + \operatorname{supp} v} = \operatorname{supp} u + \operatorname{supp} v$, since $\operatorname{supp} u$ and $\operatorname{supp} v$ are compact sets (see the Exercises). This proves part ii).

Since $(x, y) \mapsto u(x-y)v(y)$ is a uniformly continuous function on $\mathbb{R}^n \times \mathbb{R}^n$, the integration over a compact set gives a continuous function (see the Exercises). Thus $u * v$ is a continuous function of compact support.

The change of variables $y \mapsto x - z$ gives

$$\int_{\mathbb{R}^n} u(x-y)v(y)dy = \int_{\mathbb{R}^n} u(z)v(x-z)dz = (v * u)(x)$$

and proves part i). □

Corollary 7.1.1 *If $u \in C_0^m(\mathbb{R}^n)$ and $v \in C_0(\mathbb{R}^n)$, then $u * v \in C_0^m(\mathbb{R}^n)$ and $D^\alpha(u * v) = (D^\alpha u) * v$ for all $|\alpha| \leq m$; similarly, if $u \in C_0(\mathbb{R}^n)$ and $v \in C_0^m(\mathbb{R}^n)$ then $u * v \in C_0^m(\mathbb{R}^n)$ and $D^\alpha(u * v) = u * (D^\alpha v)$ for all $|\alpha| \leq m$.*

Proof. For $|\alpha| \leq m$ the function $(x, y) \mapsto (D^\alpha u)(x-y)v(y)$ is uniformly continuous on $\mathbb{R}^n \times \mathbb{R}^n$; integration with respect to y is over the compact set $\operatorname{supp} v$ and thus gives the continuous function $(D^\alpha u) * v$. Now the repeated application of the rules of differentiation of integrals with respect to parameters implies the first part. From the commutativity of the convolution product the second part is obvious. □

Naturally, the convolution of two functions u and v is defined whenever the integral in equation (7.1) exists. Obviously this is the case not only for continuous functions of compact support but for a much larger class. The following proposition looks at a number of cases for which the convolution product has convenient continuity properties in the factors and which are useful in practical problems.

Proposition 7.1.2 *Let $u, v : \mathbb{R}^n \to \mathbb{K}$ be two measurable functions. Denote $\|f\|_\infty = \operatorname{ess\,sup}_{x \in \mathbb{R}^n} |f(x)|$ and $\|f\|_1 = \int_{\mathbb{R}^n} |f(x)|dx$. Then the following holds.*

a) *If $u \in L^\infty(\mathbb{R}^n)$ and $v \in L^1(\mathbb{R}^n)$, then $u * v \in L^\infty(\mathbb{R}^n)$ and $\|u * v\|_\infty \leq \|u\|_\infty \|v\|_1$.*

b) *If $u \in L^1(\mathbb{R}^n)$ and $v \in L^\infty(\mathbb{R}^n)$, then $u * v \in L^\infty(\mathbb{R}^n)$ and $\|u * v\|_\infty \leq \|u\|_1 \|v\|_\infty$.*

c) *If $u, v \in L^1(\mathbb{R}^n)$, then $u * v \in L^1(\mathbb{R}^n)$ and $\|u * v\|_1 \leq \|u\|_1 \|v\|_1$.*

Proof. Consider the first case. One has $|\int_{\mathbb{R}^n} u(x-y)v(y)dy| \leq \int_{\mathbb{R}^n} |u(x-y)||v(y)|dy \leq \|u\|_\infty \|v\|_1$ and part a) follows. Similarly one proves b). For the third part we have to use Fubini's theorem:

$$\|u * v\|_1 = \int_{\mathbb{R}^n} |(u * v)(x)|dx \leq \int_{\mathbb{R}^n} (\int_{\mathbb{R}^n} |u(x-y)| |v(y)|dy)dx$$
$$\leq \int_{\mathbb{R}^n} (\int_{\mathbb{R}^n} |u(x-y)| |v(y)|dx)dy = \int_{\mathbb{R}^n} \int_{\mathbb{R}^n} |u(z)| |v(y)|dzdy = \|u\|_1 \|v\|_1.$$
□

Another important case where the convolution product of functions is well defined and has useful properties is the case of strongly decreasing functions. The following proposition collects the main results.

Proposition 7.1.3 1. *If $u, v \in \mathcal{S}(\mathbb{R}^n)$, then the convolution $u * v$ is a well-defined element in $\mathcal{S}(\mathbb{R}^n)$.*

2. *Equipped with the convolution $*$ as a product, the space $\mathcal{S}(\mathbb{R}^n)$ of strongly decreasing test functions is a commutative algebra.*

Proof. Recall the basic characterization of strongly decreasing test functions: $u \in \mathcal{C}^\infty(\mathbb{R}^n)$ belongs to $\mathcal{S}(\mathbb{R}^n)$ if, and only if, for all $m, l \in \mathbb{N}$ the norms $p_{m,l}(u)$ are finite. It follows that for every $\alpha \in \mathbb{N}^n$ and every $m \in \mathbb{N}$ one has
$$|D^\alpha u(x)| \leq \frac{p_{m,|\alpha|}(u)}{(1+x^2)^{\frac{m}{2}}} \qquad \forall x \in \mathbb{R}^n.$$
Thus, for $u, v \in \mathcal{S}(\mathbb{R}^n)$, for all $m, k \in \mathbb{N}$ and all $\alpha \in \mathbb{N}^n$, the estimate
$$\int_{\mathbb{R}^n} |(D^\alpha u)(x-y)v(y)|dy \leq \int_{\mathbb{R}^n} \frac{p_{m,|\alpha|}(u)}{(1+(x-y)^2)^{\frac{m}{2}}} \frac{p_{k,0}(v)}{(1+x^2)^{\frac{k}{2}}} dy$$
is available, uniformly in $x \in \mathbb{R}^n$. If we choose $k \geq n+1$, the integral on the righthand side is finite; therefore in this case the convolution $(D^\alpha u) * v$ exists. As earlier one shows $D^\alpha(u*v) = (D^\alpha u)*v$, and we deduce $u*v \in \mathcal{C}^\infty(\mathbb{R}^n)$.

In order to control the decay properties of the convolution $u*v$ observe that
$$\frac{1+x^2}{[1+(x-y)^2][1+y^2]} \leq 2 \qquad \forall x, y \in \mathbb{R}^n.$$
For $k = n+1+m$ we thus get
$$(1+x^2)^{\frac{m}{2}}|D^\alpha(u*v)(x)| \leq p_{m,|\alpha|}(u)p_{k,0}(v)2^{\frac{m}{2}} \int_{\mathbb{R}^n} \frac{dy}{(1+y^2)^{\frac{n+1}{2}}}.$$
The integral in this estimate has a finite value C. This holds for any $m \in \mathbb{N}$ and any $\alpha \in \mathbb{N}^n$. We conclude that $u*v \in \mathcal{S}(\mathbb{R}^n)$ and
$$p_{m,l}(u*v) \leq 2^{\frac{m}{2}} Cp_{m,l}(u)p_{n+1+m,0}(v). \tag{7.2}$$
This estimate also shows that the convolution is continuous on $\mathcal{S}(\mathbb{R}^n)$. As earlier the commutativity of the convolution is shown: $u*v = v*u$ for all $u, v \in \mathcal{S}(\mathbb{R}^n)$. This proves the second part and thus the proposition. □

The main application of the convolution is in the approximation of functions by smooth (i.e., $\mathcal{C}^\infty$-) functions and in the approximation of distributions by smooth functions. The basic technical preparation is provided by the following proposition.

Proposition 7.1.4 *Suppose that $(\phi_i)_{i \in \mathbb{N}}$ is a sequence of continuous function on $\mathbb{R}^n$ with support in the closed ball $\overline{B_R}(0) = \{x \in \mathbb{R}^n : \|x\| \leq R\}$. Assume furthermore*

i) $0 \leq \phi_i(x)$ for all $x \in \mathbb{R}^n$ and all $i \in \mathbb{N}$;

ii) $\int_{\mathbb{R}^n} \phi(x)dx = 1$ for all $i \in \mathbb{N}$;

iii) *For every $r > 0$ one has $\lim_{i \to \infty} \int_{B_r(0)^c} \phi_i(x)dx = 0$.*

For $u \in \mathcal{C}(\mathbb{R}^n)$ define an approximating sequence by the convolution
$$u_i = u * \phi_i \qquad i \in \mathbb{N}.$$
Then the following statements hold:

86   7. Convolution Products

a) *The sequence u_i converges to the given function u, uniformly on every compact subset $K \subset \mathbb{R}^n$;*

b) *For $u \in C^m(\mathbb{R}^n)$ and $|\alpha| \leq m$ the sequence of derivatives $D^\alpha u_i$ converges, uniformly on every compact set K, to the corresponding derivative $D^\alpha u$ of the given function u;*

c) *If in addition to the above assumptions the functions ϕ_i are of class C^∞, then the approximating functions $u_i = u * \phi_i$ are of class C^∞ and statements a) and b) hold.*

Proof. In order to prove part a) we have to show: Given a compact set $K \subset \mathbb{R}^n$ and $\epsilon > 0$ there is an $i_0 \in \mathbb{N}$ (depending on K and ϵ) such that for all $i \geq i_0$,
$$\|u - u_i\|_{K,\infty} \equiv \sup_{x \in K} |u(x) - u_i(x)| \leq \epsilon.$$

With K also the set $H = K + \overline{B_R}(0)$ is compact in $\mathbb{R}^n$. Therefore, as a continuous function on $\mathbb{R}^n$, u is bounded on H, by M let us say. Since continuous functions are uniformly continuous on compact sets, given $\epsilon > 0$ there is a $\delta > 0$ such that for all $x, x' \in H$ one has $|u(x) - u(x')| < \frac{\epsilon}{2}$ whenever $|x - x'| < \delta$. The normalization condition ii) for the functions ϕ_i allows us to write
$$u(x) - u_i(x) = \int_{\mathbb{R}^n} [u(x) - u(x-y)] \phi_i(y) dy$$
and thus, for all $x \in K$, we can estimate as follows:
$$\begin{aligned}|u(x) - u_i(x)| &\leq \int_{B_\delta(0)} |u(x) - u(x-y)| \phi_i(y) dy \\ &+ \int_{B_\delta(0)^c} |u(x) - u(x-y)| \phi_i(y) dy \\ &\leq \int_{B_\delta(0)} \tfrac{\epsilon}{2} \phi_i(y) dy + 2M \int_{B_\delta(0)^c} \phi_i(y) dy \\ &\leq \tfrac{\epsilon}{2} + 2M \int_{B_\delta(0)^c} \phi_i(y) dy.\end{aligned}$$

According to hypothesis iii) there is an $i_0 \in \mathbb{N}$ such that $\int_{B_\delta(0)^c} \phi_i(y) dy < \frac{\epsilon}{4M}$ for all $i \geq i_0$. Thus we can continue the above estimate by
$$|u(x) - u_i(x)| < \frac{\epsilon}{2} + 2M \frac{\epsilon}{4M} \qquad \forall x \in K, \forall i \geq i_0.$$

This implies statement a).

If $u \in C^m(\mathbb{R}^n)$ and $|\alpha| \leq m$, then $D^\alpha u \in C(\mathbb{R}^n)$ and by part a) we know $(D^\alpha u) * \phi_i \to D^\alpha u$, uniformly on compact sets. Corollary 7.1.1 implies that $D^\alpha u_i = D^\alpha(u * \phi_i) = (D^\alpha u) * \phi_i$. Hence part b) follows.

This corollary also implies that $u_i = u * \phi_i \in C^\infty(\mathbb{R}^n)$ whenever $\phi_i \in C^\infty(\mathbb{R}^n)$. Thus we can argue as in the previous two cases. □

Naturally the question arises how to get sequences of functions ϕ_i with the properties i)–iii) used in the above proposition. Recall the section on test function spaces. There, in the Exercises we defined a nonnegative function $\rho \in \mathcal{D}(\mathbb{R}^n)$ by equation (2.14). Denote $a = \int_{\mathbb{R}^n} \rho(x) dx$ and define
$$\phi_i(x) = \frac{i^n}{a} \rho(ix) \qquad \forall x \in \mathbb{R}^n \, \forall i \in \mathbb{N}. \tag{7.3}$$

Given $\epsilon > 0$, choose $i_0 > \frac{1}{\epsilon}$. Then for all $i \geq i_0$ one has
$$\int_{B_\epsilon(0)^c} \phi_i(x) dx = \frac{1}{a} \int_{|x| \geq \epsilon} i^n \rho(ix) dx = \frac{1}{a} \int_{|y| \geq i\epsilon} \rho(y) dy = 0,$$

since $\operatorname{supp} \rho \subseteq \{y \in \mathbb{R}^n : \|y\| \leq 1\}$. Now it is clear that this sequence satisfies the hypotheses of Proposition 7.1.4.

Corollary 7.1.2 *Suppose $\Omega \subseteq \mathbb{R}^n$ is a nonempty open set and $K \subset \Omega$ is compact. Given ϵ, $0 < \epsilon < \operatorname{dist}(\partial\Omega, K)$, denote $K_\epsilon = \{x \in \Omega : \operatorname{dist}(K, x) \leq \epsilon\}$. Then, for any continuous function f on Ω with support in K there is a sequence $(u_i)_{i \in \mathbb{N}}$ in $\mathcal{D}_{K_\epsilon}(\Omega)$ such that*

$$\lim_{i \to \infty} p_{K,0}(f - u_i) = 0.$$

If the function f is nonnegative, then also all the elements u_i of the approximating sequence can be chosen to be nonnegative.

Proof. See the Exercises. □

7.2 Regularization of distributions

This section explains how to approximate distributions by smooth functions. This approximation is understood in the sense of the weak topology on the space of distributions and is based on the convolution of distributions with test functions.

Given a test function $\phi \in \mathcal{D}(\mathbb{R}^n)$ and a point $x \in \mathbb{R}^n$, the function $y \mapsto \phi_x(y) = \phi(x - y)$ is again a test function and thus every distribution on $\mathbb{R}^n$ can be applied to it. Therefore one can define, for any $T \in \mathcal{D}'(\mathbb{R}^n)$,

$$(T * \phi)(x) = \langle T, \phi_x \rangle = \langle T(y), \phi(x - y) \rangle \qquad \forall x \in \mathbb{R}^n. \tag{7.4}$$

This function $T * \phi : \mathbb{R}^n \to \mathbb{K}$ is called the *regularization of the distribution T by the test function ϕ*, since we will learn soon that $T * \phi$ is actually a smooth function.

This definition of a convolution product between a distribution and a test function is compatible with the embedding of functions into the space of distributions. To see this take any $f \in L^1_{loc}(\mathbb{R}^n)$ and use the above definition to get

$$(I_f * \phi)(x) = \langle I_f, \phi_x \rangle = \int_{\mathbb{R}^n} f(y)\phi_x(y)dy = (f * \phi)(x) \qquad \forall x \in \mathbb{R}^n,$$

where naturally $f * \phi$ is the convolution product of functions as discussed earlier.

Basic properties of the regularization are collected in the following theorem.

Theorem 7.2.1 (Regularization) *For any $T \in \mathcal{D}'(\mathbb{R}^n)$ and any $\phi, \psi \in \mathcal{D}(\mathbb{R}^n)$ one has:*

a) $T * \phi \in \mathcal{C}^\infty(\mathbb{R}^n)$ *and, for all $\alpha \in \mathbb{N}^n$,*

$$D^\alpha(T * \phi) = T * D^\alpha \phi = D^\alpha T * \phi;$$

b) $\operatorname{supp}(T * \phi) \subseteq \operatorname{supp} T + \operatorname{supp} \phi;$

88   7. Convolution Products

c) $\langle T, \phi \rangle = (T * \check{\phi})(0)$ where $\check{\phi}(x) = \phi(-x)$ $\forall x \in \mathbb{R}^n$;

d) $(T * \phi) * \psi = T * (\phi * \psi)$.

Proof. For any test function $\phi \in \mathcal{D}(\mathbb{R}^n)$ we know that $\chi(x, y) \equiv \phi(x-y)$ belongs to $\mathcal{C}^\infty(\mathbb{R}^n \times \mathbb{R}^n)$. Given any $x_0 \in \mathbb{R}^n$ take a compact neighborhood V_{x_0} of x_0 in $\mathbb{R}^n$. Then $K = V_{x_0} - \operatorname{supp} \phi \subset \mathbb{R}^n$ is compact, and for all $x \in V_{x_0}$ we know that $\operatorname{supp} \chi_x = \{x\} - \operatorname{supp} \phi \subset K$, $\chi_x(y) = \chi(x, y)$. It follows that all hypotheses of Corollary 6.2.1 are satisfied and hence this corollary implies $T * \phi \in \mathcal{C}^\infty(\mathbb{R}^n)$ and $D^\alpha(T * \phi) = T * D^\alpha \phi$.

Now observe $D_y^\alpha \phi(x - y) = (-1)^{|\alpha|} D_x^\alpha \phi(x - y)$, hence, for all $x \in \mathbb{R}^n$,

$$\begin{aligned}(T * D^\alpha \phi)(x) &= \langle T(y), (D^\alpha \phi)(x - y) \rangle = \langle T(y), (-1)^{|\alpha|} D_y^\alpha \phi(x - y) \rangle \\ &= \langle D^\alpha T(y), \phi(x - y) \rangle = (D^\alpha T * \phi)(x).\end{aligned}$$

This proves part a).

In order that $(T * \phi)(x)$ does not vanish, the sets $\{x\} - \operatorname{supp} T$ and $\operatorname{supp} \phi$ must have a nonempty intersection, i.e., $x \in \operatorname{supp} T + \operatorname{supp} \phi$. Since $\operatorname{supp} \phi$ is compact and $\operatorname{supp} T$ is closed, the vector sum $\operatorname{supp} T + \operatorname{supp} \phi$ is closed. It follows that

$$\operatorname{supp}(T * \phi) = \overline{\{x \in \mathbb{R}^n : (T * \phi)(x) \neq 0\}} \subseteq \overline{\operatorname{supp} T + \operatorname{supp} \phi} = \operatorname{supp} T + \operatorname{supp} \phi,$$

and this proves part b).

The proof of part c) is a simple calculation.

$$(T * \check{\phi})(0) = \langle T(y), \check{\phi}(0 - y) \rangle = \langle T(y), \phi(y) \rangle = \langle T, \phi \rangle.$$

Proposition 7.1.1 and Corollary 7.1.1 together show that $\phi * \psi \in \mathcal{D}(\mathbb{R}^n)$ for all $\phi, \psi \in \mathcal{D}(\mathbb{R}^n)$. Hence, by part a) we know that $T * (\phi * \psi)$ is a well-defined $\mathcal{C}^\infty$-function on $\mathbb{R}^n$. For every $x \in \mathbb{R}^n$ it is given by

$$\langle T(y), \int_{\mathbb{R}^n} \phi(x - y - z) \psi(z) dz \rangle.$$

As we know that $T * \phi$ is a $\mathcal{C}^\infty$-function, the convolution product $(T * \phi) * \psi$ has the representation, for all $x \in \mathbb{R}^n$,

$$\int_{\mathbb{R}^n} (T * \phi)(x - z) \psi(z) dz = \int_{\mathbb{R}^n} \langle T(y), \phi(x - y - z) \rangle \psi(z) dz.$$

Hence the proof of part d) is completed by showing that the action of the distribution T with respect to the variable y and integration over $\mathbb{R}^n$ with respect to the variable z can be exchanged. This is done in the Exercises.   □

Note that in part b) the inclusion can be proper. A simple example is the constant distribution $T = I_1$ and test functions $\phi \in \mathcal{D}(\mathbb{R})$ with $\int_\mathbb{R} \phi(x) dx = 0$. Then we have $(T * \phi)(x) = \int_\mathbb{R} \phi(x) dx = 0$ for all $x \in \mathbb{R}$, thus $\operatorname{supp} T * \phi = \emptyset$ while $\operatorname{supp} I_1 = \mathbb{R}$.

As preparation for the main result of this section, namely the approximation of distributions by smooth functions, we introduce the concept of a *regularizing sequence*.

Definition 7.2.1 *A sequence of smooth functions ϕ_j on $\mathbb{R}^n$ is called a* **regularizing sequence** *if, and only if, it has the following properties.*

a) *There is a $\phi \in \mathcal{D}(\mathbb{R}^n)$, $\phi \neq 0$, such that $\phi_j(x) = j^n \phi(jx)$ for all $x \in \mathbb{R}^n$, $j = 1, 2, \ldots$;*

b) *$\phi_j \in \mathcal{D}(\mathbb{R}^n)$ for all $j \in \mathbb{N}$;*

c) $0 \leq \phi_j(x)$ for all $x \in \mathbb{R}^n$ and all $j \in \mathbb{N}$;

d) $\int_{\mathbb{R}^n} \phi_j(x)dx = 1$ for all $j \in \mathbb{N}$.

Certainly, if we choose a test function $\phi \in \mathcal{D}(\mathbb{R}^n)$ which is nonnegative and which is normalized by $\int_{\mathbb{R}^n} \phi(x)dx = 1$ and introduce the elements of the sequence as in part a), then we get a regularizing sequence. Note furthermore that every regularizing sequence converges to Dirac's delta distribution δ since regularizing sequences are special delta sequences, as discussed earlier.

Theorem 7.2.2 (Approximation of distributions) *For any $T \in \mathcal{D}'(\mathbb{R}^n)$ and any regularizing sequence $(\phi_j)_{j \in \mathbb{N}}$, the limit in $\mathcal{D}'(\mathbb{R}^n)$ of the sequence of $\mathcal{C}^\infty$-functions T_j on $\mathbb{R}^n$, defined by $T_j = T * \phi_j$ for all $j = 1, 2, \ldots$, is T, i.e.,*

$$T = \lim_{j \to \infty} T_j = \lim_{j \to \infty} T * \phi_j \quad \text{in} \quad \mathcal{D}'(\mathbb{R}^n).$$

Proof. According to Theorem 7.2.1 we know that $T * \phi_j \in \mathcal{C}^\infty(\mathbb{R}^n)$. If $\phi \in \mathcal{D}(\mathbb{R}^n)$ is the starting element of the regularizing sequence, we also know $\operatorname{supp} \phi_j \subset \operatorname{supp} \phi$ for all $j \in \mathbb{N}$. Take any $\psi \in \mathcal{D}(\mathbb{R}^n)$, then $K = \operatorname{supp} \phi - \operatorname{supp} \psi$ is compact and $\operatorname{supp}(\phi_j * \check{\psi}) \subset K$ for all $j \in \mathbb{N}$ (see part ii) of Proposition 7.1.1). Part c) of Proposition 7.1.4 implies that the sequence $D^\alpha(\phi_j * \psi)$ converges uniformly on K to $D^\alpha \check{\psi}$, for all $\alpha \in \mathbb{N}^n$, hence the sequence $(\phi_j * \check{\psi})_{j \in \mathbb{N}}$ converges to $\check{\psi}$ in $\mathcal{D}_K(\mathbb{R}^n)$. Now use part c) of Theorem 7.2.1 to conclude through the following chain of identities using the continuity of T on $\mathcal{D}_K(\mathbb{R}^n)$:

$$\lim_{j \to \infty} \int (T * \phi_j)(x)\psi(x)dx = \lim_{j \to \infty}((T * \phi_j) * \check{\psi})(0) = \lim_{j \to \infty}(T * (\phi_j * \check{\psi}))(0)$$

$$= \lim_{j \to \infty} \langle T, (\phi_j * \check{\psi}) \rangle = \langle T, \psi \rangle.$$

$\square$

Remark 7.2.1 a) *The convolution gives a bi-linear mapping $\mathcal{D}'(\mathbb{R}^n) \times \mathcal{D}(\mathbb{R}^n) \to \mathcal{C}^\infty(\mathbb{R}^n)$ defined by $(T, \phi) \mapsto T * \phi$.*

b) *Theorem 7.2.1 shows that $\mathcal{C}^\infty(\mathbb{R}^n)$ is dense in $\mathcal{D}'(\mathbb{R}^n)$. In the Exercises we show that also $\mathcal{D}(\mathbb{R}^n)$ is dense in $\mathcal{D}'(\mathbb{R}^n)$. We mention without proof that for any nonempty open set $\Omega \subset \mathbb{R}^n$ the test function space $\mathcal{D}(\Omega)$ is dense in the space $\mathcal{D}'(\Omega)$ of distributions on Ω.*

c) *The results of this section show that, and how, every distribution is the limit of a sequence of $\mathcal{C}^\infty$-functions. This observation can be used to derive another characterization of distributions. In this characterization a distribution is defined as a certain equivalence class of Cauchy sequences of $\mathcal{C}^\infty$-functions. Here a sequence of $\mathcal{C}^\infty$-functions f_j is said to be a Cauchy sequence if, and only if, $\int f_j(x)\phi(x)dx$ is a Cauchy sequence of numbers, for every test function ϕ. And two such sequences are called equivalent if, and only if, the difference sequence is a null sequence.*

d) *We mention a simple but useful observation. The convolution product is translation invariant in both factors, i.e., for $T \in \mathcal{D}'(\mathbb{R}^n)$, $\phi \in \mathcal{D}(\mathbb{R}^n)$, and every $a \in \mathbb{R}^n$ one has*

$$(T * \phi)_a = T_a * \phi = T * \phi_a.$$

For the definition of the translation of functions and distributions compare equation (4.6).

We conclude this section with an important result about the connection between differentiation in the sense of distributions and in the classical sense. The key of the proof is to use regularization.

Lemma 7.2.1 *Suppose $u, f \in C(\mathbb{R}^n)$ satisfy the equation $D_j u = f$ in the sense of distributions. Then this identity holds in the classical sense too.*

Proof. Suppose two continuous functions u, f are related by $f = D_j u = \frac{\partial u}{\partial x_j}$, in the sense of distributions. This means that for every test function ϕ the identity

$$-\int u(y) D_j \phi(y) dy = \int f(y) \phi(y) dy$$

holds. Next choose a regularizing sequence. Assume $\psi \in \mathcal{D}(\mathbb{R}^n)$ satisfies $\int \psi(y) dy = 1$. Define, for $\epsilon > 0$, $\psi_\epsilon(x) = \epsilon^{-n} \psi(\frac{x}{\epsilon})$. (With $\epsilon = \frac{1}{i}$, $i \in \mathbb{N}$, we have a regularizing sequence as above). Now approximate u and f by smooth functions:

$$u_\epsilon = u * \psi_\epsilon, \qquad f_\epsilon = f * \psi_\epsilon.$$

u_ϵ and f_ϵ are C^∞-functions, and as $\epsilon \to 0$, they converge to u, respectively f, uniformly on compact sets (see Proposition 7.1.4). A small calculation shows that

$$D_j u_\epsilon(x) = \int u(y) D_j \psi_\epsilon(x - y) dy = -\int u(y) D_{y_j} \psi_\epsilon(x - y) dy,$$

and taking the identity $D_j u = f$ in $\mathcal{D}'(\mathbb{R}^n)$ into account we find

$$D_j u_\epsilon(x) = \int f(y) \psi_\epsilon(x - y) dy = f_\epsilon(x).$$

Denote the standard unit vector in $\mathbb{R}^n$ in coordinate direction j by e_j and calculate, for $h \in \mathbb{R}$, $h \neq 0$,

$$\frac{1}{h}[u_\epsilon(x + he_j) - u_\epsilon(x)] = \int_0^1 (D_j u_\epsilon)(x + the_j) dt = \int_0^1 f_\epsilon(x + the_j) dt.$$

Take the limit $\epsilon \to 0$ of this equation. Since u_ϵ and f_ϵ converge uniformly on compact sets to u, respectively f, we get in the limit for all $|h| \leq 1$, $h \neq 0$,

$$\frac{1}{h}[u(x + he_j) - u(x)] = \int_0^1 f(x + the_j) dt.$$

It follows that we can take the limit $h \to 0$ of this equation and thus u has a partial derivative $D_j u(x)$ at the point x in the classical sense which is given by $f(x)$. Since x was arbitrary we conclude. □

7.3 Convolution of distributions

As we learned earlier, the convolution product $u * v$ is not defined for arbitrary pairs of functions (u, v). Some integrability conditions have to be satisfied. Often these integrability conditions are realized by support properties of the functions. Since the convolution product for distributions is to be defined in such a way that it is compatible with the embedding of functions, we will be able to define the convolution product for distributions under the assumption that the distributions satisfy a certain support condition which will be developed below.

7.3 Convolution of distributions

In order to motivate this support condition we calculate, for $f \in C_0(\mathbb{R}^n)$ and $g \in C(\mathbb{R}^n)$, the convolution product $f * g$ which is known to be a continuous function and thus can be considered as a distribution. For every test function ϕ the following chain of identities holds:

$$\begin{aligned}\langle I_{f*g}, \phi\rangle &= \int_{\mathbb{R}^n}(f*g)(x)\phi(x)dx = \int_{\mathbb{R}^n}(\int_{\mathbb{R}^n} f(x-y)g(y)dy)\phi(x)dx \\ &= \int_{\mathbb{R}^n}\int_{\mathbb{R}^n} f(x-y)g(y)\phi(x)dydx = \int_{\mathbb{R}^n\times\mathbb{R}^n} f(z)g(y)\phi(z+y)dydz \\ &= \langle (I_f \otimes I_g)(z,y), \phi(z+y)\rangle\end{aligned}$$

where we used Fubini's theorem for functions and the definition of the tensor product of regular distributions. Thus, in order to ensure compatibility with the embedding of functions, one has to define the convolution product for distributions $T, S \in \mathcal{D}'(\mathbb{R}^n)$ according to the formula

$$\langle T * S, \phi \rangle = \langle (T \otimes S)(x,y), \phi(x+y)\rangle \qquad \forall \phi \in \mathcal{D}(\mathbb{R}^n) \qquad (7.5)$$

whenever the right-hand side makes sense. Given $\phi \in \mathcal{D}(\mathbb{R}^n)$, the function $\psi = \psi_\phi$ defined on $\mathbb{R}^n \times \mathbb{R}^n$ by $\psi(x,y) = \phi(x+y)$, is certainly a function of class C^∞ but never has a compact support in $\mathbb{R}^n \times \mathbb{R}^n$ if $\phi \neq 0$. Thus in general the righthand side of equation (7.5) is not defined. There is an obvious and natural way to ensure the proper definition of the righthand side. Suppose supp $(T \otimes S) \cap \text{supp } \psi_\phi$ is compact in $\mathbb{R}^n \times \mathbb{R}^n$ for all $\phi \in \mathcal{D}(\mathbb{R}^n)$. Then one would expect that this definition will work. The main result of this section will confirm this. In order that this condition holds, the supports of the distributions T and S have to be in a special relation.

Definition 7.3.1 *Two distributions $T, S \in \mathcal{D}'(\mathbb{R}^n)$ are said to satisfy the* **support condition** *if, and only if, for every compact set $K \subset \mathbb{R}^n$ the set*

$$K_{T,S} = \{(x,y) \in \mathbb{R}^n \times \mathbb{R}^n : x \in \text{supp } S,\ y \in \text{supp } S,\ x+y \in K\}$$

is compact in $\mathbb{R}^n \times \mathbb{R}^n$.

Note that the set $K_{T,S}$ is always closed, but it need not be bounded. To get an idea about how this support condition can be realized, we consider several examples. Given $T, S \in \mathcal{D}'(\mathbb{R}^n)$ denote $F = \text{supp } T$ and $G = \text{supp } S$.

1. Suppose $F \subset \mathbb{R}^n$ is compact. Since $K_{T,S}$ is contained in the compact set $F \times (K - F)$ it is compact and thus the pair of distributions (T, S) satisfies the support condition.

2. Consider the case $n = 1$ and suppose $F = [a, +\infty)$ and $G = [b, +\infty)$ for some given numbers $a, b \in \mathbb{R}$. Given a compact set $K \subset \mathbb{R}$ it is contained in some closed and bounded interval $[-k, +k]$. A simple calculation shows that in this case $K_{T,S} \subseteq [a, k-b] \times [b, k-a]$, and it follows that $K_{T,S}$ is compact. Hence the support condition holds.

3. For two closed convex cones $C_1, C_2 \subset \mathbb{R}^n$, $n \geq 2$, with vertices at the origin and two points $a_j \in \mathbb{R}^n$, consider $F = a_1 + C_1$ and $G = a_2 + C_2$. Suppose

that the cones have the following property: Given any compact set $K \subset \mathbb{R}^n$ there are compact sets $K_1, K_2 \subset \mathbb{R}^n$ with the property that $x_j \in C_j$ and $x_1 + x_2 \in K$ implies $x_j \in K_j \cap C_j$ for $j = 1, 2$. Then the support condition is satisfied. The proof is given as an exercise.

4. This is a special case of the previous example. In the previous example we consider the cones $C_1 = C_2 = C = \left\{ x \in \mathbb{R}^n : x_1 \geq \theta \sqrt{\sum_{j=2}^{n} x_j^2} \right\}$ for some $\theta \geq 0$. Again we leave the proof as an exercise that the support condition holds in this case.

Theorem 7.3.1 (Definition of convolution) *If two distributions $T, S \in \mathcal{D}'(\mathbb{R}^n)$ satisfy the support condition, then the convolution product $T * S$ is a distribution on $\mathbb{R}^n$, well defined by the formula (7.5), i.e., by*

$$\langle T * S, \phi \rangle = \langle (T \otimes S)(x, y), \phi(x + y) \rangle \qquad \forall \phi \in \mathcal{D}(\mathbb{R}^n).$$

Proof. Given a compact set $K \subset \mathbb{R}^n$, there are two compact sets $K_1, K_2 \subset \mathbb{R}^n$ such that $K_{T,S} \subseteq K_1 \times K_2$, since the given distributions T, S satisfy the support condition. Now choose a test function $\psi \in \mathcal{D}(\mathbb{R}^n \times \mathbb{R}^n)$ such that $\psi(x, y) = 1$ for all $(x, y) \in K_1 \times K_2$. It follows that for all $\phi \in \mathcal{D}_K(\mathbb{R}^n)$ the function $(x, y) \mapsto (1 - \psi(x, y))\phi(x + y)$ has its support in $\mathbb{R}^n \times \mathbb{R}^n \setminus K_1 \times K_2$ and thus, because of the support condition,

$$\langle (T \otimes S)(x, y), \phi(x + y) \rangle = \langle (T \otimes S)(x, y), \psi(x, y)\phi(x + y) \rangle \qquad \forall \phi \in \mathcal{D}_K(\mathbb{R}^n).$$

By Theorem 6.2.2 we conclude that the righthand side of the above identity is a continuous linear functional on $\mathcal{D}_K(\mathbb{R}^n)$. Thus we get a well-defined continuous linear functional $(T * S)_K$ on $\mathcal{D}_K(\mathbb{R}^n)$. Let $K_i, i \in \mathbb{N}$, be a strictly increasing sequence of compact sets which exhaust $\mathbb{R}^n$. The above argument gives a corresponding sequence of functionals $(T * S)_{K_i}$. It is straightforward to show that these functionals satisfy the compatibility condition $(T * S)_{K_{i+1}} | \mathcal{D}_{K_i}(\mathbb{R}^n) = (T * S)_{K_i}, i \in \mathbb{N}$ and therefore this sequence of functionals defines a unique distribution on $\mathbb{R}^n$ (see Proposition 5.2.1) which is denoted by $T * S$ and called the convolution of T and S. □

Theorem 7.3.2 (Properties of convolution) 1. *Suppose that two distributions $T, S \in \mathcal{D}'(\mathbb{R}^n)$ satisfy the support property. Then the convolution has the following properties:*

  a) $T * S = S * T$, *i.e., the convolution product is commutative;*

  b) $\operatorname{supp}(T * S) \subseteq \overline{\operatorname{supp} T + \operatorname{supp} S}$;

  c) *For all $\alpha \in \mathbb{N}^n$ one has $D^\alpha(T * S) = D^\alpha T * S = T * D^\alpha$.*

2. *The convolution of Dirac's delta distribution δ is defined for every $T \in \mathcal{D}'(\mathbb{R}^n)$ and one has*

$$\delta * T = T.$$

3. *Suppose three distributions $S, T, U \in \mathcal{D}'(\mathbb{R}^n)$ are given whose supports satisfy the following condition: For every compact set $K \subset \mathbb{R}^n$ the set*

$$\left\{ (x, y, z) \in \mathbb{R}^{3n} : x \in \operatorname{supp} S, \ y \in \operatorname{supp} T, \ z \in \operatorname{supp} U, \ x + y + z \in K \right\}$$

is compact in $\mathbb{R}^{3n}$. Then all the convolutions $S * T$, $(S * T) * U$, $T * U$, $S * (T * U)$ are well defined and one has

$$(S * T) * U = S * (T * U).$$

Proof. Note that the pair of distributions (S, T) satisfies the support condition if, and only if, the pair (T, S) does. Thus with $T * S$ also the convolution $S * T$ is well defined by the above theorem. The righthand side of the defining formula (7.5) of the tensor product is invariant under the exchange of T and S. Therefore commutativity of the convolution follows and proves part a) of 1).

Denote $C = \overline{\operatorname{supp} T + \operatorname{supp} S}$ and consider a test function ϕ with support in $\mathbb{R}^n \setminus C$. Then $\phi(x+y) = 0$ for all $(x, y) \in \operatorname{supp} T \times \operatorname{supp} S$ and thus $\langle (T * S)(x, y), \phi(x + y) \rangle = 0$ and it follows that $\operatorname{supp}(T * S) \subseteq \mathbb{R}^n \setminus C = C$, which proves part b).

The formula for the derivatives of the convolution follows from the formula for the derivatives of tensor products (part b) of Corollary 6.2.2 and the defining identity for the convolution. The details are given in the following chain of identities, for $\phi \in \mathcal{D}(\mathbb{R}^n)$:

$$\begin{aligned}
\langle D^\alpha(T * S), \phi \rangle &= (-1)^{|\alpha|} \langle T * S, D^\alpha \phi \rangle \\
&= (-1)^{|\alpha|} \langle (T \otimes S)(x, y), (D^\alpha \phi(x + y)) \rangle \\
&= (-1)^{|\alpha|} \langle T(x), \langle S(y), D_y^\alpha \phi(x + y) \rangle \rangle \\
&= \langle T(x), \langle (D^\alpha S)(y), \phi(x + y) \rangle \rangle \\
&= \langle (T * D^\alpha S)(x, y), \phi(x + y) \rangle = \langle T * D^\alpha S, \phi \rangle.
\end{aligned}$$

Thus $D^\alpha(T * S) = T * D^\alpha S$ and in the same way $D^\alpha(T * S) = D^\alpha T * S$. This proves part c).

Dirac's delta distribution δ has the compact support $\{0\}$, hence for any distribution T on $\mathbb{R}^n$ the pair (δ, T) satisfies the support condition. Therefore the convolution $\delta * T$ is well defined. If we evaluate this product on any $\phi \in \mathcal{D}(\mathbb{R}^n)$ we find, using again Theorem 6.2.2,

$$\langle \delta * T, \phi \rangle = \langle (\delta \otimes T)(x, y), \phi(x + y) \rangle = \langle T(y), \langle \delta(x), \phi(x + y) \rangle \rangle = \langle T(y), \phi(y) \rangle$$

and we conclude $\delta * T = T$.

The proof of the third part about the three-fold convolution product is left as an exercise. $\square$

Remark 7.3.1 1. *As we have seen above, the support condition for two distributions T, S on $\mathbb{R}^n$ is* **sufficient** *for the existence of the convolution product $T * S$. Note that this condition is* **not necessary**. *This is easily seen on the level of functions. Consider two functions $f, g \in L^2(\mathbb{R}^n)$. Application of the Cauchy–Schwarz' inequality (Corollary 14.1.1) implies, for almost all $x \in \mathbb{R}^n$, $|f * g(x)| \leq \|f\|_2 \|g\|_2$ and hence the convolution product of f and g is well defined as an essentially bounded function on $L^2(\mathbb{R}^n)$.*

2. *The simple identity $D^\alpha(T * \delta) = (D^\alpha \delta) * T = D^\alpha T$ will later allow us to write linear partial differential equations with constant coefficients as a convolution identity and through this a fairly simple algebraic formalism will lead to a solution.*

3. *If either $\operatorname{supp} T$ or $\operatorname{supp} S$ is compact, then $\operatorname{supp} T + \operatorname{supp} S$ is closed and in part 1.b) of Theorem 7.3.2 the closure sign can be omitted. However, when neither $\operatorname{supp} T$ nor $\operatorname{supp} S$ is compact, then the sum $\operatorname{supp} T + \operatorname{supp} S$ is in general not closed as the folllowing simple example shows: Consider*

$T, S \in \mathcal{D}'(\mathbb{R}^2)$ with

$$\operatorname{supp} T = \{(x, y) \in \mathbb{R}^2 : 0 \le x, \; +1 \le xy\},$$
$$\operatorname{supp} S = \{(x, y) \in \mathbb{R}^2 : 0 \le x, \; xy \le -1\}.$$

Then the sum is

$$\operatorname{supp} T + \operatorname{supp} S = \{(x, y) \in \mathbb{R}^2 : 0 < x\}$$

and thus not closed.

The regularization $T * \phi$ of a distribution T by a test function ϕ is a C^∞-function by Theorem 7.2.1 and thus defines a regular distribution $I_{T*\phi}$. Certainly, the test function ϕ defines a regular distribution I_ϕ and so one can ask whether the convolution product of this regular distribution with the distribution T exists and what this convolution is. The following corollary answers this question and provides important additional information.

Corollary 7.3.1 *Let $T \in \mathcal{D}(\mathbb{R}^n)$ be a distribution on $\mathbb{R}^n$.*

a) *For all $\phi \in \mathcal{D}(\mathbb{R}^n)$ the convolution (in the sense of distributions) $T * I_\phi$ exists and one has*

$$T * I_\phi = I_{T*\phi}.$$

b) *Suppose T has a compact support. Then, for every $f \in C^\infty(\mathbb{R}^n)$, the convolution $T * I_f$ exists and is a C^∞-function. One has*

$$T * I_f = I_{T*f},$$

i.e., $T * f$ is a C^∞-function.

Proof. Since the regular distribution I_ϕ has a compact support, the support condition is satisfied for the pair (T, I_ϕ) and therefore Theorem 7.3.1 proves the existence of the convolution $T * I_\phi$, and for $\psi \in \mathcal{D}(\mathbb{R}^n)$ the following chain of identities holds.

$$\langle T * I_\phi, \psi \rangle = \langle (T \otimes I_\phi)(x, y), \psi(x + y) \rangle = \langle T(x), \langle T_\phi(y), \psi(x + y) \rangle \rangle$$
$$= \langle T(x), \int \phi(y)\psi(x + y)dy \rangle = \langle T(x), \int \phi(z - x)\psi(z)dz \rangle$$
$$= \int \langle T(x), \phi(z - x) \rangle \psi(z)dz = \int (T * \phi(z))\psi(z)dz = \langle I_{T*\phi}, \psi \rangle.$$

The key step in this chain of identities is the proof of the identity

$$\langle T(x), \int \phi(z - x)\psi(z)dz \rangle = \int \langle T(x), \phi(z - x) \rangle \psi(z)dz \qquad (7.6)$$

and this is given in the Exercises. This proves part a).

If the support K of T is compact, we know by Theorem 7.3.1 that the convolution $T * I_f$ is a well-defined distribution, for every $f \in C^\infty(\mathbb{R}^n)$. In order to show that $T * I_f$ is actually a C^∞-function, choose some $\psi \in \mathcal{D}(\mathbb{R}^n)$ such that $\psi(x) = 1$ for all $x \in K$. For all $\phi \in \mathcal{D}(\mathbb{R}^n)$ we have $\operatorname{supp}(\phi - \psi\phi) \subseteq K^c \equiv \mathbb{R}^n \setminus K$ and therefore $\langle T, \phi - \psi\phi \rangle = 0$. This shows that $T = \psi \cdot T$. Thus, for every $\phi \in \mathcal{D}(\mathbb{R}^n)$ we can define a function h_ϕ on $\mathbb{R}^n$ by

$$h_\phi(x) = \langle T(y), \phi(x + y) \rangle = \langle T(y), \psi(y)\phi(x + y) \rangle \qquad \forall x \in \mathbb{R}^n.$$

Corollary 6.2.1 implies that h_ϕ is a C^∞-function with support in $\operatorname{supp}\phi - K$. Similarly, Corollary 6.2.1 shows that the function

$$z \mapsto g(z) = \langle T(y), \psi(y) f(z-y)\rangle$$

is of class C^∞ on $\mathbb{R}^n$. Now we calculate, for all $\phi \in \mathcal{D}(\mathbb{R}^n)$,

$$\begin{aligned}\langle T * I_f, \phi\rangle &= \langle (\psi \cdot T) * I_f, \phi\rangle = \langle I_f(x), \langle (\psi \cdot T)(y), \phi(x+y)\rangle\rangle \\ &= \langle I_f(x), \langle T(y), \psi(y)\phi(x+y)\rangle\rangle = \int \langle T(y), \psi(y) f(z-y)\rangle \phi(z) dz.\end{aligned}$$

Hence $T * I_f$ is equal to I_g. Since obviously $g = T * f$, part b) follows. $\square$

From the point of view of practical applications of the convolution of distributions, it is important to know distinguished sets of distributions such that, for any pair in this set, the convolution is well defined. We present here a concrete example of such a set which later will play an important role in the *symbolic calculus*. Introduce the set of all distributions on the real line which have their support on the positive half-line:

$$\mathcal{D}'_+(\mathbb{R}) = \left\{T \in \mathcal{D}'(\mathbb{R}) \,\middle|\, \operatorname{supp} T \subseteq [0, +\infty)\right\}.$$

With regard to convolution this set has quite interesting properties as the following theorem shows.

Theorem 7.3.3 *a) $\mathcal{D}'_+(\mathbb{R})$, equipped with the convolution as a product, is an Abelian algebra with Dirac's delta distribution δ as the neutral element. It is however not a field.*

b) $(\mathcal{D}'_+(\mathbb{R}), *)$ has no divisors of zero (Theorem of Titchmarsh).

Proof. It is easily seen that any two elements $T, S \in \mathcal{D}'_+(\mathbb{R})$ satisfy the support condition (compare the second example in the discussion of this condition). Hence by Theorem 7.3.1 the convolution is well defined on $\mathcal{D}'_+(\mathbb{R})$. By Theorem 7.3.2 this product is Abelian and $T * S$ has its support in $[0, +\infty)$ for all $T, S \in \mathcal{D}'_+(\mathbb{R})$ and the neutral element is δ. $\mathcal{D}'_+(\mathbb{R})$ is not a field under the convolution since there are elements in $\mathcal{D}'_+(\mathbb{R})$ which have no inverse with respect to the convolution product though they are different from zero. Take for example a test function $\phi \in \mathcal{D}(\mathbb{R})$ with support in $\mathbb{R}_+ = [0, +\infty)$, $\phi \neq 0$. Then the regular distribution I_ϕ belongs to $\mathcal{D}'_+(\mathbb{R})$, and there is no $T \in \mathcal{D}'_+(\mathbb{R})$ such that $T * I_\phi = \delta$, since by Corollary 7.3.1 one has $T * I_\phi = I_{T*\phi}$ and by Theorem 7.2.1 it is known that $T * \phi \in C^\infty(\mathbb{R})$. This proves part a).

Statement b) means: If $T, S \in \mathcal{D}'_+(\mathbb{R})$ are given and $T * S = 0$, then either $T = 0$ or $S = 0$. The proof is somewhat involved and we refer the reader to [GŠ77]. $\square$

Remark 7.3.2 *1. The convolution product is not associative. Here is a simple example. Observe that $\delta' * \theta = D(\delta * \theta) = D\theta = \delta$, hence*

$$1 * (\delta' * \theta) = 1 * \delta = 1.$$

Similarly, $1 * \delta' = D(1 * \delta) = D1 = 0$, hence

$$(1 * \delta') * \theta = 0.$$

2. *For the proof that $(\mathcal{D}'_+(\mathbb{R}), *)$ has no divisors of zero, the support properties are essential. In $(\mathcal{D}'(\mathbb{R}), *)$ we can easily construct counterexamples. Since $\delta' \in \mathcal{D}'(\mathbb{R})$ has a compact support, we know that $\delta' * 1$ is a well-defined distribution on $\mathbb{R}$. We also know $\delta' \neq 0$ and $1 \neq 0$, but as we have seen above, $\delta' * 1 = 0$.*

3. *Fix $S \in \mathcal{D}'(\mathbb{R}^n)$ and assume that S has a compact support. Then we can consider the map $\mathcal{D}'(\mathbb{R}^n) \to \mathcal{D}'(\mathbb{R}^n)$ given by $T \mapsto T * S$. It is important to realize that this map is not continuous. Take for example the distributions $T_n = \delta_n$, $n \in \mathbb{N}$, i.e., $T_n(\phi) = \phi(n)$ for all $\phi \in \mathcal{D}(\mathbb{R})$. Then $1 * T_n = 1$ for all $n \in \mathbb{N}$, but*

$$\lim_{n \to \infty} T_n = 0 \quad \text{in } \mathcal{D}'(\mathbb{R}).$$

4. *Recall the definition of Cauchy's principal value $vp\frac{1}{x}$ in equation (3.7). It can be used to define a transformation H, called the **Hilbert transform**, by convolution $f \mapsto H(f) = vp\frac{1}{x} * f$:*

$$H(f)(x) = \lim_{\epsilon \to 0} \int_{|x-y| \geq \epsilon} \frac{f(y)}{x-y} dy \quad \forall x \in \mathbb{R}. \tag{7.7}$$

This transformation is certainly well defined on test functions. It is not difficult to show that it is also well defined on all $f \in C^1(\mathbb{R})$ with the following decay property: For every $x \in \mathbb{R}$ there are a constant C and an exponent $\alpha > 0$ such that for all $y \in \mathbb{R}$, $|y| \geq 1$, the estimate

$$|f(x-y) - f(x+y)| \leq C|y|^{-\alpha}$$

holds. This Hilbert transform is used in the formulation of "dispersion relations" which play an important role in various branches of physics (see [Thi92]).

7.4 Exercises

1. The sum $A+B$ of two subsets A and B of a vector space V is by definition the set $A+B = \{x+y \in V : x \in A, y \in B\}$. For compact subsets $A, B \subset \mathbb{R}^n$ prove that the closure of the sum is equal to the sum:

$$\overline{A+B} = A+B.$$

Give an example of two closed sets $A, B \subset \mathbb{R}^n$ such that the sum $A+B$ is not closed.

2. Fill in the details in the proof of Proposition 7.1.1.

3. Prove Corollary 7.1.2.

4. For $T \in \mathcal{D}'(\mathbb{R}^n)$ and $\phi, \psi \in \mathcal{D}(\mathbb{R}^n)$ prove the important identity (7.6)

$$\langle T(x), \int \phi(z-x)\psi(z)dz \rangle = \int \langle T(x), \phi(z-x)\rangle \psi(z)dz.$$

Hints: One can use for instance the representation theorem of distributions as weak derivatives of integrable functions (Theorem 5.2.3).

5. Prove: $\mathcal{D}(\mathbb{R}^n)$ is (sequentially) dense in $\mathcal{D}'(\mathbb{R}^n)$.

6. Prove Part 3. of Theorem 7.3.2

8
Applications of Convolution

The four sections of this chapter introduce various applications of the convolution product, for functions and distributions. The common core of these sections is a *convolution equation*, i.e., a relation of the form

$$T * X = S$$

where T, S are given distributions and X is a distribution which we want to find, in a suitable space of distributions. We will learn that various problems in mathematics can be written as convolution equations. As simple examples the case of ordinary and partial linear differential equations with constant coefficients is discussed as well as a wellknown integral equation. Naturally, in the study of convolution equations we encounter the following problems:

1. Existence of a solution: Given two distributions T, S, is there a distribution X, in a suitable space of distributions, such that $T * X = S$ holds?

2. Uniqueness: If a solution exists, is it the only solution to this equation (in a given space of distributions)?

An ideal situation would be if we could treat the convolution equation in a space of distributions which is an algebra with respect to the convolution product. Then, if T is invertible in this convolution algebra, the unique solution to the equation $T * X = S$ obviously is $X = T^{-1} * S$. If however T is not invertible in the convolution algebra the equation might not have any solution, or it might have several solutions.

Unfortunately this ideal case hardly occurs in the study of concrete problems. We discuss a few cases. Earlier we saw that the space of all distributions is not an algebra with respect to convolution. The space $L^1(\mathbb{R}^n)$ of Lebesgue integrable functions on $\mathbb{R}^n$ is an algebra for the convolution but this space is not suitable for the study of differential operators. The space $\mathcal{E}'$ of distributions with compact support is an algebra for the convolution, but not very useful since there is hardly any differential operator which is invertible in $\mathcal{E}'$. The space of distributions with support in a given cone can be shown to be an algebra for the convolution. It can be used for the study of special partial differential operators with constant coefficients. Thus we are left with the convolution algebra $\mathcal{D}'_+(\mathbb{R})$ studied in Theorem 7.3.3.

8.1 Symbolic Calculus – ordinary linear differential equations

Suppose we are given an ordinary linear differential equation with constant coefficients

$$\sum_{n=0}^{N} a_n y^{(n)} = f \qquad (8.1)$$

where the a_n are given real or complex numbers and f is a given continuous function on the positive half-line $\mathbb{R}_+$. Here $y^{(n)} = D^n y$ denotes the derivative of order n of the function y with respect to the variable x, $D = \frac{d}{dx}$.

By developing a *symbolic calculus* with the help of the convolution algebra $(\mathcal{D}'_+(\mathbb{R}), *)$ of Theorem 7.3.3 we will learn how to reduce the problem of finding solutions of equation (8.1) to a purely algebraic problem which is known to have solutions. The starting point is to consider equation (8.1) as an equation in $\mathcal{D}'_+(\mathbb{R})$ and to write it as a convolution equation

$$\left(\sum_{n=0}^{N} a_n \delta^{(n)}\right) * y = f. \qquad (8.2)$$

The rules for derivatives of convolution products and the fact that Dirac's delta distribution is the unit of the convolution algebra $(\mathcal{D}'_+(\mathbb{R}), *)$ imply that

$$y^{(n)} = \delta * y^{(n)} = \delta^{(n)} * y$$

and thus by distributivity of the convolution product equations (8.1), (8.2), and (8.3) are equivalent. In this way we assign to the differential operator

$$P(D) = \sum_{n=0}^{N} a_n D^n$$

the element

$$P(\delta) = \sum_{n=0}^{N} a_n \delta^{(n)} \qquad \in \mathcal{D}'_+(\mathbb{R})$$

8.1 Symbolic Calculus – ordinary linear differential equations

with support in $\{0\}$ such that

$$P(D)y = P(\delta) * y \tag{8.3}$$

on $\mathcal{D}'_+(\mathbb{R})$. Thus we can solve equation (8.1) by showing that the element $P(\delta)$ has an inverse in $(\mathcal{D}'_+(\mathbb{R}), *)$. We prepare the proof of this claim by a simple lemma.

Lemma 8.1.1 *The distribution $\delta' - \lambda \delta \in \mathcal{D}'_+(\mathbb{R})$, $\lambda \in \mathbb{C}$, has a unique inverse given by the regular distribution*

$$e_\lambda(x) = I_{\theta(x) e^{\lambda x}} \qquad \in \mathcal{D}'_+(\mathbb{R}).$$

Proof. The proof consists of a sequence of straightforward calculations using the rules established earlier for the convolution. We have $(\delta' - \lambda \delta) * e_\lambda = \delta' * e_\lambda - \lambda \delta * e_\lambda = \delta' * e_\lambda - \lambda e_\lambda$ and

$$\delta' * e_\lambda = D(\delta * e_\lambda) = D e_\lambda = e^{\lambda x} \delta + \lambda e_\lambda = \delta + \lambda e_\lambda$$

where we have used the differentiation rules of distributions and the fact that $D\theta = \delta$. It follows that

$$(\delta' - \lambda \delta) * e_\lambda = \delta$$

and therefore e_λ is an inverse of $\delta' - \lambda \delta$ in $\mathcal{D}'_+(\mathbb{R})$. Since $\mathcal{D}'_+(\mathbb{R})$ has no divisors of zero, the inverse is unique. □

Proposition 8.1.1 *Let $P(x) = a_0 + a_1 x + \cdots + a_n x^n$ be a polynomial of degree n $(a_n \neq 0)$ with complex coefficients a_j. Denote by $\{\lambda_1, \ldots, \lambda_p\}$ the set of zeros (roots) of P with multiplicities $\{k_1, \ldots, k_p\}$, i.e.,*

$$P(x) = a_n (x - \lambda_1)^{k_1} \cdots (x - \lambda_p)^{k_p}.$$

Then

$$P(\delta) = a_0 \delta + a_1 \delta^{(1)} + \cdots + a_n \delta^{(n)} = a_n (\delta' - \lambda_1 \delta)^{*k_1} * \cdots * (\delta' - \lambda_p \delta)^{*k_p}$$

has an inverse in $\mathcal{D}'_+(\mathbb{R})$ which is given by

$$P(\delta)^{-1} = \frac{1}{a_n} e_{\lambda_1}^{*k_1} * \cdots * e_{\lambda_p}^{*k_p} \equiv E. \tag{8.4}$$

Proof. For $\lambda, \mu \in \mathbb{C}$ calculate $(\delta' - \lambda \delta) * (\delta' - \mu \delta) = \delta' * \delta' - \lambda \delta * \delta' - \delta' * \mu \delta + \lambda \mu \delta * \delta = \delta' * \delta' - (\lambda + \mu) \delta' + \lambda \mu \delta$ where we used the distributive law for the convolution and the fact that δ is the unit in $(\mathcal{D}'_+(\mathbb{R}), *)$. Using the differentiation rules we find $\delta' * \delta' = D(\delta \delta') = D(\delta') = \delta^{(2)}$. It follows that

$$(\delta' - \lambda \delta) * (\delta' - \mu \delta) = \delta^{(2)} - (\lambda + \mu) \delta' + \lambda \mu.$$

In particular for $\lambda = \mu$ one has $(\delta' - \lambda \delta)^{*2} = \delta^{(2)} - 2\lambda \delta' + \lambda^2$ where for $S \in \mathcal{D}'_+(\mathbb{R})$ we use the notation $S^{*k} = S * \cdots * S$ (k factors). Repeated application of this argument implies (see Exercises)

$$a_0 \delta + a_1 \delta^{(1)} + \cdots + a_n \delta^{(n)} = a_n (\delta' - \lambda_1 \delta)^{*k_1} * \cdots * (\delta' - \lambda_p \delta)^{*k_p}.$$

Knowing this factorization of $P(\delta)$ it is easy to show that the given element $E \in \mathcal{D}'_+(\mathbb{R})$ is indeed the inverse of $P(\delta)$. Using the above lemma and the fact that $(\mathcal{D}'_+(\mathbb{R}), *)$ is an Abelian algebra we find

$$(\tfrac{1}{a_n} e_{\lambda_1}^{*k_1} * \cdots * e_{\lambda_p}^{*k_p}) * P(\delta) = e_{\lambda_1}^{*k_1} * (\delta' - \lambda_1 \delta)^{*k_1} * \cdots * e_{\lambda_p}^{*k_p} * (\delta' - \lambda_p \delta)^{*k_p}$$

$$= [e_{\lambda_1} * (\delta' - \lambda_1 \delta)]^{*k_1} * \cdots * [e_{\lambda_p} * (\delta' - \lambda_p \delta)]^{*k_p} = \delta^{*k_1} * \cdots * \delta^{*k_p} = \delta.$$

Hence the element $E \in \mathcal{D}'_+(\mathbb{R})$ is the inverse of $P(\delta)$. □

102   8. Applications of Convolution

After these preparations it is fairly easy to solve ordinary differential equations of the form (8.1), even for all $f \in \mathcal{D}'_+(\mathbb{R})$. To this end rewrite equation (8.1) using relation (8.2), as
$$P(\delta) * y = f,$$
and thus by Proposition 8.1.1, a solution is
$$y = P(\delta)^{-1} * f, \qquad (8.5)$$
in particular, for $f = \delta$, $E = P(\delta)^{-1}$ is a special solution of the equation
$$P(D)T = \delta \quad \text{in} \quad \mathcal{D}'_+(\mathbb{R}). \qquad (8.6)$$
This special solution E is called a *fundamental solution of the differential operator* $P(D)$. The above argument shows: Whenever we have a fundamental solution E of the differential operator $P(D)$, a solution y of the equation $P(D)y = f$ for general inhomogeneous term $f \in \mathcal{D}'_+(\mathbb{R})$ is given by
$$y = E * f.$$
Since the fundamental solution is expressed as a convolution product of explicitly known functions, one can easily derive some regularity properties of solutions.

Theorem 8.1.2 *Let $P(D) = \sum_{n=0}^{N} a_n D^n$ be an ordinary constant coefficient differential operator normalized by $a_N = 1$, $N > 1$, and $E = P(\delta)^{-1}$ the fundamental solution as determined above.*

1. *E is a function of class $\mathcal{C}^{N-2}(\mathbb{R})$ with support in $\mathbb{R}_+$.*

2. *Given an inhomogeneous term $f \in \mathcal{D}'_+(\mathbb{R})$, a solution of $P(D)y = f$ is $y = E * f$.*

3. *If the inhomogeneous term f is a continuous function on $\mathbb{R}$ with support in $\mathbb{R}_+$, then the special solution $y = E * f$ of $P(D)y = f$ is a classical solution, i.e., a function of class $\mathcal{C}^N(\mathbb{R})$ which satisfies the differential equation.*

Proof. According to Proposition 8.1.1 the fundamental solution E has the representation $E = e_{z_1} * \cdots * e_{z_N}$ where the N roots $\{z_1, \ldots, z_N\}$ are not necessarily distinct. Thus we can write $D^{N-2}E = De_{z_1} * \cdots * De_{z_{N-2}} * e_{z_{N-1}} * e_{z_N}$. Previous calculations have shown that $De_z = \delta + z e_z$. It follows that
$$D^{N-2}E = (\delta + z_1 e_{z_1}) * \cdots * (\delta + z_{N-2} e_{z_{N-2}}) * e_{z_{N-1}} * e_{z_N}$$
$$= e_{z_N} * e_{z_{N-1}} + \sum_{j=1}^{N-2} z_j e_{z_j} * e_{z_{N-1}} * e_{z_N} + \cdots + z_1 \cdots z_N e_{z_1} * \cdots * e_{z_N}.$$

Next we determine the continuity properties of the convolution product $e_z * e_w$ of the function e_z and e_w for arbitrary $z, w \in \mathbb{C}$. According to the definition of the functions e_z and the convolution we find
$$(e_z * e_w)(x) = \int e_z(y) e_w(x - y) dy$$
$$= \theta(x) \int_0^x e^{zy} e^{w(x-y)} dy = \theta(x) e^{wx} \int_0^x e^{(z-w)y} dy.$$

According to this representation $e_z * e_w$ is a continuous function on $\mathbb{R}$ with support in $\mathbb{R}_+$. It follows that also all convolution products with $m \geq 2$ factors are continuous functions on $\mathbb{R}$ with support in $\mathbb{R}_+$. Hence the formula for $D^{N-2}E$ shows that $D^{N-2}E$ is continuous and thus E has continuous derivatives up to order $N-2$ on $\mathbb{R}$ and has its support in $\mathbb{R}_+$. This proves the first part.

The second part has been shown above. In order to prove the third part we evaluate $D^N(E*f) = (D^N E) * f$. As above we find

$$\begin{aligned} D^N E &= De_{z_1} * \cdots * De_{z_N} = (\delta + z_1 e_{z_1}) * \cdots * (\delta + z_N e_{z_N}) \\ &= \delta + \sum_{j=1}^{N} z_j e_{z_j} + \sum_{i \neq j} z_i z_j e_{z_j} * e_{z_j} + \cdots + z_1 \cdots z_N e_{z_1} * \cdots * e_{z_N} \end{aligned}$$

and therefore

$$D^N y = D^n(E*f) = f + \sum_{j=1}^{N} z_j e_{z_j} * f + \cdots + z_1 \cdots z_N e_{z_1} * \cdots * e_{z_N} * f.$$

This shows that the derivative of order N of $y = E * f$, calculated in the sense of distributions, is actually a continuous function. We conclude that this solution is an N-times continuously differentiable function on $\mathbb{R}$ with support in $\mathbb{R}_+$, i.e., a classical solution. □

Remark 8.1.1 Theorem 8.1.2 reduces the problem of finding a solution of the ODE $P(D)y = f$ to the algebraic problem of finding all the roots $\{\lambda_1, \ldots, \lambda_p\}$ of the polynomial $P(x)$ and their multiplicities $\{k_1, \ldots, k_p\}$.

A simple concrete **example** will illustrate how convenient the application of Theorem 8.1.2 is in solving ordinary differential equations. Consider an electrical circuit in which a capacitor C, an inductance L, and a resistance R are put in series and connected to a power source of voltage $V(t)$. The current $I(t)$ in this circuit satisfies, according to Kirchhoff's law, the equation

$$V(t) = RI(t) + L\frac{dI(t)}{dt} + \frac{1}{C}\int_0^t I(s)ds.$$

Differentiation of this identity yields, using $D = \frac{d}{dt}$,

$$DV(t) = LP(D)I(t) \qquad P(D) = D^2 + \frac{R}{L}D + \frac{1}{LC}.$$

The roots of the polynomial $P(x) = x^2 + \frac{R}{L}x + \frac{1}{LC}$ are $\lambda_{1,2} = -\frac{R}{2L} \pm \sqrt{(\frac{R}{2L})^2 - \frac{1}{LC}}$ and therefore, according to Theorem 8.1.2,

$$I(t) = \frac{1}{L}P(\delta)^{-1} * DV(t) = \frac{1}{L}e_{\lambda_1} * e_{\lambda_2} * DV(t) = \frac{1}{L}(De_{\lambda_1}) * e_{\lambda_2} * V(t).$$

Since we know $De_z = \delta + ze_z$, a special solution of the above differential equation is

$$I(t) = \frac{1}{L}(e_{\lambda_2} * V)(t) + \frac{\lambda_1}{L}(e_{\lambda_1} * e_{\lambda_2} * V)(t).$$

8.2 Integral equation of Volterra

Given two continuous functions g, K on $\mathbb{R}_+$, we look for all functions f satisfying *Volterra's linear integral equation*

$$f(x) = \int_0^x K(x-y)f(y)dy = g(x) \qquad \forall x \in \mathbb{R}_+. \tag{8.7}$$

Integral equations of this type are for instance used in optics for the description of the distribution of brightness.

How can one solve such equations? We present here a simple method based on our knowledge of the convolution algebra $(\mathcal{D}'_+(\mathbb{R}), *)$. By identifying the functions f, g, K with the regular distributions $\theta f, \theta g, \theta K$ in $\mathcal{D}'_+(\mathbb{R})$ it is easy to rewrite equation (8.7) as a convolution equation in $\mathcal{D}'_+(\mathbb{R})$:

$$(\delta - K) * f = g. \tag{8.8}$$

In order to solve this equation we show that the element $\delta - K$ is invertible in $\mathcal{D}'_+(\mathbb{R})$. This is done in the following proposition.

Proposition 8.2.1 *If $K : \mathbb{R}_+ \to \mathbb{R}$ is a continuous function, then the element $\delta - K$ has an inverse in $\mathcal{D}'_+(\mathbb{R})$. This inverse is of the form*

$$(\delta - K)^{-1} = \delta + H$$

where H is a continuous function $\mathbb{R}_+ \to \mathbb{R}$. Volterra's integral equation has thus exactly one solution which is of the form

$$f = g + H * g.$$

Proof. We start with the well-known (in any ring with unit) identity

$$\delta - K^{*(n+1)} = (\delta - K) * (\delta + K + K^{*2} + \cdots + K^{*n}) \tag{8.9}$$

and show that the series $\sum_{i=1}^\infty K^{*i}$ converges uniformly on every compact subset of $\mathbb{R}_+$. For this it suffices to show uniform convergence on every compact interval of the form $[0, r]$ for $r > 0$. Since K is continuous we know that $M_r = \sup_{0 \le x \le r} |K(x)|$ is finite for every $r > 0$. Observe that

$$K^{*2}(x) = \int_0^x K(y)K(x-y)dy \implies |K^{*2}(x)| \le M_r^2 x,$$

and therefore by induction (see Exercises)

$$|K^{*i}(x)| \le M_r^i \frac{x^{i-1}}{(i-1)!} \qquad \forall x \in [0, r].$$

The estimate

$$\sum_{i=1}^\infty |K^{*i}(x)| \le \sum_{i=1}^\infty M_r^i \frac{x^{i-1}}{(i-1)!} = M_r e^{M_r x}$$

implies that the series $\sum_{i=1}^\infty K^{*i}(x)$ converges absolutely and uniformly on $[0, r]$, for every $r > 0$. Hence this series defines a continuous function $H : \mathbb{R}_+ \to \mathbb{R}$,

$$H = \sum_{i=1}^\infty K^{*i}.$$

With this information we can pass to the limit $n \to \infty$ in equation (8.9) and find $\delta = (\delta - K) * (\delta + H)$, hence $(\delta - K)^{-1} = \delta + H$ which proves the proposition since the convolution algebra $(\mathcal{D}'_+(\mathbb{R}), *)$ is without divisors of zeros and $\delta - K \ne 0$. □

8.3 Linear partial differential equations with constant coefficients

This section reports on one of the main achievements of the theory of distributions, namely providing a powerful framework for solving linear partial differential equations with constant coefficients. Using the multi-index notation, a linear partial differential operator with constant coefficients will generically be written as

$$P(D) = \sum_{|\alpha| \leq m} a_\alpha D^\alpha, \qquad a_\alpha \in \mathbb{C}, \quad m = 1, 2, \ldots. \tag{8.10}$$

Suppose that $\Omega \subseteq \mathbb{R}^n$ is a nonempty open set. Certainly, operators of the form (8.10) induce linear maps of the test function space over Ω into itself, and this map is continuous (see Exercises). Thus, by duality, as indicated earlier, the operators $P(D)$ can be considered as linear and continuous operators $\mathcal{D}'(\Omega) \to \mathcal{D}'(\Omega)$. Then, given $U \in \mathcal{D}'(\Omega)$, the distributional form of a linear partial differential equation with constant coefficients is

$$P(D)T = U \quad \text{in } \mathcal{D}'(\Omega). \tag{8.11}$$

Note that $T \in \mathcal{D}'(\Omega)$ is a *distributional or weak solution* of (8.11) if, and only if, for all $\phi \in \mathcal{D}(\Omega)$ one has

$$\langle U, \phi \rangle = \langle P(D)T, \phi \rangle = \langle T, P^t(D)\phi \rangle$$

where $P^t(D) = \sum_{|\alpha| \leq m}(-1)^{|\alpha|} a_\alpha D^\alpha$. In many applications however one is not so much interested in distributional solutions but in functions satisfying this partial differential equation (PDE). If the righthand side U is a continuous function, then a *classical or strong solution* of equation (8.11) is a function T on Ω which has continuous derivatives up to order m and which satisfies (8.11) in the sense of functions. As one would expect, it is easier to find solutions to equation (8.11) in the much larger space $\mathcal{D}'(\Omega)$ of distributions than in the subspace $\mathcal{C}^{(m)}(\Omega)$ of m times continuously differentiable functions. Nevertheless, the problems typically require classical and not distributional solutions and thus the question arises when, i.e., for which differential operators, a distributional solution is actually a classical solution. This is known to be the case for the so-called elliptic operators. In this *elliptic regularity theory* one shows that, for these elliptic operators, weak solutions are indeed classical solutions. This also applies to nonlinear partial differential equations. In Part III, Chapter 32 we present without proof some classes of typical examples. We mention here the earliest and quite typical result of the elliptic regularity theory, due to H. Weyl (1940), for the Laplace operator.

Lemma 8.3.1 (Lemma of Weyl) *Suppose that* $T \in \mathcal{D}'_{reg}(\Omega)$ *is a solution of* $\Delta T = 0$ *in* $\mathcal{D}'(\Omega)$, *i.e.,* $\int T(x)\Delta\phi(x)dx = 0$ *for all* $\phi \in \mathcal{D}(\Omega)$. *Then it follows that* $T \in \mathcal{C}^{(2)}(\Omega)$ *and* $\Delta T(x) = 0$ *holds in the sense of functions.*

We remark that in the special case of the Laplace operator Δ one can actually show $T \in \mathcal{C}^\infty(\Omega)$. We conclude: In order to determine classical solutions of the equation $\Delta T = 0$, $T \in \mathcal{C}^{(2)}(\Omega)$, it is sufficient to determine weak solutions in the much larger space $\mathcal{D}'_{reg}(\Omega)$.

Naturally, not all differential operators have this very convenient regularity property. As a simple example we discuss the *wave operator* $\Box_2 = \frac{\partial^2}{\partial t^2} - \frac{\partial^2}{\partial x^2}$ in two dimensions which has many weak solutions which are not strong solutions. Denote by f the characteristic function of the unit interval $[0, 1]$ and define $u(t, x) = f(x - t)$. Then $u \in \mathcal{D}'_{reg}(\mathbb{R}^2)$ and $\Box_2 u = 0$ in the sense of distributions. But u is not a strong solution.

In the context of ordinary linear differential operators we have learned already about the basic role which a fundamental solution plays in the process of finding solutions. This will be the same for linear partial differential operators with constant coefficients. Accordingly we repeat the formal definition.

Definition 8.3.1 *Given a differential operator of the form (8.10), every distribution $E \in \mathcal{D}'(\mathbb{R}^n)$ which satisfies the distributional equation*

$$P(D)E = \delta$$

is called a **fundamental solution of this differential operator**.

In the case of ordinary differential operators we saw that every constant coefficient operator has a fundamental solution and we learned how to construct them. For partial differential operators the corresponding problem is much more difficult. We indicate briefly the main reason. While for a polynomial in one variable the set of zeros (roots) is a finite set of isolated points, the set of zeros of a polynomial in $n > 1$ variables consists in general of several lower dimensional manifolds in $\mathbb{R}^n$.

It is worthwhile mentioning that some variation of the concept of a fundamental solution is used in physics under the name *Green's function*. A Green's function is a fundamental solution which satisfies certain boundary conditions. In the following section and in the sections on tempered distributions we are going to determine fundamental solutions of differential operators which are important in physics.

Despite these complications B. Malgrange (1953) and L. Ehrenpreis (1954) proved independently of each other that every constant coefficient partial differential operator has a fundamental solution.

Theorem 8.3.1 *Every partial differential operator $P(D) = \sum_{|\alpha| \leq m} a_\alpha D^\alpha$, $a_\alpha \in \mathbb{C}$, has at least one fundamental solution.*

The proof of this basic result is beyond the scope of this introduction and we have to refer to the specialized literature, for instance [Hör83b]. Knowing the existence of a fundamental solution for a PDE-operator (8.10), the problem of existence of solutions of partial differential equations of the form (8.11) has an obvious solution.

Theorem 8.3.2 *Every linear partial differential equation in $\mathcal{D}'(\mathbb{R}^n)$ with constant coefficients*

$$\sum_{|\alpha| \leq m} a_\alpha D^\alpha T = U$$

8.3 Linear partial differential equations with constant coefficients

has a solution in $\mathcal{D}'(\mathbb{R}^n)$ for all those $U \in \mathcal{D}'(\mathbb{R}^n)$ for which there is a fundamental solution $E \in \mathcal{D}'(\mathbb{R}^n)$ such that the pair (E, U) satisfies the support condition. In this case a special solution is

$$T = E * U. \tag{8.12}$$

Such a solution exists in particular for all distributions $U \in \mathcal{E}'(\mathbb{R}^n)$ of compact support.

Proof. If we have a fundamental solution E such that the pair (E, U) satisfies the support condition, then we know that the convolution $E * U$ is well defined. The rules of calculation for convolution products now yield

$$P(D)(E * U) = (P(D)E) * U = \delta * U = U,$$

hence $T = E * U$ solves the equation in the sense of distributions. If a distribution U has a compact support, then the support condition for the pair (E, U) is satisfied for every fundamental solution and thus we conclude. □

Obviously, a differential operator of the form (8.10) leaves the support of a distribution invariant: supp $(P(D)T) \subseteq$ supp T for all $T \in \mathcal{D}'(\mathbb{R}^n)$, but not necessarily the singular support as defined in Definition 3.4.1. Those constant coefficient partial differential operators which do not change the singular support of any distribution play a very important role in the solution theory for linear partial differential operators. They are called hypo-elliptic for reasons which become apparent later.

Definition 8.3.2 *A linear partial differential operator with constant coefficients $P(D)$ is called* **hypo-elliptic** *if, and only if,*

$$\text{sing supp } P(D)T = \text{sing supp } T \quad \forall T \in \mathcal{D}'(\mathbb{R}^n). \tag{8.13}$$

Since one always has sing supp $P(D)T \subseteq$ sing supp T, this definition is equivalent to the following statement: If $P(D)T$ is of class $\mathcal{C}^\infty$ on some open subset $\Omega \subset \mathbb{R}^n$, then T itself is of class $\mathcal{C}^\infty$ on Ω. With this in mind we present a detailed characterization of hypo-elliptic partial differential operators in terms of regularity properties of its fundamental solutions.

Theorem 8.3.3 *Let $P(D)$ be a linear constant coefficient partial differential operator. The following statements are equivalent:*

a) *$P(D)$ is hypo-elliptic.*

b) *$P(D)$ has a fundamental solution $E \in \mathcal{C}^\infty(\mathbb{R}^n \setminus \{0\})$.*

c) *Every fundamental solution E of $P(D)$ belongs to $\mathcal{C}^\infty(\mathbb{R}^n \setminus \{0\})$.*

Proof. We start with the observation that Dirac's delta distribution is of class $\mathcal{C}^\infty$ on $\mathbb{R}^n \setminus \{0\}$. If we now apply condition (8.13) to a fundamental solution E of the operator $P(D)$ we get

$$\text{sing supp } E = \text{sing supp } (P(D)E) = \text{sing supp } \delta = \{0\},$$

hence a) implies c). The implication c) ⇒ b) is trivial. Thus we are left with showing b) ⇒ a).

Suppose $E \in \mathcal{C}^\infty(\mathbb{R}^n \setminus \{0\})$ is a fundamental solution of the operator $P(D)$. Assume furthermore that $\Omega \subset \mathbb{R}^n$ is a nonempty open subset and $T \in \mathcal{D}'(\mathbb{R}^n)$ a distribution such that $P(D)T \in \mathcal{C}^\infty(\Omega)$ holds. Now it suffices to show that T itself is of class $\mathcal{C}^\infty$ in a neighborhood of each point x in Ω.

Given any $x \in \Omega$, there is an $r > 0$ such that the open ball $B_{2r}(x)$ is contained in Ω. There is a test function $\phi \in \mathcal{D}(\mathbb{R}^n)$ such that $\operatorname{supp} \phi \subset B_r(0)$ and such that $\phi(x) = 1$ for all x in some neighborhood V of zero.

Using Leibniz' rule we calculate

$$\begin{aligned} P(D)(\phi E) &= \sum_{|\alpha| \le m} a_\alpha \sum_{\beta \le \alpha} \tfrac{\alpha!}{\beta!(\alpha-\beta)!} D^\beta \phi D^{\alpha-\beta} E \\ &= \phi P(D)E + \sum_{|\alpha| \le m} \sum_{0 \ne \beta \le \alpha} a_\alpha \tfrac{\alpha!}{\beta!(\alpha-\beta)!} D^\beta \phi D^{\alpha-\beta} E \\ &= \phi P(D)E + \psi = \delta + \psi. \end{aligned}$$

The properties of ϕ imply that the function ψ vanishes on the neighborhood V and has its support in $B_r(0)$; by assumption b) the function ψ is of class C^∞ on $\mathbb{R}^n \setminus \{0\}$, hence $\psi \in \mathcal{D}(\mathbb{R}^n)$, and we can regularize the distribution T by ψ and find

$$T + \psi * T = (\delta + \psi) * T = [P(D)(\phi E)] * T = (\phi E) * (P(D)T),$$

or $T = \phi E * P(D)T - \psi * T$. □

8.4 Elementary solutions of partial differential operators

Theorems 8.3.2 and 8.3.3 of the previous section are the core of the solution theory for linear partial equations with constant coefficients and through them we learn that, and why, it is important to know elementary solutions of constant coefficient partial differential operators explicitly. Accordingly we determine in this section elementary solutions of differential operators which are important in physics. In some cases we include a discussion of relevant physical aspects. Later in the section on Fourier transforms and tempered distributions we learn about another method to obtain elementary solutions.

8.4.1 The Laplace operator $\Delta_n = \sum_{i=1}^n \frac{\partial^2}{\partial x_i^2}$ in $\mathbb{R}^n$

The Laplace operator occurs in a number of differential equations which play an important role in physics. After we have determined the elementary solution for this operator we discuss some of the applications in physics.

Proposition 8.4.1 *The function $E_n : \mathbb{R}^n \setminus \{0\} \to \mathbb{R}$, defined by*

$$E_n(x) = \begin{cases} \frac{1}{2\pi} \log |x| & \text{for } n = 2, \\ \frac{-1}{(n-2)|S_n|} |x|^{2-n} & \text{for } n \ge 3, \end{cases} \tag{8.14}$$

where $|S_n| = 2\pi^{\frac{n+1}{2}} \Gamma(\frac{n+1}{2})$ is the area of the unit sphere S_n in $\mathbb{R}^n$, has the following properties:

a) $E_n \in L^1_{loc}(\mathbb{R}^n) \cap C^\infty(\mathbb{R}^n \setminus \{0\})$;

b) $\Delta_n E_n(x) = 0$ for all $x \in \mathbb{R}^n \setminus \{0\}$;

c) E_n is the elementary solution of the Laplace operator Δ_n in $\mathbb{R}^n$ which is thus hypo-elliptic.

Proof. Using polar coordinates it is an elementary calculation to show that E_n is locally integrable in $\mathbb{R}^n$. Similarly, standard differentiation rules imply that E_n is of class C^∞ on $\mathbb{R}^n \setminus \{0\}$. This proves part a). The elementary proof of part b) is left as an exercise. Uniqueness of the elementary solution for the Laplace operator follows from Hörmander's theorem (see Theorem 10.4.1). Thus we are left with proving that the function E_n is an elementary solution.

For any test function $\phi \in \mathcal{D}(\mathbb{R}^n)$ we calculate

$$\langle \Delta_n E_n, \phi \rangle = \int E_n(x) \Delta_n \phi(x) dx = \lim_{r \to 0} \int_{[r \leq |x| \leq R]} E_n(x) \Delta_n(x) \phi(x) dx$$

since E_n is locally integrable and where R is chosen such that $\operatorname{supp} \phi \subset B_R(0)$. Here $[r \leq |x| \leq R]$ denotes the set $\{x \in \mathbb{R}^n : r \leq |x| \leq R\}$. Observe that ϕ vanishes in a neighborhood of the boundary of the ball $B_R(0)$ and that $\Delta_n E_n(x) = 0$ in $[r \leq |x| \leq R]$. Therefore, applying partial integration and Gauss' Theorem twice, we get for the integral under the limit

$$-\int_{|x|=r} \phi(x) \nabla_n E_n(x) \cdot dS(x) + \int_{|x|=r} E_n(x) \nabla_n \phi(x) \cdot dS(x).$$

In the Exercises one shows that the limit $r \to 0$ of the first integral gives $\phi(0)$ while the limit of the second integral is zero. It follows $\langle \Delta_n, \phi \rangle = \phi(0) = \langle \delta, \phi \rangle$ for all $\phi \in \mathcal{D}(\mathbb{R}^n)$ and thus $\Delta_n E_n = \delta$.
□

The case $n = 1$ is elementary. We claim $E_1(x) = x\theta(x)$ is the elementary solution of $\Delta_1 = \frac{d^2}{dx^2}$. The proof is a straightforward differentiation in the sense of distributions and is left as an exercise.

Now we discuss the case $n = 3$ which is of particular importance for physics. The fundamental solution for Δ_3 is

$$E_3(x) = -\frac{1}{4\pi} \frac{1}{|x|} \qquad \forall x \in \mathbb{R}^3 \setminus \{0\}. \tag{8.15}$$

This solution is well known in physics in connection with the *Poisson equation*

$$\Delta_3 U = \rho \tag{8.16}$$

where ρ is a given density (of masses or electrical charges), and one is looking for the potential U generated by this density. In physics we learn that this potential is given by the formula

$$U(x) = -\frac{1}{4\pi} \int_{\mathbb{R}^3} \frac{\rho(x)}{|x-y|} d^3y \qquad \forall x \in \mathbb{R}^3 \tag{8.17}$$

whenever ρ is an integrable function. One easily recognizes that this solution formula is just the convolution formula for this special case:

$$U(x) = E_3 * \rho(x).$$

Certainly, the formula $U = E_3 * \rho$ gives the solution of equation (8.16) for all $\rho \in \mathcal{E}'(\mathbb{R}^3)$, not just for integrable densities.

8.4.2 The PDE operator $\frac{\partial}{\partial t} - \Delta_n$ of the heat equation in $\mathbb{R}^{n+1}$

We proceed as in the case of the Laplace operator but refer for a discussion of the physical background of this operator to the physics literature.

Proposition 8.4.2 *The function E_n defined on the set $(0, +\infty) \times \mathbb{R}^n$ by the formula*

$$E_n(t, x) = (\frac{1}{2\sqrt{\pi t}})^n \theta(t) e^{-\frac{|x|^2}{4t}} \tag{8.18}$$

has the following properties:

a) $E_n \in L^1_{loc}(\mathbb{R}^{n+1}) \cap C^\infty((0, +\infty) \times \mathbb{R}^n)$;

b) $(\frac{\partial}{\partial t} - \Delta_n) E_n(t, x) = 0$ for all $(t, x) \in (0, +\infty) \times \mathbb{R}^n$;

c) E_n is the elementary solution of the operator $\frac{\partial}{\partial t} - \Delta_n$ which is thus hypoelliptic.

Proof. Since the statements of this proposition are quite similar to the result on the elementary solution of the Laplace operator, it is natural that we can use nearly the same strategy of proof. Certainly, the function (8.18) is of class C^∞ on $(0, +\infty) \times \mathbb{R}^n$. In order to show that this function is locally integrable on $\mathbb{R}^{n+1}$, it suffices to show that the integral

$$I(t) = \int_0^t \int_{|x| \leq R} E_n(s, x) dx dt$$

is finite, for every $t > 0$ and every $R > 0$. For every $s > 0$ the integral with respect to x can be estimated in absolute value by 1, after the change of variables $x = 2\sqrt{s} y$. Thus it follows that $I(t) \leq t$ and therefore $E_n \in L^1_{loc}(\mathbb{R}^{n+1})$. Elementary differentiation shows that part b) holds. Again, uniqueness of the elementary solution follows from Hörmander's theorem (Theorem 10.4.1).

Now take any $\phi \in \mathcal{D}(\mathbb{R}^{n+1})$; since E_n is locally integrable it follows, using $\partial_t = \frac{\partial}{\partial t}$, that

$$\langle (\partial_t - \Delta_n) E_n, \phi \rangle = -\langle E_n, (\partial_t + \Delta_n)\phi \rangle$$
$$= \lim_{r \to 0} \int_r^\infty \int_{\mathbb{R}^n} E_n(t, x)(\partial + \Delta_n)\phi(t, x) dx dt = \lim_{r \to 0} I_r(\phi).$$

Since ϕ has a compact support, repeated partial integration in connection with Gauß' theorem yields

$$\int_{\mathbb{R}^n} E_n(t, x)(\Delta_n \phi)(t, x) dx = \int_{\mathbb{R}^n} (\Delta_n E_n)(t, x) \phi(t, x) dx$$

for every $t > 0$. Therefore, by partial integration with respect to t, we find

$$I_r(\phi) = \int_r^\infty \int_{\mathbb{R}^n} (\Delta_n E_n)(t, x) \phi(t, x) dx dt + \int_{\mathbb{R}^n} E_n(t, x) \phi(t, x)|_{t=r}^{t=+\infty} dx$$
$$- \int_r^\infty \int_{\mathbb{R}^n} (\partial_t E_n)(t, x) \phi(t, x) dx dt$$
$$= \int_r^\infty \int_{\mathbb{R}^n} ((-\partial_t + \Delta_n) E_n)(t, x) \phi(t, x) dx dt - \int_{\mathbb{R}^n} E_n(r, x) \phi(r, x) dx$$
$$= -\int_{\mathbb{R}^n} E_n(r, x) \phi(r, x) dx.$$

Here we have used Fubini's theorem for integrable functions to justify the exchange of the order of integration and in the last identity we have used part b). This allows the conclusion

$$\langle (\partial_t - \Delta_n) E_n, \phi \rangle = (\frac{1}{2\sqrt{\pi}})^n \lim_{r \to 0} \int_{\mathbb{R}^n} r^{-\frac{n}{2}} e^{-\frac{|x|^2}{4r}} \phi(r, x) dx$$
$$= (\frac{1}{2\sqrt{\pi}})^n \lim_{r \to 0} \int_{\mathbb{R}^n} e^{-\frac{|y|^2}{4}} \phi(r, \sqrt{r} y) dy$$
$$= \phi(0, 0),$$

where we used the new integration variable $y = \frac{x}{\sqrt{r}}$, Lebesgue's theorem of dominated convergence, and the fact that

$$\int_{\mathbb{R}^n} e^{-\frac{y^2}{4}} dy = (2\sqrt{\pi})^n.$$

Since $\phi \in \mathcal{D}(\mathbb{R}^{n+1})$ is arbitrary, this shows that $(\partial_t - \Delta_n)E_n = \delta$ and hence the given function is indeed the elementary solution of the operator $\partial - \Delta_n$. □

8.4.3 The wave operator $\Box_4 = \partial_0^2 - \Delta_3$ in $\mathbb{R}^4$

Here we use the notation $\partial_0 = \frac{\partial}{\partial x_0}$. In applications to physics the variable x_0 has the interpretation of $x_0 = ct$, c being the velocity of light and t the time variable. The variable $x \in \mathbb{R}^3$ stands for the space coordinate. For the wave operator Hörmander's theorem does not apply and accordingly several elementary solutions for the wave operator are known. We mention two:

$$E_{r,a}(x_0, x) = \frac{1}{2\pi} \theta(\pm x_0) \delta(x_0^2 - x^2). \tag{8.19}$$

These distributions are defined as follows:

$$\langle \theta(\pm x_0) \delta(x_0^2 - x^2), \phi(x_0, x) \rangle = \int_{\mathbb{R}^3} \phi(\pm |x|, x) \frac{dx}{2|x|}.$$

Since the function $x \mapsto \frac{1}{|x|}$ is integrable over compacts sets in $\mathbb{R}^3$ these are indeed well defined distributions.

Proposition 8.4.3 *The distributions (8.19) are two elementary solutions of the wave operator $\Box_4$ in dimension 4. Their support properties are:*

$$\operatorname{supp} E_r = \left\{ (x_0, x) \in \mathbb{R}^4 : x_0 \geq 0,\ x_0^2 - x^2 = 0 \right\},$$

$$\operatorname{supp} E_a = \left\{ (x_0, x) \in \mathbb{R}^4 : x_0 \leq 0,\ x_0^2 - x^2 = 0 \right\}.$$

Proof. The obvious invariance of the wave equation under rotations in $\mathbb{R}^3$ can be used to reduce the number of dimensions which have to be considered. This can be done by averaging over the unit sphere S^2 in $\mathbb{R}^3$. Accordingly, to every $\phi \in \mathcal{D}(\mathbb{R}^4)$ we assign a function $\tilde{\phi} : \mathbb{R} \times \mathbb{R}_+$ by the formula

$$\tilde{\phi}(t, s) = \int_{S^2} \phi(t, s\omega) d\omega$$

where $d\omega$ denotes the normalized surface measure on S^2. Introducing polar coordinates in $\mathbb{R}^3$ we thus see

$$\langle E_{r,a}, \phi \rangle = \int_0^\infty s \tilde{\phi}(\pm s, s) ds.$$

In the Exercises it is shown that

$$\widetilde{\Box_4 \phi}(t, s) = \left[\frac{\partial^2}{\partial t^2} - \frac{\partial^2}{\partial s^2} - \frac{2}{t} \frac{\partial}{\partial s} \right] \tilde{\phi}(t, s).$$

Thus we get

$$\langle \Box_4 E_{r,a}, \phi \rangle = \langle E_{r,a}, \Box_4 \phi \rangle = \int_0^\infty t \widetilde{\Box_4 \phi}(t, t) dt.$$

Introducing, for $t > 0$, the auxiliary function

$$u(t) = t\frac{\partial \tilde{\phi}}{\partial t}(t,t) - t\frac{\partial \tilde{\phi}}{\partial s}(t,t) - \tilde{\phi}(t,t)$$

which has the derivative

$$u'(t) = t\left[\frac{\partial^2 \tilde{\phi}}{\partial t^2} - \frac{\partial^2 \tilde{\phi}}{\partial s^2} - \frac{2}{t}\frac{\partial \tilde{\phi}}{\partial s}\right](t,t)$$

it follows that

$$\langle \Box_4 E_r, \phi \rangle = \int_0^\infty u'(t)dt = -u(0) = \tilde{\phi}(0,0) = \phi(0) = \langle \delta, \phi \rangle$$

and thus we conclude that E_r is an elementary solution of the wave operator. The argument for E_a is quite similar. □

Remark 8.4.1   1. *Though the wave operator $\Box_4$ is not hypo-elliptic, it can be shown that it is hypo-elliptic in the variable x_0. This means that every weak solution $u(x_0, x)$ of the wave equation is a C^∞-function in x_0 (see [Hör83b]).*

2. *Later with the help of Fourier transformation for tempered distributions we will give another proof for $E_{r,a}$ being elementary solutions of the wave operator.*

3. *In particular in applications to physics the support properties will play an important role. According to these support properties one calls E_r a **retarded** and E_a an **advanced** elementary solution. The reasoning behind these names is apparent from the following discussion of solutions of Maxwell's equation.*

Maxwell's equation in vacuum.

Introducing the abbreviations $x_0 = ct$ and $\partial_0 = \frac{\partial}{\partial x_0}$, Maxwell's equation in vacuum can be written as follows (see [Thi92]):

$$\operatorname{curl} \underline{E} + \partial_0 \underline{B} = 0 \quad \text{Faraday's law}$$
$$\operatorname{div} \underline{B} = 0 \quad \text{source-free magnetic field}$$
$$\operatorname{curl} \underline{B} + \partial_0 \underline{E} = \underline{j} \quad \text{Maxwell's form of Ampère's law}$$
$$\operatorname{div} \underline{E} = \rho \quad \text{Coulomb's law}$$

In courses on electrodynamics it is shown: Given a density ρ of electric charges and a density $\underline{j}$ of electric currents, the electric field $\underline{E}$ and the magnetic field $\underline{B}$ are given by

$$\underline{B} = \operatorname{curl} \underline{A}, \quad \underline{E} = -\nabla \Phi - \partial_0 \underline{A}$$

where $(\Phi, \underline{A})$ are the electromagnetic potentials. In the Lorenz gauge, i.e., $\partial_0 \Phi + \operatorname{div} \underline{A} = 0$, these electromagnetic potentials are solutions of the inhomogeneous wave equations

$$\Box_4 \Phi = \rho, \qquad \Box_4 \underline{A} = \underline{j}.$$

(The last equation is understood component-wise, i.e., $\Box_4 A_i = j_i$ for $i = 1, 2, 3$.)

Thus the problem of solving Maxwell's equations in vacuum has been put into a form to which our previous results apply since we know elementary solutions of the wave operator.

In concrete physical situations the densities of charges and currents are switched on at a certain moment which we choose to be our time reference point $t = 0$. Then one knows $\operatorname{supp} \rho$, $\operatorname{supp} \underline{j} \subseteq \{(x_0, x) \in \mathbb{R}^4 : x_0 \geq 0\}$.

It follows that the pairs (E_r, ρ) and $(E_r, \underline{j})$ satisfy the support condition and thus the convolution products $E_r * \rho$ and $E_r * \underline{j}$ are well defined. We conclude that the electromagnetic potentials are given by

$$(\Phi, \underline{A}) = (E_r * \rho, E_r * \underline{j})$$

which in turn give the electromagnetic field as mentioned above. Because of the known support properties of E_r and ρ and the formula for the support of a convolution we know: $\operatorname{supp} \Phi \subseteq \{(x_0, x) \in \mathbb{R}^4 : x_0 \geq 0\}$ and similarly for $\underline{A}$. Hence our solution formula shows causality, i.e., no electromagnetic field before the charge and current densities are switched on! The other elementary solution E_a of the wave operator does not allow this conclusion.

Note that the above formula gives a solution for Maxwell's equation not only for proper densities ($\rho \in L^1(\mathbb{R}^3)$) but also for the case where ρ is any distribution with the support property used earlier. The same applies to $\underline{j}$.

Under well-known decay properties for ρ and $\underline{j}$ for $|x| \to +\infty$ one can show that the electromagnetic field $(\underline{E}, \underline{B})$ determined above is the only solution to Maxwell's equation in vacuum.

8.5 Exercises

1. Let $f : \mathbb{R}_+ \to \mathbb{R}$ be a continuous function. In $\mathcal{D}'_+(\mathbb{R})$ find a special solution of the ordinary differential equation

$$y^{(4)} - 8y^{(2)} + 16y = f$$

and verify that it is actually a classical solution.

2. Let $K : \mathbb{R}_+ \to \mathbb{R}$ be a continuous function. For $n = 2, 3, \ldots$ define

$$K^{*n}(x) = \int_0^x K^{*(n-1)}(y) K(x - y) dy, \qquad \forall x \in \mathbb{R}_+$$

and show that for every $0 < r < \infty$ one has

$$|K^{*n}(x)| \leq M_r^n \frac{x^{n-1}}{(n-1)!} \qquad \forall x \in [0, r].$$

Here $M_r = \sup_{0 \leq x \leq r} |K(x)|$.

3. Let Δ_n be the Laplace operator in $\mathbb{R}^n$ ($n = 2, 3, \ldots$). For $\alpha \in \mathbb{N}^n$ solve the partial differential equation

$$\Delta_n u = \delta^{(\alpha)}.$$

4. For the function E_n of equation (8.18) show $(\frac{\partial}{\partial t} - \Delta_n) E_n(t, x) = 0$ for all $(t, x) \in (0, +\infty) \times \mathbb{R}^n$.

5. Find the causal solution of Maxwell's equations in vacuum.

   Hints: Use the retarded elementary solution of the wave operator and calculate $\underline{E}$ and $\underline{B}$ according to the formulae given in the text.

9
Holomorphic Functions

This chapter gives a brief introduction to the theory of holomorphic functions of one complex variable from a special point of view which defines holomorphic functions as elements of the kernel or null space of a certain hypo-elliptic differential operator. Thus this chapter offers a new perspective of some aspects on the theory of functions of one complex variable. A comprehensive modern presentation of this classical subject is [Rem98].

Our starting point will be the observation that the differential operator in $\mathcal{D}'(\mathbb{R}^2)$

$$\bar{\partial} = \frac{1}{2}(\frac{\partial}{\partial x} + i\frac{\partial}{\partial y}) \tag{9.1}$$

is hypo-elliptic and some basic results about convergence in the sense of distributions. Then holomorphic functions will be defined as elements in the null-space in $\mathcal{D}'(\mathbb{R}^2)$ of this differential operator. Relative to the theory of distributions developed thus far, this approach to the theory of holomorphic functions is fairly easy, though certainly this is neither a standard nor a too direct approach.

9.1 Hypo-ellipticity of $\bar{\partial}$

We begin by establishing several basic facts about the differential operator $\bar{\partial}$ in $\mathcal{D}'(\mathbb{R}^2)$.

9. Holomorphic Functions

Lemma 9.1.1 *The regular distribution on $\mathbb{R}^2$, $(x, y) \mapsto \frac{1}{\pi(x+iy)}$ is an elementary solution of the differential operator $\bar{\partial}$ in $\mathcal{D}'(\mathbb{R}^2)$, i.e., in $\mathcal{D}'(\mathbb{R}^2)$ one has*

$$\bar{\partial} \frac{1}{\pi(x+iy)} = \delta.$$

Proof. It is easy to see that the function $(x, y) \mapsto \frac{1}{\pi(x+iy)}$ is locally integrable on $\mathbb{R}^2$ and thus it defines a regular distribution. On $\mathbb{R}^2 \setminus \{0\}$ a straightforward differentiation shows $\bar{\partial} \frac{1}{\pi(x+iy)} = 0$. Now take any $\phi \in \mathcal{D}(\mathbb{R}^2)$ and calculate

$$\langle \bar{\partial} \frac{1}{\pi(x+iy)}, \phi \rangle = -\langle \frac{1}{\pi(x+iy)}, \bar{\partial}\phi \rangle = -\int_{\mathbb{R}^2} \frac{\bar{\partial}\phi(x, y)}{\pi(x+iy)} dx dy.$$

Since the integrand is absolutely integrable this integral can be represented as

$$= -\lim_{r \to 0} \int_{r \leq \sqrt{x^2+y^2} \leq R} \frac{\bar{\partial}\phi(x, y)}{\pi(x+iy)} dx dy$$

where R is chosen large enough such that $\operatorname{supp}\phi \subset B_R(0)$. For any $0 < r < R$ we observe that

$$\int_{r \leq \sqrt{x^2+y^2} \leq R} \frac{\bar{\partial}\phi(x, y)}{\pi(x+iy)} dx dy = \int_{r \leq \sqrt{x^2+y^2} \leq R} \bar{\partial} \left[\frac{\phi(x, y)}{\pi(x+iy)} \right] dx dy$$

since $\bar{\partial} \frac{1}{\pi(x+iy)} = 0$ in $\mathbb{R}^2 \setminus \{0\}$. Recall the formula of Green–Riemann for a domain $\Omega \subset \mathbb{R}^2$ with smooth boundary $\Gamma = \partial \Omega$ (see [Hör67]):

$$\int_\Omega \bar{\partial} u \, dx dy = \frac{1}{2i\pi} \oint_\Gamma u(x, y)(dx + idy) \tag{9.2}$$

which we apply to the function $u(x, y) = \frac{1}{\pi} \frac{\phi(x,y)}{x+iy}$ to obtain

$$\langle \bar{\partial} \frac{1}{\pi(x+iy)}, \phi \rangle = -\frac{1}{2i\pi} \lim_{r \to 0} \int_{\sqrt{x^2+y^2}=r} \frac{\phi(x, y)}{x+iy} (dx + idy).$$

Introducing polar coordinates $x = r\cos\theta$, $y = r\sin\theta$, this limit becomes

$$\lim_{r \to 0} \frac{1}{2\pi} \int_0^{2\pi} \phi(r\cos\theta, r\sin\theta) d\theta = \phi(0, 0) = \langle \delta, \phi \rangle$$

and thus we conclude. $\square$

Corollary 9.1.1 *The differential operator $\bar{\partial}$ in $\mathcal{D}'(\mathbb{R}^2)$ is hypo-elliptic, i.e., every distribution $T \in \mathcal{D}'(\mathbb{R}^2)$ for which $\bar{\partial}T$ is of class $\mathcal{C}^\infty$ on some open set $\Omega \subset \mathbb{R}^2$ is itself of class $\mathcal{C}^\infty$ on Ω.*

Proof. Lemma 9.1.1 gives an elementary solution of $\bar{\partial}$ which is of class $\mathcal{C}^\infty$ on $\mathbb{R}^2 \setminus \{0\}$. Thus by Theorem 8.3.3 we conclude. $\square$

If we apply this corollary to a distribution T on $\mathbb{R}^2$ which satisfies $\bar{\partial}T = 0$, it follows immediately that T is equal to a $\mathcal{C}^\infty$-function g, $T = I_g$, for some $g \in \mathcal{C}^\infty(\mathbb{R}^2)$, since obviously the zero function is of class $\mathcal{C}^\infty$ everywhere. Therefore the null space of the operator $\bar{\partial}$ on $\mathcal{D}'(\mathbb{R}^2)$ can be described as

$$\ker \bar{\partial} = \left\{ T \in \mathcal{D}'(\mathbb{R}^2) : \bar{\partial}T = 0 \right\} = \left\{ g \in \mathcal{C}^\infty(\mathbb{R}^2) : \bar{\partial}g = 0 \right\}$$

where as usual we identify the function g and the regular distribution I_g.

9.1 Hypo-ellipticity of $\bar{\partial}$

Now let $\Omega \subset \mathbb{R}^2$ be a nonempty open set. Similarly one deduces

$$\ker(\bar{\partial}|\mathcal{D}'(\Omega)) = \{T \in \mathcal{D}'(\Omega) : \bar{\partial}T = 0\} = \{g \in C^\infty(\Omega) : \bar{\partial}g = 0\}.$$

This says in particular that a complexvalued function g in $L^1_{loc}(\Omega)$ which satisfies $\bar{\partial}g = 0$ in the sense of distributions is actually a C^∞-function on Ω.

As usual we identify the point $(x, y) \in \mathbb{R}^2$ with the complex number $z = x + iy$. Under this identification we introduce

$$H(\Omega) = \left\{g \in L^1_{loc}(\Omega) : \bar{\partial}g = 0 \text{ in } \mathcal{D}'(\Omega)\right\} = \{u \in C^\infty(\Omega) : \bar{\partial}u = 0\}. \quad (9.3)$$

Elements in $H(\Omega)$ are called *holomorphic functions on* Ω. The following theorem lists the basic properties of the space of holomorphic functions.

Theorem 9.1.1 *Let $\Omega \subset \mathbb{C}$ be a nonempty open set. The space $H(\Omega)$ of holomorphic functions on Ω has the following properties:*

1. $H(\Omega)$ *is a complex algebra;*

2. $H(\Omega)$ *is complete for the topology of uniform convergence on all compact subsets of Ω;*

3. *If $u \in H(\Omega)$ does not vanish on Ω, then $\frac{1}{u} \in H(\Omega)$;*

4. *If a function u is holomorphic on Ω and a function v is holomorphic on an open set $\tilde{\Omega}$ which contains $u(\Omega)$, then the composition $v \circ u$ is holomorphic on Ω.*

Proof. As the nullspace of a linear operator on a vector space $H(\Omega)$ is certainly a complex vector space. The product rule of differentiation easily implies that with $u, v \in H(\Omega)$ also the (pointwise) product $u \cdot v$ belongs to $H(\Omega)$. The verification that with this product $H(\Omega)$ is indeed an algebra is straightforward and is left as an exercise.

Suppose that (u_n) is a Cauchy sequence in $H(\Omega)$ for the topology of uniform convergence on all compact sets $K \subset \Omega$. It follows that there is some continuous function u on Ω such that the sequence u_n converges uniformly to u, on every compact set $K \subset \Omega$. Take any $\phi \in \mathcal{D}(\Omega)$. It follows, as $n \to \infty$, that

$$\int u_n(x+iy)\phi(x, y)dxdy \to \int u(x+iy)\phi(x, y)dxdy,$$

thus $u_n \to u$ in $\mathcal{D}'(\Omega)$. As a linear differential operator with constant coefficients the operator $\bar{\partial} : \mathcal{D}'(\Omega) \to \mathcal{D}'(\Omega)$ is continuous and therefore $\bar{\partial}u = \lim_{n \to \infty} \bar{\partial}u_n = \lim_{n \to \infty} 0 = 0$. We conclude $u \in H(\Omega)$. This proves the second part.

If $u \in H(\Omega)$ has no zeroes in Ω, then $\frac{1}{u}$ is a well-defined continuous function on Ω and the differentiation rules imply $\bar{\partial}\frac{1}{u} = -\frac{1}{u^2}\bar{\partial}u = 0$, hence $\frac{1}{u} \in H(\Omega)$.

The final part follows by a straightforward application of the chain rule and is in the Exercises. □

It is easy to give many *examples of holomorphic functions*. Naturally, every constant function $u = a \in \mathbb{C}$ satisfies $\bar{\partial}a = 0$ and thus all constants belong to $H(\Omega)$. Next consider the function $z \mapsto z$. It follows that $\bar{\partial}z = \frac{1}{2}(1 - 1) = 0$, hence this function belongs to $H(\Omega)$ too. Since we learned in Theorem 9.1.1 that $H(\Omega)$ is a complex algebra, it follows immediately that all polynomials $P(z) = \sum_{n=0}^{m} a_n z^n$, $a_n \in \mathbb{C}$, belong to $H(\Omega)$.

According to Theorem 9.1.1 the algebra $H(\Omega)$ is complete for the topology of uniform convergence on all compact sets $K \subset \Omega$. Therefore all functions $u : \Omega \to \mathbb{C}$ belong to $H(\Omega)$ which are the limit of a sequence of polynomials for this topology. We investigate this case in more detail.

Recall some properties of power series (see for instance [Rem98]). A power series $\sum_{n=0}^{\infty} a_n (z-c)^n$ with center $c \in \mathbb{C}$ and coefficients $a_n \in \mathbb{C}$ has a unique disk of convergence $B_R(c) = \{z \in \mathbb{C} : |z - c| < R\}$ where the radius of convergence R is determined by the coefficients $\{a_n : n = 0, 1, 2, \ldots\}$. On every compact subset K of the disk of convergence the power series converges uniformly and thus defines a complex valued function u on $B_R(c)$. From our earlier considerations it follows that u is holomorphic on this disk.

Let $\Omega \subset \mathbb{C}$ be a nonempty open set. A function $u : \Omega \to \mathbb{C}$ is said to be *analytic* on Ω if, and only if, for every point $c \in \Omega$ there is some disk $B_r(c) \subset \Omega$ such that on this disk the function u is given by some power series, i.e., $u(z) = \sum_{n=0}^{\infty} a_n(z-c)^n$ for all $z \in B_r(c)$. Since every compact subset $K \subset \Omega$ can be covered by a finite number of such disks of convergence, it follows that every analytic function is holomorphic, i.e.,

$$A(\Omega) \subseteq H(\Omega)$$

where $A(\Omega)$ denotes the set of all analytic functions on Ω. In the following section we will learn that actually every holomorphic function is analytic so that these two sets of functions are the same.

9.2 Cauchy theory

According to our definition a holomorphic function u on an open set is a function which solves the differential equation $\bar{\partial} u = 0$. If this is combined with a well-known result from classical analysis, the Green–Riemann formula, the basic result of the Cauchy theory follows easily.

Theorem 9.2.1 (Theorem of Cauchy) *Let $\Omega \subset \mathbb{C}$ be a nonempty open set and B be an open set such that the closure $\bar{B}$ of B is contained in Ω. Assume that the boundary ∂B of B is piecewise smooth (i.e., piecewise of class $\mathbb{C}^1$). Then, for all $u \in H(\Omega)$,*

$$\oint_{\partial B} u(z) dz = 0. \tag{9.4}$$

Proof. The proof of Cauchy's theorem is a simple application of the Green–Riemann formula (9.2). □

Theorem 9.2.2 (Cauchy's integral formula I) *Let $\Omega \subset \mathbb{C}$ be a nonempty open set and $K \subset \Omega$ a compact subset whose boundary (with standard orientation) $\Gamma = \partial K$ is piecewise smooth. Then, for every $u \in H(\Omega)$, one has*

$$\frac{1}{2i\pi} \oint_\Gamma \frac{u(z)}{z - z_0} dz = \begin{cases} 0 & \text{if } z_0 \notin K, \\ u(z_0) & \text{if } z_0 \in K \setminus \Gamma. \end{cases} \tag{9.5}$$

9.2 Cauchy theory

Proof. Denote by $\dot{K} = K \setminus \Gamma$ the interior of the compact set K and by χ the characteristic function of $\dot{K}$. Now, given any $u \in H(\Omega)$, introduce the regular distribution $T = \chi u$. Using $\bar{\partial} u = 0$ and again the Green–Riemann formula (9.2) we find, for any $\phi \in \mathcal{D}(\Omega)$,

$$\langle \bar{\partial} T, \phi \rangle = -\langle T, \bar{\partial} \phi \rangle = -\int_{\dot{K}} u(z) \bar{\partial} \phi(z) dx dy$$
$$= -\int_{\dot{K}} \bar{\partial}(u\phi)(z) dx dy = -\frac{1}{2i\pi} \oint_\Gamma u(z) \phi(z) dz.$$

Lemma 9.1.1 says that $\bar{\partial} \frac{1}{\pi z} = \delta$. Since T has a compact support in K, the convolution with $\frac{1}{\pi z}$ exists and the identity $T = \frac{1}{\pi} \bar{\partial} T * \frac{1}{z}$ holds. Take a test function ϕ which satisfies $\phi(z) = 1$ for all $z \in K$ and which has its support in a sufficiently small neighborhood U of K. For $z_0 \in \Omega \setminus \Gamma$ the combination of these identities yields

$$T(z_0) = \frac{1}{\pi}(\bar{\partial} T * \frac{1}{z})(z_0) = \frac{1}{\pi} \langle (\bar{\partial} T)(z), \frac{\phi(z)}{z - z_0} \rangle$$
$$= -\frac{1}{2i\pi} \oint_\Gamma u(z) \frac{\phi(z)}{z - z_0} dz = \frac{1}{2i\pi} \oint_\Gamma \frac{u(z)}{z - z_0} dz,$$

and thus Cauchy's integral formula follows. $\square$

Cauchy's integral formula (9.5) has many applications, practical and theoretical. We discuss now one of the most important applications which shows that every holomorphic function has locally a power series expansion and thus is analytic.

Theorem 9.2.3 (Cauchy's integral formulae II) *Let $\Omega \subset \mathbb{C}$ be a nonempty open set. For every $c \in \Omega$ define*

$$R = R(c) = \sup \left\{ R' > 0 : \overline{B'_R(c)} \subset \Omega \right\}. \tag{9.6}$$

Then, for every $u \in H(\Omega)$ and every $r \in (0, R)$ the following statements hold:

1. *u has a power series expansion in $B_r(c)$,*

$$u(z) = \sum_{n=0}^{\infty} a_n (z - c)^n \quad \forall z \in B_r(c);$$

   and this power series expansion converges uniformly on every compact subset $K \subset B_r(c)$;

2. *the coefficients a_n of this power series expansion are given by **Cauchy's integral formulae***

$$a_n = a_n(c) = \frac{1}{2i\pi} \oint_{|z-c|=r} \frac{u(z)}{(z-c)^{n+1}} dz \quad n = 0, 1, 2, \ldots; \tag{9.7}$$

   and these coefficients depend on c and naturally on the function u but not on the radius $r \in (0, R)$ which is used to calculate them.

Proof. Take any $c \in \Omega$ and determine $R = R(c)$ as in the theorem. Then for every $r \in (0, R)$ we know $\overline{B_r(c)} \subset \Omega$. Thus Theorem 9.2.2 applies to $K = \overline{B_r(c)}$ and $\Gamma = \partial \overline{B_r(c)} = \{z \in \mathbb{C} : |z - c| = r\}$ and hence for all $z \in B_r(c)$ one has

$$u(z) = \frac{1}{2i\pi} \oint_{|z-c|=r} \frac{u(\xi)}{\xi - z} d\xi.$$

Take any compact set $K \subset B_r(c)$. Since $|z - c| < r = |\xi - c|$ we can expand the function $\xi \mapsto \frac{1}{\xi - z}$ into a geometric series:

$$\frac{1}{\xi - z} = \frac{1}{\xi - c} \frac{1}{1 - \frac{z-c}{\xi - c}} = \frac{1}{\xi - c} \sum_{n=0}^{\infty} \left(\frac{z - c}{\xi - c}\right)^n.$$

This series converges uniformly in $\xi \in \partial B_r(c)$ and $z \in K$. Hence we can exchange the order of integration and summation in the above formula to get

$$u(z) = \sum_{n=0}^{\infty} \left[\frac{1}{2i\pi} \oint_{|\xi-c|=r} \frac{u(\xi)d\xi}{(\xi - c)^{n+1}}\right] (z - c)^n.$$

Hence the function u has a power series expansion in $B_r(c)$ with coefficients a_n given by formula (9.7). The proof that the coefficients do not depend on $r \in (0, R)$ is left as an exercise. □

Corollary 9.2.1 *Let $\Omega \subset \mathbb{C}$ be a nonempty open set.*

1. *A function u on Ω is holomorphic if, and only if, it is analytic:*

$$H(\Omega) = A(\Omega).$$

2. *The power series expansion of a holomorphic function u on Ω is unique in a given disk $B_r(c) \subset \Omega$ of convergence.*

3. *Given $c \in \Omega$ determine $R = R(c)$ according to (9.6) and choose $r \in (0, R)$. Then, for every $u \in H(\Omega)$, the following holds:*

   a) *For $n = 0, 1, 2, \ldots$ the coefficient a_n of the power series expansion of u at the point c and the nth complex derivative of u at c are related by*

   $$u^{(n)}(c) = n! a_n; \tag{9.8}$$

   b) *The nth derivative of u at c is bounded in terms of the values of u on the boundary of the disk $B_r(c)$ according to the following formula* **(Cauchy estimates)**:

   $$|a_n| r^n \leq \sup_{|z-c|=r} |u(z)|. \tag{9.9}$$

Proof. In the discussion following Theorem 9.1.1 we saw that every analytic function is holomorphic. The previous theorem shows that conversely every holomorphic function is analytic. The uniqueness of the power series expansion of a holomorphic function at a point $c \in \Omega$ was shown in Theorem 9.2.3. The Cauchy estimates are a straightforward consequence of the Cauchy formulae (9.7):

$$|a_n| \leq \frac{1}{2\pi} \oint_{|z-c|=r} \frac{|u(z)||dz|}{|z - c|^{n+1}} \leq \frac{1}{2\pi} \frac{1}{r^{n+1}} \sup_{|z-c|=r} |u(z)| 2\pi r$$

This estimate implies (9.9) and thus we conclude. □

9.3 Some properties of holomorphic functions

As a consequence of the Cauchy theory we derive some very important properties of holomorphic functions which themselves have many important applications.

Corollary 9.3.1 (Theorem of Liouville) *The only bounded functions in $H(\mathbb{C})$ are the constants, i.e., if a function u is holomorphic on all of $\mathbb{C}$ and bounded there, then u is a constant function.*

Proof. Suppose $u \in H(\mathbb{C})$ is bounded on $\mathbb{C}$ by M, i.e., $\sup_{z \in \mathbb{C}} |u(z)| = M < +\infty$. Since u is holomorphic on $\mathbb{C}$ the value of $R = R(0)$ in (9.6) is $+\infty$. Hence, in the Cauchy estimates we can choose r as large as we wish. Therefore in this case we have $|a_n| \leq \frac{M}{r^n}$ for every $r > 0$. It follows that $a_n = 0$ for $n = 1, 2, \ldots$, and thus Theorem 9.2.3 shows that u is constant. □

Corollary 9.3.2 (Fundamental Theorem of Algebra) *Suppose P is a polynomial of degree $N \geq 1$ with coefficients $a_n \in \mathbb{C}$, i.e., $P(z) = \sum_{n=0}^{N} a_n z^n$, $a_N \neq 0$. Then there are complex numbers $\{z_1, \ldots, z_N\}$, the roots of P, which are unique up to ordering such that*

$$P(z) = a_n(z - z_1) \cdots (z - z_N) \quad \forall z \in \mathbb{C}.$$

If all the coefficients of the polynomial P are real, then P has either only real roots or if complex roots exist, they occur as pairs of complex numbers which are complex conjugate to each other and have the same multiplicity; in such a case the polynomial factorizes as

$$P(z) = a_n(z - x_1) \cdots (z - x_m)|z - z_1|^2 \cdots |z - z_k|^2 \quad \forall z \in \mathbb{C}.$$

Here $x_1, \ldots, x_m$ are the real roots of P and $z_1, \overline{z_1}, \ldots, z_k, \overline{z_k}$ are the complex roots of P; hence $m + 2k = N$.

Proof. In a first and basic step we show that a polynomial which is not constant has at least one root. Suppose P is a polynomial of degree $N \geq 1$ which has no roots in $\mathbb{C}$. Then we know that the function $z \mapsto \frac{1}{P(z)}$ is holomorphic on $\mathbb{C}$.

We write the polynomial in the form

$$P(z) = a_N z^N \left[1 + \frac{a_{N-1}}{a_N z} + \cdots + \frac{a_0}{a_N z^N} \right]$$

and choose R so large that

$$\left| \frac{a_{N-1}}{a_N z} + \cdots + \frac{a_0}{a_N z^N} \right| \leq \frac{1}{2}$$

for all $|z| \geq R$. It follows that

$$|P(z)| \geq \frac{1}{2} |a_N| R^N \quad \forall |z| \geq R.$$

On the compact set $K_R = \{z \in \mathbb{C} : |z| \leq R\}$ the continuous function $|P|$ is strictly positive (since we have assumed that P has no roots), i.e.,

$$b = b_R = \inf_{z \in K_R} |P(z)| > 0.$$

It thus follows that $\frac{1}{P(z)}$ is bounded on $\mathbb{C}$:

$$\frac{1}{|P(z)|} \leq \max \left\{ \frac{1}{b}, \frac{2}{|a_N| R^N} \right\} \quad \forall z \in \mathbb{C}.$$

By Liouville's theorem (Corollary 9.3.1) we conclude that $\frac{1}{P(z)}$ and thus $P(z)$ is constant which is a contradiction to our hypothesis that the degree N of P is larger than or equal to 1. We deduce that a polynomial of degree $N \geq 1$ has at least one root, i.e., for at least one $z_0 \in \mathbb{C}$, one has $P(z_0) = 0$.

In order to complete the proof, a proof by induction with respect to the degree N has to be done. For details we refer to the Exercises where also the special case of polynomials with real coefficients is considered. □

Holomorphic functions differ from functions of class C^∞ in a very important way: If all derivatives of two holomorphic functions agree in one point, then these functions agree everywhere, if the domain is 'connected'. As we have seen earlier this is not all the case for C^∞-functions which are not holomorphic.

Theorem 9.3.1 (Identity theorem) *Suppose that $\Omega \subset \mathbb{C}$ is a nonempty open and connected set and $f, g : \Omega \to \mathbb{C}$ are two holomorphic functions. The following statements are equivalent:*

a) $f = g$;

b) The set of all points in Ω at which f and g agree, i.e., the set

$$\{z \in \Omega : f(z) = g(z)\}$$

has an accumulation point $c \in \Omega$;

c) There is a point $c \in \Omega$ in which all complex derivatives of f and g agree: $f^{(n)}(c) = g^{(n)}(c)$ for all $n = 0, 1, 2, \ldots$.

Proof. The implication a) ⇒ b) is trivial. In order to show that b) implies c), introduce the holomorphic function $h = f - g$ on Ω. According to b) the set $M = \{z \in \Omega : h(z) = 0\}$ of zeros of h has an accumulation point $c \in \Omega$. Suppose that $h^{(m)}(c) \neq 0$ for some $m \in \mathbb{N}$. We can assume that m is the smallest number with this property. Then in some open disk around c we can write $h(z) = (z-c)^m h_m(z)$ with $h_m(z) = \sum_{i=m}^\infty \frac{h^{(i)}(c)}{i!}(z-c)^{i-m}$ and $h_m(c) \neq 0$. Continuity of h_m implies that $h_m(z) \neq 0$ for all points z in some neighborhood U of c, $U \subset B$. It follows that the only point in U in which h vanishes is the point c, hence this point is an isolated point of M. This contradiction implies $h^{(n)}(c) = 0$ for all $n = 0, 1, 2 \ldots$ and statement c) holds.

For the proof of the implication c) ⇒ a) we introduce again the holomorphic function $h = f - g$ and consider, for $k = 0, 1, 2, \ldots$, the sets $N_k = \{z \in \Omega : h^{(k)}(z) = 0\}$. Since the function $h^{(k)}$ is continuous, the set N_k is closed in Ω. Hence the intersection $N = \cap_{k=0}^\infty N_k$ of these sets is closed too. But N is at the same time open: Take any $z \in N$. Since $h^{(k)}$ is holomorphic in Ω its Taylor series at z converges in some open nonempty disk B with centre z. Since $z \in N$, all Taylor coefficients of this series vanish and it follows that $h^{(k)}|B = 0$ for all $k \in \mathbb{N}$. This implies $B \subset N$ and we conclude that N is open. Since Ω is assumed to be connected and N is not empty ($c \in N$ because of c)) we conclude $N = \Omega$ and thus $f = g$. □

There are other versions and some extensions of the identity theorem for holomorphic functions, see [Rem98].

Another important application of Cauchy's integral formula (9.5) is the classification of isolated singularities of a function and the corresponding series representation. Here one says that a complex function u has an *isolated singularity at*

9.3 Some properties of holomorphic functions

a point $c \in \mathbb{C}$ if, and only if, there is some $R > 0$ such that u is holomorphic in the set
$$K_{0,R}(c) = \{z \in \mathbb{C} : 0 < |z - c| < R\}$$
which is a disk of radius R from which the center c is removed. If a function is holomorphic in such a set it allows a characteristic series representation which gives the classification of isolated singularities. This series representation is in terms of powers and inverse powers of $z - c$ and is called the Laurent expansion of u.

Theorem 9.3.2 (Laurent expansion) *For $0 \leq r < R \leq +\infty$ consider the annulus $K_{r,R}(c) = \{z \in \mathbb{C} : r < |z - c| < R\}$ with center c and radii r and R. Every function u which is holomorphic in $K_{r,R}(c)$ has the unique **Laurent expansion***

$$u(z) = \sum_{n=-\infty}^{+\infty} a_n (z - c)^n \qquad \forall z \in K_{r,R}(c) \tag{9.10}$$

which converges uniformly on every compact subset $K \subset K_{r,R}(c)$. The coefficients a_n of this expansion are given by

$$a_n = \frac{1}{2i\pi} \oint_{|t-c|=\rho} \frac{u(t)}{(t-c)^{n+1}} dt \qquad \forall n \in \mathbb{Z} \tag{9.11}$$

where $\rho \in (r, R)$ is arbitrary. These coefficients depend only on the function u and on the annulus but not on the radius $\rho \in (r, R)$.

Proof. Consider any compact set $K \subset K_{r,R}(c)$. There are radii r_i such that for all $z \in K$,
$$r < r_1 < |z - c| < r_2 < R.$$
Apply Cauchy's integral formula (9.5) to the annulus $K_{r,R}(c)$ and a given function $u \in H(K_{r,R}(c))$. This yields
$$u(z) = \frac{1}{2i\pi} \oint_{|t-c|=r_1} \frac{u(t)}{t-z} dt + \frac{1}{2i\pi} \oint_{|t-c|=r_2} \frac{u(t)}{t-z} dt \qquad \forall z \in K.$$
Uniformly in $z \in K$ and $|t - c| = r_1$, respectively $|t - c| = r_2$, one has
$$\frac{|t-c|}{|z-c|} = \frac{r_1}{|z-c|} \leq \alpha < 1 \quad \text{respectively} \quad \frac{|z-c|}{|t-c|} = \frac{|z-c|}{r_2} \leq \beta < 1.$$
The convergence of the geometric series $\sum_{n=0}^{\infty} q^n$ for $0 \leq q < 1$ ensures the uniform convergence of the series
$$\frac{1}{t-z} = -\frac{1}{z-c} \sum_{n=0}^{\infty} \left(\frac{t-c}{z-c}\right)^n \qquad \forall |t-c|=r_1; \ \forall z \in K,$$

$$\frac{1}{t-z} = \frac{1}{t-c} \sum_{n=0}^{\infty} \left(\frac{z-c}{t-c}\right)^n \qquad \forall |t-c|=r_2; \ \forall z \in K.$$

Therefore we may exchange the order of summation and integration in the above integral representation of u and obtain uniformly in $z \in K$,
$$u(z) = \sum_{n=0}^{\infty} \left[\frac{1}{2i\pi} \oint_{|t-c|=r_2} \frac{u(t) dt}{(t-c)^{n+1}} \right] (z-c)^n +$$
$$+ \sum_{n=0}^{\infty} \left[\frac{1}{2i\pi} \oint_{|t-c|=r_1} u(t)(t-c)^n dt \right] (z-c)^{-n-1}.$$

9. Holomorphic Functions

If we choose $-n - 1$ as new summation index in the second series, we arrive at the Laurent expansion (9.10) with coefficients given by (9.11). A straightforward application of (9.5) shows that the integrals

$$\oint_{|t-c|=\rho} \frac{u(t)dt}{(t-c)^m} \qquad \forall m \in \mathbb{Z}$$

are independent of the choice of $\rho \in (r, R)$ and thus we conclude. □

The announced classification of isolated singularities of a function u is based on the Laurent expansion of u at the singularities and classifies these singularities according to the number of coefficients $a_n \neq 0$ for $n \leq 0$ in the Laurent expansion. In detail one proceeds in the following way.

Suppose $c \in \mathbb{C}$ is an isolated singularity of a function u. Then there is an $R = R(u, c) > 0$ such that u is holomorphic in the annulus $K_{0,R}(c)$ and thus has a unique Laurent expansion there.

$$u(z) = \sum_{n=-\infty}^{+\infty} a_n(z-c)^n \qquad \forall z \in K_{0,R}(c).$$

One distinguishes three cases:

a) $a_n = 0$ for all $n < 0$. Then c is called a *removable singularity*. Initially u is not defined at $z = c$, but the limit $\lim_{z \to c} u(z)$ exists and is used to define the value of u at $z = c$. In this way u becomes defined and holomorphic in the disk $\{z \in \mathbb{C} : |z - c| < R\}$. A well-known example is $u(z) = \frac{\sin z}{z}$ for all $z \in \mathbb{C}, z \neq 0$. Using the power series expansion for $\sin z$ we find easily the Laurent series for u at $z = 0$ and see that $\lim_{z \to 0} u(z)$ exists.

b) There is $k \in \mathbb{N}, k > 0$, such that $a_n = 0$ for all $n \in \mathbb{Z}, n < -k$ and $a_k \neq 0$. Then the point $z = c$ is called a *pole of order k* of the function u. One has $|u(z)| \to +\infty$ as $z \to c$. A simple example is the function $u(z) = z^{-3}$ for $z \in \mathbb{C}, z \neq 0$. It has a pole of order 3 in $z = 0$.

c) $a_n \neq 0$ for infinitely many $n \in \mathbb{Z}, n < 0$. In this case the point c is called an *essential singularity* of u. As an example we mention the function $u(z) = e^{\frac{1}{z}}$ defined for all $z \in \mathbb{C}\setminus\{0\}$. The well-known power series expansion of the exponential function shows easily that the Laurent series of u at $z = 0$ is given by $\sum_{n=0}^{\infty} \frac{1}{n!} \frac{1}{z^n}$ and thus u has an essential singularity at $z = 0$.

Assume that a function u has an isolated singularity at a point c. Then, in a certain annulus $K_{0,R}(c)$ it has a unique Laurent expansion (9.10) where the coefficients a_n have the explicit integral representation (9.11). For $n = -1$ this integral representation is

$$a_{-1} = \frac{1}{2i\pi} \oint_{|z-c|=\rho} u(z)dz \qquad (9.12)$$

for a suitable radius ρ. This coefficient is called the *residue of the function u at the isolated singularity c*, usually denoted as

$$a_{-1} = \text{Res}(u, c).$$

9.3 Some properties of holomorphic functions

If c is a pole of order 1, the Laurent expansion shows that the residue can be calculated in a simple way as

$$\text{Res}(u, c) = \lim_{z \to c}(z - c)u(z). \tag{9.13}$$

In most cases it is fairly easy to determine this limit and thus the residue. This offers a convenient way to determine the value of the integral in (9.12) and is the starting point for a method which determines the values of similar path integrals.

Theorem 9.3.3 (Theorem of residues) *Suppose $\Omega \subset \mathbb{C}$ is a nonempty open set and $D \subset \Omega$ a discrete subset (this means that in every open disk $K_r(z) = \{\xi \in \mathbb{C} : |\xi - z| < r\}$ there are only a finite number of points from D). Furthermore assume that K is a compact subset of Ω such that the boundary $\Gamma = \partial K$ of K with standard mathematical orientation is piecewise smooth and does not contain any point from D. Then, for every $u \in H(\Omega \setminus D)$, the following holds:*

a) *The number of isolated singularities of u in K is finite.*

b) *Suppose $\{z_0, z_1, \ldots, z_N\}$ are the isolated singularities of u in K, then one has*

$$\oint_\Gamma u(z)dz = 2\pi i \sum_{n=0}^{N} \text{Res}(u, z_n). \tag{9.14}$$

Proof. Given a point $z \in K$, there is an open disk in Ω which contains at most one point from D since D is discrete. Since K is compact a finite number of such disks cover K. This proves part a).

Suppose that $z_0, z_1, \ldots, z_N$ are the isolated singularities of u in K. One can find radii $r_0, r_1, \ldots, r_N$ such that the closed disks $\overline{K_{r_j}}(z_j)$ are pairwise disjoint. Now choose the orientation of the boundaries $\partial K_{r_j}(z_j) = -\gamma_j$ of these disks in such a way that $\Gamma \cup \cup_{j=0}^N \gamma_j$ is the oriented boundary of some compact set $K' \subset \Omega$. By construction the function u is holomorphic in some open neighborhood of K' and thus (9.5) applies to give

$$\oint_\Gamma u(t)dt + \sum_{j=0}^N \oint_{\gamma_j} u(t)dt = 0,$$

i.e., by (9.12)

$$\oint_\Gamma u(t)dt = \sum_{j=0}^N \oint_{\partial K_{r_j}(z_j)} u(t)dt = 2\pi i \sum_{j=0}^N \text{Res}(u, z_j)$$

and we conclude. □

Remark 9.3.1 1. *Only in the case of a pole of order 1, can we calculate the residue by the simple formula (9.13). In general one has to use the Laurent series. A discussion of some other special cases in which it is relatively easy to find the residue without going to the Laurent expansion is explained in most textbooks on complex analysis.*

2. *In the case of u being the quotient of two polynomials P and Q, $u(z) = \frac{P(z)}{Q(z)}$, one has a pole of order 1 at a point $z = c$ if $Q(c) = 0$, $Q'(c) \neq 0$, and*

$P(c) \neq 0$. Then the residue of u at the point c can be calculated by formula (9.13). The result is a convenient formula

$$\text{Res}(u, c) = \lim_{z \to c} (z - c) u(z) = \lim_{z \to c} \frac{P(z)}{\frac{Q(z) - Q(c)}{z - c}} = \frac{P(c)}{Q'(c)}.$$

9.4 Exercises

1. Write a complex valued function $f : \Omega \to \mathbb{C}$ on some open set $\Omega \subset \mathbb{C}$ in terms of its real and imaginary parts, $f(x + iy) = u(x, y) + iv(x, y)$ for all $z = x + iy \in \Omega$ where u and v are real valued functions. Show: If $\bar{\partial} f(z) = 0$ on Ω, then the functions u, v satisfy the **Cauchy–Riemann equations**

$$\frac{\partial u}{\partial x}(x, y) = +\frac{\partial v}{\partial y}(x, y),$$
$$\frac{\partial u}{\partial y}(x, y) = -\frac{\partial v}{\partial x}(x, y).$$

2. Prove Part 4 of Theorem 9.1.1.

3. Show: In Cauchy's integral formula (9.7) the right-hand side is independent of r, $0 < r < R$.

4. Complete the proof of Corollary 9.3.2.

   Hints: For the case of a real polynomial prove first that $P(z) = 0$ implies $P(\bar{z}) = 0$ and observe that a complex root z and its complex conjugate $\bar{z}$ have the same multiplicity.

10
Fourier Transformation

Our goal in this chapter is to define the Fourier transformation in a setting which is as general as possible and to discuss the most important properties of this transformation. This is followed by some typical and important applications, mainly in the theory of partial differential operators with constant coefficients as they occur in physics.

If one wants to introduce the Fourier transformation on the space $\mathcal{D}'(\mathbb{R}^n)$ of all distributions on $\mathbb{R}^n$, one encounters a natural difficulty which has its origin in the fact that general distributions are not restricted in their growth when one approaches the boundary of their domain of definition. It turns out that the growth restrictions which control tempered distributions are sufficient to allow a convenient and powerful Fourier transformation on the space $\mathcal{S}'(\mathbb{R}^n)$ of all tempered distributions. Actually, the space of tempered distributions was introduced for this purpose.

The starting point of the theory of Fourier transformation is very similar to that of the theory of Fourier series. Under well-known conditions a periodic complex valued function can be represented as the sum of exponential functions of the form $a_n e^{in\kappa x}$, $n \in \mathbb{Z}$, $a_n \in \mathbb{C}$, where κ is determined by the period of the function in question. The theory of Fourier transformation aims at a similar representation without assuming periodicity, but allowing that the summation index n might have to vary continuously so that the sum is replaced by an integral.

On a formal level the transition between the two representations is achieved in the following way. Suppose that $f : \mathbb{R} \to \mathbb{C}$ is an integrable continuous function. For each $T > 0$ introduce the auxiliary function f_T with period $2T$ which is equal to f on the interval $[-T, T]$. Then f_T has a representation in terms of a Fourier series

$$f_T(x) = \frac{1}{2T} \sum_{n \in \mathbb{Z}} c_n e^{in\frac{\pi}{T}x} \quad \text{with} \quad c_n = \int_{-T}^{+T} f(x) e^{-in\frac{\pi}{T}x} dx.$$

Now introduce $\nu = n\frac{\pi}{T}$ and $a_\nu = c_n$ and rewrite the above representation as

$$f_T(x) = \frac{1}{2T} \sum_{\nu} a_\nu e^{i\nu x} \quad \text{with} \quad a_\nu = \int_{-T}^{+T} f(x) e^{-i\nu x} dx.$$

Two successive values of the summation index differ by $\frac{\pi}{T}$; thus formally we get in the limit $T \to +\infty$,

$$f(x) = \frac{1}{2\pi} \int_{\mathbb{R}} a_\nu e^{i\nu x} d\nu \quad \text{with} \quad a_\nu = \int_{\mathbb{R}} f(x) e^{-i\nu x} dx.$$

The following section will give a precise meaning to these relations.

In order to be able to define and to study the Fourier transformation for distributions, we begin by establishing the basic properties of the Fourier transformation on various spaces of functions. In the first section we introduce and study the Fourier transformation on the space $L^1(\mathbb{R}^n)$ of Lebesgue integrable functions. Recall that $L^1(\mathbb{R}^n)$ denotes the space (of equivalence classes) of measurable functions $f : \mathbb{R}^n \to \mathbb{C}$ which are absolutely integrable, i.e., for which

$$\|f\|_1 = \int_{\mathbb{R}^n} |f(x)| dx$$

is finite. The main result of Section 2 is that the Fourier transformation is an isomorphism of the topological vector space $\mathcal{S}(\mathbb{R}^n)$ which is the test function space for tempered distributions. This easily implies that the Fourier transform can be defined on the space of tempered distributions by duality. Section 3 then establishes the most important properties of the Fourier transformation for tempered distributions. In the final section on applications we come back to the study of linear partial differential operators with constant coefficients and the improvements of the solution theory one has in the context of tempered distributions. There we will learn, among other things, that with the help of the Fourier transformation it is often fairly easy to find elementary solutions of linear partial differential operators with constant coefficients.

10.1 Fourier transformation for integrable functions

For $x = (x_1, \ldots, x_n) \in \mathbb{R}^n$ and $p = (p_1, \ldots, p_n) \in \mathbb{R}^n$, denote by $p \cdot x = p_1 x_1 + \cdots + p_n x_n$ the Euclidean inner product. Since for all $x, p \in \mathbb{R}^n$ one has $|e^{ip \cdot x}| = 1$, all the functions $x \mapsto e^{ip \cdot x} f(x)$, $p \in \mathbb{R}^n$, $f \in L^1(\mathbb{R})$, are integrable and thus we get a well-defined function $\tilde{f} : \mathbb{R}^n \to \mathbb{C}$ by defining, for all $p \in \mathbb{R}^n$,

$$\tilde{f}(p) = (2\pi)^{-\frac{n}{2}} \int_{\mathbb{R}^n} e^{-ip \cdot x} f(x) dx. \tag{10.1}$$

This function $\tilde f$ is called the *Fourier transform of f* and the map defined on $L^1(\mathbb{R}^n)$ by $\mathcal{F}f \equiv \mathcal{F}(f) = \tilde f$ is called the *Fourier transformation* (on $L^1(\mathbb{R}^n)$).

Remark 10.1.1 *The choice of the normalization factor $(2\pi)^{-\frac{n}{2}}$ and the choice of the sign of the argument of the exponential function in the definition of the Fourier transform are not uniform in the literature (see for instance [Cha89, Don69, DS58, Hör83a, MS57]). Each choice has some advantage and some disadvantage.*

In our normalization the Fourier transform of Dirac's delta distribution on $\mathbb{R}^n$ will be $\mathcal{F}\delta = (2\pi)^{-\frac{n}{2}}$.

The starting point of our investigation of the properties of the Fourier transform is the following basic result.

Lemma 10.1.1 (Riemann–Lebesgue) *The Fourier transform $\tilde f = \mathcal{F}f$ of $f \in L^1(\mathbb{R}^n)$ has the following properties:*

a) *$\tilde f$ is a continuous and bounded function on $\mathbb{R}^n$.*

b) *$\mathcal{F} : L^1(\mathbb{R}^n) \to L^\infty(\mathbb{R}^n)$ is a continuous linear map. One has the following bound:*
$$\|\mathcal{F}f\|_\infty \equiv \sup_{p\in\mathbb{R}^n} |\tilde f(p)| \leq (2\pi)^{-\frac{n}{2}} \|f\|_1.$$

c) *$\tilde f$ vanishes at infinity, i.e.,*
$$\lim_{|p|\to\infty} \tilde f(p) = 0.$$

Proof. The bound given in part b) is evident from the definition (10.1) of the Fourier transformation. The basic rules of Lebesgue integration imply that $\mathcal{F}$ is a linear map from $L^1(\mathbb{R}^n)$ into $L^\infty(\mathbb{R}^n)$. In order to prove continuity of $\tilde f$ at any point $p \in \mathbb{R}^n$, for any $f \in L^1(\mathbb{R}^n)$, consider any sequence of points p_k which converges to p. It follows that
$$\lim_{k\to\infty} \tilde f(p_k) = (2\pi)^{-\frac{n}{2}} \lim_{k\to\infty} \int_{\mathbb{R}^n} e^{-ip_k\cdot x} f(x)dx = (2\pi)^{-\frac{n}{2}} \int_{\mathbb{R}^n} e^{-ip\cdot x} f(x)dx = \tilde f(p)$$
since $e^{-ip_k\cdot x} f(x) \to e^{-ip\cdot x} f(x)$ as $k \to \infty$, for almost all $x \in \mathbb{R}^n$ and $|e^{-ip_k\cdot x} f(x)| \leq |f(x)|$ for all $x \in \mathbb{R}^n$ and all $k \in \mathbb{N}$, so that Lebesgue's theorem on dominated convergence implies the convergence of the integrals. The sequence test for continuity now proves continuity of $\tilde f$ at p. Thus continuity of $\tilde f$ follows. This proves parts a) and b).

The proof of part c) is more involved. We start with the observation $e^{-i\pi} = -1$ and deduce, for all $p \in \mathbb{R}^n$, $p \neq 0$:
$$(2\pi)^{\frac{n}{2}} \tilde f(p) = -\int_{\mathbb{R}^n} e^{-ip\cdot(x+\frac{\pi p}{p^2})} f(x)dx = -\int_{\mathbb{R}^n} e^{-ip\cdot x} f(x - \frac{\pi p}{p^2})dx.$$

Recall the definition of translation by a vector a of a function f, $f_a(x) = f(x - a)$ for all $x \in \mathbb{R}^n$. Then with $a = \frac{\pi p}{p^2}$ for $p \neq 0$, we can write
$$(\mathcal{F}f)(p) = \frac{1}{2}[(\mathcal{F}f)(p) - (\mathcal{F}f_a)(p)]|_{a=\frac{\pi p}{p^2}},$$
hence, using linearity of $\mathcal{F}$ and the estimate of part b), it follows that
$$|(\mathcal{F}f)(p)| \leq \frac{1}{2}(2\pi)^{-\frac{n}{2}} \|f - f_{\frac{\pi p}{p^2}}\|_1.$$

This shows that one can prove part c) by showing

$$\lim_{a \to 0} \|f - f_a\|_1 = 0 \quad \forall f \in L^1(\mathbb{R}^n),$$

i.e., translations act continuously on $L^1(\mathbb{R}^n)$. This is a well known result in the theory of Lebesgue integrals. In the Exercises one is asked to prove this result, first for continuous functions with compact support and then for general elements in $L^1(\mathbb{R}^n)$. This concludes the proof. □

In general it is not so easy to calculate the Fourier transform $\tilde{f}$ of a function f in $L^1(\mathbb{R}^n)$ explicitly. We give now a few examples where this calculation is straightforward. A more comprehensive list will follow at the end of this chapter.

Example 10.1.1 1. Denote by $\chi_{[-a,a]}$ the characteristic function of the symmetric interval $[-a, a]$, that is, $\chi_{[-a,a]}(x) = 1$ for $x \in [-a, a]$ and $\chi_{[-a,a]}(x) = 0$ otherwise. Clearly this function is integrable and for $\mathcal{F}\chi_{[-a,a]}$ we find, for any $p \in \mathbb{R} \setminus \{0\}$:

$$\begin{aligned}\tilde{\chi}_{[-a,a]}(p) &= (2\pi)^{-\frac{1}{2}} \int_{\mathbb{R}} e^{-ipx} \chi_{[-a,a]}(x) dx = (2\pi)^{-\frac{1}{2}} \int_{-a}^{+a} e^{-ipx} dx \\ &= (2\pi)^{-\frac{1}{2}} \frac{e^{-ipx}}{-ip} \Big|_{-a}^{+a} = \frac{2}{\sqrt{2\pi}} \frac{\sin ap}{p}.\end{aligned}$$

It is easy to see that the apparent singularity at $p = 0$ is removable.

2. Consider the function $f(x) = e^{-x^2}$. f is certainly integrable and one has $\int_{\mathbb{R}} e^{-x^2} dx = \sqrt{\pi}$. In order to calculate the Fourier transform of this function we have to rely on Cauchy's Integral Theorem 9.2.1 applied to the function $z \mapsto e^{-z^2}$ which is holomorphic on the complex plane $\mathbb{C}$.

$$\begin{aligned}(2\pi)^{\frac{1}{2}} \tilde{f}(p) &= \int_{-\infty}^{+\infty} e^{-ipx} e^{-x^2} dx = e^{-\frac{p^2}{4}} \int_{-\infty}^{+\infty} e^{-(x+ip)^2} dx \\ &= e^{-\frac{p^2}{4}} \int_{C_p} e^{-z^2} dz \quad C_p = \{z = x + ip : x \in \mathbb{R}\} \\ &= e^{-\frac{p^2}{4}} \int_{C_0} e^{-z^2} dz \quad C_0 = \{z = x : x \in \mathbb{R}\} \\ &= e^{-\frac{p^2}{4}} \sqrt{\pi},\end{aligned}$$

and thus we conclude that

$$\mathcal{F}(e^{-x^2})(p) = \frac{1}{\sqrt{2}} e^{-\frac{p^2}{4}}.$$

3. For some number $a > 0$ define the integrable function $f(x) = e^{-a|x|}$ for $x \in \mathbb{R}$. Its L^1-norm is $\|f\|_1 = \frac{2}{a}$. Its Fourier transform $\tilde{f}$ can be calculated as follows, for all $p \in \mathbb{R}$:

$$\begin{aligned}(2\pi)^{\frac{1}{2}} \tilde{f}(p) &= \int_{-\infty}^{+\infty} e^{-ipx} e^{-a|x|} dx = \int_{-\infty}^{0} e^{-ipx+ax} dx + \int_{0}^{+\infty} e^{-ipx-ax} dx \\ &= \frac{e^{ax-ipx}}{a-ip} \Big|_{-\infty}^{0} + \frac{e^{-ax-ipx}}{-a-ip} \Big|_{0}^{+\infty} = \frac{1}{a-ip} + \frac{1}{a+ip}.\end{aligned}$$

We rewrite this as

$$\mathcal{F}\left(e^{-a|\cdot|}\right)(p) = \frac{1}{\sqrt{2\pi}} \frac{2a}{a^2 + p^2}.$$

The following proposition collects a number of basic properties of the Fourier transformation. These properties say how the Fourier transformation acts on the translation, scaling, multiplication and differentiation of functions. In addition we learn that the Fourier transformation transforms a convolution product into an ordinary pointwise product. These properties are the starting point of the analysis of the Fourier transformation on the test function space $\mathcal{S}(\mathbb{R}^n)$ addressed in the next section and are deduced from the Riemann–Lebesgue lemma in a straightforward way.

Proposition 10.1.1 1. *For $f \in L^1(\mathbb{R}^n)$ and $a \in \mathbb{R}^n$ the translation by a is defined as $f_a(x) = f(x-a)$ for almost all $x \in \mathbb{R}^n$. These translations and the multiplication by a corresponding exponential function are related under the Fourier transformation according to the following formulae:*

  a) $\mathcal{F}(e^{iax} f)(p) = (\mathcal{F}f)_a(p) \quad \forall p \in \mathbb{R}^n,$
  b) $(\mathcal{F}f_a)(p) = e^{iap}(\mathcal{F}f)(p) \quad \forall p \in \mathbb{R}^n.$

2. *For any $\lambda > 0$ define the scaled function f_λ by $f_\lambda(x) = f(\frac{x}{\lambda})$ for almost all $x \in \mathbb{R}^n$. Then, for $f \in L^1(\mathbb{R}^n)$, one has*

$$(\mathcal{F}f_\lambda)(p) = \lambda^n (\mathcal{F}f)(p) \quad \forall p \in \mathbb{R}^n.$$

3. *For all $f, g \in L^1(\mathbb{R}^n)$ one has $f * g \in L^1(\mathbb{R}^n)$ and*

$$\mathcal{F}(f * g) = (2\pi)^{\frac{n}{2}} (\mathcal{F}g) \cdot (\mathcal{F}g).$$

4. *Suppose that $f \in L^1(\mathbb{R}^n)$ satisfies $x_j \cdot f \in L^1(\mathbb{R}^n)$ for some $j \in \{1, 2, \ldots, n\}$. Then the Fourier transform $\mathcal{F}f$ of f is continuously differentiable with respect to the variable p_j and one has*

$$\frac{\partial}{\partial p_j}(\mathcal{F}f)(p) = \mathcal{F}(-ix_j \cdot f)(p) \quad \forall p \in \mathbb{R}^n.$$

5. *Suppose that $f \in L^1(\mathbb{R}^n)$ has a derivative with respect to the variable x_j which is integrable, $\frac{\partial f}{\partial x_j} \in L^1(\mathbb{R}^n)$ for some $j \in \{1, 2, \ldots, n\}$. Then the following holds:*

$$\mathcal{F}(\frac{\partial f}{\partial x_j})(p) = ip_j (\mathcal{F}f)(p) \quad \text{and} \quad |p_j (\mathcal{F}f)(p)| \le \|\frac{\partial f}{\partial x_j}\|_1 \quad \forall p \in \mathbb{R}^n.$$

Proof. The proof of the first two properties is straightforward and is done in the Exercises.

To prove the relation for the convolution product we apply Fubini's theorem on the exchange of the order of integration to conclude $\|f * g\|_1 \leq \|f\|_1 \|g\|_1$ for all $f, g \in L^1(\mathbb{R}^n)$, hence $f * g \in L^1(\mathbb{R}^n)$. The same theorem and the first property justify the following calculations for all fixed $p \in \mathbb{R}^n$:

$$(2\pi)^{\frac{n}{2}} \mathcal{F}(f * g)(p) = \int_{\mathbb{R}^n} e^{-ip \cdot x} (f * g)(x) dx$$

$$= \int_{\mathbb{R}^n} \left(\int_{\mathbb{R}^n} f(x-y) g(y) dy \right) dx = \int_{\mathbb{R}^n} \left(\int_{\mathbb{R}^n} e^{-ip \cdot x} f(x-y) dx \right) g(y) dy$$

$$= (2\pi)^{\frac{n}{2}} \int_{\mathbb{R}^n} e^{-ip \cdot y} (\mathcal{F}f)(p) g(y) dy = (2\pi)^n (\mathcal{F}f)(p)(\mathcal{F}g)(p).$$

Now the third property follows easily.

In order to prove differentiability of $\mathcal{F}f$ under the assumptions stated above, take any $p \in \mathbb{R}^n$ and denote by $e_j = (0, \ldots, 0, 1, 0, \ldots, 0)$ the standard unit vector in $\mathbb{R}^n$ in coordinate direction j. By definition, for all $h \in \mathbb{R}$, $h \neq 0$, we find

$$(2\pi)^{\frac{n}{2}} \frac{\tilde{f}(p + h e_j) - \tilde{f}(p)}{h} = \int_{\mathbb{R}^n} \frac{e^{-i(p+h e_j) \cdot x} - e^{-ip \cdot x}}{h} f(x) dx.$$

For arbitrary but fixed $x \in \mathbb{R}^n$ we know

$$\lim_{h \to 0} \frac{e^{-i(p+h e_j) \cdot x} - e^{-ip \cdot x}}{h} = -i x_j e^{-ip \cdot x}.$$

Furthermore the estimate

$$\left| \frac{e^{-i(p+h e_j) \cdot x} - e^{-ip \cdot x}}{h} \right| \leq |x_j| \qquad \forall x, p \in \mathbb{R}^n$$

is well known. Thus a standard application of Lebesgue's theorem of dominated convergence implies, taking the hypothesis $x_j f \in L^1(\mathbb{R}^n)$ into account,

$$\lim_{h \to 0} \frac{\tilde{f}(p + h e_j) - \tilde{f}(p)}{h} = (2\pi)^{-\frac{n}{2}} \int_{\mathbb{R}^n} (-i x_j) e^{-ip \cdot x} f(x) dx,$$

and we conclude

$$\frac{\partial \tilde{f}}{\partial p_j}(p) = \mathcal{F}(-i x_j f)(p) \qquad \forall p \in \mathbb{R}^n.$$

This partial derivative is continuous by the Riemann–Lebesgue lemma and thus the fourth property follows.

In order to prove the fifth property we start with the observation

$$f \in L^1(\mathbb{R}) \quad \text{and} \quad f' \in L^1(\mathbb{R}) \quad \Rightarrow \quad \lim_{|x| \to \infty} f(x) = 0.$$

This is shown in the Exercises. Now we calculate

$$i p_j (\mathcal{F}f)(p) = (2\pi)^{-\frac{n}{2}} \int_{\mathbb{R}^n} i p_j e^{-ip \cdot x} f(x) dx = -(2\pi)^{-\frac{n}{2}} \int_{\mathbb{R}^n} \frac{\partial}{\partial x_j} (e^{-ip \cdot x}) f(x) dx$$

and perform a partial integration with respect to x_j. By the above observation the boundary terms vanish under our hypotheses and thus this partial integration yields

$$(2\pi)^{-\frac{n}{2}} \int_{\mathbb{R}^n} e^{-ip \cdot x} \frac{\partial f}{\partial x_j}(x) dx = \mathcal{F}(\frac{\partial f}{\partial x_j})(p).$$

We conclude by Lemma 10.1.1. □

Denote by $\mathcal{C}_b(\mathbb{R}^n)$ the space of all bounded continuous functions $f : \mathbb{R}^n \to \mathbb{C}$ which vanish at infinity as expressed in the Riemann–Lebesgue lemma. Then this lemma shows that the Fourier transformation $\mathcal{F}$ maps the space $L^1(\mathbb{R}^n)$ into $\mathcal{C}_b(\mathbb{R}^n)$. A natural and very important question is whether this map has an inverse and what this inverse is. In order to answer these questions some preparations are necessary.

Lemma 10.1.2 1. For all $f, g \in L^1(\mathbb{R}^n)$ the following identity holds:

$$\int_{\mathbb{R}^n} f(x)(\mathcal{F}g)(x)dx = \int_{\mathbb{R}^n} (\mathcal{F}f)(y)g(y)dy.$$

2. Suppose $f, g \in L^1(\mathbb{R}^n)$ are continuous and bounded and their Fourier transforms $\mathcal{F}f$, $\mathcal{F}g$ belong to $L^1(\mathbb{R}^n)$ too. Then one has

$$f(0) \int_{\mathbb{R}^n} (\mathcal{F}g)(x)dx = g(0) \int_{\mathbb{R}^n} (\mathcal{F}f)(y)dy.$$

Proof. If $f, g \in L^1(\mathbb{R}^n)$, then the function $(x, y) \mapsto e^{-ix \cdot y} f(x)g(y)$ belongs to $L^1(\mathbb{R}^n \times \mathbb{R}^n)$ and thus Fubini's theorem implies

$$I_1 = (2\pi)^{-\frac{n}{2}} \int_{\mathbb{R}^n} \left(\int_{\mathbb{R}^n} e^{-ix \cdot y} f(x)g(y)dy \right) dx = (2\pi)^{-\frac{n}{2}} \int_{\mathbb{R}^n} \int_{\mathbb{R}^n} e^{-ix \cdot y} f(x)g(y)dxdy$$

$$= (2\pi)^{-\frac{n}{2}} \int_{\mathbb{R}^n} \left(\int_{\mathbb{R}^n} e^{-ix \cdot y} f(x)g(y)dx \right) dy = I_2.$$

According to the definition of the Fourier transformation we have

$$I_1 = \int_{\mathbb{R}^n} f(x)(\mathcal{F}g)(x)dx, \qquad I_2 = \int_{\mathbb{R}^n} (\mathcal{F}f)(y)g(y)dy.$$

Thus the identity $I_1 = I_2$ proves the first part.

Next apply the identity of the first part to $f, g_\lambda \in L^1(\mathbb{R}^n)$, for $g \in L^1(\mathbb{R}^n)$ and $g_\lambda(y) = g(\frac{y}{\lambda})$, $\lambda > 0$, to get

$$\int_{\mathbb{R}^n} f(x)(\mathcal{F}g_\lambda)(x)dx = \int_{\mathbb{R}^n} (\mathcal{F}f)(y)g_\lambda(y)dy \qquad \forall \lambda > 0.$$

The second part of Proposition 10.1.1 says $(\mathcal{F}g_\lambda)(x) = \lambda^n (\mathcal{F}g)(\lambda x)$. This implies

$$\int_{\mathbb{R}^n} f(x)\lambda^n (\mathcal{F}g)(\lambda x)dx = \int_{\mathbb{R}^n} (\mathcal{F}f)(y)g(\frac{y}{\lambda})dy \qquad \forall \lambda > 0.$$

Now we use the additional assumptions on f, g to determine the limit $\lambda \to \infty$ of this identity. Since f is continuous and bounded and since $\mathcal{F}g \in L^1(\mathbb{R}^n)$, a simple application of Lebesgue's dominated convergence theorem proves, by changing variables, $\xi = \lambda x$,

$$\lim_{\lambda \to \infty} \int_{\mathbb{R}^n} f(x)\lambda^n (\mathcal{F}g)(\lambda x)dx = \lim_{\lambda \to \infty} \int_{\mathbb{R}^n} f(\frac{\xi}{\lambda})(\mathcal{F}g)(\xi)d\xi = \int_{\mathbb{R}^n} f(0)(\mathcal{F}g)(\xi)d\xi.$$

Similarly, the limit of the right-hand side is determined:

$$\lim_{\lambda \to \infty} \int_{\mathbb{R}^n} (\mathcal{F}f)(y)g(\frac{y}{\lambda})dy = \int_{\mathbb{R}^n} (\mathcal{F}f)(y)g(0)dy.$$

Thus the identity of the second part follows. □

Theorem 10.1.2 (Inverse Fourier transformation) 1. On $L^1(\mathbb{R}^n)$ define a map $\mathcal{L}$ by

$$(\mathcal{L}f)(x) = (2\pi)^{-\frac{n}{2}} \int_{\mathbb{R}^n} e^{ix \cdot p} f(p)dp \qquad \forall p \in \mathbb{R}^n.$$

This map $\mathcal{L}$ maps $L^1(\mathbb{R}^n)$ into $\mathcal{C}_b(\mathbb{R}^n)$ and satisfies

$$(\mathcal{L}f)(x) = (\mathcal{F}f)(-x) \qquad \forall x \in \mathbb{R}^n. \tag{10.2}$$

2. On the space of continuous bounded functions f such that f and $\mathcal{F}f$ belong to $L^1(\mathbb{R}^n)$ one has

$$\mathcal{L}\mathcal{F}f = f \quad \text{and} \quad \mathcal{F}\mathcal{L}f = f;$$

hence on this space of functions, $\mathcal{L}$ is the inverse of the Fourier transformation $\mathcal{F}$.

Proof. The proof of the first part is obvious. For the proof of the second part we observe that for every $x \in \mathbb{R}^n$ the translated function f_{-x} has the same properties as the function f and that the relation $\mathcal{F}(f_{-x}) = e_x \cdot (\mathcal{F}f)$ holds where e_x denotes the exponential function $e_x(p) = e^{ix \cdot p}$. Now apply the second part of the Lemma to the function f_{-x} and any $g \in L^1(\mathbb{R}^n)$ which is bounded and continuous and for which $\mathcal{F}g$ belongs to $L^1(\mathbb{R}^n)$ to obtain

$$f_{-x}(0) \int_{\mathbb{R}^n} (\mathcal{F}g)(p)dp = g(0) \int_{\mathbb{R}^n} (\mathcal{F}f_{-x})(p)dp = g(0) \int_{\mathbb{R}^n} e_x(p)(\mathcal{F}f)(p)dp,$$

or, by taking $f_{-x}(0) = f(x)$ into account,

$$f(x) \int_{\mathbb{R}^n} (\mathcal{F}g)(p)dp = g(0) \int_{\mathbb{R}^n} e^{ix \cdot p}(\mathcal{F}f)(p)dp = g(0)(2\pi)^{\frac{n}{2}}(\mathcal{L}(\mathcal{F}f))(x).$$

Next choose a special function g which satisfies all our hypotheses and for which we can calculate the quantities involved explicitly: We choose for instance ($x = (x_1, \ldots, x_n)$)

$$g(x) = \prod_{k=1}^{n} e^{-a|x_k|} \quad a > 0.$$

In the Exercises we show $I(g) = \int (\mathcal{F}g)(p)dp = (2\pi)^{\frac{n}{2}}$ and thus we deduce $f(x) = (\mathcal{L}(\mathcal{F}f))(x)$ for all $x \in \mathbb{R}^n$. With the help of the first part the second identity follows easily: For all $p \in \mathbb{R}^n$ one has

$$\begin{aligned}(\mathcal{F}(\mathcal{L}f))(p) &= (2\pi)^{-\frac{n}{2}} \int_{\mathbb{R}^n} e^{-ip \cdot x}(\mathcal{L}f)(x)dx \\ &= (2\pi)^{-\frac{n}{2}} \int_{\mathbb{R}^n} e^{ip \cdot x}(\mathcal{L}f)(-x)dx = (\mathcal{L}(\mathcal{F}f))(p).\end{aligned}$$

$\square$

10.2 Fourier transformation on $\mathcal{S}(\mathbb{R}^n)$

As indicated earlier our goal in this chapter is to extend the definition and the study of the Fourier transformation on a suitable space of distributions. Certainly, this extension has to be done in such a way that it is compatible with the embedding of integrable functions into the space of distributions and the Fourier transformation on integrable functions we have studied in the previous section. From the Riemann–Lebesgue lemma it follows that $\tilde{f} = \mathcal{F}f \in L^1_{loc}(\mathbb{R}^n)$ whenever $f \in L^1(\mathbb{R}^n)$. Thus the regular distribution $I_{\mathcal{F}f}$ is well defined. In the Exercises we show

$$\langle I_{\mathcal{F}f}, \phi \rangle = \langle I_f, \mathcal{F}\phi \rangle \quad \forall \phi \in \mathcal{D}(\mathbb{R}^n).$$

If $\mathcal{F}'$ denotes the Fourier transformation on distributions we want to define, the compatibility with the embedding requires

$$\mathcal{F}' I_f = I_{\mathcal{F}f} \quad \forall f \in L^1(\mathbb{R}^n).$$

Accordingly one should define $\mathcal{F}'$ as follows:

$$\langle \mathcal{F}'T, \phi \rangle = \langle T, \mathcal{F}\phi \rangle \qquad \forall \phi \in \mathcal{T}(\mathbb{R}^n), \quad \forall T \in \mathcal{T}'(\mathbb{R}^n) \qquad (10.3)$$

where $\mathcal{T}(\mathbb{R}^n)$ denotes the test function space of the distribution space $\mathcal{T}'(\mathbb{R}^n)$ on which one can define the Fourier transformation naturally.

In the Exercises we show: If $\phi \in \mathcal{D}(\mathbb{R}^n)$, $\phi \neq 0$, then $\mathcal{F}\phi$ is an entire analytic function different from 0 and thus does not belong to $\mathcal{D}(\mathbb{R}^n)$ so that the right-hand side of equation (10.3) is not defined in general in this case. We conclude that we cannot define the Fourier transformation $\mathcal{F}'$ naturally on $\mathcal{D}'(\mathbb{R}^n)$.

Equation (10.3) also indicates that the test function space $\mathcal{T}(\mathbb{R}^n)$ should have the property that the Fourier transformation maps this space into itself and is continuous in order that this definition be effective. In this section we will learn that this is the case for the test function space $\mathcal{T}(\mathbb{R}^n) = \mathcal{S}(\mathbb{R}^n)$ and thus the space of tempered distributions becomes the natural and effective distribution space on which one studies the Fourier transformation.

Recall that the elements of the test function space $\mathcal{S}(\mathbb{R}^n)$ of strongly decreasing C^∞-functions are characterized by condition (2.10). An equivalent way is to say: A function $\phi \in C^\infty(\mathbb{R}^n)$ belongs to $\mathcal{S}(\mathbb{R}^n)$ if and only if

$$\forall \alpha \in \mathbb{N}^n \; \forall \beta \in \mathbb{N}^n \; \exists C_{\alpha,\beta} \in \mathbb{R}_+ \; \forall x \in \mathbb{R}^n \; |x^\beta D^\alpha \phi(x)| \leq C_{\alpha,\beta}. \qquad (10.4)$$

Recall furthermore that the topology on $\mathcal{S}(\mathbb{R}^n)$ is defined by the norms $p_{m,l}$, $m, l = 0, 1, 2, \ldots$, where

$$p_{m,l}(\phi) = \sup\left\{ (1+x^2)^{\frac{m}{2}} |D^\alpha \phi(x)| : x \in \mathbb{R}^n, \; |\alpha| \leq l \right\}.$$

An easy consequence is the following invariance property of $\mathcal{S}(\mathbb{R}^n)$:

$$\phi \in \mathcal{S}(\mathbb{R}^n), \; \alpha, \beta \in \mathbb{N}^n \Rightarrow x^\beta D^\alpha \phi \in \mathcal{S}(\mathbb{R}^n)$$

and

$$p_{m,l}(x^\beta D^\alpha \phi) \leq p_{m+|\beta|,l+|\alpha|}(\phi). \qquad (10.5)$$

In the previous section we learned that the Fourier transformation is invertible on a certain subspace of $L^1(\mathbb{R}^n)$. Here we are going to show that the test function space $\mathcal{S}(\mathbb{R}^n)$ is contained in this subspace. As a first step we observe that $\mathcal{S}(\mathbb{R}^n)$ is continuously embedded into $L^1(\mathbb{R}^n)$ by the identity map:

$$\mathcal{S}(\mathbb{R}^n) \subset L^1(\mathbb{R}^n), \qquad \|\phi\|_1 \leq C p_{n+1,0}(\phi) \quad \forall \phi \in \mathcal{S}(\mathbb{R}^n). \qquad (10.6)$$

Here the embedding constant C depends only on the dimension n:

$$C = \int_{\mathbb{R}^n} \frac{dx}{(1+x^2)^{\frac{n+1}{2}}}.$$

This is shown in the Exercises.

Theorem 10.2.1 (Fourier transformation on $\mathcal{S}(\mathbb{R}^n)$) 1. *The Fourier transformation $\mathcal{F}$ is an isomorphism on $\mathcal{S}(\mathbb{R}^n)$, i.e., a continuous bijective mapping with continuous inverse.*

2. *The inverse of $\mathcal{F}$ is the map $\mathcal{L}$ introduced in equation (10.2).*

3. *The following relations hold for all $\phi \in \mathcal{S}(\mathbb{R}^n)$, $p \in \mathbb{R}^n$ and $\alpha \in \mathbb{N}^n$:*

   a) $D^\alpha (\mathcal{F}\phi)(p) = \mathcal{F}((-ix)^\alpha \phi)(p);$

   b) $\mathcal{F}(D^\alpha \phi)(p) = (ip)^\alpha (\mathcal{F}\phi)(p).$

Proof. In a first step we show that the Fourier Transformation $\mathcal{F}$ is a continuous linear map from $\mathcal{S}(\mathbb{R}^n)$ into $\mathcal{S}(\mathbb{R}^n)$. Take any $\phi \in \mathcal{S}(\mathbb{R}^n)$ and any $\alpha, \beta \in \mathbb{N}^n$. Then we know $x^\beta D^\alpha \phi \in \mathcal{S}(\mathbb{R}^n)$ and the combination of the estimates (10.5) and (10.6) implies

$$\|x^\beta D^\alpha \phi\|_1 \leq C p_{n+1+|\beta|,|\alpha|}(\phi). \tag{10.7}$$

Hence parts 4) and 5) of Proposition 10.1.1 can be applied repeatedly, to every order, and thus it follows that

$$D^\alpha (\mathcal{F}\phi)(p) = \mathcal{F}((-ix)^\alpha \phi)(p) \quad \forall p \in \mathbb{R}^n, \quad \forall \alpha \in \mathbb{N}^n.$$

We deduce $\mathcal{F}\phi \in \mathcal{C}^\infty(\mathbb{R}^n)$ and relation a) of part 3) holds.

Similarly one shows for all $\alpha, \beta \in \mathbb{N}^n$ and all $p \in \mathbb{R}^n$,

$$p^\beta D^\alpha (\mathcal{F}\phi)(p) = p^\beta \mathcal{F}((-ix)^\alpha \phi)(p) = \mathcal{F}((-iD)^\beta [(-ix)^\alpha \phi])(p). \tag{10.8}$$

Choosing $\alpha = 0$ in equation (10.8) implies relation b) of part 3). Equation (10.8) also implies

$$|p^\beta D^\alpha (\mathcal{F}\phi)(p)| \leq \|D^\beta (x^\alpha \phi)\|_1$$

and therefore by estimate (10.7), for all $m, l = 0, 1, 2, \ldots$ and all $\phi \in \mathcal{S}(\mathbb{R}^n)$,

$$p_{m,l}(\mathcal{F}\phi) \leq C p_{n+1+l,m}(\phi) \tag{10.9}$$

where the constant C depends only on m, n, l. This estimate implies $\mathcal{F}\phi \in \mathcal{S}(\mathbb{R}^n)$. It follows easily that $\mathcal{F}$ is linear. Hence this estimate also implies that $\mathcal{F}$ is bounded and thus continuous.

Since we know $(\mathcal{L}\phi)(p) = (\mathcal{F}\phi)(-p)$ on $\mathcal{S}(\mathbb{R}^n)$, it follows that the map $\mathcal{L}$ has the same properties as $\mathcal{F}$. The estimate above shows in addition that $\mathcal{S}(\mathbb{R}^n)$ is contained in the subspace of $L^1(\mathbb{R}^n)$ on which $\mathcal{F}$ is invertible. We conclude that the continuous linear map $\mathcal{L}$ on $\mathcal{S}(\mathbb{R}^n)$ is the inverse of the Fourier transformation on this space. This concludes the proof of the theorem. □

On the test function space $\mathcal{S}(\mathbb{R}^n)$ we have introduced two products, the standard pointwise product and the convolution product. As one would expect on the basis of part 3) of Proposition 10.1.1 the Fourier transformation transforms the convolution product into the pointwise product and conversely. More precisely we have the following.

Corollary 10.2.1 1. *The Fourier transformation $\mathcal{F}$ and its inverse $\mathcal{L}$ are related on $\mathcal{S}(\mathbb{R}^n)$ as follows, $u \in \mathcal{S}(\mathbb{R}^n)$:*

$$\mathcal{L}u = \mathcal{F}\check{u} = (\mathcal{F}u)\check{} \quad \mathcal{L}\mathcal{F}u = \mathcal{F}\mathcal{L}u = u \quad \mathcal{F}\mathcal{F}u = \check{u} = \mathcal{L}\mathcal{L}u$$

where $\check{u}(x) = u(-x)$ for all $x \in \mathbb{R}^n$.

2. For all $\phi, \psi \in \mathcal{S}(\mathbb{R}^n)$ the following relations hold:

$$\mathcal{F}(\phi * \psi) = (2\pi)^{\frac{n}{2}} (\mathcal{F}\phi) \cdot (\mathcal{F}\psi),$$
$$\mathcal{F}(\phi \cdot \psi) = (2\pi)^{-\frac{n}{2}} (\mathcal{F}\phi) * (\mathcal{F}\psi).$$

Proof. The first identity in the first part is immediate from the definitions of the maps involved. The second repeats the fact that $\mathcal{L}$ is the inverse of $\mathcal{F}$, on $\mathcal{S}(\mathbb{R}^n)$. The third identity is a straightforward consequence of the first two.

In order to prove the second part, recall that by part 3) of Proposition 10.1.1 the first identity is known for functions in $L^1(\mathbb{R}^n)$, and we know that $\mathcal{S}(\mathbb{R}^n)$ is continuously embedded into $L^1(\mathbb{R}^n)$. Furthermore we know from Proposition 7.1.3 that $\phi * \psi \in \mathcal{S}(\mathbb{R}^n)$. This proves that the first identity is actually an identity in $\mathcal{S}(\mathbb{R}^n)$ and not only in $L^1(\mathbb{R}^n)$.

Now replace in the first identity of the second part the function ϕ with $\mathcal{L}\phi$ and the function ψ with $\mathcal{L}\psi$ to obtain $\mathcal{F}((\mathcal{L}\phi) * (\mathcal{L}\psi)) = (2\pi)^{\frac{n}{2}} (\mathcal{F}(\mathcal{L}\phi)) \cdot (\mathcal{F}(\mathcal{L}\psi)) = (2\pi)^{\frac{n}{2}} \phi \cdot \psi$. It follows that $\mathcal{F}(\phi \cdot \psi) = (2\pi)^{-\frac{n}{2}} \mathcal{F}(\mathcal{F}((\mathcal{L}\phi) * (\mathcal{L}\psi)))$ and thus, taking the first part into account $= (2\pi)^{-\frac{n}{2}} ((\mathcal{L}\phi) * (\mathcal{L}\psi))\check{\,} = (2\pi)^{-\frac{n}{2}} (\mathcal{L}\phi)\check{\,} * (\mathcal{L}\psi)\check{\,} = (2\pi)^{-\frac{n}{2}} (\mathcal{F}\phi) * (\mathcal{F}\psi)$, hence $\mathcal{F}(\phi \cdot \psi) = (2\pi)^{-\frac{n}{2}} \mathcal{F}\phi * \mathcal{F}\psi$. □

10.3 Fourier transformation for tempered distributions

According to the previous section the Fourier transformation is an isomorphism of the test function space $\mathcal{S}(\mathbb{R}^n)$, hence it can be extended to the space of tempered distributions $\mathcal{S}'(\mathbb{R}^n)$ by the standard duality method. After the formal definition has been given we look at some simple examples to illustrate how this definition works in practice. Then several important general results about the Fourier transformation on $\mathcal{S}'(\mathbb{R}^n)$ are discussed.

Definition 10.3.1 *The Fourier transform $\tilde{T} = \mathcal{F}'T$ of a tempered distribution $T \in \mathcal{S}'(\mathbb{R}^n)$ is defined by the relation*

$$\langle \mathcal{F}'T, \phi \rangle = \langle T, \mathcal{F}\phi \rangle \quad \forall \phi \in \mathcal{S}(\mathbb{R}^n). \tag{10.10}$$

Example 10.3.1 1. *Dirac's delta distribution is obviously tempered and thus it has a Fourier transform according to the definition given above. The actual calculation is very simple: For all $\phi \in \mathcal{S}(\mathbb{R}^n)$ one has*

$$\langle \mathcal{F}'\delta, \phi \rangle = \langle \delta, \mathcal{F}\phi \rangle = (\mathcal{F}\phi)(0) = (2\pi)^{-\frac{n}{2}} \int_{\mathbb{R}^n} \phi(x) dx = (2\pi)^{-\frac{n}{2}} \langle I_1, \phi \rangle,$$

hence

$$\mathcal{F}'\delta = (2\pi)^{-\frac{n}{2}} I_1,$$

i.e., the Fourier transform of Dirac's delta distribution is the constant distribution. This is often written as $\mathcal{F}'\delta = (2\pi)^{-\frac{n}{2}}$.

2. *Next we calculate the Fourier transform of a constant distribution $I_c, c \in \mathbb{C}$. According to the previous example we expect it to be proportional to Dirac's*

delta distribution. Indeed one finds for all $\phi \in \mathcal{S}(\mathbb{R}^n)$,

$$\langle \mathcal{F}'I_c, \phi\rangle = \langle I_c, \mathcal{F}\phi\rangle = \int_{\mathbb{R}^n} c(\mathcal{F}\phi)(p)dp = c(2\phi)^{\frac{n}{2}}(\mathcal{L}\mathcal{F}\phi)(0)$$
$$= c(2\pi)^{\frac{n}{2}}\phi(0) = \langle c(2\pi)^{\frac{n}{2}}\delta, \phi\rangle,$$

i.e.,

$$\mathcal{F}'I_c = c(2\pi)^{\frac{n}{2}}\delta.$$

3. Another simple example of a tempered distribution is the Heaviside function θ. It certainly has no Fourier transform in the classical sense. We determine here its Fourier transform in the sense of tempered distributions. The calculations contain a new element, namely a suitable limit procedure. For all $\phi \in \mathcal{S}(\mathbb{R})$ we find

$$\mathcal{F}'\theta, \phi\rangle = \langle \theta, \mathcal{F}\phi\rangle = \int_0^\infty (\mathcal{F}\phi)(p)dp = \lim_{r\to 0, r>0} \int_0^\infty e^{-rp}(\mathcal{F}\phi)(p)dp.$$

For fixed $r > 0$ we apply Fubini's theorem to exchange the order of integration so that one of the integrals can be calculated explicitly. The result is

$$\int_0^\infty e^{-rp}(\mathcal{F}\phi)(p)dp = \int_0^\infty e^{-rp}\left(\int_{-\infty}^{+\infty} e^{-ip\cdot x}\phi(x)\frac{dx}{\sqrt{2\pi}}\right)dp,$$
$$\int_{-\infty}^{+\infty} \phi(x)\left(\int_0^\infty e^{-rp-ipx}dp\right)\frac{dx}{\sqrt{2\pi}} = (2\pi)^{-\frac{1}{2}}\int_\mathbb{R} \frac{i}{x-ir}\phi(x)dx,$$

hence

$$(\mathcal{F}'\theta)(x) = \lim_{r\to 0, r>0} \frac{i}{\sqrt{2\pi}}\frac{1}{x-ir} \equiv \frac{i}{\sqrt{2\pi}}\frac{1}{x-io}.$$

By duality the properties of the Fourier transformation on $\mathcal{S}(\mathbb{R}^n)$ as expressed in Theorem 10.2.1 are easily translated into similar properties of the Fourier transformation on the space of tempered distributions $\mathcal{S}'(\mathbb{R}^n)$.

Theorem 10.3.1 (Fourier transformation on $\mathcal{S}'(\mathbb{R}^n)$) 1. *The Fourier transformation $\mathcal{F}'$ is an isomorphism of $\mathcal{S}'(\mathbb{R}^n)$. It is compatible with the embedding of integrable functions: For all $f \in L^1(\mathbb{R}^n)$ we have*

$$\mathcal{F}'I_f = I_{\mathcal{F}f}.$$

2. *The inverse of $\mathcal{F}'$ is the dual $\mathcal{L}'$ of the inverse $\mathcal{L}$ of $\mathcal{F}$, i.e., $\mathcal{F}'^{-1} = \mathcal{L}'$.*

3. *The following rules hold, $\alpha \in \mathbb{N}^n$:*

$$\mathcal{F}'(D_x^\alpha T)(p) = (ip)^\alpha(\mathcal{F}'T)(p), \qquad D_p^\alpha(\mathcal{F}'T)(p) = \mathcal{F}'((-ix)^\alpha T)(p).$$

10.3 Fourier transformation for tempered distributions

Proof. In the Exercises we show: If I is an isomorphism of the HLCTVS E, then its dual I' is an isomorphism of the topological dual space E' equipped with the topology of pointwise convergence (weak topology σ). Thus we deduce from Theorem 10.2.1 that $\mathcal{F}'$ is an isomorphism of $\mathcal{S}'(\mathbb{R}^n)$.

Next consider any $f \in L^1(\mathbb{R}^n)$. We know that its Fourier transform $\mathcal{F}f$ is a bounded continuous and thus locally integrable function which defines the tempered distribution $I_{\mathcal{F}f}$. For all $\phi \in \mathcal{S}(\mathbb{R}^n)$ a simple application of Fubini's theorem shows that

$$\begin{aligned}\langle \mathcal{F}'I_f, \phi \rangle &= \langle I_f, \mathcal{F}\phi \rangle = \int_{\mathbb{R}^n} f(x)(\mathcal{F}\phi)(x)dx = \int_{\mathbb{R}^n} f(x)\left(\int_{\mathbb{R}^n} e^{-ix\cdot p}\phi(p)\frac{dp}{(2\pi)^{\frac{n}{2}}}\right)dx\\ &= \int_{\mathbb{R}^n}\left(\int_{\mathbb{R}^n} f(x)e^{-ip\cdot x}\frac{dp}{(2\pi)^{\frac{n}{2}}}\right)\phi(p)dp = \int_{\mathbb{R}^n}(\mathcal{F}f)(p)\phi(p)dp = \langle I_{\mathcal{F}f}, \phi \rangle.\end{aligned}$$

This implies compatibility of the Fourier transformations on $L^1(\mathbb{R}^n)$ and on $\mathcal{S}'(\mathbb{R}^n)$ and thus part 1) has been shown.

In order to prove part 2) take any $T \in \mathcal{S}'(\mathbb{R}^n)$ and calculate for all $\phi \in \mathcal{S}(\mathbb{R}^n)$ using Theorem 10.2.1 $\langle \mathcal{L}'\mathcal{F}'T, \phi \rangle = \langle \mathcal{F}'T, \mathcal{L}\phi \rangle = \langle T, \mathcal{F}\mathcal{L}\phi \rangle = \langle T, \phi \rangle$, thus $\mathcal{L}'\mathcal{F}' = id$. It follows that $\mathcal{L}'$ is the inverse of $\mathcal{F}'$.

Finally we establish the rules of part 3) relying on the corresponding rules as stated in Theorem 10.2.1: Take any $T \in \mathcal{S}'(\mathbb{R}^n)$ and any $\phi \in \mathcal{S}(\mathbb{R}^n)$ and use the definitions, respectively the established rules, to get

$$\langle \mathcal{F}'(D_x^\alpha T), \phi \rangle = \langle D_x^\alpha T, \mathcal{F}\phi \rangle = (-1)^{|\alpha|}\langle T, D_x^\alpha(\mathcal{F}\phi)\rangle$$
$$= (-1)^{|\alpha|}\langle T, \mathcal{F}((-ip)^\alpha \phi)\rangle = \langle \mathcal{F}'T, (ip)^\alpha \phi \rangle = \langle (ip)^\alpha \mathcal{F}T, \phi \rangle.$$

Since this identity holds for every $\phi \in \mathcal{S}(\mathbb{R}^n)$ the first relation is proven. Similarly we proceed with the second.

$$\langle D_p^\alpha(\mathcal{F}'T), \phi \rangle = (-1)^{|\alpha|}\langle \mathcal{F}'T, D_p^\alpha \phi \rangle = (-1)^{|\alpha|}\langle T, \mathcal{F}(D_p^\alpha \phi)\rangle$$
$$= (-1)^{|\alpha|}\langle T, (ix)^\alpha \mathcal{F}\phi \rangle = \langle (-ix)^\alpha T, \mathcal{F}\phi \rangle = \langle \mathcal{F}'((-ix)^\alpha T), \phi \rangle.$$

$\square$

As a simple illustration of the rules in part 3) we mention the following. Apply the first rule to Dirac's delta distribution. Recalling the relation $\mathcal{F}'\delta = (2\pi)^{-\frac{n}{2}}$ we get

$$\mathcal{F}'(D^\alpha \delta)(p) = (2\pi)^{-\frac{n}{2}}(ip)^\alpha. \tag{10.11}$$

Similarly, applying the second rule to the constant distribution $T = I_1$ produces the relation

$$\mathcal{F}'((-ix)^\alpha)(p) = (2\pi)^{\frac{n}{2}} D^\alpha \delta(p). \tag{10.12}$$

Certainly, these convenient rules have no counterpart in the classical theory of Fourier transformation. Further applications are discussed in the Exercises.

In Corollary 10.2.1 we learned that the Fourier transformation $\mathcal{F}$ transforms a convolution of test functions ϕ, ψ into a pointwise product: $\mathcal{F}(\phi * \psi) = (2\pi)^{\frac{n}{2}}(\mathcal{F}\phi) \cdot (\mathcal{F}\psi)$. Since we have also learned that the convolution and the pointwise product of distributions is naturally defined only in special cases, we cannot expect this relation to hold for distributions in general. However there is an important class for which one can show this relation to hold for distributions too: One distribution is tempered and the other has a compact support. As preparation we show that the Fourier transform of a distribution of compact support is a multiplier for tempered distributions, i.e., a $\mathcal{C}^\infty$-function with polynomially bounded derivatives.

To begin we note

Lemma 10.3.1 *For $p \in \mathbb{R}^n$ define a function $e_p : \mathbb{R}^n \to \mathbb{C}$ by $e_p(x) = \frac{e^{-ip \cdot x}}{(2\pi)^{n/2}}$. Suppose $T \in \mathcal{D}'(\mathbb{R}^n)$ is a distribution with support contained in the compact set $K \subset \mathbb{R}^n$. For any function $u \in \mathcal{D}(\mathbb{R}^n)$ define a function $T_u : \mathbb{R}^n \to \mathbb{C}$ by*

$$T_u(p) = \langle T, e_p \cdot u \rangle \quad \forall p \in \mathbb{R}^n.$$

Then the following holds.

1. *$T_u \in \mathcal{O}_m(\mathbb{R}^n)$, i.e., T_u is a C^∞-function with polynomially bounded derivatives.*

2. *If $u, v \in \mathcal{D}(\mathbb{R}^n)$ satisfy $u(x) = v(x) = 1$ for all $x \in K$, then $T_u = T_v$.*

Proof. Since for each $p \in \mathbb{R}^n$ the function $e_p \cdot u$ belongs to $\mathcal{D}(\mathbb{R}^n)$ if u does, the function T_u is well defined for $u \in \mathcal{D}(\mathbb{R}^n)$. As in Theorem 7.2.1 it follows that T_u is a C^∞-function and

$$D^\alpha T_u(p) = \langle T, D_p^\alpha(e_p \cdot u) \rangle = \langle T, e_p \cdot (-ix)^\alpha \cdot u \rangle \quad \forall \alpha \in \mathbb{N}^n.$$

Since T has its support in the compact set K, there are $m \in \mathbb{N}$ and a constant C such that $|\langle T, \phi \rangle| \leq C p_{K,m}(\phi)$ for all $\phi \in \mathcal{D}_K(\mathbb{R}^n)$. It follows that, for all $p \in \mathbb{R}^n$,

$$|D^\alpha T_u(p)| \leq C p_{K,m}(e_p \cdot (-ix)^\alpha \cdot u). \tag{10.13}$$

As we show in the Exercises, the right-hand side of this inequality is a polynomially bounded function of $p \in \mathbb{R}^n$. It follows that $T_u \in \mathcal{O}_m(\mathbb{R}^n)$. This proves the first part.

If two functions $u, v \in \mathcal{D}(\mathbb{R}^n)$ are equal to 1 on K, then, for every $p \in \mathbb{R}^n$, the function $e_p \cdot (u - v)$ vanishes on a neighborhood of the support of the distribution T and hence $\langle T, e_p \cdot (u - v) \rangle = 0$. Linearity of T implies $T_u = T_v$. □

Theorem 10.3.2 *A distribution $T \in \mathcal{D}'(\mathbb{R}^n)$ of compact support is tempered and its Fourier transform $\tilde{T} = \mathcal{F}'(T)$ is a C^∞-function such that all derivatives $D^\alpha \tilde{T}(p)$ are polynomially bounded, i.e., $\tilde{T} \in \mathcal{O}_m(\mathbb{R}^n)$. (See also Proposition 4.2.2).*

Proof. A distribution T with compact support is an element of the dual $\mathcal{E}'(\mathbb{R}^n)$ of the test function space $\mathcal{E}(\mathbb{R}^n)$, according to Theorem 3.5.3. Since $\mathcal{S}(\mathbb{R}^n) \subset \mathcal{E}(\mathbb{R}^n)$ with continuous identity map, it follows that $\mathcal{E}'(\mathbb{R}^n) \subset \mathcal{S}'(\mathbb{R}^n)$. Therefore a distribution with compact support is tempered and thus has a well defined Fourier transform.

Suppose $T \in \mathcal{D}'(\mathbb{R}^n)$ has its support in the compact set $K \subset \mathbb{R}^n$. Choose any $u \in \mathcal{D}(\mathbb{R}^n)$ with the property $u(x) = 1$ for all $x \in K$ and define the function T_u as in the previous lemma. It follows that $T_u \in \mathcal{O}_m(\mathbb{R}^n)$ and we claim

$$\mathcal{F}'T = I_{T_u},$$

i.e., for all $\phi \in \mathcal{S}(\mathbb{R}^n)$,

$$\langle \mathcal{F}'T, \phi \rangle = \int_{\mathbb{R}^n} T_u(p) \phi(p) dp.$$

According to the specification of u we know

$$\langle u \cdot T, \mathcal{F}\phi \rangle = \langle T, \mathcal{F}\phi \rangle = \langle \mathcal{F}'T, \phi \rangle.$$

Now observe $\mathcal{F}\phi(x) = \int_{\mathbb{R}^n} e_p(x) \phi(p) dp$ and thus

$$\langle u \cdot T, \mathcal{F}\phi \rangle = \langle (u \cdot T)(x), \int_{\mathbb{R}^n} e_p(x) \phi(p) dp \rangle = \langle T(x), u(x) \int_{\mathbb{R}^n} e_p(x) \phi(p) dp \rangle$$

$$= \int_{\mathbb{R}^n} \langle T(x), u(x) e_p(x) \rangle \phi(p) dp = \int_{\mathbb{R}^n} T_u(p) \phi(p) dp.$$

In the second but last step we used equation (7.6). This gives $\langle \mathcal{F}'T, \phi \rangle = \int_{\mathbb{R}^n} T_u(p) \phi(p) dp$ for all $\phi \in \mathcal{S}(\mathbb{R}^n)$ and thus proves $\mathcal{F}'T = T_u$. The previous lemma now gives the conclusion. □

10.3 Fourier transformation for tempered distributions

As further preparation we present a result which is also of considerable interest in itself since it controls the convolution of distributions, in $\mathcal{S}'(\mathbb{R}^n)$ and in $\mathcal{E}'(\mathbb{R}^n)$, with test functions in $\mathcal{S}(\mathbb{R}^n)$.

Proposition 10.3.3 *The convolution of a tempered distribution $T \in \mathcal{S}'(\mathbb{R}^n)$ with a test function $\psi \in \mathcal{S}(\mathbb{R}^n)$ is a tempered distribution $T * \psi$ which has the Fourier transform*

$$\mathcal{F}'(T * \psi) = (2\pi)^{\frac{n}{2}} (\mathcal{F}'T) \cdot (\mathcal{F}\psi).$$

In particular, if $T \in \mathcal{E}'(\mathbb{R}^n)$, then $T * \psi \in \mathcal{S}(\mathbb{R}^n)$.

Proof. The convolution $T * \psi$ is defined by

$$\langle T * \psi, \phi \rangle = \langle T, \check{\psi} * \phi \rangle \qquad \forall \phi \in \mathcal{S}(\mathbb{R}^n).$$

Since we have learned that, for fixed $\psi \in \mathcal{S}(\mathbb{R}^n)$, $\phi \mapsto \check{\psi} * \phi$ is a continuous linear map from $\mathcal{S}(\mathbb{R}^n)$ into itself (see Proposition 7.1.3) it follows that $T * \psi$ is well defined as a tempered distribution. For its Fourier transform we find, using Corollary 10.2.1 and Theorem 10.3.1,

$$\begin{aligned}\langle \mathcal{F}'(T*\psi), \phi \rangle &= \langle T*\psi, \mathcal{F}\phi \rangle = \langle T, \check{\psi} * \mathcal{F}\phi \rangle \\ &= \langle \mathcal{F}'T, \mathcal{L}(\check{\psi} * \mathcal{F}\phi) \rangle = (2\pi)^{\frac{n}{2}} \langle \mathcal{F}'T, \mathcal{L}(\check{\psi}) \cdot (\mathcal{L}\mathcal{F}\phi) \rangle \\ &= (2\pi)^{\frac{n}{2}} \langle \mathcal{F}'T, (\mathcal{F}\psi) \cdot \phi \rangle = (2\pi)^{\frac{n}{2}} \langle (\mathcal{F}\psi) \cdot (\mathcal{F}'T), \phi \rangle.\end{aligned}$$

This implies

$$\mathcal{F}'(T*\psi) = (2\pi)^{\frac{n}{2}} (\mathcal{F}'T) \cdot (\mathcal{F}\psi) \qquad \forall T \in \mathcal{S}'(\mathbb{R}^n), \forall \psi \in \mathcal{S}(\mathbb{R}^n).$$

If $T \in \mathcal{E}'(\mathbb{R}^n)$, then $\mathcal{F}'T \in \mathcal{O}_m(\mathbb{R}^n)$ by Theorem 10.3.2 and thus $(\mathcal{F}'T) \cdot (\mathcal{F}\psi) \in \mathcal{S}(\mathbb{R}^n)$, hence $\mathcal{F}'(T*\psi) = \mathcal{F}(T*\psi) \in \mathcal{S}(\mathbb{R}^n)$. $\square$

Theorem 10.3.4 (Convolution theorem) *The convolution $T * S$ of a tempered distribution $T \in \mathcal{S}'(\mathbb{R}^n)$ and a compactly supported distribution $S \in \mathcal{E}'(\mathbb{R}^n)$ is a tempered distribution whose Fourier transform is*

$$\mathcal{F}'(T * S) = (2\pi)^{\frac{n}{2}} (\mathcal{F}'T) \cdot (\mathcal{F}'S). \qquad (10.14)$$

Proof. Since $S \in \mathcal{E}'(\mathbb{R}^n)$ Proposition 10.3.3 ensures that

$$x \mapsto \langle S(y), \phi(x+y) \rangle = (\check{S} * \phi)(x)$$

belongs to $\mathcal{S}(\mathbb{R}^n)$ for every $\phi \in \mathcal{S}(\mathbb{R}^n)$. Using Corollary 10.2.1 we calculate its inverse Fourier transform:

$$\mathcal{L}(\check{S} * \phi) = (2\pi)^{\frac{n}{2}} (\mathcal{F}'S) \cdot (\mathcal{L}\phi)$$

with $\mathcal{F}'S \in \mathcal{O}_m(\mathbb{R}^n)$ according to Theorem 10.3.2. Observe now that the definition of the convolution of two distributions can be rewritten as $\langle T * S, \phi \rangle = \langle T, \check{S} * \phi \rangle$, for all $\phi \in \mathcal{S}(\mathbb{R}^n)$. Hence $T * S$ is a well-defined tempered distribution.

The inverse of $\mathcal{F}'$ is $\mathcal{L}'$. This implies

$$\langle T * S, \phi \rangle = \langle \mathcal{F}'T, \mathcal{L}(\check{S} * \phi) \rangle = (2\pi)^{\frac{n}{2}} \langle \mathcal{F}'T, (\mathcal{F}'S) \cdot \mathcal{L}\phi \rangle = (2\pi)^{\frac{n}{2}} \langle (\mathcal{F}'S) \cdot (\mathcal{F}'T), \mathcal{L}\phi \rangle$$

and therefore $T * S = (2\pi)^{\frac{n}{2}} \mathcal{L}'((\mathcal{F}'S) \cdot (\mathcal{F}T))$. Now equation (10.14) follows and we conclude. $\square$

142   10. Fourier Transformation

We started the study of the Fourier transformation on the space $L^1(\mathbb{R}^n)$. We found that the domain and the range of $\mathcal{F}$ are not symmetric. However when we restricted $\mathcal{F}$ to the test function space $\mathcal{S}(\mathbb{R}^n)$ we could prove that the domain and the range are the same; actually we found that $\mathcal{F}$ is an isomorphism of topological vector spaces and used this to extend the definition of the Fourier transformation to the space of all tempered distributions $\mathcal{S}'(\mathbb{R}^n)$, using duality. Certainly, the space $L^1(\mathbb{R}^n)$ is contained in $\mathcal{S}'(\mathbb{R}^n)$, in the sense of the embedding $L^1(\mathbb{R}^n) \ni f \mapsto I_f \in \mathcal{S}'(\mathbb{R}^n)$. In this sense there are many other function spaces contained in $\mathcal{S}'(\mathbb{R}^n)$, for instance the space $L^2(\mathbb{R}^n)$ of (equivalence classes of) square integrable functions which is known to be a Hilbert space with inner product

$$\langle f, g \rangle_2 = \int_{\mathbb{R}^n} \overline{f(x)} g(x) dx \qquad \forall f, g \in L^2(\mathbb{R}^n).$$

This is discussed in Section 14.1. There we also learn that the test function space $\mathcal{S}(\mathbb{R}^n)$ is dense in $L^2(\mathbb{R}^n)$. Since $L^2(\mathbb{R}^n)$ is 'contained' in $\mathcal{S}'(\mathbb{R}^n)$, the restriction of the Fourier transformation $\mathcal{F}'$ to $L^2(\mathbb{R}^n)$ gives a definition of the Fourier transformation on $L^2(\mathbb{R}^n)$. More precisely this means the following: Denote the Fourier transformation on $L^2(\mathbb{R}^n)$ by $\mathcal{F}_2$; it is defined by the identity

$$\mathcal{F}' I_f = I_{\mathcal{F}_2 f} \qquad \forall f \in L^2(\mathbb{R}^n).$$

In order to get a more concrete representation of $\mathcal{F}_2$ and to study some of its properties we use our results on the Fourier transform on $\mathcal{S}(\mathbb{R}^n)$ and combine them with Hilbert space methods as developed in Part II.

To begin we show that the restriction of the inner product of $L^2(\mathbb{R}^n)$ to $\mathcal{S}(\mathbb{R}^n)$ is invariant under $\mathcal{F}$. First we observe that for all $\phi, \psi \in \mathcal{S}(\mathbb{R}^n)$ one has

$$\langle \mathcal{F}\phi, \mathcal{F}\psi \rangle_2 = \langle I_1, \overline{\mathcal{F}\phi} \cdot \mathcal{F}\psi \rangle.$$

Express the complex conjugate of the Fourier transform of ϕ as $\overline{\mathcal{F}\phi} = \mathcal{L}(\overline{\phi}) = \mathcal{F}(\check{\overline{\phi}})$ and apply Corollary 10.2.1 to get $\overline{\mathcal{F}\phi} \cdot \mathcal{F}\psi = (2\pi)^{-\frac{n}{2}} \mathcal{F}(\check{\overline{\phi}} * \psi)$. It follows that, using $\mathcal{F}' I_1 = (2\pi)^{\frac{n}{2}} \delta$, $\langle \mathcal{F}\phi, \mathcal{F}\psi \rangle_2 = \langle I_1, (2\pi)^{-\frac{n}{2}} \mathcal{F}(\check{\overline{\phi}} * \psi) \rangle = \langle (2\pi)^{-\frac{n}{2}} \mathcal{F}' I_1, \check{\overline{\phi}} * \psi \rangle = \langle \delta, \check{\overline{\phi}} * \psi \rangle = (\check{\overline{\phi}} * \psi)(0) = \int_{\mathbb{R}^n} \overline{\phi(x)} \psi(x) dx = \langle \phi, \psi \rangle_2$, and thus we get the announced invariance

$$\langle \mathcal{F}\phi, \mathcal{F}\psi \rangle_2 = \langle \phi, \psi \rangle_2 \qquad \forall \phi, \psi \in \mathcal{S}(\mathbb{R}^n).$$

This nearly proves

Theorem 10.3.5 (Plancherel) *The Fourier transformation $\mathcal{F}_2$ on $L^2(\mathbb{R}^n)$ can be obtained as follows: Given any $f \in L^2(\mathbb{R}^n)$ choose a sequence $(u_j)_{j \in \mathbb{N}}$ in $\mathcal{S}(\mathbb{R}^n)$ which converges to f (in $L^2(\mathbb{R}^n)$). Then the sequence $(\mathcal{F} u_j)_{j \in \mathbb{N}}$ is a Cauchy sequence in $L^2(\mathbb{R}^n)$ which thus converges to some element $g \in L^2(\mathbb{R}^n)$ which defines $\mathcal{F}_2 f$, i.e.,*

$$\mathcal{F}_2 f = \lim_{j \to \infty} \mathcal{F} u_j.$$

$\mathcal{F}_2$ is a well-defined unitary map of the Hilbert space $L^2(\mathbb{R}^n)$.

Proof. Since we know that $\mathcal{S}(\mathbb{R}^n)$ is dense in $L^2(\mathbb{R}^n)$ and that the inner product $\langle \cdot, \cdot \rangle_2$ is invariant under the Fourier transformation $\mathcal{F}$ on $\mathcal{S}(\mathbb{R}^n)$, this follows easily from Proposition 22.2.2. □

The relation of the Fourier transformation on the various spaces can be summarized by the following diagram:

$$\begin{array}{ccc} \mathcal{S}'(\mathbb{R}^n) & \xrightarrow{\mathcal{F}'} & \mathcal{S}'(\mathbb{R}^n) \\ I \uparrow & & I \uparrow \\ L^2(\mathbb{R}^n) & \xrightarrow{\mathcal{F}_2} & L^2(\mathbb{R}^n) \\ id \uparrow & & id \uparrow \\ \mathcal{S}(\mathbb{R}^n) & \xrightarrow{\mathcal{F}} & \mathcal{S}(\mathbb{R}^n) \end{array}$$

All maps in the diagram are continuous and linear. $\mathcal{F}_2$ is unitary.

Remark 10.3.1 *The fact that the Fourier transformation $\mathcal{F}_2$ is a unitary map of the Hilbert space $L^2(\mathbb{R}^n)$ is of particular importance to the quantum mechanics of localized systems since it allows us to pass from the coordinate representation $L^2(\mathbb{R}_x^n)$ of the state space to the momentum representation $L^2(\mathbb{R}_p^n)$ without changing expectation values.*

10.4 Some applications

This section deals with several aspects of the solution theory for linear partial differential operators with constant coefficients in the framework of tempered distributions, which arise from the fact that for tempered distributions the Fourier transformation is available. The results will be considerably stronger.

Central to the solution theory for linear partial differential operators with constant coefficients in the space of tempered distributions is the following result by L. Hörmander, see reference [Hör83b].

Theorem 10.4.1 (L. Hörmander) *Suppose P is a polynomial in n variables with complex coefficients, $P \neq 0$. Then the following holds:*

a) *For every $T \in \mathcal{S}'(\mathbb{R}^n)$ there is an $S \in \mathcal{S}'(\mathbb{R}^n)$ such that*

$$P \cdot S = T.$$

b) *If the polynomial P has no real roots, then the equation $P \cdot S = T$ has exactly one solution S.*

The proof of this core result is far beyond the scope of our elementary introduction, and we have to refer to the book [Hör83b]. But we would like to give a few comments indicating the difficulties involved.

Introduce the set of roots or zeros of the polynomial:

$$N(P) = \{x \in \mathbb{R}^n : P(x) = 0\}.$$

If the polynomial P has no real roots, then it is easy to see that $\frac{1}{P}$ belongs to the multiplier space $\mathcal{O}_m(\mathbb{R}^n)$ of tempered distributions and thus the equation $P \cdot S = T$ has the unique solution $S = \frac{1}{P} \cdot T$.

But we know that in general $N(P)$ is not empty. In the case of one variable $N(P)$ is a discrete set (see the fundamental theorem of algebra, Corollary 9.3.2). For $n \geq 2$ the set of roots of a polynomial can be a fairly complicated set embedded in $\mathbb{R}^n$; in some cases it is a differentiable manifold of various dimensions, in other cases it is more complicated than a differentiable manifold. In the Exercises we consider some examples.

On the set $\mathbb{R}^n \setminus N(P)$ the solution S has to be of the form $\frac{1}{P} \cdot T$, in some way. But $\frac{1}{P}$ can fail to be locally integrable. Accordingly the problem is: Define a distribution $[\frac{1}{P}] \in \mathcal{S}'(\mathbb{R}^n)$ with the properties

$$P \cdot [\frac{1}{P}] = I_1$$

and the product of the two tempered distributions

$$[\frac{1}{P}] \cdot T$$

is a well-defined tempered distribution.

As an illustration we look at the simplest nontrivial case, i.e., $n = 1$ and $P(x) = x$. In the section on the convergence of sequences of distributions we have already encountered tempered distributions $[\frac{1}{P}]$ which satisfy $x \cdot [\frac{1}{P}] = 1$, namely the distributions

$$\frac{1}{x \pm io}, \quad \text{vp}\,\frac{1}{x}.$$

Then, given $T \in \mathcal{S}'(\mathbb{R})$, it is not clear whether we can multiply T with these distributions. Hörmander's theorem resolves this problem.

Naturally, in the general case where the structure of the set of roots of P is much more complicated these two steps are much more involved. There are a number of important consequences of Hörmander's theorem.

Corollary 10.4.1 *Suppose that $P(D) = \sum_{|\alpha| \leq N} a_\alpha D^\alpha$, $a_\alpha \in \mathbb{C}$ is a constant coefficient partial differential operator, $P \neq 0$. Then the following holds.*

a) $P(D)$ has a tempered elementary solution $E_P \in \mathcal{S}'(\mathbb{R}^n)$;

b) If $P(ix)$ has no real roots, then there is exactly one tempered elementary solution E_P;

c) For every $T \in S'(\mathbb{R}^n)$ there is an $S \in S'(\mathbb{R}^n)$ such that

$$P(D)S = T,$$

i.e., every linear partial differential equation with constant coefficients $P(D)S = T$, $T \in S'(\mathbb{R}^n)$, has at least one tempered solution.

Proof. We discuss only the easy part of the proof. For $S \in S'(\mathbb{R}^n)$ we calculate first

$$\mathcal{F}'(P(D)S) = \sum_{|\alpha| \leq N} a_\alpha \mathcal{F}'(D^\alpha S) = \sum_{|\alpha| \leq N} a_\alpha (ip)^\alpha \mathcal{F}'S$$

where in the last step we used the third part of Theorem 10.3.1. This implies: Given $T \in S'(\mathbb{R}^N)$, a distribution $S \in S'(\mathbb{R}^n)$ solves the partial differential equation $P(D)S = T$ if, and only if, $\tilde{S} = \mathcal{F}'S$ solves the algebraic equation

$$P(ip)\tilde{S} = \tilde{T}$$

with $\tilde{T} = \mathcal{F}'T$. Now recall $\mathcal{F}'\delta = (2\pi)^{-\frac{n}{2}}I_1$. According to Theorem 10.4.1 there is $[\frac{1}{P(ip)}] \in S'(\mathbb{R}^n)$ such that $P(ip)[\frac{1}{P(ip)}] = (2\pi)^{-\frac{n}{2}}I_1$ and $[\frac{1}{P(ip)}]$ is unique if $P(ip)$ has no real roots. By applying the inverse Fourier transformation we deduce that a (exactly one) tempered elementary solution

$$E_P = \mathcal{L}'[\frac{1}{P(ip)}]$$

exists. This proves parts a) and b). For the proof of the third part we have to refer to Hörmander. In many cases one can find a tempered elementary solution E_P such that the convolution product $E_P * T$ exists. Then a solution is

$$S = E_P * T.$$

As we know this is certainly the case if T has compact support. □

10.4.1 Examples of tempered elementary solutions

For several simple partial differential operators with constant coefficients, which play an important role in physics, we calculate the tempered elementary solution explicitly.

The Laplace operator $\triangle_3$ in $\mathbb{R}^3$.

A fundamental solution E_3 for the Laplace operator $\triangle_3$ satisfies the equation $\triangle_3 E_3 = \delta$. By taking the Fourier transform of this equation $-p^2 \mathcal{F}'E_3 = \mathcal{F}'\delta = (2\pi)^{-\frac{n}{2}}I_1$, we find

$$\mathcal{F}'E_3(p) = (2\pi)^{\frac{3}{2}} \frac{-1}{p^2}, \qquad p^2 = p_1^2 + p_2^2 + p_3^2.$$

Since $p \mapsto \frac{1}{p^2}$ is locally integrable on $\mathbb{R}^3$, $\mathcal{F}'E_3$ is a regular distribution and its inverse Fourier transform can be calculated explicitly. For $\phi \in S(\mathbb{R}^3)$ we proceed as follows:

$$\begin{aligned}
\langle E_3, \phi \rangle &= \langle \mathcal{F}'E_3, \mathcal{L}\phi \rangle = \tfrac{-1}{(2\pi)^3} \int_{\mathbb{R}^3} \tfrac{1}{p^2} \left(\int_{\mathbb{R}^3} e^{ip\cdot x} \phi(x) dx \right) dp \\
&= \tfrac{-1}{(2\pi)^3} \lim_{R \to \infty} \int_{|p| \le R} \tfrac{1}{p^2} \left(\int_{\mathbb{R}^3} e^{ip\cdot x} \phi(x) dx \right) dp \\
&= \tfrac{-1}{(2\pi)^3} \lim_{R \to \infty} \int_{\mathbb{R}^3} \left(\int_{|p| \le R} \tfrac{e^{ip\cdot x} dp}{p^2} \right) \phi(x) dx \\
&= \tfrac{-1}{(2\pi)^3} \lim_{R \to \infty} \int_{\mathbb{R}^3} \left(\int_0^R 2\pi \int_0^\pi \sin\theta \, \tfrac{e^{i|x|\rho \cos\theta}}{\rho^2} \rho^2 d\rho \right) \phi(x) dx \\
&= \tfrac{-1}{(2\pi)^2} \lim_{R \to \infty} \int_{\mathbb{R}^3} \left(\int_0^R \tfrac{e^{i|x|\rho} - e^{-i|x|\rho}}{i|x|\rho} d\rho \right) \phi(x) dx \\
&= \tfrac{-1}{(2\pi)^2} \int_{\mathbb{R}^3} \left(\int_0^\infty \tfrac{e^{i\lambda} - e^{-i\lambda}}{i\lambda} d\lambda \right) \tfrac{\phi(x)}{|x|} dx.
\end{aligned}$$

The exchange of the order of integration is justified by Fubini's theorem. Recalling the integral

$$\int_0^\infty \frac{e^{i\lambda} - e^{-i\lambda}}{i\lambda} d\lambda = 2 \int_0^\infty \frac{\sin\lambda}{\lambda} d\lambda = 2\frac{\pi}{2} = \pi$$

we thus get

$$\langle E_3, \phi \rangle = \frac{-1}{4\pi} \int_{\mathbb{R}^3} \frac{\phi(x)}{|x|} dx \quad \text{i.e.,} \quad E_3(x) = -\frac{1}{4\pi |x|}.$$

Helmholtz' differential operator $\triangle_3 - \mu^2$

Again, by Fourier transformation the partial differential equation for the fundamental solution E_H of this operator is transformed into an algebraic equation for the Fourier transform: $(\triangle_3 - \lambda) E_H = \delta$ implies $(-p^2 - \lambda)\tilde{E}(p) = (2\pi)^{-\frac{3}{2}}$ with $\tilde{E} = \mathcal{F}'E_H$. Hence for $\lambda = \mu^2 > 0$ one finds that $P(ip) = -(p^2 + \mu^2)$ has no real roots and thus the division problem has a simple unique solution

$$\tilde{E}(p) = \frac{-1}{(2\pi)^{\frac{3}{2}}} \frac{1}{p^2 + \mu^2} \in L^1_{loc}(\mathbb{R}^3).$$

The unique (tempered) fundamental solution of Helmholtz' operator thus is, for all $x \in \mathbb{R}^3 \setminus \{0\}$,

$$E_H(x) = \mathcal{L}\tilde{E}(x) = \frac{-1}{(2\pi)^3 2} \int_{\mathbb{R}^3} \frac{e^{ip\cdot x}}{p^2 + \mu^2} dp = \frac{-1}{4\pi} \frac{e^{-|\mu||x|}}{|x|}.$$

The details of this calculation are given in the Exercises.

The Wave operator $\square_4$ in $\mathbb{R}^4$.

In Proposition 8.4.3 it was shown that the distribution

$$E_r(x_0, x) = \frac{1}{2\pi} \theta(x_0) \delta(x_0^2 - x^2)$$

is an elementary solution of the wave operator. Using the Fourier transformation we give here another proof of this fact. It is easy to see that the assignment

$$S(\mathbb{R}^4) \ni \phi \mapsto \frac{1}{2\pi} \int_{\mathbb{R}^3} \phi(|x|, x) \frac{d^3x}{2|x|} \equiv \langle E_r, \phi \rangle$$

defines a tempered distribution on $\mathbb{R}^4$. For any $\phi \in S(\mathbb{R}^4)$ we calculate

$$\langle \mathcal{F}' E_r, \phi \rangle = \langle E_r, \mathcal{F}\phi \rangle = \frac{1}{4\pi} \int_{\mathbb{R}^3} (\mathcal{F}\phi)(|x|, x) \frac{d^3x}{2|x|}$$

and observe that this integral equals

$$\frac{1}{4\pi} \lim_{t \to 0} \int_{\mathbb{R}^3} e^{-t|x|} (\mathcal{F}\phi)(|x|, x) \frac{d^3x}{|x|} = \lim_{t \to 0} \int_{\mathbb{R}^4} I_t(p_0, p) \phi(p_0, p) dp_0 d^3 p$$

where

$$I_t(p_0, p) = \int_{\mathbb{R}^3} e^{-i|x|(p_0-it)-ip\cdot x} \frac{d^3x}{4\pi|x|}$$

$$= \frac{1}{(2\pi)^2} \left(\frac{1}{p_0+|p|-it} - \frac{1}{p_0-|p|-it} \right).$$

It follows that

$$\langle (-p_0^2 + p^2) \mathcal{F}' E_r, \phi \rangle = \langle \mathcal{F}' E_r, (-p_0^2 + p^2)\phi \rangle$$
$$= \lim_{t \to 0} \int_{\mathbb{R}^4} I_t(p_0, p)(-p_0^2 + p^2)\phi(p_0, p) dp_0 d^3 p$$
$$= \lim_{t \to 0} \int_{\mathbb{R}^4} \frac{\phi(p_0, p)}{|p|} \left(\frac{1}{p_0+|p|-it} - \frac{1}{p_0-|p|-it} \right) (|p| + p_0)(|p| - p_0) dp_0 d^3 p$$
$$= (2\pi)^{-2} \int_{\mathbb{R}^4} \phi(p_0, p) dp_0 d^3 p = (2\pi)^{-2} \langle I_1, \phi \rangle = \langle \mathcal{F}' \delta, \phi \rangle,$$

and thus $\Box_4 E_r = \delta$.

Operator of Heat Conduction, Heat Equation

Suppose $E(t, x) \in S'(\mathbb{R}^{n+1})$ satisfies the partial differential equation

$$(\partial_t - \Delta_n) E = \delta$$

where $\partial_t = \frac{\partial}{\partial t}$, i.e., E is an elementary solution of the differential operator of heat conduction. Per Fourier transform one obtains the algebraic equation

$$(ip_0 + p^2)(\mathcal{F}' E)(p_0, p) = (2\pi)^{-\frac{n+1}{2}} I_1 \qquad p_0 \in \mathbb{R}, \quad p \in \mathbb{R}^n.$$

Since $\frac{1}{ip_0+p^2} \in L^1_{loc}(\mathbb{R}^{n+1})$, the solution of this equation is the regular distribution given by the function

$$\tilde{E}(p_0, p) = (2\pi)^{-\frac{n+1}{2}} (ip_0 + p^2)^{-1}.$$

Now consider the function

$$E(t, x) = \theta(t)(4\pi t)^{-\frac{n}{2}} e^{-\frac{x^2}{4t}} \qquad t \in \mathbb{R}, \quad x \in \mathbb{R}^n.$$

Its Fourier transform is easily calculated (in the sense of functions).

$$\begin{aligned}(\mathcal{F}E)(p_0, p) &= (2\pi)^{-\frac{n+1}{2}} (4\pi)^{-\frac{n}{2}} \int_0^\infty \int_{\mathbb{R}^n} e^{-ip_0 t} e^{-ip\cdot x} t^{-\frac{n}{2}} e^{-\frac{x^2}{4t}} dx dt \\ &= (2\pi)^{-\frac{n+1}{2}} (4\pi)^{-\frac{n}{2}} \int_0^\infty e^{-ip_0 t} t^{-\frac{n}{2}} (4\pi t)^{\frac{n}{2}} e^{-tp^2} dt \\ &= (2\pi)^{-\frac{n+1}{2}} \frac{1}{ip_0 + p^2}.\end{aligned}$$

We conclude that E is a tempered elementary solution of the operator $\partial_t - \Delta_n$.

Free Schrödinger operator in $\mathbb{R}^n$

The partial differential operator

$$i\frac{\partial}{\partial t} - \Delta_n$$

is called *the free Schrödinger operator* of dimension n. In the Exercises it is shown that the function

$$E_S(t, x) = \theta(t)(4\pi i t)^{-\frac{n}{2}} e^{-\frac{x^2}{4it}}$$

defines a tempered distribution which solves the equation $(i\frac{\partial}{\partial t} - \Delta_n)E_S = \delta$ in $\mathcal{S}'(\mathbb{R}^{n+1+})$ and therefore it is a tempered elementary solution.

Other examples of elementary solutions and Green functions are given in the book [YCB82].

Some comments

There is an important difference in the behaviour of solutions of the heat equation and the wave equation: The propagation speed of solutions of the wave equation is finite and is determined by the 'speed parameter' in this equation. However the propagation speed of heat according to the heat equation is infinite! Certainly this is physically not realistic. Nevertheless, the formula $u(t, x) = E * U_0$ (E is the elementary solution given above) implies that an initial heat source U_0 localized in the neighborhood of some point x_0 will cause an effect $u(t, x) \neq 0$ at a point x which is at an arbitrary distance from x_0, within a time $t > 0$ which is arbitrarily small.

10.4.2 Summary of properties of the Fourier transformation

In a short table we summarize the basic properties and some important relations for the Fourier transformation. Following the physicists convention, we denote the variables for the functions in the domain of the Fourier transformation by x and the variables for functions in the range of the Fourier transformation by p. Though all statements have a counterpart in the general case, we present the one dimensional case in our table.

For a function f we denote by $\tilde{f}$ its Fourier transform $\mathcal{F}f$. In the table we use the words 'strongly decreasing' to express that a function f satisfies the condition defined by Equation 2.10.

As a summary of the table one can mention the following rule of thumb: If f or $T \in \mathcal{S}'(\mathbb{R})$ decays sufficiently rapidly at "infinity", then $\tilde{f}$, respectively $\tilde{T}$, is smooth, i.e., is a differentiable function, and conversely.

In the literature there are a good number of books giving detailed tables where the Fourier transforms of explicitly given functions are calculated. We mention the book by F. Oberhettinger, entitled " Fourier transforms of distributions and their inverses : a collection of tables", Academic Press, New York, 1973.

Properties of Fourier transformation $\mathcal{F}$	
Properties in x-space	Properties in p-space
decay for $\lvert x \rvert \to \infty$	local regularity
1) $f \in L^1(\mathbb{R})$ 2) f strongly decreasing 3) $f \in \mathcal{S}(\mathbb{R}_x)$ 4) $f(x) = e^{-ax^2}, a > 0$ 5) $f \in L^1(\mathbb{R})$, supp $f \subseteq [-a, a], a > 0$ 6) $T \in \mathcal{S}'(\mathbb{R})$, supp $T \subseteq [-a, a], a > 0$	1) $\tilde{f} \in \mathcal{C}(\mathbb{R})$ and $\lim_{\lvert p \rvert \to \infty} \tilde{f}(p) = 0$ 2) $\tilde{f} \in \mathcal{C}^\infty(\mathbb{R}) \cap L^1(\mathbb{R})$ 3) $\tilde{f} \in \mathcal{S}(\mathbb{R}_p)$ 4) $\tilde{f}(p) = e^{-\frac{p^2}{4a}}$ 5) $\tilde{f}$ analytic on $\mathbb{C}$ bounded by const $e^{a\lvert p \rvert}$ 6) $\tilde{T}$ analytic on $\mathbb{C}$, bounded by $Q(p)e^{a\lvert p \rvert}$, Q polynomial
7) $T \in \mathcal{S}'(\mathbb{R})$	7) $\tilde{T} \in \mathcal{S}'(\mathbb{R})$
growth for $\lvert x \rvert \to \infty$	local singularity
8) $1, e^{iax}$ 9) x^m 10) multiplication with $(ix)^m$	8) $\delta(p), \delta(p-a)$ 9) $\delta^{(m)}(p)$ 10) differential operator $(\frac{d}{dp})^m$
11) $\theta(\pm x)$ 12) sign x 13) $\theta(x)x^{m-1}$	11) $\frac{1}{p \pm io}$ 12) vp $\frac{1}{p}$ 13) $\frac{1}{(p+io)^m}$

10.5 Exercises

1. For $f \in L^1(\mathbb{R}^n)$ show:

$$\lVert f - f_a \rVert_1 \to 0 \quad \text{as} \quad a \to 0.$$

   Hints: Consider first continuous functions of compact support. Then approximate elements of $L^1(\mathbb{R}^n)$ accordingly.

2. Prove the first two properties of the Fourier transformation mentioned in Proposition 10.1.1:

(a) For $f \in L^1(\mathbb{R}^n)$ and $a \in \mathbb{R}^n$ the translation by a is defined as $f_a(x) = f(x-a)$ for almost all $x \in \mathbb{R}^n$. These translations and the multiplication by a corresponding exponential function are related under the Fourier transformation according to the following formulae:
   a) $\mathcal{F}(e^{iax}f)(p) = (\mathcal{F}f)_a(p)$ $\forall p \in \mathbb{R}^n$;
   b) $(\mathcal{F}f_a)(p) = e^{iap}(\mathcal{F}f)(p)$ $\forall p \in \mathbb{R}^n$.

(b) For any $\lambda > 0$ define the scaled function f_λ by $f_\lambda(x) = f(\frac{x}{\lambda})$ for almost all $x \in \mathbb{R}^n$. Then, for $f \in L^1(\mathbb{R}^n)$ one has

$$(\mathcal{F}f_\lambda)(p) = \lambda^n(\mathcal{F}f)(p) \qquad \forall p \in \mathbb{R}^n.$$

3. Prove: If $f \in L^1(\mathbb{R})$ has a derivative $f' \in L^1(\mathbb{R})$, then $f(x) \to 0$ as $|x| \to \infty$.

4. Show the embedding relation 10.6.

5. Prove: If I is an isomorphism of the Hausdorff locally convex topological vector space E, then the adjoint I' is an isomorphism of the topological dual equipped with weak topology σ.

6. Show that the right-hand of inequality (10.13) is a polynomially bounded function of $p \in \mathbb{R}^n$.

7. Prove the following relation:

$$\langle I_{\mathcal{F}f}, \phi \rangle = \langle I_f, \mathcal{F}\phi \rangle \qquad \forall \phi \in \mathcal{D}(\mathbb{R}^n), \quad \forall f \in L^1(\mathbb{R}^n).$$

8. Show that the Fourier transform $\mathcal{F}\phi$ of a test function $\phi \in \mathcal{D}(\mathbb{R}^n)$ is the restriction of an entire function to $\mathbb{R}^n$, i.e., $\mathcal{F}\phi$ is the restriction to $\mathbb{R}^n$ of the function

$$\mathbb{C}^n \ni z = (z_1, \ldots, z_n) \mapsto (2\pi)^{-\frac{n}{2}} \int_{\mathbb{R}^n} e^{iz \cdot \xi} \phi(\xi) d\xi$$

which is holomorphic on $\mathbb{C}^n$. Conclude: If $\phi \neq 0$, then $\mathcal{F}\phi$ cannot be a test function in $\mathcal{D}(\mathbb{R}^n)$ (it cannot have a compact support).

9. For any $a > 0$ introduce the function $g(x) = \prod_{k=1}^n e^{-a|x_k|}$ and show that

$$I(\mathcal{F}g) = \int (\mathcal{F}g)(p) dp = (2\pi)^{\frac{n}{2}}.$$

10. Assume that we know $\mathcal{F}'I_1 = (2\pi)^{\frac{n}{2}} \delta$. Then show that $\mathcal{L}(\mathcal{F}\phi)(x) = \phi(x)$ for all $x \in \mathbb{R}^n$ and all $\phi \in \mathcal{S}(\mathbb{R}^n)$, i.e., $\mathcal{L}\mathcal{F} = id$ on $\mathcal{S}(\mathbb{R}^n)$.

Hint: In a straightforward calculation use the relation $e^{ip \cdot x}(\mathcal{F}\phi)(p) = (\mathcal{F}\phi_{-x})(p)$.

11. Define the action of a rotation R of $\mathbb{R}^n$ on tempered distributions T on $\mathbb{R}^n$ by $R \cdot T = T \circ R^{-1}$ where $T \circ R^{-1}$ is defined by equation (4.6). Prove that the Fourier transformation $\mathcal{F}'$ commutes with this action of rotations:

$$R \cdot (\mathcal{F}'T) = \mathcal{F}'(R \cdot T) \qquad \forall T \in \mathcal{S}'(\mathbb{R}^n).$$

Conclude: If a distribution T is invariant under a rotation R (i.e., $R \cdot T = T$), so is its Fourier transform.

Hints: Show first that $(\mathcal{F}\phi) \circ R = \mathcal{F}(\phi \circ R)$ for all test functions ϕ.

12. In the notation of Lemma 10.3.1 show that the function $\mathbb{R}^n \ni p \mapsto p_{K,m}(e_p \cdot (-ix)^\alpha \cdot u)$ is polynomially bounded, for any $\alpha \in \mathbb{N}^n$ and any $u \in \mathcal{D}(\mathbb{R}^n)$.

13. Calculate the integral

$$\frac{1}{(2\pi)^3 2} \int_{\mathbb{R}^3} \frac{e^{ip \cdot x}}{p^2 + \mu^2} dp = \frac{1}{4\pi} \frac{e^{-|\mu||x|}}{|x|}.$$

Hints: Introduce polar co-ordinates and apply the Theorem of Residues 9.3.3.

14. Find a tempered elementary solution of the free Schrödinger operator.

11
Distributions and Analytic Functions

For reasons explained earlier we introduced various classes of distributions as elements of the topological dual of suitable test function spaces. Later we learned that distributions can also be defined as equivalence classes of certain Cauchy sequences of smooth functions or, locally, as finite order weak derivatives of continuous functions. In this chapter we learn that distributions have another characterization, namely as finite sums of boundary values of analytic functions.

11.1 Distributions as boundary values of analytic functions

This section introduces the subject for the case of one variable. We begin by considering a simple example discussed earlier from a different perspective. The function $z \mapsto \frac{1}{z}$ is analytic in $\mathbb{C}\setminus\{0\}$ and thus in particular in the upper and lower half planes H_+ and H_-,

$$H_\pm = \{z = x + iy \in \mathbb{C} : x \in \mathbb{C}, \pm y > 0\}.$$

The limits in $\mathcal{D}'(\mathbb{R})$ of the function $f(z) = \frac{1}{z} = \frac{1}{x+iy}$ exist for $y \to 0$, $z = x + iy \in H_\pm$ as we saw earlier,

$$\lim_{y \to 0,\, y>0} \frac{1}{x \pm iy} = \frac{1}{x \pm io}.$$

The distributions $\frac{1}{x \pm io}$ are called the *boundary values* of the analytic function $z \mapsto \frac{1}{z}$ restricted to the half planes $H_\pm$. In Section 3.2 we established the following

11. Distributions and Analytic Functions

relations between these boundary values with two other distributions, namely

$$\frac{1}{x+io} - \frac{1}{x-io} = -2\pi i\, \delta, \qquad \frac{1}{x+io} + \frac{1}{x-io} = 2\mathrm{vp}\,\frac{1}{x},$$

i.e., Dirac's delta distribution and Cauchy's principal value are represented as finite sums of boundary values of the function $z \mapsto \frac{1}{z}$, $z \in \mathbb{C}\setminus\{0\}$. In this chapter we will learn that every distribution can be represented as a finite sum of boundary values of analytic functions.

Recall that $\mathcal{A}(H_\pm)$ stands for the algebra of functions which are analytic on $H_\pm$. Every $F \in \mathcal{A}(H_+)$ defines naturally a family F_y, $y > 0$, of regular distributions on $\mathbb{R}$ according to the formula

$$\langle F_y, \phi \rangle = \int_\mathbb{R} F(x+iy)\phi(x)dx \qquad \forall \phi \in \mathcal{D}(\mathbb{R}). \tag{11.1}$$

The basic definition of a boundary value now reads as follows.

Definition 11.1.1 *A function $F \in \mathcal{A}(H_+)$ is said to have a **boundary value** $F_+ \in \mathcal{D}'(\mathbb{R})$ if, and only if, the family of regular distributions F_y, $y > 0$, has a limit in $\mathcal{D}'(\mathbb{R})$, for $y \to 0$, i.e., for every $\phi \in \mathcal{D}(\mathbb{R})$,*

$$\lim_{y \to 0} \langle F_y, \phi \rangle$$

exists in $\mathbb{C}$. The boundary value $F_+ \in \mathcal{D}'(\mathbb{R})$ is usually denoted by $F(x+io) \equiv F_+$.

The following result is a concrete characterization of those analytic functions which have a boundary value in the space of distributions.

Theorem 11.1.1 *A holomorphic function $F_\pm \in \mathcal{A}(H_\pm)$ has a boundary value in $\mathcal{D}'(\mathbb{R})$ if it satisfies the following condition:*

For every compact set $K \subset \mathbb{R}$ there are a positive constant C and an integer $m \in \mathbb{N}$ such that for all $x \in K$ and all $|y| \in (0, 1]$ the estimate

$$|F_\pm(x+iy)| \leq \frac{C}{|y|^m} \tag{11.2}$$

holds.

Proof. Consider the case of the upper halfplane. In order to show that the above condition is sufficient for the existence of a boundary value of F one has to show that under this condition, for each $\phi \in \mathcal{D}(\mathbb{R})$, the auxiliary function

$$g(y) = \langle F_y, \phi \rangle = \int_\mathbb{R} F(x+iy)\phi(x)dx, \qquad y > 0$$

has a limit for $y \to 0$.

It is clear (for instance by Corollary 6.2.1) that this function g is of class $C^\infty((0, \infty))$. Since F is holomorphic, it satisfies the Cauchy–Riemann equations $\partial_y F = i\partial_x F$ for all $z = x+iy \in H_+$. (Recall that ∂_x stands for $\frac{\partial}{\partial x}$). This allows us to express derivatives of g as follows, for $n \in \mathbb{N}$ and any $y > 0$:

$$\begin{aligned} g^{(n)}(y) &= \int_\mathbb{R} (\partial_y)^n F(x+iy)\phi(x)dx \\ &= \int_\mathbb{R} (i\partial_x)^n F(x+iy)\phi(x)dx = (-i)^n \int_\mathbb{R} F(x+iy)\phi^{(n)}(x)dx. \end{aligned}$$

11.1 Distributions as boundary values of analytic functions

The Taylor expansion of g at $y=1$ reads

$$g(y) = \sum_{j=0}^{n} \frac{g^{(j)}(1)}{j!}(y-1)^j + R_n(y), \quad R_n(y) = \frac{1}{n!}\int_1^y (y-t)^n g^{(n+1)}(t)dt.$$

This expansion shows that $g(y)$ has a limit for $y \to 0$ if, and only if, the remainder $R_n(y)$ does, for some $n \in \mathbb{N}$. Apply the hypothesis on F for the compact set $K = \operatorname{supp}\phi$. This then gives a constant C and an integer $m \in \mathbb{N}$ such that the estimate of our hypothesis holds. For this integer we deduce ($|K|$ denotes the Lebesgue measure of the set K)

$$|g^{(m+1)}(t)| \leq \int_{\mathbb{R}} |F(x+it)||\phi^{(m+1)}(x)|dx \leq \frac{C}{t^m}|K|p_{K,m+1}(\phi)$$

for all $0 < y \leq t \leq 1$, and this implies that for $n = m$ the remainder has a limit for $t \to 0$, and this limit is

$$\lim_{y \to 0, y > 0} R_m(y) = \frac{(-1)^{m+1}}{m!} \int_0^1 t^m g^{(m+1)}(t)dt.$$

Since $\phi \in \mathcal{D}(\mathbb{R})$ is arbitrary, we conclude by Theorem 3.3.1 that F has a boundary value in $\mathcal{D}'(\mathbb{R})$. □

The restriction of the function $z \mapsto \frac{1}{z}$ to $H_\pm$ certainly belongs to $\mathcal{A}(H_\pm)$ and clearly these two analytic functions satisfy the condition (11.2), hence by Theorem 11.1.1 they have boundary values $\frac{1}{x \pm io}$. Thus we find on the basis of a general result what we have shown earlier by direct estimates. There we have also shown that the difference of these two boundary values equals $2\pi i \delta$. In the section on convolution we learned that $T * \delta = T$ for all $T \in \mathcal{D}'(\mathbb{R})$. Thus one would conjecture that every distribution on $\mathbb{R}$ is the difference of boundary values of analytic functions on the upper, respectively lower, half plane. This conjecture is indeed true. We begin with the easy case of distributions of compact support.

Theorem 11.1.2 *If $T \in \mathcal{E}'(\mathbb{R})$ has the compact support K, then there is a holomorphic function $\hat{T}$ on $\mathbb{C}\setminus K$ such that for all $f \in \mathcal{D}(\mathbb{R})$,*

$$T(f) = \lim_{\epsilon \searrow 0} \int_{\mathbb{R}} [\hat{T}(x+i\epsilon) - \hat{T}(x-i\epsilon)]f(x)dx. \tag{11.3}$$

Proof. For every $z \in \mathbb{R}^c = \mathbb{C}\setminus\mathbb{R}$ the Cauchy kernel $t \mapsto \frac{1}{2i\pi}\frac{1}{t-z}$ belongs to $\mathcal{E}(\mathbb{R})$. Hence a function $\hat{T}: \mathbb{R}^c \to \mathbb{C}$ is well defined by

$$\hat{T}(z) = \frac{1}{2i\pi}\langle T(t), \frac{1}{t-z}\rangle.$$

Since there is an $m \in \mathbb{N}$ such that T satisfies the estimate

$$|T(f)| \leq C \sup_{t \in K, \nu \leq m} |D^\nu f(t)|,$$

we find immediately the estimate

$$|\hat{T}(x+iy)| \leq C|y|^{-m-1} \quad \forall x \in \mathbb{R}, \, \forall y \neq 0.$$

Furthermore, the estimate for T implies that $\hat{T}$ can be analytically continued to $K^c = \mathbb{C}\setminus K$. For all $z, \zeta \in K^c$ one has, for $z \neq \zeta$,

$$\frac{1}{z-\zeta}\left[\frac{1}{t-z} - \frac{1}{t-\zeta}\right] = \frac{1}{(t-z)(t-\zeta)}.$$

As $\zeta \to z$ the right-hand side converges to $\frac{1}{(t-z)^2}$ in $\mathcal{E}(\mathbb{R})$. We conclude that

$$\frac{\hat{T}(z) - \hat{T}(\zeta)}{z - \zeta} = \frac{1}{2i\pi}\langle T(t), \frac{1}{(t-z)(t-\zeta)}\rangle \to \frac{1}{2i\pi}\langle T(t), \frac{1}{(t-z)^2}\rangle,$$

hence $\hat{T}$ is complex differentiable on K^c.

Now for $z = x + iy$, $y > 0$, we calculate $\hat{T}(z) - \hat{T}(\bar{z}) = \langle T(t), \chi_y(t-x)\rangle$ where $\chi_y(t) = \frac{1}{\pi}\frac{y}{t^2+y^2}$.
This allows us to write, for $f \in \mathcal{D}(\mathbb{R})$,

$$\int_\mathbb{R} [\hat{T}(x+iy) - \hat{T}(x-iy)]f(x)dx = \int_\mathbb{R} \langle T(t), \chi_y(t-x)\rangle f(x)dx.$$

In the Exercises of the chapter on convolution products (Section 7.4) we have shown that this equals

$$\langle T(t), (\chi_y * f)(t)\rangle.$$

According to the Breit–Wigner Formula (3.9) $\chi_y \to \delta$ as $y \searrow 0$, hence $(\chi_y * f) \to f$ in $\mathcal{D}(\mathbb{R})$ as $y \searrow 0$, and it follows that $\langle T(t), (\chi_y * f)(t)\rangle \to \langle T(t), f(t)\rangle = T(f)$ as $y \searrow 0$. We conclude that the formula (11.3) holds. □

Note that in Theorem 11.1.2 the condition $f \in \mathcal{D}(\mathbb{R})$ cannot be replaced by $f \in \mathcal{E}(\mathbb{R})$. A careful inspection of the proof however shows that formula (11.3) can be extended to all $f \in \mathcal{E}(\mathbb{R})$ which are bounded and which have bounded derivatives. In this case the convolution products occurring in the proof are well defined too.

When one wants to extend Theorem 11.1.2 to the case of general distributions $T \in \mathcal{D}'(\mathbb{R})$ one faces the problem that the Cauchy kernel belongs to $\mathcal{E}(\mathbb{R})$ but not to $\mathcal{D}(\mathbb{R})$. Thus a suitable approximation of T by distributions with compact support is needed. As shown in Theorem 5.9 of the book [Bre65] this strategy is indeed successful.

Theorem 11.1.3 *For every $T \in \mathcal{D}'(\mathbb{R})$ there is an analytic function F on K^c, $K = \mathrm{supp}\, T$, satisfying the growth condition (11.2) on $H_\pm$ such that*

$$T(f) = \lim_{\epsilon \searrow 0} \int_\mathbb{R} [F(x+i\epsilon) - F(x-i\epsilon)]f(x)dx \qquad (11.4)$$

for all $f \in \mathcal{D}(\mathbb{R})$. One writes $T(x) = F(x+io) - F(x-io)$.

Similar results are available for distributions of more than one variable. This case is much more difficult than the one-dimensional case for a variety of reasons. Let us mention the basic ones. 1) One has to find an appropriate generalization of the process of taking boundary values from above and below the real line. 2) In the theory of analytic functions of more than one complex variable one encounters a number of subtle difficulties absent in the one-dimensional theory.

We sketch the solution due to A. Martineau [Mar64]. Suppose that $U \subset \mathbb{C}^n$ is a pseudo-convex open set (for the definition of this concept we have to refer to Definition 2.6.8 of the book [Hör67]) and $\Gamma \subset \mathbb{R}^n$ an open convex cone. Suppose furthermore that F is a holomorphic function on $U_\Gamma = (\mathbb{R}^n + i\Gamma) \cap U$ which satisfies the following condition: For every compact subset $K \subset \Omega = \mathbb{R}^n \cap U$ and every closed subcone $\Gamma' \subset \Gamma$ there are positive constants C and k such that

$$\sup_{x \in K} |F(x+iy)| \leq C|y|^{-k} \qquad \forall y \in \Gamma'. \qquad (11.5)$$

Then $F(x+iy)$ has the boundary value $F(x+i\Gamma 0)$ which is a distribution on Ω and, as y tends to zero in a closed subcone $\Gamma' \subset \Gamma$,

$$F(x+iy) \to F(x+i\Gamma 0) \qquad \text{in} \quad \mathcal{D}'(\Omega). \qquad (11.6)$$

For the converse suppose that a distribution $T \in \mathcal{D}'(\Omega)$ is given on $\Omega = \mathbb{R}^n \cap U$. Then there are open convex cones $\Gamma_1, \ldots, \Gamma_m$ in $\mathbb{R}^n$ such that their dual cones $\Gamma_1^o, \ldots, \Gamma_m^o$ cover the dual space of $\mathbb{R}^n$ ($\Gamma_j^o = \{\xi \in \mathbb{R}^n : \xi \cdot x \geq 0 \, \forall \, x \in \Gamma_j\}$) and holomorphic functions F_j on U_{Γ_j}, $j = 1, \ldots, m$, each satisfying the growth condition (11.5), such that T is the sum of the boundary values of these holomorphic functions:

$$T(x) = F_1(x + i\Gamma_1 0) + \cdots + F_m(x + i\Gamma_m 0). \tag{11.7}$$

11.2 Exercises

1. For $n = 1, 2, \ldots$ define $f_n(z) = \frac{1}{z^n}$, $z \in \mathbb{C}\setminus\{0\}$ and show that the functions $f_n^\pm = f_n|H_\pm$ have boundary values $\frac{1}{(x \pm io)^n}$ in $\mathcal{D}'(\mathbb{R})$. Then prove the formula
$$\frac{1}{(x+io)^{n+1}} = \frac{(-1)^n}{n!} D^n \frac{1}{x+io}$$
where D denotes the distributional derivative.

2. For $f \in L^1(\mathbb{R})$ define two functions $F_\pm$ on $H_\pm$ by the formula
$$F_\pm(z) = \frac{1}{2i\pi} \int_\mathbb{R} \frac{f(x)}{x-z} dx \qquad \forall \, z \in H_\pm.$$

   Show:

   (a) $F_\pm$ is well defined and is estimated by
   $$|F_\pm(x+iy)| \leq \frac{1}{2\pi|y|} \|f\|_1 \qquad \forall \, z \in H_\pm.$$

   (b) $F_\pm$ is holomorphic on $H_\pm$.
   (c) $F_\pm$ has a boundary value $f_\pm \in \mathcal{D}'(\mathbb{R})$.
   (d) For a Hölder-continuous function $f \in L^1(\mathbb{R})$ show that the boundary values are given by
   $$f_\pm = \pm\frac{1}{2}f + \frac{1}{2i\pi}(\text{vp}\frac{1}{x}) * f$$
   and deduce $f = f_+ - f_-$.

3. a) Suppose a function $f \in L^1_{\text{loc}}(\mathbb{R})$ has its support in $\mathbb{R}_+$ and there are some constants a, C such that $|f(\xi)| \leq Ce^{a\xi}$ for almost all $\xi \in \mathbb{R}_+$. Introduce the half plane $H_a = \{z \in \mathbb{C} : \text{Re}\, z > a\}$ and show that
$$\hat{f}(z) = \int_0^\infty e^{-z\xi} f(\xi) d\xi \tag{11.8}$$
is a well defined analytic function on H_a.

11. Distributions and Analytic Functions

b) Suppose a distribution $u \in \mathcal{E}'(\mathbb{R})$ has its support in the interval $[-a, a]$, for some $a > 0$. Prove that

$$\hat{u}(z) = \langle u(\xi), e^{-z\xi} \rangle \qquad (11.9)$$

is a well defined analytic function on the complex plane $\mathbb{C}$ and show that there is a constant C such that

$$|\hat{u}(z)| \leq C e^{a \operatorname{Re} z} \qquad \forall z \in \mathbb{C}.$$

The function $\hat{f}$ is called the *Laplace transform of the function f* usually written as $\hat{f}(z) = (\mathcal{L}f)(z)$ and similarly the function $\hat{u}$ is called the *Laplace transform of the distribution $u \in \mathcal{E}'(\mathbb{R})$*, also denoted usually by $\hat{u}(z) = (\mathcal{L}u)(z)$. For further details on the Laplace transform and related transformations see [Wid71, Dav02].

Hints: For the proof of the second part one can use the representation of distributions as weak derivatives of functions.

12
Other Spaces of Generalized Functions

For a nonempty open set $\Omega \subseteq \mathbb{R}^n$ we have introduced three classes of distributions or generalized functions, distributions with compact support $\mathcal{E}'(\Omega)$, tempered distributions $\mathcal{S}'(\Omega)$ and general distributions $\mathcal{D}'(\Omega)$ and we have found that these spaces of distributions are related by the inclusions

$$\mathcal{E}'(\Omega) \subset \mathcal{S}'(\Omega) \subset \mathcal{D}'(\Omega).$$

These distributions are often called *Schwartz distributions*. They have found numerous applications in mathematics and physics. One of the most prominent areas of successful applications of Schwartz distributions and their subclasses has been the solution theory of linear partial differential operators with constant coefficients as it is documented in the monograph of L. Hörmander ([Hör83a, Hör83b]). Though distributions do not admit in general a product, certain subclasses have been successfully applied in solving many important classes of nonlinear partial differential equations. These classes of distributions are the *Sobolev spaces* $W^{m,p}(\Omega)$, $m \in \mathbb{N}$, $1 \leq p < \infty$, $\Omega \subseteq \mathbb{R}^n$ open and nonempty and related spaces. We will use them in solving some nonlinear partial differential equations through the variational approach in Part III.

In physics, mainly tempered distributions are used, since there Fourier transformation is a very important tool in connecting the position representation with the momentum representation of the theory, and the class of tempered distributions is the only class of Schwartz distributions which is invariant under the Fourier transformation. General relativistic quantum field theory in the sense of Gårding and Wightman [WG64, SW64, Jos65, BLOT90] is based on the theory of tempered distributions.

160   12. Other Spaces of Generalized Functions

All Schwartz distributions are *localizable* and the notion of *support* is well defined for them via duality and the use of compactly supported test functions. Furthermore all these distributions are locally of finite order. This gives Schwartz distributions a relatively simple structure but limits their applicability in an essential way. Another severe limitation for the use of tempered distributions in physics is the fact that they are polynomially bounded, since in physics one often has to deal with exponential functions, for instance e^x, $x \in \mathbb{R}$, which is a distribution but not a tempered distribution on $\mathbb{R}$. These are some very important reasons to look for more general classes of generalized functions than the Schwartz distributions. And certainly a systematic point of view invites a study of other classes of generalized functions too. Accordingly we discuss the most prominent spaces of generalized functions which are known today from the point of view of their applicability to a solution theory of more general partial differential operators and in physics. However we do not give proofs in this chapter since its intention is just to inform about the existence of these other spaces of generalized functions and to stimulate some interest.

The first section presents the generalized functions with test function spaces of Gelfand type $\mathcal{S}$. The next section introduces *hyperfunctions* and in particular *Fourier hyperfunctions* and the final section explains *ultra-distributions* according to Komatsu.

12.1 Generalized functions of Gelfand type $\mathcal{S}$

The standard reference for this section is Chapter IV of [GŠ72] in which one finds all the proofs for the statements.

Denote $\mathbb{N}_0 = \{0, 1, 2, 3, \ldots\}$ and introduce for $0 < a, L < \infty$, $j \in \mathbb{N}$, $m \in \mathbb{N}_0$ the following functions on $\mathcal{C}^\infty(\mathbb{R}^n)$:

$$q(f; a, m, L, j) = \sup \left\{ \frac{|x^k D^\alpha f(x)|}{(L + \frac{1}{j})^{|k|} k^{ak}} : x \in \mathbb{R}^n, k, \alpha \in \mathbb{N}_0^n, |\alpha| \leq m \right\} \quad (12.1)$$

The set of all functions $f \in \mathcal{C}^\infty(\mathbb{R}^n)$ for which $q(f; a, m, L, j)$ is finite for all $m \in \mathbb{N}_0$ and all $j \in \mathbb{N}$ is denoted by

$$\mathcal{S}_{a,L}(\mathbb{R}^n). \quad (12.2)$$

Equipped with the system of norms $q(\cdot; a, m, L, j)$, $m \in \mathbb{N}_0$, $j \in \mathbb{N}$, it is a Fréchet space. Finally we take the inductive limit of these spaces with respect to L to get

$$\mathcal{S}_a(\mathbb{R}^n) = \text{ind lim}_{L>0} \, \mathcal{S}_{a,L}(\mathbb{R}^n). \quad (12.3)$$

For a nonempty open subset $\Omega \subset \mathbb{R}^n$ the spaces $\mathcal{S}_a(\Omega)$ are defined in the same way by replacing $\mathcal{C}^\infty(\mathbb{R}^n)$ by $\mathcal{C}^\infty(\Omega)$ and by taking the supremum over $x \in \Omega$ instead of $x \in \mathbb{R}^n$.

Some basic properties of the class of spaces $\mathcal{S}_a(\Omega)$, $a > 0$, are collected in

12.1 Generalized functions of Gelfand type $\mathcal{S}$

Proposition 12.1.1 *Suppose $0 < a \leq a'$ and consider any open nonempty subset $\Omega \subseteq \mathbb{R}^n$, then*

$$\mathcal{D}(\Omega) \subset \mathcal{S}_a(\Omega) \subset \mathcal{S}_{a'}(\Omega) \subset \mathcal{S}(\Omega). \tag{12.4}$$

In this chain each space is densely contained in its successor and all the embeddings are continuous.

Similarly we introduce another class of test function spaces $\mathcal{S}^b(\mathbb{R}^n)$ distinguished by a parameter $b > 0$. For $f \in C^\infty(\mathbb{R}^n)$, $m \in \mathbb{N}_0$ and $j \in \mathbb{N}$ define

$$p(f; b, m, M, j) = \sup \left\{ \frac{|x^k D^\alpha f(x)|}{(M + \frac{1}{j})^{|\alpha|} \alpha^{b\alpha}} : x \in \mathbb{R}^n, k, \alpha \in \mathbb{N}_0^n, |k| \leq m \right\}. \tag{12.5}$$

The set of all $f \in C^\infty(\mathbb{R}^n)$ for which $p(f; b, m, M, j)$ is finite for all $m \in \mathbb{N}_0$ and all $j \in \mathbb{N}$ is denoted by $\mathcal{S}^{b,M}(\mathbb{R}^n)$. Equipped with the system of norms $p(\cdot; b, m, L, j)$, $j \in \mathbb{N}$, $m \in \mathbb{N}_0$, the spaces

$$\mathcal{S}^{b,M}(\mathbb{R}^n) \tag{12.6}$$

are Fréchet spaces (see Chapter IV of [GŠ72]). Again we take the inductive limit of these spaces with respect to $M > 0$ to obtain

$$\mathcal{S}^b(\mathbb{R}^n) = \text{ind}\lim_{M>0} \mathcal{S}^{b,M}(\mathbb{R}^n). \tag{12.7}$$

Note the important difference in the definition of the spaces $\mathcal{S}^b(\mathbb{R}^n)$ and the spaces $\mathcal{S}_a(\mathbb{R}^n)$: In the definition of the continuous norms for these spaces the rôles of multiplication with powers of the variable x and the derivative monomials D^α have been exchanged and therefore according to the results on the Fourier transformation (see Proposition 10.1.1) one would expect that the Fourier transform maps these spaces into each other. Indeed the precise statement about this connection is contained in the following proposition.

Proposition 12.1.2 *The Fourier transformation $\mathcal{F}$ is a homeomorphism $\mathcal{S}^b(\mathbb{R}^n) \to \mathcal{S}_b(\mathbb{R}^n)$.*
Suppose $0 < b \leq b'$, then

$$\mathcal{S}^b(\mathbb{R}^n) \subset \mathcal{S}^{b'}(\mathbb{R}^n) \subset \mathcal{S}(\mathbb{R}^n). \tag{12.8}$$

In this chain each space is densely contained in its successor and all the embeddings are continuous.

The elements of $\mathcal{S}^1(\mathbb{R}^n)$ are analytic functions and those of $\mathcal{S}^b(\mathbb{R}^n)$ for $0 < b < 1$ are entire analytic.

A third class of test function spaces of type $\mathcal{S}$ is the intersection of the spaces defined above. They can be defined directly as an inductive limit of spaces $\mathcal{S}_{a,L}^{b,M}(\mathbb{R}^n)$ with respect to $L, M > 0$. To this end consider the following system of norms on

$\mathcal{C}^\infty(\mathbb{R}^n)$, for $L, M > 0$ and $j, m \in \mathbb{N}$:

$$q(f; a, b, m, j, L, M) =$$

$$= \sup \left\{ \frac{|x^k D^\alpha f(x)|}{(L + \frac{1}{j})^{|k|} k^{ak} (M + \frac{1}{m})^{|\alpha|} \alpha^{b\alpha}} : x \in \mathbb{R}^n, k, \alpha \in \mathbb{N}_0^n \right\}. \qquad (12.9)$$

Denote the set of functions $f \in \mathcal{C}^\infty(\mathbb{R}^n)$ for which $q(f; a, b, m, j, L, M)$ is finite for all $m, j \in \mathbb{N}$ by

$$\mathcal{S}_{a,L}^{b,M}(\mathbb{R}^n). \qquad (12.10)$$

Equipped with the system of norms $q(\cdot; a, b, m, j, L, M)$, $m, j \in \mathbb{N}$, the space $\mathcal{S}_{a,L}^{b,M}(\mathbb{R}^n)$ is a Fréchet space. The third class of test function spaces is now defined by

$$\mathcal{S}_a^b(\mathbb{R}^n) = \text{ind} \lim_{M>0, L>0} \mathcal{S}_{a,L}^{b,M}(\mathbb{R}^n) \qquad (12.11)$$

for $a, b > 0$. For a function $f \in \mathcal{C}^\infty(\mathbb{R}^n)$ to be an element of $\mathcal{S}_a^b(\mathbb{R}^n)$, it has to satisfy the constraints both from $\mathcal{S}_a(\mathbb{R}^n)$ and $\mathcal{S}^b(\mathbb{R}^n)$ with the effect that for certain values of the parameters $a, b > 0$ only the trivial function $f = 0$ is allowed.

Proposition 12.1.3 *The spaces $\mathcal{S}_a^b(\mathbb{R}^n)$ are not trivial if, and only if,*

$$a + b \geq 1, \ a > 0, b > 0 \quad \text{or} \quad a = 0, b > 1 \quad \text{or} \quad a > 1, b = 0.$$

The Fourier transformation $\mathcal{F}$ is a homeomorphism $\mathcal{S}_a^b(\mathbb{R}^n) \to \mathcal{S}_b^a(\mathbb{R}^n)$.

Suppose $0 < a \leq a'$ and $0 < b \leq b'$ such that the space $\mathcal{S}_a^b(\mathbb{R}^n)$ is not trivial, then $\mathcal{S}_a^b(\mathbb{R}^n)$ is densely contained in $\mathcal{S}_{a'}^{b'}(\mathbb{R}^n)$ and the natural embedding is continuous.

In addition we have the following continuous embeddings:

$$\mathcal{S}_a^b(\mathbb{R}^n) \subset \mathcal{S}(\mathbb{R}^n), \quad \mathcal{S}_a^b(\mathbb{R}^n) \subset \mathcal{S}_a(\mathbb{R}^n), \quad \mathcal{S}_a^b(\mathbb{R}^n) \subset \mathcal{S}^b(\mathbb{R}^n). \qquad (12.12)$$

The elements in $\mathcal{S}_a^1(\mathbb{R}^n)$ are analytic functions and those in $\mathcal{S}_a^b(\mathbb{R}^n)$ for $0 < b < 1$ are entire analytic, i.e., they have extensions to analytic, respectively to entire analytic, functions.

The topological dual $\mathcal{S}_a^b(\mathbb{R}^n)'$ of $\mathcal{S}_a^b(\mathbb{R}^n)$ defines the *class of generalized functions of Gelfand type $\mathcal{S}_a^b$*. Thus we get a two-parameter family of spaces of generalized functions. Since $\mathcal{S}_a^b(\mathbb{R}^n) \subset \mathcal{S}(\mathbb{R}^n)$ with continuous embedding we know that these new classes of generalized functions contain the space of tempered distributions:

$$\mathcal{S}'(\mathbb{R}^n) \subset \mathcal{S}_a^b(\mathbb{R}^n)'.$$

There are three important aspects under which one can look at these various spaces of generalized functions:

a) Does this space of generalized functions admit the Fourier transformation as a homeomorphism (isomorphism)?

b) Are the generalized functions of this space localizable?

c) Are the Fourier transforms of the generalized functions of the space localizable?

These questions are relevant in particular for applications to the theory of partial differential operators and in mathematical physics (relativistic quantum field theory).

One can show that the spaces $\mathcal{S}_a^b(\mathbb{R}^n)$, $1 < b$, contain test functions of compact support. Thus for generalized functions over these test function spaces the concept of a support can be defined as usual. Since the Fourier transformation maps the space $\mathcal{S}_a^b(\mathbb{R}^n)$ into $\mathcal{S}_b^a(\mathbb{R}^n)$, all three questions can be answered affirmatively for the spaces $\mathcal{S}_a^b(\mathbb{R}^n)$ $1 < a, b < \infty$.

According to Proposition 12.1.3 the smaller the parameters $a, b > 0$ are the smaller is the test function space $\mathcal{S}_a^b(\mathbb{R}^n)$ and thus the larger is the corresponding space of generalized functions. Therefore it is worthwhile to consider generalized functions over the spaces $\mathcal{S}_a^b(\mathbb{R}^n)$ with $0 < a \leq 1$ and/or $0 < b \leq 1$ too. However according to Proposition 12.1.3 elements of the spaces $\mathcal{S}_a^b(\mathbb{R}^n)$, $0 < b \leq 1$ are analytic functions. Since there are no nontrivial analytic functions with compact support, the localization of the generalized functions with this test function space cannot be defined through compactly supported test functions as in the case of Schwartz distributions. Thus it is not obvious how to define the concept of support in this case.

The topological dual of a space of analytic functions is called a space of *analytic functionals*. As we are going to indicate, analytic functionals admit the concept of a *carrier* which is the counterpart of the concept of support of a Schwartz distribution.

Let $\Omega \subset \mathbb{C}^n$ be a nonempty open set and consider the space $\mathcal{O}(\Omega)$ of holomorphic functions on Ω equipped with the system of semi-norms

$$|f|_K = \sup_{z \in K} |f(z)|, \qquad K \subset \Omega \quad \text{compact.} \tag{12.13}$$

Since Ω can be exhausted by a sequence of compact sets, the space $\mathcal{O}(\Omega)$ is actually a Fréchet space. For $T \in \mathcal{O}(\Omega)'$ there are a constant $C, 0 \leq C < \infty$, and a compact set $K \subset \Omega$ such that

$$|T(f)| \leq C|f|_K \qquad \forall f \in \mathcal{O}(\Omega). \tag{12.14}$$

The compact set K of relation (12.14) is called a *carrier of the analytic functional T*. Naturally one would like to proceed to define the *support of an analytic functional* as the smallest of its carriers. But in general this does not exist and thus the concept of support is not always available. In this context it is worthwhile to recall the definition $\mathcal{E}'$ of Schwartz distributions of compact support where the same type of topology is used.

With regard to our three questions the space $\mathcal{S}_1^1(\mathbb{R}^n)$ plays a distinguished rôle since it is invariant under the Fourier transform and elements of its topological

12.2 Hyperfunctions and Fourier hyperfunctions

Recall the representation

$$T(x) = F_1(x + i\Gamma_1 0) + \cdots + F_m(x + i\Gamma_m 0) \tag{12.15}$$

of a distribution $T \in \mathcal{D}'(\Omega)$ as a finite sum of boundary values of certain holomorphic functions $F_1, \ldots, F_m$, each of which satisfyies a growth condition of the form (11.5). In a series of articles [Sat58, Sat59, Sat60] M. Sato has shown how to give a precise mathematical meaning to a new class of generalized functions when in the above representation of distributions as a sum of boundary values of analytic functions all growth restrictions are dropped. For this he used a cohomological method and called these new generalized functions *hyperfunctions on* Ω. In this way a hyperfunction T on Ω is identified with a class of m-tuples of holomorphic functions. When equation (12.15) holds, one calls $\{F_1, \ldots, F_m\}$ *defining functions* of the hyperfunction T.

The space of all hyperfunctions on Ω is denoted by $\mathcal{B}(\Omega)$. From the above definition it is evident that it contains all Schwartz distributions on Ω:

$$\mathcal{D}'(\Omega) \subset \mathcal{B}(\Omega).$$

It has to be emphasized that in contrast to the other spaces of generalized functions we have discussed thus far the space $\mathcal{B}(\Omega)$ is not defined as the topological dual of some test function space.

Spaces of hyperfunctions are well suited for a solution theory of linear differential operators with real analytic coefficients (see [Kom73a]). Consider for example the ordinary differential operator

$$P(x, D) = a_m(x)D^m + \cdots + a_1(x)D + a_0(x), \qquad D = \frac{d}{dx}$$

with a_j, $j = 1, \ldots, m$, real analytic functions on some open interval $\Omega \subset \mathbb{R}$, $a_m \neq 0$. In [Kom73a] it is shown how a comprehensive and transparent solution theory for

$$P(x, D)u(x) = T(x) \tag{12.16}$$

can be given in the space $\mathcal{B}(\Omega)$ of all hyperfunctions on Ω, for any given $T \in \mathcal{B}(\Omega)$.

As in the case of Schwartz distributions, one can characterize the subspace of those hyperfunctions which admit the Fourier transformation as an isomorphism (for this appropriate growth restrictions at infinity are needed). This subspace is called the space of *Fourier hyperfunctions*. Later the space of Fourier hyperfunctions on $\mathbb{R}^n$ was recognized as the topological dual of the test function space of rapidly decreasing analytic functions $\mathcal{Q}(D^n)$ which is isomorphic to the space

12.2 Hyperfunctions and Fourier hyperfunctions

$\mathcal{S}_1^1(\mathbb{R}^n)$ introduced in the previous section. Briefly the space $\mathcal{O}(\boldsymbol{D}^n)$ can be described as follows (see[Kan88]).

First we recall the radial compactification $\boldsymbol{D}^n$ of $\mathbb{R}^n$. Let S_∞^{n-1} be the $(n-1)$-dimensional sphere at infinity, which is homeomorphic to the unit sphere $S^{n-1} = \{x \in \mathbb{R}^n; |x| = 1\}$ by the mapping $x \to x_\infty$, where the point $x_\infty \in S_\infty^{n-1}$ lies on the ray connecting the origin with the point $x \in S^{n-1}$. The set $\mathbb{R}^n \cup S_\infty^{n-1}$, equipped with its natural topology (a fundamental system of neighborhoods of x_∞ is the set of all the sets $O_{\Omega, R}(x_\infty)$ given by:

$$O_{\Omega, R}(x_\infty) = \{\xi \in \mathbb{R}^n; \xi/|\xi| \in \Omega, \ |\xi| > R\} \cup \{\xi_\infty; \xi \in \Omega\}$$

for every neighborhood Ω of x in S^{n-1} and $R > 0$), is denoted by $\boldsymbol{D}^n$, called the radial compactification of $\mathbb{R}^n$. Equip the space $Q^n = \boldsymbol{D}^n \times i\mathbb{R}^n$ with its natural product topology. Clearly, $\mathbb{C}^n = \mathbb{R}^n \times i\mathbb{R}^n$ is embedded in Q^n. Let K be a compact set in $\boldsymbol{D}^n$, $\{U_m\}$ a fundamental system of neighborhoods of K in Q^n and $\mathcal{O}_c^m(U_m)$ the Banach space of functions f analytic in $U_m \cap \mathbb{C}^n$ and continuous on $\bar{U}_m \cap \mathbb{C}^n$ which satisfy

$$\|f\|_m = \sup_{z \in U_m \cap \mathbb{C}^n} |f(z)| e^{|z|/m} < \infty.$$

Finally we introduce the inductive limit of these Banach spaces of analytic functions

$$\mathcal{O}(K) = \text{ind} \lim_{m \to \infty} \mathcal{O}_c^m(U_m).$$

It has the following properties:

Proposition 12.2.1 *Let $K \subseteq \boldsymbol{D}^n$ be compact. Then the space $\mathcal{O}(K)$ is a DFS-space (a dual Fréchet – Schwartz space), i.e., all the embedding mappings*

$$\mathcal{O}_c^m(U_m) \to \mathcal{O}_c^{m+1}(U_{m+1}), \quad m = 1, 2, \ldots,$$

are compact.
The space $\mathcal{O}(\boldsymbol{D}^n)$ is dense in $\mathcal{O}(K)$.
The Fourier transform $\mathcal{F}$ is well defined on $\mathcal{O}(\boldsymbol{D}^n)$ by the standard formula

$$(\mathcal{F}f)(p) = (2\pi)^{-n/2} \int e^{ip \cdot x} f(x) dx.$$

It is an isomorphism of the topological vector space $\mathcal{O}(\boldsymbol{D}^n)$.

Note that this inductive limit is not strict. Since $\mathcal{O}(\boldsymbol{D}^n)$ is dense in $\mathcal{O}(K)$, continuous extensions from $\mathcal{O}(\boldsymbol{D}^n)$ to $\mathcal{O}(K)$ are unique if they exist at all.

The topological dual $\mathcal{O}(\boldsymbol{D}^n)'$ of $\mathcal{O}(\boldsymbol{D}^n)$ is called the space of *Fourier hyperfunctions* on $\mathbb{R}^n$.

Suppose $T \in \mathcal{O}(\boldsymbol{D}^n)'$ is a Fourier hyperfunction. Introduce the class $C(T)$ of all those compact subsets $K \subseteq \boldsymbol{D}^n$ such that T has a continuous extension T_K to $\mathcal{O}(K)$. As we have mentioned above each $K \in C(T)$ is called a *carrier of T*.

On the basis of the Mittag–Leffler theorem for rapidly decreasing analytic functions (see [Kan88, NN90]) one proves the nontrivial

166   12. Other Spaces of Generalized Functions

Lemma 12.2.1 *For any $T \in \mathcal{O}(D^n)'$ one has*

$$K_1, K_2 \in C(T) \Rightarrow K_1 \cap K_2 \in C(T).$$

Corollary 12.2.1 *Fourier hyperfunctions T admit the concept of* **support**, *defined as the smallest carrier of T:*

$$\operatorname{supp} T = \cap_{K \in C(T)} K.$$

The *localization of Fourier hyperfunctions* means that for every open nonempty subset $\Omega \subset \mathbb{R}^n$ one has the space of Fourier hyperfunctions on Ω. This is summarized by stating that Fourier hyperfunctions form a (flabby) sheaf over $\mathbb{R}^n$ [Kom73a, Kan88].

Fourier hyperfunctions have an interesting and quite useful integral representation which uses analyticity of the test functions in a decisive way. For $j = 1, \ldots, n$ introduce the open set $W_j = \{z \in Q^n : \operatorname{Im} z_j \neq 0\}$. The intersection $W = \cap_{j=1}^n W_j$ of all these sets consists of 2^n open connected components of Q^n separated by the 'real points'. For every $z \in W$ introduce the function h_z defined by

$$h_z(t) = \prod_{j=1}^n \frac{e^{-(t_j - z_j)^2}}{2\pi i (t_j - z_j)}.$$

One shows $h_z \in \mathcal{O}(D^n)$ for every $z \in W$. Hence, for every $T \in \mathcal{O}(D^n)'$, we can define a function $\hat{T} : W \to \mathbb{C}$ by $\hat{T}(z) = T(h_z)$. It follows that $\hat{T}$ actually is a 'slowly increasing' analytic function on W. Now given $f \in \mathcal{O}(D^n)$ there is an $m \in \mathbb{N}$ such that $f \in \mathcal{O}_c^m(U_m)$. Hence we can find $\delta_m > 0$ such that $\Gamma_1 \times \cdots \times \Gamma_n \subset U_m \cap W \cap \mathbb{C}^n$ where $\Gamma_j = \Gamma_j^+ + \Gamma_j^-$ and $\Gamma_j^\pm = \{z_j = \pm x_j \pm i\delta_m : -\infty < x_j < \infty\}$. Since h_z is a modified Cauchy kernel with appropriate decay properties at infinity, an application of Cauchy's integral theorem implies

$$\int_{\Gamma_1 \times \cdots \times \Gamma_n} f(z) h_z(\cdot) dz = f(\cdot).$$

Now applying $T \in \mathcal{O}(D^n)'$ to this identity we get

$$\int_{\Gamma_1 \times \cdots \times \Gamma_n} f(z) \hat{T}(z) dz = T(f). \tag{12.17}$$

The integral on the lefthand side exists since $\hat{T}(z)$ is slowly increasing and $f(z)$ is 'rapidly decreasing'. Certainly one has to prove that the application of the Fourier hyperfunction T 'commutes' with integration so that T can be applied to the integrand of this path integral. Then in equation (12.17) one has a very useful structure theorem for Fourier hyperfunctions: Every Fourier hyperfunction is represented by a path integral over a slowly increasing analytic function on W. In this way the powerful theory of analytic functions can be used in the analysis of Fourier hyperfunctions.

Most results known for (tempered) distributions have been extended to (Fourier) hyperfunctions. And certainly there are a number of interesting results which are characteristic for (Fourier) hyperfunctions and which are not available for distributions. From a structural point of view and for applications the most important difference between Schwartz distributions and hyperfunctions is that hyperfunctions can locally be of *infinite order*. For instance the infinite series

$$\sum_{n=1}^{\infty} a_n \delta^{(n)}, \qquad \lim_{n \to \infty} (|a_n| n!)^{1/n} = 0$$

has a precise meaning as a (Fourier) hyperfunction. Actually all hyperfunctions with support in $\{0\}$ are of this form. Hence the set of hyperfunctions with support in $\{0\}$ is much larger than the set of distributions with support in a point (compare Proposition 4.4.3).

As an example consider the function $e^{-\frac{1}{z}}$ which is defined and holomorphic on $\mathbb{C}\setminus\{0\}$. Hence one can consider $e^{-\frac{1}{z}}$ as a defining function of a hyperfunction $[e^{-\frac{1}{z}}]$ with support in $\{0\}$ and one shows (see [Kan88])

$$[e^{-\frac{1}{z}}] = \sum_{n=0}^{\infty} \frac{2\pi i}{n!(n+1)!} \delta^{(n)}.$$

In mathematical physics, Fourier hyperfunctions have been used successfully to extend the Gårding–Wightman formulation of relativistic quantum field theory considerably (see [NM76, BN89, NB01]). For other applications of hyperfunctions we refer to the books [Kom73a, Kan88].

12.3 Ultradistributions

The standard reference for this section is the article [Kom73b]. The theory of ultradistributions has been developed further in [Kom77, Kom82]. Ultradistributions are special hyperfunctions and the space of all ultradistributions on an open set $\Omega \subset \mathbb{R}^n$ is the strong dual of a test function space which is defined in terms of a sequence $(M_p)_{p \in \mathbb{N}_0}$ of positive numbers M_p satisfying the following conditions:

(M1) logarithmic convexity: $M_p^2 \leq M_{p-1} M_{p+1}$ for all $p \in \mathbb{N}$;

(M2) stability under ultradifferential operators (defined later): There are constants $C > 0$, $L > 1$ such that for all $p \in \mathbb{N}_0$,

$$M_p \leq CL^p \min_{0 \leq q \leq p} M_q M_{p-q};$$

(M3) strong non-quasianalyticity: There is a constant $C > 0$ such that for all $p \in \mathbb{N}$,

$$\sum_{q=p+1}^{\infty} \frac{M_{q-1}}{M_q} \leq Cp \frac{M_p}{M_{p+1}}.$$

For special purposes some weaker conditions suffice. Examples of sequences satisfying these conditions are the Gevrey sequences

$$M_p = (p!)^s \quad \text{or} \quad p^{ps} \quad \text{or} \quad \Gamma(1+ps)$$

for $s > 1$.

Now let $\Omega \subset \mathbb{R}^n$ be a nonempty open set. A function $f \in C^\infty(\Omega)$ is called an *ultradifferentiable function of class* M_p if, and only if, on each compact set $K \subset \Omega$ the derivatives of f are bounded according to the estimate

$$\|D^\alpha f\|_K = \sup_{x \in K} |D^\alpha f(x)| \leq C r^{|\alpha|} M_{|\alpha|}, \qquad \alpha \in \mathbb{N}_0^n \tag{12.18}$$

for some positive constants C and r. In order to make such a class of functions invariant under affine coordinate transformations, there are two ways to choose the constant r and accordingly we get two classes of ultradifferentiable functions: $f \in C^\infty(\Omega)$ is called an *ultradifferentiable function of class* (M_p) *(respectively of class* $[M_p]$*)* if condition (12.18) holds for every $r > 0$ (respectively for some $r > 0$).

$\mathcal{E}^{(M_p)}(\Omega)$ ($\mathcal{E}^{[M_p]}(\Omega)$) denotes the space of all ultradifferentiable functions of class (M_p) (of class $[M_p]$) on Ω. The corresponding subspaces of all ultradifferentiable functions with compact support are denoted by $\mathcal{D}^{(M_p)}(\Omega)$, respectively $\mathcal{D}^{[M_p]}(\Omega)$. All these spaces can be equipped with natural locally convex topologies, using the construction of inductive and projective limits.

Under these topologies the functional analytic properties of these spaces are well known (Theorem 2.6 of [Kom73b]), and we can form their strong duals $\mathcal{E}^{(M_p)}(\Omega)'$, $\mathcal{E}^{[M_p]}(\Omega)'$, $\mathcal{D}^{(M_p)}(\Omega)'$, $\mathcal{D}^{[M_p]}(\Omega)'$.

$\mathcal{D}^{(M_p)}(\Omega)'$ ($\mathcal{D}^{[M_p]}(\Omega)'$) is called the space of *ultradistributions of class* M_p *of Beurling type (of Roumieu type) or of class* (M_p) *(of class* $[M_p]$*)*.

Ultradistributions of class (M_p) (of class $[M_p]$) each form a (soft) sheaf over $\mathbb{R}^n$. Multiplication by a function in $\mathcal{E}^{(M_p)}(\Omega)$ (in $\mathcal{E}^{[M_p]}(\Omega)$) acts as a sheaf homomorphism.

These spaces of ultradistributions have been studied as comprehensively as Schwartz distributions but they have found up to now nearly no applications in physics or mathematical physics. The spaces of ultradistributions are invariant under a by far larger class of partial differential operators than the corresponding spaces of Schwartz distributions, and this was one of the major motivations for the construction of the spaces of ultradistributions. Consider a differential operator of the form

$$P(x, D) = \sum_{|\alpha| \leq m} a_\alpha(x) D^\alpha \qquad a_\alpha \in \mathcal{E}^*(\Omega). \tag{12.19}$$

It defines a linear partial differential operator $P(x, D) : \mathcal{D}^*(\Omega)' \to \mathcal{D}^*(\Omega)'$ as the dual of the formal adjoint $P'(x, D)$ operator of the operator $P(x, D)$ which is a continuous linear operator $\mathcal{D}^*(\Omega) \to \mathcal{D}^*(\Omega)$. Here $*$ stands for either (M_p) or $[M_p]$. In addition certain partial differential operators of *infinite order* leave the spaces of ultradistributions invariant and thus provide the appropriate setting for a study of such operators.

A partial differential operator of the form

$$P(D) = \sum_{|\alpha|=0}^{\infty} a_\alpha D^\alpha, \qquad a_\alpha \in \mathbb{C} \qquad (12.20)$$

is called an *ultradifferential operator of class* (M_p) *(of class* $[M_p]$*)* if there are constants r and C (for every $r > 0$ there is a constant C) such that

$$|a_\alpha| \leq C r^{|\alpha|} / M_{|\alpha|}, \qquad |\alpha| = 0, 1, 2, \ldots.$$

An ultradifferential operator of class $*$ maps the space of ultradistributions $\mathcal{D}^*(\Omega)'$ continuously into itself.

Part II
Hilbert Space Operators

13
Hilbert Spaces: A Brief Historical Introduction

13.1 Survey: Hilbert spaces

The eigenvalue problem in finite dimensional spaces was completely solved at the end of the 19th century. At the beginning of the 20th century the focus shifted to eigenvalue problems for certain linear partial differential operators of second order (e.g., Sturm–Liouville problems) and one realized quickly that these are eigenvalue problems in infinite dimensional spaces, which presented completely new properties and unexpected difficulties.

In an attempt to use, by analogy, the insight gathered in the finite dimensional case, also in the infinite dimensional case, one started with the problem of expanding 'arbitrary functions' in terms of systems of known functions according to the requirements of the problem under consideration, for instance exponential functions, Hermite functions, spherical functions, etc. The coefficients of such an expansion were viewed as the coordinates of the unknown function with respect to the given system of functions (V. Volterra, I. Fredholm, E. Schmidt). Clearly, in this context many mathematical problems had to be faced, for instance:

1. Which sequences of numbers can be interpreted as the sequence of coefficients of which functions?

2. Which notion of convergence is suitable for such an expansion procedure?

3. Which systems of functions, besides exponential and Hermite functions, can be used for such an expansion?

4. Given a differential operator of the type mentioned above how do we choose the system of functions for this expansion?

Accordingly we start our introduction into the theory of Hilbert spaces and their operators with some remarks on the history of this subject. The answers to the first two questions were given at the beginning of the 20th century by D. Hilbert in his studies of linear integral equations. They became the paradigm for this type of problems. Hilbert suggested using the space $\ell^2(\mathbb{R})$ of all sequences $x = (x_i)_{i \in \mathbb{N}}$ of real numbers x_i which are square summable and introduced new topological concepts which turned out to be very important later. Soon afterwards E. Schmidt, M. Fréchet, and F. Riesz gave Hilbert's theory a more geometrical form which emphasized the analogy with the finite dimensional Euclidean spaces $\mathbb{R}^n$ and $\mathbb{C}^n$, $n = 1, 2, \ldots$. This analogy is supported by the concept of an *inner product* or *scalar product* which depends on the dimension of the space and which provides the connection between the metric and geometric structures on the space. This is well known for Euclidean spaces and one expects that the notions and results known from Euclidean space are valid in general. Indeed, this turned out to be the case. We mention here the concepts of length, of angles, as well as orthogonality and results such as the theorem of Pythagoras, the theorem of diagonals, and Schwarz' inequality. This will be discussed in the section on the geometry of Hilbert spaces. However we will follow more the axiomatic approach to the theory of Hilbert spaces which was developed later, mainly by J. von Neumann and F. Riesz. In this approach a Hilbert space is defined as a vector space on which an inner product is defined in such a way that the space is complete with respect to the norm induced by the inner product. For details see the Chapter 14, "Inner product spaces and Hilbert spaces".

After the basic concepts of the theory of Hilbert spaces have been introduced a systematic study of the consequences of the concept of orthogonality follows in the section on the geometry of Hilbert spaces. The main results are the 'Projection Theorem' 15.1.1 and its major consequences. Here it is quite useful to keep the analogy with the Euclidean spaces in mind. Recall the direct orthogonal decomposition $\mathbb{R}^n = \mathbb{R}^p \oplus \mathbb{R}^q$, $p + q = n$. This decomposition has a direct counterpart in a general Hilbert space $\mathcal{H}$ and reads $\mathcal{H} = M \oplus M^\perp$ where M is any closed linear subspace of $\mathcal{H}$ and $M^\perp$ its 'orthogonal complement'.

A very important consequence of this decomposition is the characterization of the continuous linear functionals on a Hilbert space (Theorem of Riesz–Fréchet 15.3.1). According to this theorem a Hilbert space $\mathcal{H}$ and its topological dual space $\mathcal{H}'$ (as the space of all continuous linear functionals on $\mathcal{H}$) are 'isometrically anti-isomorphic'. Thus, in sharp contrast to the 'duality theory' of a general complete normed space, the 'duality theory' of a Hilbert space is nearly as simple as that of the Euclidean spaces. The reason is that the norm of a Hilbert space has a special form since it is defined by the inner product.

The expansion problem mentioned above receives a comprehensive solution in the 'theory of separable Hilbert spaces' which is based on the notions of an 'orthonormal basis' and 'Hilbert space basis' (Chapter 16, "Separable Hilbert spaces"). Certainly, in this context it is important to have a characterization of an orthonormal basis and a method to construct such a basis (Gram–Schmidt orthonormalization procedure).

Besides the sequence spaces $\ell^2(\mathbb{K})$, $\mathbb{K} = \mathbb{R}$ or $\mathbb{C}$, examples of Hilbert spaces which are important for us, are the Lebesgue spaces $L^2(\Omega, dx)$ and the Sobolev spaces $H^k(\Omega)$, $k = 1, 2, \ldots$, where Ω is a closed or an open subset of a Euclidean space $\mathbb{R}^n$, $n = 1, 2, \ldots$. For some of the Lebesgue spaces the problem of constructing an orthonormal basis is discussed in detail. It turns out that the system of exponential functions e_n,

$$e_n(x) = \frac{1}{\sqrt{2\pi}} e^{inx}, \qquad x \in [0, 2\pi], \quad n \in \mathbb{Z}$$

is an orthonormal basis of the Hilbert space $\mathcal{H} = L^2([0, 2\pi), dx)$. This means that every 'function' $f \in L^2([0, 2\pi]), dx)$ has an expansion with respect to these basis functions (Fourier expansion):

$$f = \sum_{n \in \mathbb{Z}} c_n e_n, \qquad c_n = \langle e_n, f \rangle_2 \equiv \int_0^{2\pi} \overline{e_n(x)} f(x) dx.$$

Here, naturally, the series converges with respect to the topology of the Hilbert space $L^2([0, 2\pi), dx)$. This shows that Fourier series can be dealt with in a simple and natural way in the theory of Hilbert spaces.

Next we construct an orthonormal basis for several 'weighted Lebesgue spaces' $L^2(I, \rho dx)$, for an interval $I = [a, b]$ and a weight function $\rho : I \to \mathbb{R}_+$. By specializing the interval and the weight function one thus obtains several well-known orthonormal systems of polynomials, namely the Hermite-, Laguerre- and Legendre polynomials.

We proceed with some remarks related to the second question. For the Euclidean spaces $\mathbb{R}^n$ one has a characterization of compact sets which is simple and convenient in applications: A subset $K \subset \mathbb{R}^n$ is compact if, and only if, it is bounded and closed. However in an infinite dimensional Hilbert space, as for instance the sequence space $\ell^2(\mathbb{R})$, a closed and bounded subset is not necessarily compact, with respect to the 'strong' or norm topology. This fact creates a number of new problems unknown in finite dimensional spaces. D. Hilbert had recognized this, and therefore he was looking for a weaker topology on the sequence space with respect to which the above convenient characterization of compact sets would still be valid. He introduced the 'weak topology' and studied its main properties. We will discuss the basic topological concepts for this weak topology and their relation to the corresponding concepts for the strong topology. It turns out that a subset of a Hilbert space is 'weakly bounded', i.e., bounded with respect to the weak topology, if, and only if, it is 'strongly bounded', i.e., bounded with respect to the strong or norm topology. This important result is based on the fundamental 'principle of uniform boundedness' which is discussed in good detail in the Appendix (Section 34.4). An immediate important consequence of the equivalence of weakly and strongly bounded sets is that (strongly) bounded subsets of a Hilbert space are relatively sequentially compact for the weak topology and this implies sequential completeness of Hilbert spaces for the weak topology.

After we have learned the basic facts about the geometrical and topological structure of Hilbert spaces we study mappings between Hilbert spaces which are compatible with the linear structure. These mappings are called 'linear operators'. A linear operator is specified by a linear subspace D of a Hilbert space $\mathcal{H}$ and an assignment A which assigns to each point x in D a unique point Ax in a Hilbert space $\mathcal{K}$. This linear subspace D is called the 'domain of the operator'. If $\mathcal{K} = \mathcal{H}$ one speaks about a 'linear operator in the Hilbert space $\mathcal{H}$', otherwise about a 'linear operator from $\mathcal{H}$ into $\mathcal{K}$'. In order to indicate explicitly the dependence of a linear operator on its domain we write $\hat{A} = (D, A)$ for a linear operator with domain D and assignment A. In this notation it is evident that the same assignment on different linear subspaces D_1 and D_2 defines different linear operators.

Observe that in the above definition of a linear operator no continuity requirements enter. If one takes also the topological structure of Hilbert spaces into account one is lead to the distinction of different classes of linear operators. Accordingly we discuss in Chapter 19 'Linear operators' the definition and the characterization of the following classes of linear operators: Bounded, unbounded, closed, closable and densely defined operators; for densely defined linear operators one proves the existence of a unique 'adjoint operator' which allows one to distinguish between the classes of 'symmetric', 'essentially self-adjoint' and 'self-adjoint' operators. In applications, for instance in quantum mechanics, it is often important to decide whether a given linear operator is self-adjoint or not. Thus some criteria for self-adjointness are presented and these are illustrated in a number of examples which are of interest in quantum mechanics.

If for two linear operators $\hat{A}_i = (D_i, A_i)$, $i = 1, 2$, one knows $D_1 \subseteq D_2$ and $A_1 x = A_2 x$ for all $x \in D_1$, one says that the linear operator $\hat{A}_2$ is an 'extension' of the linear operator $\hat{A}_1$, respectively that $\hat{A}_1$ is a 'restriction' of $\hat{A}_2$. A standard problem which occurs quite frequently is the following: Given a linear differential operator on a space of 'smooth' functions, construct all self-adjoint extensions of this differential operator. Ideally one would like to prove that there is exactly one self-adjoint extension (which one then could call the natural self-adjoint extension). For the construction of self-adjoint extensions (for instance of a linear differential operator) one can often use the 'method of quadratic forms' since there is a fundamental result which states that 'semi-bounded self-adjoint operators' and 'closed semi-bounded densely defined quadratic forms' are in a one-to-one correspondence (see Representation Theorem 20.2.2 and 20.2.3 of T. Kato). The method of quadratic forms is also applied successfully to the definition of the sum of two unbounded self-adjoint operators, even in some cases when the intersection of the domains of the two operators is trivial, i.e., only contains the null vector. In this way one gets the 'form sum' of two unbounded operators.

Naturally, most of the problems addressed above do not occur for the class of 'bounded' linear operators. Two bounded linear operators can be added in the standard way since they are defined on the whole space, and they can be multiplied by scalars, i.e., by numbers in $\mathbb{K}$. Furthermore one can define a product of two such operators by the composition for mappings. Thus it turns out that the class of all bounded linear operators on a Hilbert space $\mathcal{H}$ is an algebra $\mathfrak{B}(\mathcal{H})$, in a natural way. This algebra $\mathfrak{B}(\mathcal{H})$ has a number of additional properties which make it the standard example of a 'C^*-algebra'. On $\mathfrak{B}(\mathcal{H})$ we consider three different topologies, the 'uniform' or 'operator-norm' topology, the 'strong' topology, and the 'weak' topology and look at the relations between these topologies.

The algebra $\mathfrak{B}(\mathcal{H})$ contains several important classes of bounded linear operators. Thus we discuss the class of 'projection operators' or 'projectors', the class of 'isometries', and the class of 'unitary operators'. Projectors are in one-to-one correspondence with closed subspaces of the Hilbert space. Isometric operators between two Hilbert spaces do not change the metric properties of these spaces. The class of unitary operators can be considered as the class of those operators between Hilbert spaces which respect the linear, the metric, and the geometric structures. This can be expressed by saying that unitary operators are those bijective linear operators which do not change the inner products. As we will learn there is an important connection between self-adjoint operators and 'strongly continuous one-parameter groups of unitary operators $U(t)$, $t \in \mathbb{R}$: Such groups are 'generated by self-adjoint operators', in analogy to the unitary group of complex numbers $z(t) = e^{iat}$, $t \in \mathbb{R}$, which is 'generated' by the real number a. The unitary groups and their relation to self-adjoint operators play a very important role in quantum mechanics (time evolutions, symmetries). Another class of bounded linear operators are the 'trace class' operators which are used in the form of 'density matrices' in the description of states for a quantum mechanical system. As an important application we present here the 'general uncertainty relations of Heisenberg'.

In more concrete terms and in greater detail we will discuss the above concepts and results in the following section which is devoted to those self-adjoint operators which play a fundamental role in the description of quantum systems, i.e., position, momentum and energy or Hamilton operators. As in classical mechanics the Hamilton operator of an interacting system is the 'sum' of the operator corresponding to the kinetic energy, the free Hamilton operator, and the operator describing the interaction. Typically both operators are unbounded and we are here in a concrete situation of the problem of defining the 'sum' of two unbounded self-adjoint operators. The solution of this problem is due to T. Kato who suggested considering the potential operator or interaction energy as a certain perturbation of the free Hamilton operator (nowadays called 'Kato perturbation'). In this way many self-adjoint Hamilton operators can be constructed which are of great importance to quantum mechanics.

The final sections of the part 'Hilbert Spaces' come back to the class of problems from which the theory of Hilbert spaces originated, namely finding 'eigenvalues' of linear operators in Hilbert spaces. It turns out that in infinite dimensional Hilbert spaces the concept of an eigenvalue is too narrow for the complexity of the problem. As the suitable generalization of the set of all eigenvalues of linear maps in the finite dimensional case to the infinite dimensional setting, the concept of 'spectrum' is used. In an infinite dimensional Hilbert space the spectrum of a self-adjoint operator can have a much richer structure than in the finite dimensional situation where it equals the set of all eigenvalues: Besides 'eigenvalues of finite multiplicity' there can be 'eigenvalues of infinite multiplicity' and a 'continuous part', i.e., a nonempty open interval can be contained in the spectrum. Accordingly the spectrum of a linear operator is divided into two parts, the 'discrete spectrum' and the 'essential spectrum'. H. Weyl found a powerful characterization of the discrete and the essential spectrum and he observed a remarkable stability of the essential spectrum under certain perturbations of the operator: If the difference of the 'resolvents' of two closed linear operators is a 'compact operator', then both operators have the same essential spectrum.

Recall the 'spectral representation' of a symmetric $n \times n$ matrix. If $\sigma(A) = \{\lambda_1, \ldots, \lambda_n\}$ are the eigenvalues of A and $\{e_1, \ldots, e_n\} \subset \mathbb{R}^n$ the corresponding orthonormal eigenvectors, the matrix A has the spectral representation

$$A = \sum_{\lambda \in \sigma(A)} \lambda P_\lambda = \sum_{j=1}^{n} \lambda_j |e_j\rangle\langle e_j|$$

where $P_{\lambda_j} = |e_j\rangle\langle e_j|$ is the orthogonal projector onto the space spanned by the eigenvector e_j, i.e., $P_{\lambda_j} x = \langle e_j, x\rangle e_j$ for all $x \in \mathbb{K}^n$.

For a self-adjoint operator in an infinite dimensional Hilbert space one must take into account that the operator might have a nonempty continuous spectrum and accordingly the general version of the spectral representation of a self-adjoint operator A should be, in analogy with the finite dimensional case,

$$A = \int_{\sigma(A)} \lambda \, dP_\lambda. \tag{13.1}$$

The proof of the validity of such a spectral representation for general self-adjoint operators needs a number of preparations which we will give in considerable detail. The proof of the spectral representation which we present has the advantage that it relies completely on Hilbert space intrinsic concepts and methods, namely the 'geometric characterization of self-adjointness'. This approach has the additional advantage that it allows us to prove the fact that every closed symmetric operator has a 'maximal self-adjoint part', without any additional effort.

Early results in the 'spectral theory' of self-adjoint operators concentrated on the case where the operator is 'compact'. Such operators do not have a continuous spectrum. We discuss here briefly the main results in this area, the 'Riesz–Schauder theory' including the 'Fredholm alternative' and several examples.

The spectral representation of a self-adjoint operator A (13.1) has many applications some of which we discuss in detail, others we just mention briefly. From the point of view of applications to quantum mechanics the following consequences are very important. Starting from the spectral representation (13.1) the classification of the different parts of the spectrum $\sigma(A)$ of the operator A can be done in terms of properties of the measures

$$dm_\psi(\lambda) = d\langle \psi, P_\lambda \psi\rangle, \qquad \psi \in \mathcal{H}$$

relative to the Lebesgue measure $d\lambda$. Here the most important distinction is whether the measure dm_ψ is absolutely continuous with respect to the Lebesgue measure or not. In this way one gets a decomposition of the Hilbert space $\mathcal{H}$ into different 'spectral subspaces'. This spectral decomposition plays an important role in the 'scattering theory' for self-adjoint 'Schrödinger operators' $H = H_0 + V$ in the Hilbert space $\mathcal{H} = L^2(\mathbb{R}^3)$, for instance. According to physical intuition one expects that every state of such a system is either a 'bound state', i.e., stays essentially localized in a bounded region of $\mathbb{R}^3$, or a 'scattering state', i.e., a state which 'escapes to infinity'. The finer spectral analysis shows that this expectation is not always correct. The final section of this part discusses when precisely this statement is correct and how it is related to the different spectral subspaces of the Schrödinger operator H.

13.2 Some historical remarks

We sketch a few facts which led to the development of the theory of Hilbert spaces. For those readers who are interested in further details of the history of this theory and of functional analysis in general we recommend the book [Die69]. As mentioned above the theory of Hilbert spaces has its origin in the theory of expansion of arbitrary functions with respect to certain systems of orthogonal functions (with respect to a given inner product). Such systems of orthogonal functions usually were systems of eigenfunctions of certain linear differential operators.

In the second half of the 19th century, under the influence of mathematical physics, the focus of much research was on the linear partial differential equation

$$\Delta_3 u(x) + \lambda u(x) = 0 \quad \forall x \in \Omega, \quad u|\partial\Omega = 0, \qquad (13.2)$$

where $\Omega \subset \mathbb{R}^3$ is a nonempty domain with smooth boundary and where Δ_3 is the Laplace operator in three dimensions. In this context the concept of *Green's function* or elementary solution was introduced by Schwarz, as a predecessor of the concept of elementary solution as introduced and discussed in the the first part on distribution theory (Section 8.4). Around 1894, H. Poincaré proved the existence and the main properties of the *eigenfunctions* of the *eigenvalue problem* (13.2).

As we will learn later, these results are closely related to the emergence of the theory of linear *integral equations*, i.e., equations of the form

$$u(x) + \int_a^b K(x, y)u(y)dy = f(x), \qquad (13.3)$$

in the case of one dimension, for an unknown function u, for a given kernel function K and a given source term f. And this theory of linear integral equations in turn played a decisive role in the development of those ideas which shaped *functional analysis*, as we know it today. Many well-known mathematicians of that period, e.g., C. Neumann, H. Poincaré, I. Fredholm, V. Volterra, and E. Schmidt studied this type of equations and obtained many interesting results. Eventually, at the beginning of the 20th century, D. Hilbert introduced a good number of new and very fruitful ideas. In his famous papers of 1906, he showed that solving the integral equation (13.3) is equivalent, under certain conditions on K and f, to solving the infinite linear system for the unknown real sequence u_i, $i = 1, 2, \ldots$, for a given infinite matrix with real coefficients K_{ij} and a given real sequence f_i:

$$u_i + \sum_{j=1}^{\infty} K_{ij} u_j = f_i \quad i = 1, 2, \ldots. \tag{13.4}$$

Furthermore he succeeded in showing that the only relevant solutions of this system are those which satisfy the condition

$$\sum_{j=1}^{\infty} u_j^2 < \infty. \tag{13.5}$$

The set of all real sequences $(u_i)_{i \in \mathbb{N}}$ satisfying condition (13.5), i.e., the set of all square summable real sequences, is denoted by $\ell^2(\mathbb{R})$. We will learn later that it is a real vector space with an inner product so that this space is complete with respect to the norm defined by this inner product. Thus $\ell^2(\mathbb{R})$ is an example of a *Hilbert space*. Naturally one would expect that this space plays a prominent role in the theory of Hilbert spaces and this expectation will be confirmed later when we learn that every *separable* Hilbert space is isomorphic to $\ell^2(\mathbb{R})$ or $\ell^2(\mathbb{C})$.

All the Euclidean spaces $\mathbb{R}^n$, $n = 1, 2, \ldots$, are naturally embedded into $\ell^2(\mathbb{R})$ by assigning to the point $\underline{x} = (x_1, \ldots, x_n) \in \mathbb{R}^n$ the sequence whose components with index $i > n$ all vanish. In this sense we can consider the space $\ell^2(\mathbb{R})$ as the natural generalization of the Euclidean space $\mathbb{R}^n$ to the case of infinite dimensions. On the space $\ell^2(\mathbb{R})$, D. Hilbert introduced two important notions of convergence which are known today as *strong* and *weak convergence*. These will be studied later in considerable detail. These two notions of convergence correspond to two different topologies on this infinite dimensional vector space. Linear mappings, linear functionals and bilinear forms were classified and studied by Hilbert on the basis of their continuity with respect to these two topologies. In such a space the meaning and interpretation of many concepts of Euclidean geometry were preserved. This is the case in particular for theory of diagonalization of quadratic forms which is well established in Euclidean spaces. Hilbert proved that also in the space $\ell^2(\mathbb{R})$ every quadratic form can be given a normal (i.e., diagonal) form by a 'rotation of the coordinate system'. In his theory of diagonalization of quadratic forms in the infinite dimensional case, Hilbert discovered a number of new mathematical structures, e.g., the possibility of a 'continuous spectrum'.

Hilbert's new theory was of great importance for the emerging *quantum mechanics* since it offered, through Hilbert's new concept of a 'mathematical spectrum', the possibility of interpreting and understanding the energy spectra of atoms as they were observed experimentally. Since then the theory of Hilbert spaces grew enormously, mainly through its interaction with quantum physics.

The next important step in the development of the theory of Hilbert spaces came through the ideas of M. Fréchet, E. Schmidt, and F. Riesz who introduced in the years 1907 to 1908 the concepts of Euclidean geometry (length, angle, orthogonality, basis, etc) to the theory of Hilbert spaces. A remarkable early observation in these studies by F. Riesz and M. Fréchet was the following: The Lebesgue space $L^2(\mathbb{R})$ of all equivalence classes of square integrable functions on $\mathbb{R}$ has a very similar geometry to the Hilbert space $\ell^2(\mathbb{R})$. Several months later the analogy between the two spaces $L^2(\mathbb{R})$ and $\ell^2(\mathbb{R})$ was established completely when F. Riesz and E. Fischer proved the completeness of the space $L^2(\mathbb{R})$ and the isomorphy of these spaces. Soon one realized that many classical function spaces were also isomorphic to $\ell^2(\mathbb{R})$. Thus most of the important properties of Hilbert spaces were already known at that period.

Later, around 1920, the abstract and axiomatic presentation of the theory of Hilbert spaces emerged, mainly through the efforts of *J. von Neumann* [vN67] and *R. Riesz* who also started major developments of the theory of linear operators on Hilbert spaces.

Certainly there many other interesting aspects of history of the theory of Hilbert spaces and their operators. These are addressed, for instance in J. Dieudonné's book "History of Functional analysis" [Die69] which we highly recommend.

13.3 Hilbert spaces and Physics

In our context *Physics* refers for the most part to 'Quantum Physics'. In quantum physics a system, for instance a particle or several particles in some force field, is described in terms of 'states, 'observables', and 'expectation values'. States are given in terms of vectors in a Hilbert space, more precisely in terms of 'unit rays' generated by a nonvanishing vector in a Hilbert space. The set of all states of a system is called the state space. Observables are realized by self-adjoint operators in this Hilbert space while expectation values are calculated in terms of the inner product of the Hilbert space.

In quantum physics, a particle is considered as an object which is localizable in (physical) space, i.e., in the Euclidean space $\mathbb{R}^3$. Its state space is the Hilbert space $L^2(\mathbb{R}^3)$. The motivation for this choice is as follows. If the particle is in the state given by $\psi \in L^2(\mathbb{R}^3)$, the quantity $|\psi(x)|^2$ has the interpretation of the probability density of finding the particle at the point $x \in \mathbb{R}^3$. This interpretation which is due to M. Born obviously requires

$$\int_{\mathbb{R}^3} |\psi(x)|^2 d^3x = 1.$$

Thus the choice of $L^2(\mathbb{R}^3)$ as the state space of one localizable particle is consistent with the probability interpretation of the 'wave function' ψ. Observables are then self-adjoint operators in $L^2(\mathbb{R}^3)$ and the expectation value $E_A(\psi)$ of an observable described by the self-adjoint operator A when the particle is in the state $\psi \in L^2(\mathbb{R}^3)$

is
$$E_A(\psi) = \frac{\langle \psi, A\psi \rangle}{\langle \psi, \psi \rangle}.$$

The self-adjoint operators of quantum mechanics are typically unbounded and thus not continuous. Therefore Hilbert's original version of the theory of Hilbert spaces and their operators could not cope with many important aspects and problems arising in quantum mechanics. Thus, in order to provide quantum mechanics with a precise mathematical framework, R. Riesz, M. H. Stone, and in particular J. von Neumann developed around 1930 an axiomatic approach to the theory of Hilbert spaces and their operators. While in Hilbert's understanding quadratic forms (or operators) were given in terms of concrete quantities, J. von Neumann defined this concept abstractly, i.e., in terms of precise mathematical relation to previously defined concepts. This step in abstraction allowed him to overcome the limitation of Hilbert's original theory and it enabled this abstract theory of Hilbert spaces to cope with all mathematical demands from quantum physics. A more recent example of the successful use of operator methods in quantum mechanics is the book [Sch81].

Earlier we presented L. Schwartz' theory of distributions as part of modern functional analysis, i.e., the unification in terms of concepts and methods of linear algebra and analysis. It is worthwhile mentioning here that the deep results of D. Hilbert, F. and R. Riesz, M. Fréchet, E. Fischer, J. von Neumann, and E. Schmidt were historically the starting point of modern functional analysis.

Now we recall several applications of the theory of Hilbert spaces in classical physics. There this theory is used mainly in the form of Hilbert spaces $L^2(\Omega)$ of square integrable functions on some measurable set $\Omega \subset \mathbb{R}^n$, $n = 1, 2, 3, \ldots$. If for instance Ω is some interval in time and if $|f(t)|^2 \Delta t$ denotes the energy radiating off some system, the total energy which is radiated off the system during this period in time, is

$$E = \int_\Omega |f(t)|^2 dt = \|f\|^2_{L^2(\Omega)}.$$

In such a context physicists prefer to call the square integrable functions the 'functions with finite total energy'. Theorem 10.3.5 of Parseval–Plancherel states for this case

$$\int |f(t)|^2 dt = \int |\tilde{f}(\nu)|^2 d\nu$$

where $\tilde{f}$ denotes the Fourier transform of the function f. In this way one has two equivalent expressions for the total energy of the system. The quantity $|\tilde{f}(\nu)|^2$ has naturally the interpretation of the radiated energy during a unit interval in frequency space. The second integral in the above equation thus corresponds to a decomposition into harmonic components. It says that the total energy is the sum of the energies of all its harmonic components. This important result which is easily derived from the theory of the L^2 spaces was originally proposed by the physicist Lord Rayleigh.

The conceptual and technical aspects of the development of quantum theory are well documented in the book [Jam74] of M. Jammer. A quite comprehensive account of the development of quantum theory can be found in the six volumes of Mehra and Rechenberg [MR01].

14 Inner Product Spaces and Hilbert Spaces

In close analogy with the Euclidean spaces we develop in this short chapter the basis of the theory of inner product spaces or 'pre-Hilbert spaces' and of 'Hilbert spaces'. Recall that a Euclidean space is a finite dimensional real or complex vector space equipped with an inner product (also called a scalar product). In the theory of Euclidean space we have the important concepts of the length of a vector, of orthogonality between two vectors, of an orthonormal basis etc. Through the inner product it is straightforward to introduce these concepts in the infinite dimensional case too. In particular we will learn in a later chapter that, and how, a Hilbert space can be identified with its topological dual space. This, together with the fact that Hilbert spaces can be considered as the natural extension of the concept of a Euclidean space to the infinite dimensional situation, gives Hilbert spaces a distinguished role in mathematical physics, in particular in quantum physics, and in functional analysis in general.

14.1 Inner product spaces

Before we turn our attention to the definition of abstract inner product spaces and Hilbert spaces we recall some basic facts about Euclidean spaces. We hope that thus the reader gets some intuitive understanding of Hilbert spaces.

The distinguishing geometrical properties of the three dimensional Euclidean space $\mathbb{R}^3$ is the existence of the concept of the 'angle between two vectors' of this space, which has a concrete meaning. As is well known, this can be expressed in terms of the inner product of this space. For $x = (x_1, x_2, x_3) \in \mathbb{R}^3$ and $y =$

$(y_1, y_2, y_3) \in \mathbb{R}^3$ one defines

$$\langle x, y \rangle = \sum_{j=1}^{3} x_j y_j.$$

Then

$$\|x\| = +\sqrt{\langle x, x \rangle}$$

is the Eucidean length of the vector $x \in \mathbb{R}^3$, and the angle θ between two vectors $x, y \in \mathbb{R}^3$ is determined by the equation

$$\langle x, y \rangle = \|x\| \|y\| \cos\theta.$$

θ is unique in the interval $[0, \pi]$.

For the finite dimensional Euclidean spaces $\mathbb{R}^n$ and $\mathbb{C}^n$, $n \in \mathbb{N}$, the situation is very similar; with the inner product

$$\langle x, y \rangle = \sum_{j=1}^{n} \overline{x_j} y_j \qquad x, y \in \mathbb{C}^n$$

the angle θ between x, y is defined in the same way.

Thus we have three fundamental concepts at our disposal in these spaces together with their characteristic relation: Vectors (linear structure), length of vectors (metric structure), angle between two vectors (geometric structure).

14.1.1 Basic definitions and results

The concept of an inner product space or pre-Hilbert space is obtained by abstraction, by disregarding the restriction in the dimension of the underlying vector space. As in the finite dimensional case, the metric and geometric structures are introduced through the concept of an 'inner' or 'scalar' product.

Definition 14.1.1 *For a vector space V over the field $\mathbb{K}$ (of complex or real numbers) every mapping*

$$\langle \cdot, \cdot \rangle : V \times V \to \mathbb{K}$$

is called an **inner product** *or a* **scalar product** *if this mapping satisfies the following conditions:*

(IP1) $\langle x, x \rangle \geq 0 \; \forall \; x \in V$, and $\langle x, x \rangle = 0$ implies $x = 0 \in V$;

(IP2) $\langle x, y + z \rangle = \langle x, y \rangle + \langle x, z \rangle$ for all $x, y, z \in V$;

(IP3) $\langle x, \alpha y \rangle = \alpha \langle x, y \rangle$ for all $x, y \in V$ and all $\alpha \in \mathbb{K}$;

(IP4) $\langle x, y \rangle = \overline{\langle y, x \rangle}$ for all $x, y \in V$;

i.e., $\langle \cdot, \cdot \rangle$ is a **positive definite sesquilinear form** on V.

A vector space equipped with an inner product is called an **inner product space** or a **pre-Hilbert space**.

There is an immediate consequence of this definition: For all $x, y, z \in V$ and all $\alpha, \beta \in \mathbb{K}$ one has

$$\langle x, \alpha y + \beta z \rangle = \alpha \langle x, y \rangle + \beta \langle x, z \rangle,$$
$$\langle \alpha x, y \rangle = \overline{\alpha} \langle x, y \rangle.$$

Note that in Definition 14.1.1 we have used the convention which is most popular among physicists in requiring that an inner product is linear in the second argument while it is antilinear in the first argument. Among mathematicians, linearity in the first argument seems to be more popular.

We recall two well-known examples of inner products:

1) On the Euclidean space $\mathbb{K}^n$ the standard inner product is

$$\langle x, y \rangle = \sum_{j=1}^{n} \overline{x_j} y_j$$

for all $x = (x_1, \ldots, x_n) \in \mathbb{K}^n$ and all $y = (y_1, \ldots, y_n) \in \mathbb{K}^n$.

2) On the vector space $V = \mathcal{C}(I, \mathbb{K})$ of continuous functions on the interval $I = [a, b]$ with values in $\mathbb{K}$, the following formula defines an inner product as one easily proves:

$$\langle f, g \rangle = \int_a^b \overline{f(x)} g(x) dx \qquad \forall\, f, g \in V.$$

As in the Euclidean spaces the concept of orthogonality can be defined in any inner product space.

Definition 14.1.2 *Suppose that $(V, \langle \cdot, \cdot \rangle)$ is an inner product space. One calls*

a) *an element $x \in V$ **orthogonal** to an element $y \in V$, denoted $x \perp y$, if, and only if, $\langle x, y \rangle = 0$;*

b) *a system $(x_\alpha)_{\alpha \in A} \subset V$ **orthonormal** or an **orthonormal system** if, and only if, $\langle x_\alpha, x_\beta \rangle = 0$ for $\alpha \neq \beta$ and $\langle x_\alpha, x_\alpha \rangle = 1$ for all $\alpha, \beta \in A$. Here A is any index set;*

c) $\|x\| = +\sqrt{\langle x, x \rangle}$ *the **length** of the vector $x \in V$.*

A simple and well-known example of an orthonormal system in the inner product space $V = \mathcal{C}(I, \mathbb{C})$, $I = [0, 2\pi]$, mentioned above, is the system of functions f_n, $n \in \mathbb{Z}$, defined by $f_n(x) = \frac{1}{\sqrt{2\pi}} e^{inx}$, $x \in I$. By an elementary integration one finds

$$\langle f_n, f_m \rangle = \frac{1}{2\pi} \int_0^{2\pi} e^{i(m-n)x} dx = \delta_{nm}.$$

In elementary geometry we learn the theorem of Pythagoras. The following lemma shows that this result holds in any inner product space.

Lemma 14.1.1 (Theorem of Pythagoras) *If $\{x_1, \ldots, x_N\}$, $N \in \mathbb{N}$, is an orthonormal system in an inner product space $(V, \langle \cdot, \cdot \rangle)$, then, for every $x \in V$ the following identity holds:*

$$\|x\|^2 = \sum_{n=1}^{N} |\langle x_n, x \rangle|^2 + \|x - \sum_{n=1}^{N} \langle x_n, x \rangle x_n \|^2.$$

Proof. Given any $x \in V$ introduce the vectors $y = \sum_{n=1}^{N} \langle x_n, x \rangle x_n$ and $z = x - y$. Now we calculate, for $j \in \{1, \ldots, N\}$:

$$\begin{aligned}\langle x_j, z \rangle &= \langle x_j, x - \sum_{n=1}^{N} \langle x_n, x \rangle x_n \rangle = \langle x_j, x \rangle - \sum_{n=1}^{N} \overline{\langle x_n, x \rangle} \langle x_j, x_n \rangle \\ &= \langle x_j, x \rangle - \sum_{n=1}^{N} \langle x_n, x \rangle \delta_{jn} = 0.\end{aligned}$$

It follows that $\langle y, z \rangle = 0$. This shows that $x = y + z$ is the decomposition of the vector x into a vector y which is contained in the space spanned by the orthonormal system and a vector z which is orthogonal to this space. This allows us to calculate

$$\|x\|^2 = \langle y + z, y + z \rangle = \langle y, y \rangle + \langle z, y \rangle + \langle y, z \rangle + \langle z, z \rangle = \|y\|^2 + \|z\|^2.$$

And a straightforward calculation shows that $\langle y, y \rangle = \sum_{n=1}^{N} |\langle x_n, x \rangle|^2$ and thus Pythagoras' theorem follows. □

Pythagoras' theorem has two immediate consequences which are used in many estimates.

Corollary 14.1.1   *1.* **Bessel's inequality:** *If $\{x_1, \ldots, x_N\}$ is a countable orthonormal system (i.e., $N \in \mathbb{N}$ or $N = +\infty$) in a pre-Hilbert space V, then, for every $x \in V$, the following estimate holds:*

$$\sum_{n=1}^{N} |\langle x_n, x \rangle|^2 \leq \|x\|^2.$$

2. **Schwarz' inequality:** *For any two vectors x, y in a pre-Hilbert space V one has*

$$|\langle x, y \rangle| \leq \|x\| \cdot \|y\|.$$

Proof. To prove the first part take $L \in \mathbb{N}$, $L \leq N$. Pythagoras' theorem implies

$$S_L = \sum_{n=1}^{L} |\langle x_n, x \rangle|^2 \leq \|x\|^2.$$

Thus, for $N \in \mathbb{N}$ the first part follows. If $N = +\infty$ one observes that $(S_L)_L$ is a monotone increasing sequence which is bounded by $\|x\|^2$. Therefore this sequence converges to a number which is smaller than or equal to $\|x\|^2$:

$$\sum_{n=1}^{\infty} |\langle x_n, x \rangle|^2 = \lim_{L \to \infty} S_L \leq \|x\|^2$$

14.1 Inner product spaces

which proves Bessel's inequality in the second case.

Schwarz' inequality is an easy consequence of Bessel's inequality. Take any two vectors $x, y \in V$. If for instance $x = 0$, then $\langle x, y \rangle = \langle 0, y \rangle = 0$ and $\|x\| = 0$, and Schwarz' inequality holds in this case. If $x \neq 0$, then $\|x\| > 0$ and thus $\left\{ \frac{x}{\|x\|} \right\}$ is an orthonormal system in V. Hence for any $y \in V$ Bessel's inequality implies

$$\left| \langle \frac{x}{\|x\|}, y \rangle \right|^2 \leq \|y\|^2.$$

Now Schwarz' inequality follows easily. □

Remark 14.1.1 1. *In the literature Schwarz' inequality is often called* **Cauchy–Schwarz–Bunjakowski inequality**. *It generalizes the classical Cauchy inequality*

$$\left| \sum_{j=1}^n a_j b_j \right|^2 \leq \sum_{j=1}^n a_j^2 \sum_{j=1}^n b_j^2 \qquad \forall a_j, b_j \in \mathbb{R}, \quad \forall n \in \mathbb{N}.$$

2. *Later in the section on the geometry of Hilbert spaces we will learn about a powerful generalization of Schwarz' inequality, in the form of the 'Gram determinants'.*

3. *Suppose that* $(V, \langle \cdot, \cdot \rangle)$ *is a real inner product space and suppose that* $x, y \in V$ *are two nonzero vectors. Then Schwarz' inequality says*

$$-1 \leq \frac{\langle x, y \rangle}{\|x\| \|y\|} \leq +1.$$

It follows that in the interval $[0, \pi]$ *there is exactly one number* $\theta = \theta(x, y)$ *such that* $\cos \theta = \frac{\langle x, y \rangle}{\|x\| \|y\|}$. *This number* θ *is called the* **angle between the vectors x and y**.

Finally we study the concept of *length* in a general inner product space, in analogy to the Euclidean spaces.

Proposition 14.1.1 *If* $(V, \langle \cdot, \cdot \rangle)$ *is a pre-Hilbert space, then the function* $V \ni x \mapsto \|x\| = +\sqrt{\langle x, x \rangle} \in \mathbb{R}_+$ *has the following properties:*

(N1) $\|x\| \geq 0$ *for all* $x \in V$;

(N2) $\|\lambda x\| = |\lambda| \|x\|$ *for all* $x \in V$ *and all* $\lambda \in \mathbb{K}$;

(N3) $\|x + y\| \leq \|x\| + \|y\|$ *for all* $x, y \in V$ (**triangle inequality**);

(N4) $\|x\| = 0$ *if, and only if,* $x = 0 \in V$.

This function $\| \cdot \| = \sqrt{\langle \cdot, \cdot \rangle}$ *is thus a norm on* V; *it is called the* **norm induced by the inner product**.

Proof. It is a straightforward calculation to verify properties (N1), (N2), and (N3) using the basic properties of an inner product. This is done in the Exercises. Property (N3) follows from Schwarz' inequality, as the following calculations show:

$$\begin{aligned}\|x+y\|^2 &= \langle x+y, x+y\rangle = \langle x, x\rangle + \langle y, x\rangle + \langle x, y\rangle + \langle y, y\rangle \\ &= \|x\|^2 + \|y\|^2 + 2\Re\langle x, y\rangle \\ &\leq \|x\|^2 + \|y\|^2 + 2|\langle x, y\rangle| \\ &\leq \|x\|^2 + \|y\|^2 + 2\|x\|\|y\| = (\|x\| + \|y\|)^2,\end{aligned}$$

hence the triangle inequality (N3) follows. □

14.1.2 Basic topological concepts

Every inner product space $(V, \langle \cdot, \cdot \rangle)$ is a normed space under the induced norm $\|\cdot\| = \sqrt{\langle \cdot, \cdot \rangle}$ and thus all results from the theory of normed spaces apply which we have discussed in the first part. Here we recall some basic results.

The system of neighborhoods of a point $x \in V$ is the system of all subsets of V which contain some open ball $B_r(x) = \{y \in V : \|y - x\| < r\}$ with centre x and radius $r > 0$. This system of neighborhoods defines the *norm topology* on V. For this topology one has: The addition $(x, y) \mapsto x + y$ is a continuous map $V \times V \to V$. The scalar multiplication $(\lambda, x) \mapsto \lambda x$ is a continuous map $\mathbb{K} \times V \to V$. In the following definition we recall the basic concepts related to convergence of sequences with respect to the norm topology and express them explicitly in terms of the induced norm.

Definition 14.1.3 *Equip the inner product space $(V, \langle \cdot, \cdot \rangle)$ with its norm topology. One says:*

a) *A sequence $(x_n)_{n \in \mathbb{N}} \subset V$ is a **Cauchy sequence** if, and only if, for every $\epsilon > 0$ there is $N \in \mathbb{N}$ such that $\|x_n - x_m\| < \epsilon$ for all $n, m \geq N$;*

b) *A sequence $(x_n)_{n \in \mathbb{N}} \subset V$ **converges** if, and only if, there is $x \in V$ such that for every $\epsilon > 0$ there is $N \in \mathbb{N}$ such that $\|x - x_n\| < \epsilon$ for all $n \geq N$;*

c) *The inner product space $(V, \langle \cdot, \cdot \rangle)$ is **complete** if, and only if, every Cauchy sequence in V converges.*

Some immediate important consequences of these definitions are

Corollary 14.1.2 *1. Every convergent sequence is a Cauchy sequence.*

2. If a sequence $(x_n)_{n \in \mathbb{N}}$ converges to $x \in V$, then

$$\lim_{n \to \infty} \|x_n\| = \|x\|. \tag{14.1}$$

Proof. The first part is obvious and is left as exercise. Concerning the proof of the second statement observe the basic estimate

$$|(\|x\| - \|y\|)| \leq \|x \pm y\| \quad \forall x, y \in V \tag{14.2}$$

which follows easily from the triangle inequality (see Exercises). □

14.1 Inner product spaces

Remark 14.1.2   *1. The axiom of completeness of the space of real numbers $\mathbb{R}$ plays a very important role in (real and complex) analysis. There are two equivalent ways to formulate the completeness of the set of real numbers. a) Every nonempty subset $M \subset \mathbb{R}$ which is bounded from above (from below) has a supremum (an infimum). b) Every Cauchy sequence of real numbers converges. The first characterization of completeness relies on the order structure of real numbers. Such an order is not available in general. Therefore we have defined completeness of an inner product space in terms of convergence of Cauchy sequences.*

2. *Finite dimensional and infinite dimensional pre-Hilbert spaces differ in a very important way: Every finite dimensional pre-Hilbert space is complete, but there are infinite dimensional pre-Hilbert spaces which are not complete. This is discussed in some detail in the Exercises.*

3. *If a space is complete we can deal with convergence of a sequence without a priori knowledge of the limit. It suffices to verify that the sequence is a Cauchy sequence. Thus completeness often plays a decisive role in existence proofs.*

4. *In the Appendix 34.1 it is shown that every metric space can be 'completed' by 'adding' certain 'limit elements'. This applies to pre-Hilbert spaces as well. We illustrate this here for the pre-Hilbert space $\mathbb{Q}$ of rational numbers with the ordinary product as inner product. In this case these limit elements are those real numbers which are not rational and thus rather different from the original rational numbers. In this process of completion the (inner) product is extended by continuity to all real numbers.*

   We mention another example illustrating the fact that these limit elements generated in the process of completion are typically very different from the elements of the original space. If one completes the inner product space V of all polynomials on an interval $I = [a, b]$, $a, b \in \mathbb{R}$, $a < b$, with the inner product $\langle f, g \rangle = \int_I f(t) g(t) dt$, one obtains the Lebesgue space $L^2(I, dt)$ whose elements differ in many ways from polynomials.

One has to distinguish clearly the concepts complete and closed for a space. A topological space V which is not complete is closed, as is every topological space. However as part of its completion $\tilde{V}$, the space V is not closed. The closure of V in the space $\tilde{V}$ is just $\tilde{V}$. This will be evident when we look at the construction of the completion of a metric space in some detail in the Appendix 34.1.

The basic definition of this part is

Definition 14.1.4 (J. von Neumann, 1925) *A **Hilbert space** is an inner product space which is complete (with respect to its norm topology).*

Thus, in order to verify whether a given inner product space is a Hilbert space, one has to show that every Cauchy sequence in this space converges. Therefore, Hilbert space are examples of complete normed spaces, i.e., of *Banach spaces*.

It is interesting and important to know when a given Banach space is actually a Hilbert space, i.e., when its norm is induced by an inner product. The following subsection addresses this question.

14.1.3 On the relation between normed spaces and inner product spaces

As we know, for instance from the example of Euclidean spaces, one can define on a vector many different norms. The norm induced by the inner product satisfies a characteristic identity which is well known from elementary Euclidean geometry.

Lemma 14.1.2 (Parallelogram law) *In an inner product space* $(V, \langle \cdot, \cdot \rangle)$ *the norm induced by the inner product satisfies the identity*

$$\|x+y\|^2 + \|x-y\|^2 = 2\|x\|^2 + 2\|y\|^2 \quad \forall x, y \in V. \quad (14.3)$$

Proof. The simple proof is done in the Exercises. □

The intuitive meaning of the parallelogram law is as in elementary geometry. To see this recall that $x + y$ and $x - y$ are the two diagonals of the parallelogram spanned by the vectors x and y.

According to Lemma 14.1.2 the parallelogram law (14.3) is a necessary condition for a norm to be induced by a scalar product, i.e., to be a *Hilbertian norm*. Naturally, not every norm satisfies the parallelogram law as the following simple example shows.

Consider the vector space $V = \mathcal{C}([0, 3], \mathbb{R})$ of continuous real functions on the interval $I = [0, 3]$. We know that

$$\|f\| = \sup_{x \in I} |f(x)|$$

defines a norm on V. In the Exercises it is shown that there are functions $f, g \in V$ such that $\|f\| = \|g\| = \|f + g\| = \|f - g\| = 1$. It follows that this norm does not satisfy (14.3). Hence this norm is not induced by an inner product.

The following proposition shows that the parallelogram law is not only necessary but also sufficient for a norm to be a Hilbertian norm.

Proposition 14.1.2 (Fréchet–von Neumann–Jordan) *If in a normed space* $(V, \|\cdot\|)$ *the parallelogram law holds, then there is an inner product on V such that* $\|x\|^2 = \langle x, x \rangle$ *for all $x \in V$.*

If V is a real vector space, then the inner product is defined by the **polarization identity**

$$\langle x, y \rangle = \frac{1}{4}(\|x+y\|^2 - \|x-y\|^2) \quad \forall x, y \in V; \quad (14.4)$$

if V is a complex vector space the inner product is given by the **polarization identity**

$$\langle x, y \rangle = \frac{1}{4}(\|x+y\|^2 - \|x-y\|^2 + i\|x+iy\|^2 - i\|x-iy\|^2) \quad \forall x, y \in V. \quad (14.5)$$

Proof. The proof is left as an exercise. □

Without proof (see however [Kak39, dFK67]) we mention two other criteria which ensure that a norm is actually a Hilbertian norm. Here we have to use some concepts which are only introduced in later sections.

Proposition 14.1.3 (Kakutani, 1939) *Suppose that $(V, \|\cdot\|)$ is a normed space of dimension ≥ 3. If every subspace $F \subset V$ of dimension 2 has a projector of norm 1, then the norm is Hilbertian.*

Proposition 14.1.4 (de Figueiredo–Karlovitz, 1967) *Let $(V, \langle\cdot, \cdot\rangle)$ be a normed space of dimension ≥ 3; define a map T from V into the closed unit ball $B = \{x \in V : \|x\| \leq 1\}$ by*

$$Tx = \begin{cases} x & \text{if } \|x\| \leq 1, \\ \frac{x}{\|x\|} & \text{if } \|x\| \geq 1. \end{cases}$$

If $\|Tx - Ty\| \leq \|x - y\|$ for all $x, y \in V$, then $\|\cdot\|$ is a Hilbertian norm.

It is worthwhile to mention that in a normed space one always has $\|Tx - Ty\| \leq 2\|x - y\|$ for all $x, y \in V$. In general the constant 2 in this estimate cannot be improved.

14.1.4 Examples of Hilbert spaces

We discuss a number of concrete examples of Hilbert spaces which are used in many applications of the theory of Hilbert spaces. The generic notation for a Hilbert space is $\mathcal{H}$.

1. **The Euclidean spaces:** As mentioned before, the Euclidean spaces $\mathbb{K}^n$ are Hilbert spaces when they are equipped with the inner product $\langle x, y \rangle = \sum_{j=1}^{n} \overline{x_j} y_j$ for all $x, y \in \mathbb{K}^n$. Since vectors in $\mathbb{K}^n$ have a finite number of components, completeness of the inner product space $(\mathbb{K}^n, \langle\cdot, \cdot\rangle)$ follows easily from that of $\mathbb{K}$.

2. **Matrix spaces:** Denote by $\mathcal{M}_n(\mathbb{K})$ the set of all $n \times n$ matrices with coefficients in $\mathbb{K}$, $n = 2, 3, \ldots$. Addition and scalar multiplication are defined component-wise. This gives $\mathcal{M}_n(\mathbb{K})$ the structure of a vector space over the field $\mathbb{K}$. In order to define an inner product on this vector space recall the definition of the *trace of a matrix* $A \in \mathcal{M}_n(\mathbb{K})$. If $A_{ij} \in \mathbb{K}$ are the components of A, the trace of A is defined as $\text{Tr}\, A = \sum_{j=1}^{n} A_{jj}$. The transpose of the complex conjugate matrix $\overline{A}$ is called the adjoint of A and denoted by $A^* = \overline{A}^t$. It is easy to show that $\langle A, B \rangle = \text{Tr}\,(A^*B) = \sum_{i,j=1}^{n} \overline{A_{ij}} B_{ij}$, $A, B \in \mathcal{M}_n(\mathbb{K})$, defines a scalar product. Again, completeness of this inner product space follows easily from completeness of $\mathbb{K}$ since matrices have a finite number of coefficients (see Exercises).

3. **The sequence space:** Recall that the space $\ell^2(\mathbb{K})$ of all sequences in $\mathbb{K}$ which are square summable was historically the starting point of the theory of Hilbert spaces. This space can be considered as the natural generalization of the Euclidean spaces $\mathbb{K}^n$ for $n \to \infty$. Here we show that this set is a Hilbert space in the sense of the axiomatic definition given above. Again, addition and scalar multiplication are defined component-wise. If $x = (x_n)_{n \in \mathbb{N}}$ and $y = (y_n)_{n \in \mathbb{N}}$ are elements in $\ell^2(\mathbb{K})$, then the estimate $|x_n + y_n|^2 \leq 2(|x_n|^2 + |y_n|^2)$ implies that the sequence $x + y = (x_n + y_n)_{n \in \mathbb{N}}$ is square summable too and thus addition is well defined. Similarly it follows that scalar multiplication is well defined and therefore $\ell^2(\mathbb{K})$ is a vector space over the field $\mathbb{K}$. The estimate $|\overline{x_n} y_n| \leq \frac{1}{2}(|x_n|^2 + |y_n|^2)$ implies that the series $\sum_{n=1}^{\infty} \overline{x_n} y_n$ converges absolutely for $x, y \in \ell^2(\mathbb{K})$ and thus can be used to define

$$\langle x, y \rangle = \sum_{n=1}^{\infty} \overline{x_n} y_n \tag{14.6}$$

as a candidate for an inner product on the vector space $\ell^2(\mathbb{K})$. In the Exercises we show that equation (14.6) defines indeed a scalar product on this space. Finally we show the completeness of the inner product space $(\ell^2(\mathbb{K}), \langle \cdot, \cdot \rangle)$.

Suppose that $(x^i)_{i \in \mathbb{N}}$ is a Cauchy sequence in this inner product space. Then each x^i is a square summable sequence $(x_n^i)_{n \in \mathbb{N}}$ and for every $\epsilon > 0$ there is an $i_0 \in \mathbb{N}$ such that for all $i, j \geq i_0$ we have

$$\|x^i - x^j\|^2 = \sum_{n=1}^{\infty} |x_n^i - x_n^j|^2 < \epsilon^2.$$

It follows that for every $n \in \mathbb{N}$ the sequence $(x_n^i)_{i \in \mathbb{N}}$ is actually a Cauchy sequence in $\mathbb{K}$. Completeness of $\mathbb{K}$ implies that these Cauchy sequences converge, i.e., for all $n \in \mathbb{N}$ the limits $x_n = \lim_{i \to \infty} x_n^i$ exist in $\mathbb{K}$.

Next we prove that the sequence $x = (x_n)_{n \in \mathbb{N}}$ of these limits is square summable. Given $\epsilon > 0$ choose $i_0 \in \mathbb{N}$ as in the basic Cauchy estimate above. Then, for all $i, j \geq i_0$ and for all $m \in \mathbb{N}$,

$$\sum_{n=1}^{m} |x_n^i - x_n^j|^2 \leq \|x^i - x^j\|^2 < \epsilon^2;$$

and we deduce, since limits can be taken in finite sums,

$$\lim_{i \to \infty} \sum_{n=1}^{m} |x_n^i - x_n^j|^2 = \sum_{n=1}^{m} |x_n - x_n^j|^2 \leq \epsilon^2$$

for all $j \geq i_0$ and all $m \in \mathbb{N}$. Therefore, for each $j \geq i_0$, $s_m = \sum_{n=1}^{m} |x_n - x_n^j|^2$ is a monotone increasing sequence with respect to $m \in \mathbb{N}$ which is bounded by ϵ^2. Hence this sequence has a limit, with the same upper bound:

$$\sum_{n=1}^{\infty} |x_n - x_n^j|^2 = \lim_{m \to \infty} s_m \leq \epsilon^2,$$

i.e., for each $j \geq i_0$, we know $\|x - x^j\| \leq \epsilon$. Since $\|x\| = \|x - x^j + x^j\| \leq \|x - x^j\| + \|x^j\| \leq \epsilon + \|x^j\|$, for fixed $j \geq i_0$, the sequence x belongs to $\ell^2(\mathbb{K})$ and the given Cauchy sequence $(x^i)_{i \in \mathbb{N}}$ converges (with respect to the induced norm) to x. It follows that every Cauchy sequence in $\ell^2(\mathbb{K})$ converges, thus this space is complete.

Proposition 14.1.5 *The space $\ell^2(\mathbb{K})$ of square summable sequences is a Hilbert space.*

4. **The Lebesgue space:** For this example we have to assume familiarity of the reader with the basic aspects of Lebesque's integration theory. Here we concentrate on the Hilbert space aspects.

Denote by $\mathcal{L}(\mathbb{R}^n)$ the set of Lebesgue measurable functions $f : \mathbb{R}^n \to \mathbb{K}$ which are square integrable, i.e., for which the Lebesgue integral

$$\int_{\mathbb{R}^n} |f(x)|^2 dx$$

is finite. Since for almost all $x \in \mathbb{R}^n$ one has $|f(x) + g(x)|^2 \leq 2(|f(x)|^2 + |g(x)|^2)$, it follows easily that $\mathcal{L}(\mathbb{R}^n)$ is a vector space over $\mathbb{K}$. Similarly one has $2|f(x)g(x)| \leq |f(x)|^2 + |g(x)|^2$, for almost all $x \in \mathbb{R}^n$, and therefore $2\int_{\mathbb{R}^n} |f(x)g(x)|dx \leq \int_{\mathbb{R}^n} |f(x)|^2 dx + \int_{\mathbb{R}^n} |g(x)|^2 dx$, for all $f, g \in \mathcal{L}(\mathbb{R}^n)$. Thus a function $\mathcal{L}(\mathbb{R}^n) \times \mathcal{L}(\mathbb{R}^n) \ni (f, g) \mapsto \langle f, g \rangle_2 \equiv \int_{\mathbb{R}^n} \overline{f(x)}g(x)dx \in \mathbb{K}$ is well defined. The basic rules for the Lebesgue integral imply that this function satisfies conditions (IP2) – (IP4) of Definition 14.1.1. It also satisfies $\langle f, f \rangle_2 \geq 0$ for all $f \in \mathcal{L}(\mathbb{R}^n)$. However $\langle f, f \rangle_2 = 0$ does not imply $f = 0 \in \mathcal{L}(\mathbb{R}^n)$. Therefore one introduces the 'kernel' $\mathcal{N} = \{f \in \mathcal{L}(\mathbb{R}^n) : \langle f, f \rangle_2 = 0\}$ of $\langle \cdot, \cdot \rangle_2$ which consists of all those functions in $\mathcal{L}(\mathbb{R}^n)$ which vanish almost everywhere on $\mathbb{R}^n$. As above it follows that $\mathcal{N}$ is a vector space over $\mathbb{K}$. Now introduce the quotient space

$$L^2(\mathbb{R}^n) = \mathcal{L}(\mathbb{R}^n)/\mathcal{N}$$

with respect to this kernel which consists of all equivalence classes

$$[f] = f + \mathcal{N}, \qquad f \in \mathcal{L}(\mathbb{R}^n).$$

On this quotient space we define

$$\langle [f], [g] \rangle_2 = \langle f, g \rangle_2$$

where $f, g \in \mathcal{L}(\mathbb{R}^n)$ are any representatives of their respective equivalence class. It is straightforward to show that now $\langle \cdot, \cdot \rangle_2$ is a scalar product on $L^2(\mathbb{R}^n)$. Hence $\mathcal{H} = (L^2(\mathbb{R}^n), \langle \cdot, \cdot \rangle_2)$ is an inner product space. That it is actually a Hilbert space follows from the important theorem

Theorem 14.1.6 (Riesz–Fischer) *The inner product space*

$$\mathcal{H} = (L^2(\mathbb{R}^n), \langle \cdot, \cdot \rangle_2)$$

is complete.

Following tradition we identify the equivalence class $[f] \equiv f$ with its representative in $\mathcal{L}(\mathbb{R}^n)$ in the rest of the book. Similarly one introduces the Lebesgue spaces $L^2(\Omega)$ for measurable subsets $\Omega \subset \mathbb{R}^n$ with nonempty interior. They too are Hilbert spaces.

5. **The Sobolev spaces:** For an open nonempty set $\Omega \subset \mathbb{R}^n$ denote by $W_k^2(\Omega)$ the space of all $f \in L^2(\Omega)$ which have 'weak' or distributional derivatives $D^\alpha f$ of all orders α, $|\alpha| \leq k$, for $k = 0, 1, 2, \ldots$, which again belong to $L^2(\Omega)$. Obviously one has

$$W_{k+1}^2(\Omega) \subset W_k^2(\Omega) \subset \cdots \subset W_0^2(\Omega) = L^2(\Omega), \qquad k = 0, 1, 2, \ldots.$$

On $W_k^2(\Omega)$ the natural inner product is

$$\langle f, g \rangle_{2,k} = \sum_{|\alpha| \leq k} \int_\Omega \overline{D^\alpha f(x)} g(x) dx \qquad \forall\, f, g \in W_k^2(\Omega).$$

It is fairly easy to verify that this function defines indeed a scalar product on the Sobolev space $W_k^2(\Omega)$. Finally, completeness of the Hilbert space $L^2(\Omega)$ implies completeness of the Sobolev spaces. Details of the proof are considered in the Exercises.

14.2 Exercises

1. Let $(V, \langle \cdot, \cdot \rangle)$ be an inner product space. For $x \in V$ define $\|x\| = +\sqrt{\langle x, x \rangle}$ and show that $V \ni x \mapsto \|x\|$ is a norm on V.

2. Give an example of a pre-Hilbert space which is not complete.
   Hints: Consider for instance the space $V = \mathcal{C}(I; \mathbb{R})$ of continuous real valued functions on the unit interval $I = [-1, 1]$ and equip this infinite dimensional space with the inner product $\langle f, g \rangle_2 = \int_I f(t) g(t) dt$. Then show that the sequence $(f_n)_{n \in \mathbb{N}}$ in V defined by

$$f_n(t) = \begin{cases} 0 & -1 \leq t \leq 0, \\ nt & 0 < t \leq \frac{1}{n}, \\ 1 & \frac{1}{n} < t \leq 1, \end{cases}$$

is a Cauchy sequence in this inner product space which does not converge to an element in V.

3. Show: For any two vectors a, b in a normed space $(V, \|\cdot\|)$ one has
$$\pm(\|a\| - \|b\|) \leq \|a \pm b\|$$
for any combination of the $\pm$ signs.

4. Prove the parallelogram law (14.3).

5. On the vector space $V = \mathcal{C}([0, 3], \mathbb{R})$ of continuous real functions on the interval $I = [0, 3]$ define the norm
$$\|f\| = \sup_{x \in I} |f(x)|$$
and show that there are functions $f, g \in V$ such that $\|f\| = \|g\| = \|f + g\| = \|f - g\| = 1$. It follows that this norm does not satisfy the identity (14.3). Hence this norm is not induced by an inner product.

Hints: Consider functions $f, g \in V$ with disjoint supports.

6. Prove Proposition 14.1.2.

7. Show that the space $\mathcal{M}_n(\mathbb{K})$ of $n \times n$ matrices with coefficients in $\mathbb{K}$ is a Hilbert space under the inner product $\langle A, B \rangle = \text{Tr}(A^*B)$, $A, B \in \mathcal{M}_n(\mathbb{K})$.

8. Prove: Equation 14.6 defines an inner product on the sequence space $\ell^2(\mathbb{K})$.

9. Prove that the Sobolev spaces $W_k^2(\Omega)$, $k \in \mathbb{N}$, are Hilbert spaces.

Hints: Use completenes of the Lebesgue space $L^2(\Omega)$.

15
Geometry of Hilbert Spaces

According to its definition a Hilbert space differs from a general Banach space in the important aspect that the norm is derived from an inner product. This inner product provides additional structure, mainly of geometric nature. This short chapter looks at basic and mostly elementary consequences of the presence of an inner product in a (pre-) Hilbert space.

15.1 Orthogonal complements and projections

In close analogy to the corresponding concepts in Euclidean spaces, the concepts of orthogonal complement and projections are introduced and basic properties are studied. This analogy helps to understand these results in the general infinite dimensional setting. Only very few additional difficulties occur in the infinite dimensional case as will become apparent later.

Definition 15.1.1 *For any subset M in a pre-Hilbert space $(V, \langle \cdot, \cdot \rangle)$ the **orthogonal complement** of M in V is defined as*

$$M^\perp = \{ y \in V : \langle y, x \rangle = 0 \quad \forall x \in M \}.$$

There are a number of elementary but important consequences of this definition.

Lemma 15.1.1 *Suppose that $(V, \langle \cdot, \cdot \rangle)$ is an inner product space. Then the following holds:*

1. *$V^\perp = \{0\}$ and $\{0\}^\perp = V$.*

2. *For any two subset $M \subset N \subset V$ one has $N^\perp \subseteq M^\perp$.*

3. *The orthogonal complement $M^\perp$ of any subset $M \subset V$ is a linear subspace of V.*

4. *If $0 \in M \subset V$, then $M \cap M^\perp = \{0\}$.*

The simple proof is done in the Exercises.

The following definition and subsequent discussion take into account that linear subspaces are not necessarily closed in the infinite dimensional setting.

Definition 15.1.2 1. *A **closed subspace** of a Hilbert space $\mathcal{H}$ is a linear subspace of $\mathcal{H}$ which is closed.*

2. *If M is any subset of a Hilbert space $\mathcal{H}$ the **span or linear hull of M** is defined by*

$$\lin M = \left\{ x = \sum_{j=1}^{n} a_j x_j \,:\, x_j \in M,\ a_j \in \mathbb{K},\ n \in \mathbb{N} \right\}.$$

3. *The **closed subspace generated by a set** M is the closure of the linear hull; it is denoted by $[M]$, i.e.,*

$$[M] = \overline{\lin M}.$$

That these definitions, respectively notations, are consistent is the contents of the next lemma.

Lemma 15.1.2 *For a subset M in a Hilbert space $\mathcal{H}$ the following holds:*

1. *The linear hull $\lin M$ is the smallest linear subspace of $\mathcal{H}$ which contains M.*

2. *The closure of a linear subspace is again a linear subspace.*

3. *The orthogonal complement $M^\perp$ of a subset M is a closed subspace.*

4. *The orthogonal complement of a subset M and the orthogonal complement of the closed subspace generated by M are the same:*

$$M^\perp = [M]^\perp.$$

Proof. The proof of the first two items is left as an exercise.

For the proof of the third point observe first that according to Lemma 15.1.1 $M^\perp$ is a linear subspace. In order to show that this linear subspace is closed, take any $y \in \mathcal{H}$ in the closure of $M^\perp$. Then there is a sequence $(y_n)_{n \in \mathbb{N}} \subset M^\perp$ which converges to y. Therefore, for any $x \in M$, we know $\langle y, x \rangle = \lim_{n \to \infty} \langle y_n, x \rangle$, because of continuity of the inner product. $y_n \in M^\perp$ implies $\langle y_n, x \rangle = 0$ and thus $\langle y, x \rangle = 0$. We conclude that $y \in M^\perp$. This proves $\overline{M^\perp} \subseteq M^\perp$ and thus $M^\perp = \overline{M^\perp}$ is closed and the third point follows.

In a first step of the proof of the fourth part we show: $M^\perp = (\operatorname{lin} M)^\perp$. Since $M \subset \operatorname{lin} M$ the first part of Lemma 15.1.1 proves $(\operatorname{lin} M)^\perp \subseteq M^\perp$. If now $y \in M^\perp$ and $x = \sum_{j=1}^n a_j x_j \in \operatorname{lin} M$ are given, it follows that $\langle y, x \rangle = \sum_{j=1}^n a_j \langle y, x_j \rangle = 0$ since all x_j belong to M, thus $y \in (\operatorname{lin} M)^\perp$ and therefore the equality $M^\perp = (\operatorname{lin} M)^\perp$.

Since $M \subseteq [M]$ we know by the first part of Lemma 15.1.1 that $[M]^\perp$ is contained in $M^\perp$. In order to show the converse $M^\perp \subseteq [M]^\perp$, take any $y \in M^\perp$. Every point $x \in [M]$ can be represented as a limit $x = \lim_{j \to \infty} x_j$ of points $x_j \in \operatorname{lin} M$. Since we know $M^\perp = (\operatorname{lin} M)^\perp$ we deduce as above that $\langle y, x \rangle = \lim_{j \to \infty} \langle y, x_j \rangle = 0$. This proves $y \in [M]^\perp$ and we conclude. □

From elementary Euclidean geometry we know that given any line in $\mathbb{R}^3$, i.e., a one dimensional subspace of $\mathbb{R}^3$, we can write any vector $x \in \mathbb{R}^3$ in precisely one way as a sum of two vectors where one vector is the projection of x onto this line and the other vector is perpendicular to it. A similar statements holds if a two dimensional plane is given in $\mathbb{R}^3$. The following important result extends this orthogonal decomposition to any Hilbert space.

Theorem 15.1.1 (Projection theorem) *Suppose M is a closed subspace of a Hilbert space $\mathcal{H}$. Every vector $x \in \mathcal{H}$ has the unique representation*

$$x = u + v, \quad u = u(x) \in M, \quad v = v(x) \in M^\perp,$$

and one has

$$\|v\| = \inf_{y \in M} \|x - y\| = d(x, M)$$

where $d(x, M)$ denotes the **distance of the vector x from the subspace M**. *Or equivalently,*

$$\mathcal{H} = M \oplus M^\perp,$$

i.e., the Hilbert space $\mathcal{H}$ is the **direct orthogonal sum** *of the closed subspace M and its orthogonal complement $M^\perp$.*

Proof. Given any $x \in \mathcal{H}$ we have, for $u \in M$,

$$\|x - u\| \geq \inf_{v \in M} \|x - v\| \equiv d(x, M).$$

There is a sequence $(u_n)_{n \in \mathbb{N}} \subset M$ such that

$$d \equiv d(x, M) = \lim_{n \to \infty} \|x - u_n\|.$$

The parallelogram law (Lemma 14.1.2) implies that this sequence is actually a Cauchy sequence as the following calculation shows:

$$\begin{aligned}
\|u_n - u_m\|^2 &= \|(u_n - x) + (x - u_m)\|^2 \\
&= 2\|u_n - x\|^2 + 2\|x - u_m\|^2 - \|(u_n - x) - (x - u_m)\|^2 \\
&= 2\|u_n - x\|^2 + 2\|x - u_m\|^2 - 4\|\tfrac{1}{2}(u_n + u_m) - x\|^2.
\end{aligned}$$

Since u_n and u_m belong to M their convex combination $\tfrac{1}{2}(u_n + u_m)$ is an element of M too and thus, by definition of d, $\|\tfrac{1}{2}(u_n + u_m) - x\|^2 \geq d^2$. It follows that

$$0 \leq \|u_n - u_m\|^2 \leq 2\|u_n - x\|^2 + 2\|x - u_m\|^2 - 4d^2 \to_{n,m \to \infty} 0.$$

Hence $(u_n)_{n\in\mathbb{N}} \subset M$ is a Cauchy sequence in the Hilbert space $\mathcal{H}$ and thus converges to a unique element $u \in \overline{M} = M$ since M is closed. By construction one has
$$d = \lim_{n\to\infty} \|x - u_n\| = \|x - u\|.$$
Next we show that the element $v = x - u$ belongs to the orthogonal complement of M. For $y \in M$, $y \neq 0$, introduce $\alpha = -\frac{\langle y,v \rangle}{\langle y,y \rangle}$. For arbitrary $z \in M$, we know $z - \alpha y \in M$ and thus in particular $d^2 \leq \|x - (u - \alpha y)\|^2 = \|v + \alpha y\|^2 = \|v\|^2 + |\alpha|^2 \|y\|^2 + \langle v, \alpha y \rangle + \langle \alpha y, v \rangle$, hence $d^2 \leq d^2 - \frac{|\langle y,v\rangle|^2}{\|y\|^2}$ and this estimate implies $\langle y, v \rangle = 0$. Since this argument applies to every $y \in M$, $y \neq 0$, we deduce that v belongs to the orthogonal complement $M^\perp$ of M.

Finally we show uniqueness of the decomposition of elements $x \in \mathcal{H}$ into a component u parallel to the closed subspace M and a component v orthogonal to it. Assume that $x \in \mathcal{H}$ has two such decompositions:
$$x = u_1 + v_1 = u_2 + v_2, \qquad u_i \in M, \quad v_i \in M^\perp, \quad i = 1, 2.$$
It follows that $u_1 - u_2 = v_2 - v_1 \in M \cap M^\perp$. By part 4 of Lemma 15.1.1 we conclude $u_1 - u_2 = v_2 - v_1 = 0 \in \mathcal{H}$, hence this decomposition is unique. $\square$

Recall that in the Euclidean space $\mathbb{R}^2$ the shortest distance between a point x and a line M is given by the distance between x and the point u on the line M which is the intersection of the line M and the line perpendicular to M, through the point x. The projection theorem says that this result holds in any Hilbert space and for any closed linear subspace M.

As an easy consequence of the projection theorem one obtains a detailed description of the bi-orthogonal complement $M^{\perp\perp}$ of a set M which is defined as the orthogonal complement of the orthogonal complement of M, i.e., $M^{\perp\perp} = (M^\perp)^\perp$.

Corollary 15.1.1 *For any subset M in a Hilbert space $\mathcal{H}$ one has*
$$M^{\perp\perp} = [M] \quad \text{and} \quad M^{\perp\perp\perp} \equiv (M^{\perp\perp})^\perp = M^\perp. \tag{15.1}$$
In particular, if M is a linear subspace, $M^{\perp\perp} = \overline{M}$, and if M is a closed linear subspace $M^{\perp\perp} = M$.

Proof. Obviously one has $M \subset M^{\perp\perp}$. By Lemma 15.1.2 the bi-orthogonal complement of a set M is known to be a closed linear subspace of $\mathcal{H}$, hence the closed linear hull $[M]$ of M is contained in $M^{\perp\perp}$: $[M] \subseteq M^{\perp\perp}$.

Given any $x \in M^{\perp\perp}$ there are $u \in [M]$ and $v \in [M]^\perp$ such that $x = u + v$ (projection theorem). Since $x - u \in M^{\perp\perp} - [M] \subseteq M^{\perp\perp}$ and $[M]^\perp = M^\perp$ (Lemma 15.1.2), it follows that $v = x - u \in M^{\perp\perp} \cap M^\perp$. But this intersection is trivial by Lemma 15.1.1, therefore $x = u \in [M]$; this proves $M^{\perp\perp} \subseteq [M]$ and together with the opposite inclusion shown above, $M^{\perp\perp} = [M]$. In order to show the second part we take the orthogonal complement of the identity we have just shown and apply Lemma 15.1.2 to conclude $M^{\perp\perp\perp} = [M]^\perp = M^\perp$. $\square$

Remark 15.1.1 *Naturally one can ask whether the assumptions in the projection theorem can be weakened. By considering examples we see that this is not possible in the case of infinite dimensional spaces.*

a) *For a linear subspace of an infinite dimensional Hilbert space which is not closed or for a closed linear subspace of an infinite dimensional inner product space which is not complete, one can construct examples which show that in these cases the projection theorem does not hold (see Exercises).*

b) The projection theorem also does not hold for closed linear subspaces of a Banach space which are not Hilbert spaces. In these cases the uniqueness statement in the projection theorem is not assured (see Exercises).

There is however a direction in which the projection theorem can be generalized. One is allowed to replace the closed linear subspace M by a closed convex set M. Recall that a subset M of a vector space is called convex if all the convex combinations $\lambda x + (1-\lambda)y$, $0 \leq \lambda \leq 1$, belong to M whenever x, y do. Thus one arrives at the projection theorem for closed convex sets. According to the methods used for its proof we present this result in Part C on variational methods, Theorem 32.1.1.

Recall that a subset D of a Hilbert space $\mathcal{H}$ is called dense in $\mathcal{H}$ if every open ball $B_r(x) \subset \mathcal{H}$ has a nonempty intersection with D, i.e., if the closure $\overline{D}$ of D is equal to $\mathcal{H}$. Closely related to dense subsets are the 'total' subsets, i.e., those sets whose linear hull is dense. They play an important role in the study of linear functions. The formal definition reads:

Definition 15.1.3 *A subset M of a Hilbert space $\mathcal{H}$ is called* **total** *if, and only if, the closed linear hull of M equals $\mathcal{H}$, i.e., in the notation introduced earlier, if, and only if, $[M] = \mathcal{H}$.*

The results on orthogonal complements and their relation to the closed linear hull give a very convenient and much used characterization of total sets.

Corollary 15.1.2 *A subset M of a Hilbert space $\mathcal{H}$ is total if, and only if, the orthogonal complement of M is trivial: $M^\perp = \{0\}$.*

Proof. If M is total, then $[M] = \mathcal{H}$ and thus $[M]^\perp = \mathcal{H}^\perp = \{0\}$. Lemma 15.1.2 implies $M^\perp = \{0\}$. If conversely $M^\perp = \{0\}$, then $M^{\perp\perp} = \{0\}^\perp = \mathcal{H}$. But by Corollary 15.1.1 we know $[M] = M^{\perp\perp}$. Thus we conclude. □

15.2 Gram determinants

If we are given a closed subspace M of a Hilbert space $\mathcal{H}$ and a point $x \in \mathcal{H}$ there is, according to the projection theorem, a unique *element of best approximation* $u \in M$, i.e., an element $u \in M$ such that $\|x - u\|$ is minimal. In concrete applications one often has to calculate this element of best approximation explicitly. In general this is a rather difficult task. However, if M is a finite dimensional subspace of a Hilbert space, there is a fairly simple solution to this problem, based on the concept of *Gram determinants*.

Suppose that M is a subspace of dimension n and with basis $\{x_1, \ldots, x_n\}$. The projection theorem implies: Given $x \in \mathcal{H}$ there are a unique $u = u(x) \in M$ and a unique $v = v(x) \in M^\perp$ such that $x = u + v$. $u \in M$ has a unique representation in terms of the elements of the basis: $u = \lambda_1 x_1 + \cdots + \lambda_n x_n$, $\lambda_j \in \mathbb{K}$. Since $v = x - u \in M^\perp$ we know for $k = 1, \ldots, n$ that $\langle x_k, x - u \rangle = 0$

or $\langle x_k, x \rangle = \langle x_k, u \rangle$. Inserting the above representation of $u \in M$ we get a linear system for the unknown coefficients $\lambda_1, \ldots, \lambda_n$:

$$\sum_{j=1}^{n} \lambda_j \langle x_k, x_j \rangle = \langle x_k, x \rangle \qquad k = 1, \ldots, n. \tag{15.2}$$

The determinant of this linear system is called the *Gram determinant*. It is defined in terms of the inner products of basis elements $x_1, \ldots, x_n$:

$$G(x_1, \ldots, x_n) = \det \begin{pmatrix} \langle x_1, x_1 \rangle & \langle x_1, x_2 \rangle & \cdots & \langle x_1, x_n \rangle \\ \langle x_2, x_1 \rangle & \langle x_2, x_2 \rangle & \cdots & \langle x_2, x_n \rangle \\ \vdots & \vdots & \ddots & \vdots \\ \langle x_n, x_1 \rangle & \langle x_n, x_2 \rangle & \cdots & \langle x_n, x_n \rangle \end{pmatrix}. \tag{15.3}$$

Certainly, the function G is well defined for any finite number of vectors of an inner product space.

Next we express the distance $d = d(x, M) = \|x - u\| = \|v\|$ of the point x from the subspace M in terms of the coefficients λ_j: A straightforward calculation gives:

$$\begin{aligned} d^2 &= \|x - u\|^2 = \langle x - u, x - u \rangle = \langle x, x - u \rangle = \langle x, x \rangle - \langle x, u \rangle \\ &= \|x\|^2 - \sum_{j=1}^{n} \lambda_j \langle x, x_j \rangle. \end{aligned}$$

This identity and the linear system (15.2) is written as one homogeneous linear system for the coefficients $(\lambda_0 = 1, \lambda_1, \ldots, \lambda_n)$:

$$\begin{aligned} (d^2 - \|x\|^2)\lambda_0 + \sum_{j=1}^{n} \lambda_j \langle x, x_j \rangle &= 0, \\ -\langle x_k, x \rangle \lambda_0 + \sum_{j=1}^{n} \lambda_j \langle x_k, x_j \rangle &= 0, \qquad k = 1, \ldots, n. \end{aligned} \tag{15.4}$$

By the projection theorem it is known that this homogeneous linear system has a nontrivial solution $(\lambda_0, \lambda_1, \ldots, \lambda_n) \neq (0, 0, \ldots, 0)$. Hence the determinant of this system vanishes, i.e.,

$$\det \begin{pmatrix} \langle d^2 - \langle x, x \rangle & \langle x, x_1 \rangle & \cdots & \langle x, x_n \rangle \\ -\langle x_1, x \rangle & \langle x_1, x_1 \rangle & \cdots & \langle x_1, x_n \rangle \\ \vdots & \vdots & \ddots & \vdots \\ -\langle x_n, x \rangle & \langle x_n, x_1 \rangle & \cdots & \langle x_n, x_n \rangle \end{pmatrix} = 0.$$

Elementary properties of determinants thus give

$$d^2 G(x_1, \ldots, x_n) - G(x, x_1, \ldots, x_n) = 0. \tag{15.5}$$

The Gram determinant of two vectors is:

$$G(x_1, x_2) = \|x_1\|^2 \|x_2\|^2 - |\langle x_1, x_2 \rangle|^2.$$

Schwarz' inequality shows $G(x_1, x_2) \geq 0$. Now an induction with respect to $n \geq 2$, using equation (15.5), proves the following theorem which gives in particular an explicit way to calculate the distance of a point from a finite dimensional subspace.

Theorem 15.2.1 (Gram determinants) *In a Hilbert space $\mathcal{H}$ define the Gram determinants by equation (15.3). Then the following holds:*

1. $G(x_1, \ldots, x_n) \geq 0$ *for all* $x_1, \ldots, x_n \in \mathcal{H}$;

2. $G(x_1, \ldots, x_n) = 0 \Leftrightarrow \{x_1, \ldots, x_n\}$ *is a linearly independent set;*

3. *If $x_1, \ldots, x_n$ are linearly independent vectors in $\mathcal{H}$, denote by $[\{x_1, \ldots, x_n\}]$ the closed linear subspace generated by $\{x_1, \ldots, x_n\}$. Then the distance of any point $x \in \mathcal{H}$ from the subspace $[\{x_1, \ldots, x_n\}]$ is*

$$d = d(x, [\{x_1, \ldots, x_n\}]) = \sqrt{\frac{G(x, x_1, \ldots, x_n)}{G(x_1, \ldots, x_m)}}. \quad (15.6)$$

The proof of this result and some generalizations of Schwarz' inequality given by part 1 of Theorem 15.2.1 are discussed in the Exercises.

15.3 The dual of a Hilbert space

Recall that the (topological) dual of a topological vector space V is defined as the space of all continuous linear functions $T : V \to \mathbb{K}$. In general it is not known how to determine the form of the elements of the topological dual explicitly, even in the case of Banach spaces. However, in the case of a Hilbert space $\mathcal{H}$ the additional information that the norm is induced by an inner product suffices to easily determine the explicit form of continuous linear functions $T : \mathcal{H} \to \mathbb{K}$. Recall that a linear function $T : \mathcal{H} \to \mathbb{K}$ is continuous if, and only if, it is bounded, i.e., if, and only if, there is some constant C_T such that $|T(x)| \leq C_T \|x\|$ for all $x \in \mathcal{H}$. Recall furthermore that under pointwise addition and scalar multiplication the set of all linear functions $T : \mathcal{H} \to \mathbb{K}$ is a vector space over the field $\mathbb{K}$ (see Exercises). For bounded linear functions $T : \mathcal{H} \to \mathbb{K}$ one defines

$$\|T\|' = \sup\{|T(x)| : x \in \mathcal{H}, \|x\| \leq 1\}. \quad (15.7)$$

In the Exercises it is shown that this defines a norm on the space $\mathcal{H}'$ of all bounded linear functions on $\mathcal{H}$. Explicit examples of elements of $\mathcal{H}'$ are all T_u, $u \in \mathcal{H}$, defined by

$$T_u(x) = \langle u, x \rangle \qquad \forall x \in \mathcal{H}. \quad (15.8)$$

The properties of inner products easily imply that the functions T_u, $u \in \mathcal{H}$, are linear. Schwarz' inequality $|\langle u, x \rangle| \leq \|u\| \|x\|$ shows that these linear functions are bounded and thus continuous. And it follows immediately that $\|T_u\|' \leq \|u\|$, for all $u \in \mathcal{H}$. Since $T_u(u) = \|u\|^2$ one actually has equality in this estimate:

$$\|T_u\|' = \|u\| \qquad \forall u \in \mathcal{H}. \tag{15.9}$$

The following theorem characterizes continuous linear functions on a Hilbert space explicitly. This representation theorem says that all elements of $\mathcal{H}'$ are of the form T_u with some $u \in \mathcal{H}$.

Theorem 15.3.1 (Riesz–Fréchet) *Let $\mathcal{H}$ be a Hilbert space over the field $\mathbb{K}$. A linear function $T : \mathcal{H} \to \mathbb{K}$ is continuous if, and only if, there is a $u \in \mathcal{H}$ such that $T = T_u$, and one has $\|T\|' = \|u\|$.*

Proof. According to the discussion preceding the theorem we have to show that every continuous linear functional T on the Hilbert space $\mathcal{H}$ is of the form T_u for some $u \in \mathcal{H}$. If $T = 0$ is the null functional, choose $u = 0$. If $T \neq 0$, then

$$M = \ker T = T^{-1}(0) = \{x \in \mathcal{H} : T(x) = 0\}$$

is a closed linear subspace of $\mathcal{H}$ which is not equal to $\mathcal{H}$. This is shown in the Exercises. It follows that the orthogonal complement $M^\perp$ of M is not trivial (Corollary 15.1.1) and the projection theorem states that the Hilbert space has a decomposition into two nontrivial closed linear subspaces: $\mathcal{H} = M \oplus M^\perp$. Hence there is a $v \in M^\perp$ such that $T(v) \neq 0$ and thus, for every $x \in \mathcal{H}$ we can define the element $u = u(x) = x - \frac{T(x)}{T(v)} v$. Linearity of T implies $T(u) = 0$ and thus $u \in M$, therefore $\langle v, u(x) \rangle = 0$ for all $x \in \mathcal{H}$, or

$$T(x) = \frac{T(v)}{\langle v, v \rangle} \langle v, x \rangle \qquad \forall x \in \mathcal{H}.$$

This proves $T = T_u$ for the element

$$u = \frac{\overline{T(v)}}{\langle v, v \rangle} v \in (\ker T)^\perp.$$

It follows that $\|T\|' = \|T_u\| = \|u\|$.

Finally we show uniqueness of the element $u \in \mathcal{H}$ which defines the given continuous linear function T by $T = T_u$. Suppose $u, v \in \mathcal{H}$ define T by this relation. Then $T_u = T_v$ or $\langle u, x \rangle = \langle v, x \rangle$ for all $x \in \mathcal{H}$ and hence $u - v \in \mathcal{H}^\perp = \{0\}$, i.e., $u = v$. □

Corollary 15.3.1 *A Hilbert space $\mathcal{H}$ and its (topological) dual $\mathcal{H}'$ are isometrically anti-isomorphic, i.e., there is an isometric map $J : \mathcal{H} \to \mathcal{H}'$ which is antilinear.*

Proof. Define a map $J : \mathcal{H} \to \mathcal{H}'$ by $J(u) = T_u$ for all $u \in \mathcal{H}$ where T_u is defined by equation (15.8). Thus J is well defined and we know $\|J(u)\|' = \|T_u\|' = \|u\|$. Hence the map J is isometric. The definition of T_u easily implies

$$J(\alpha u + \beta v) = \overline{\alpha} J(u) + \overline{\beta} J(v) \qquad \forall \alpha, \beta \in \mathbb{K}, \ \forall u, v \in \mathcal{H},$$

i.e., the map J is antilinear. As an isometric map J is injective, and by the Riesz–Fréchet Theorem we know that it is surjective. Hence J is an isometric antiisomorphism. □

Remark 15.3.1 1. *The theorem of Riesz and Fréchet relies in a decisive way on the assumption of completeness. This theorem does not hold in inner product spaces which are not complete. An example is discussed in the Exercises.*

2. *The duality property of a Hilbert space $\mathcal{H}$, i.e., $\mathcal{H} \simeq \mathcal{H}'$ is used in the* **bra- and ket- vector notation of Dirac.** *For vectors $u \in \mathcal{H}$ Dirac writes a ket vector $|u>$ and for elements $T_v \in \mathcal{H}'$ he writes the bra vector $<v|$. Bra vectors act on ket vectors according to the relation $<v|u> = T_v(u)$, $u, v \in \mathcal{H}$. In this notation the projector P_ψ onto the subspace spanned by the vector ψ is $P_\psi = |\psi><\psi|$.*

Every continuous linear function $T : \mathcal{H} \to \mathbb{K}$ is of the form $T = T_u$ for a unique element u in the Hilbert space $\mathcal{H}$, by Theorem 15.3.1. This implies the following orthogonal decomposition of the Hilbert space: $\mathcal{H} = \ker T \oplus \mathbb{K} u$, i.e., the kernel or null space of a continuous linear functional on a Hilbert space is a closed linear subspace of co-dimension 1. This says in particular that a continuous linear functional "lives" on the one dimensional subspace $\mathbb{K} u$. This is actually the case in the general setting of locally convex topological vector spaces as the Exercises show.

The Theorem of Riesz and Fréchet has many other applications. We discuss here an easy solution of the extension problem, i.e., the problem of finding a continuous linear functional T on the Hilbert space $\mathcal{H}$ which agrees with a given continuous linear functional T_0 on a linear subspace M of $\mathcal{H}$ and which has the same norm as T_0.

Theorem 15.3.2 (Extension theorem) *Let M be a linear subspace of a Hilbert space $\mathcal{H}$ and $T_0 : M \to \mathbb{K}$ a continuous linear functional, i.e., there is some constant C such that $|T_0(x)| \leq c \|x\|$ for all $x \in M$. Then there is exactly one continuous linear functional $T : \mathcal{H} \to \mathbb{K}$ such that $T|M = T_0$ and $\|T\|' = \|T_0\|'$ where the definition*

$$\|T_0\|' = \sup \{|T_0(x)| : x \in M, \|x\| \leq 1\}$$

is used.

Proof. The closure $\overline{M}$ of the linear subspace M is itself a Hilbert space, when we use the restriction of the inner product $\langle \cdot, \cdot \rangle$ of $\mathcal{H}$ to $\overline{M}$. This is shown as an exercise. We show next that T_0 has a unique extension T_1 to a continuous linear function $\overline{M} \to \mathbb{K}$. Given $x \in \overline{M}$ there is a sequence $(x_n)_{n \in \mathbb{N}}$ in M which converges to x. Define $T_1(x) = \lim_{n \to \infty} T_0(x_n)$. This limit exists since the field $\mathbb{K}$ is complete and $(T_0(x_n))_{n \in \mathbb{N}}$ is a Cauchy sequence in $\mathbb{K}$: We have the estimate $|T_0(x_n) - T_0(x_m)| = |T_0(x_n - x_m)| \leq C\|x_n - x_m\|$, and we know that $(x_n)_{n \in \mathbb{N}}$ is a Cauchy sequence in the Hilbert space $\overline{M}$. If we take another sequence $(y_n)_{n \in \mathbb{N}}$ in M with limit x we know $|T_0(x_n) - T_0(y_n)| = |T_0(x_n - y_n)| \leq C\|x_n - y_n\| \to 0$ as $n \to \infty$ and thus both sequences give the same limit $T_1(x)$. It follows that

$$\|T_1\|' = \sup \{|T_1(x)| : x \in \overline{M}, \|x\| \leq 1\} = \sup \{|T_0(x)| : x \in M, \|x\| \leq 1\} = \|T_0\|' \leq C.$$

The second identity is shown in the Exercises.

The Theorem of Riesz–Fréchet implies: There is exactly one $v \in \overline{M}$ such that $T_1(u) = \langle v, u \rangle$ for all $u \in \overline{M}$ and $\|T_1\|' = \|v\|$. Since the inner product is actually defined on all of $\mathcal{H}$, we get an easy extension T of T_1 to the Hilbert space $\mathcal{H}$ by defining $T(x) = \langle v, x \rangle$ for all $x \in \mathcal{H}$ and it follows that $\|T\|' = \|v\| = \|T_0\|'$.

This functional T is an extension of T_0 since for all $u \in M$ one has $T(u) = \langle v, u \rangle = T_1(u) = T_0(u)$, by definition of T_1.

Suppose that S is a continuous linear extension of T_0. As a continuous linear functional on $\mathcal{H}$ this extension is of the form $S(x) = \langle y, x \rangle$, for all $x \in \mathcal{H}$, with a unique $y \in \mathcal{H}$. And, since S is an extension of T_0, we know $S(u) = T_0(u) = \langle v, u \rangle$ for all $u \in M$ and thus for all $u \in \overline{M}$. This shows $\langle y - v, u \rangle = 0$ for all $u \in M$, hence $y - v \in M^\perp$, and we deduce $\|S\|' = \|y\| = \sqrt{\|v\|^2 + \|y - v\|^2}$. Hence this extension S satisfies $\|S\|' = \|T_0\|' = \|v\|$ if, and only if, $y - v = 0$, i.e., if, and only if, $S = T = T_v$, and we conclude. □

Methods and results from the theory of Hilbert spaces and their operators are used in various areas of mathematics. We present here an application of Theorem 15.3.2 to a problem from distribution theory, namely to prove the existence of a fundamental solution for a special constant coefficient partial differential operator. Earlier we had used Fourier transformation for distributions to find a fundamental solution for this type of differential operator. The proof of the important Theorem 8.3.1 follows a similar strategy.

Corollary 15.3.2 *The linear partial differential operator with constant coefficients*

$$1 - \Delta_n \quad \text{in} \quad \mathbb{R}^n$$

has a fundamental solution in $S'(\mathbb{R}^n)$.

Proof. Consider the subspace $M = (1 - \Delta_n)\mathcal{D}(\mathbb{R}^n)$ of the Hilbert space $\mathcal{H} = L^2(\mathbb{R}^n)$ and define a linear functional $T_0 : M \to \mathbb{K}$ by

$$T_0((1 - \Delta_n)\phi) = \phi(0) \quad \forall \phi \in \mathcal{D}(\mathbb{R}^n).$$

Applying Lemma 10.1.1 to the inverse Fourier transformation one has the estimate

$$|\phi(0)| \leq \|\phi\|_\infty \leq (2\pi)^{-\frac{n}{2}} \|\mathcal{F}\phi\|_1.$$

If $2m > n$, then the function $p \mapsto (1 + p^2)^{-m}$ belongs to the Hilbert space $L^2(\mathbb{R}^n)$, and we can use Schwarz' inequality to estimate the norm of $\tilde{\phi} = \mathcal{F}\phi$ as follows:

$$\|\tilde{\phi}\|_1 = \|(1 + p^2)^{-m} \cdot (1 + p^2)^m \tilde{\phi}\|_1 \leq \|(1 + p^2)^{-m}\|_2 \|(1 + p^2)^m \tilde{\phi}\|_2.$$

By theorems 10.3.5 and 10.2.1 we know $\|(1 + p^2)^m \tilde{\phi}\|_2 = \|(1 - \Delta_n)^m \phi\|_2$ and thus the estimate

$$|\phi(0)| \leq C\|(1 - \Delta_n)^m \phi\|_2 \quad \forall \phi \in \mathcal{D}(\mathbb{R}^n)$$

follows, with a constant C which is given by the above calculations. This estimate shows first that the functional T_0 is well defined on M. It is easy to see that T_0 is linear. Now the above estimate also implies that T_0 is continuous. Hence the above extension theorem can be applied, and thus there is $u \in L^2(\mathbb{R}^n)$ such that

$$T_0((1 - \Delta_n)^m \phi) = \langle u, (1 - \Delta_n)^m \phi \rangle_2 = \int_{\mathbb{R}^n} \overline{u(x)}(1 - \Delta_n)^m \phi(x) dx$$

for all $\phi \in \mathcal{D}(\mathbb{R}^n)$. By definition of T_0 this shows that

$$\int_{\mathbb{R}^n} \overline{u(x)}(1 - \Delta_n)^m \phi(x) dx = \phi(0),$$

i.e., the distribution $E = (1 - \Delta_n)^{m-1} u$ is a fundamental solution of the operator $1 - \Delta_n$. Since $u \in L^2(\mathbb{R}^n)$ the distribution $E = (1 - \Delta_n)^{m-1} u$ is tempered, and we conclude. □

15.4 Exercises

1. Prove Lemma 15.1.1.

2. Prove the first two parts of Lemma 15.1.2.

3. Find an example supporting the first part of Remark 15.1.1.

4. Consider the Euclidean space $\mathbb{R}^2$, but equipped with the norm $\|x\| = |x_1| + |x_2|$ for $x = (x_1, x_2) \in \mathbb{R}^2$. Show that this is a Banach space but not a Hilbert space. Consider the point $x = (-\frac{r}{2}, \frac{r}{2})$ for some $r > 0$ and the closed linear subspace $M = \{x \in \mathbb{R}^2 : x_1 = x_2\}$. Prove that this point has the distance r from the subspace M, i.e., $\inf\{\|x - u\| : u \in M\} = r$ and that there are infinitely many points $u \in M$ such that $\|x - u\| = r$. Conclude that the projection theorem does not hold for Banach spaces which are not Hilbert spaces (compare with part b) of Remark 15.1.1).

5. Prove Theorem 15.2.1.

6. For three vectors x, y, z in a Hilbert space $\mathcal{H}$, calculate the Gram determinant $G(x, y, z)$ explicitly and discuss in detail the inequality $0 \leq G(x, y, z)$. Consider some special cases: $x \perp y$, $x \perp z$, or $y \perp z$.

7. For a nontrivial continuous linear function $T : \mathcal{H} \to \mathbb{K}$ on a Hilbert space $\mathcal{H}$ show that its null-space $\ker T$ is a proper closed linear subspace of $\mathcal{H}$.

8. Consider the space of all terminating sequences of elements in $\mathbb{K}$:
$$\ell_e^2(\mathbb{K}) = \{x = (x_1, x_2, \ldots, x_N, 0, 0, \ldots) : x_j \in \mathbb{K}, N = N(x) \in \mathbb{N}\}.$$

Obviously one has $\ell_e^2(\mathbb{K}) \subset \ell^2(\mathbb{K})$ and it is naturally a vector space over the field $\mathbb{K}$; as an inner product on $\ell_e^2(\mathbb{K})$ we take $\langle x, y \rangle_2 = \sum_{j=1}^{\infty} \overline{x_j} y_j$. Consider the sequence $u = (1, \frac{1}{2}, \frac{1}{3}, \ldots, \frac{1}{n}, \ldots) \in \ell^2(\mathbb{K})$ and use it to define a linear function $T = T_u : \ell_e^2(\mathbb{K}) \to \mathbb{K}$ by
$$T(x) = \langle u, x \rangle_2 = \sum_{n=1}^{\infty} \frac{1}{n} x_n.$$

This function is continuous by Schwarz' inequality: $|T(x)| \leq \|u\|_2 \|x\|_2$ for all $x \in \ell_e^2(\mathbb{K})$. Conclude that the theorem of Riesz – Fréchet does not hold for the inner product space $\ell_e^2(\mathbb{K})$.

9. For a linear subspace M of a Hilbert space $\mathcal{H}$ with inner product $\langle \cdot, \cdot \rangle$ show: The closure $\overline{M}$ of M is a Hilbert space when equipped with the restriction of the inner product $\langle \cdot, \cdot \rangle$ to $\overline{M}$.

15. Geometry of Hilbert Spaces

10. Prove the identity

    $$\sup\{|T_1(x)| : x \in \overline{M}, \|x\| \leq 1\} = \sup\{|T_0(x)| : x \in M, \|x\| \leq 1\}$$

    used in the proof of Theorem 15.3.2.

11. Give an example of a linear functional which is not continuous.

    Hints: Consider the real vector space V of all real polynomials P on the interval $I = [0, 1]$, take a point $a \notin I$, for instance $a = 2$, and define $T_a : V \to \mathbb{R}$ by $T_a(P) = P(a)$ for all $P \in V$. Show that T_a is not continuous with respect to the norm $\|P\| = \sup_{x \in I} |P(x)|$ on V.

16
Separable Hilbert Spaces

Up to now we have studied results which are available in any Hilbert space. Now we turn our attention to a very important subclass which one encounters in many applications, in mathematics as well as in physics. This subclass is characterized by the property that the Hilbert space has a countable basis defined in a way suitable for Hilbert spaces. Such a 'Hilbert space basis' plays the same role as a coordinate system in a finite dimensional vector space.

Recall that two finite dimensional vector spaces are isomorphic if, and only if, they have the same dimension. Similarly, Hilbert spaces are characterized up to isomorphy by the cardinality of their Hilbert space basis. Those Hilbert spaces which have a countable Hilbert space basis are called *separable*.

In a first section we introduce and discuss the basic concepts and results in the theory of separable Hilbert spaces. Then a special class of separable Hilbert spaces is investigated. For this subclass the Hilbert space basis is defined in an explicit way through a given weight function and an orthogonalization procedure. These spaces play an important role in the study of differential operators, in particular in quantum mechanics.

16.1 Basic facts

As indicated above the concept of a Hilbert space basis differs from the concept of a basis in a vector space. The point which distinguishes these two concepts is that for the definition of a Hilbert space basis a limit process is used.

16. Separable Hilbert Spaces

We begin by recalling the concept of a basis in a vector space V over the field $\mathbb{K}$. A nonempty subset $A \subset V$ is called *linearly independent* if, and only if, every finite subset $\{x_1, \ldots, x_n\} \subset A$, $n \in \mathbb{N}$, is linearly independent. A finite subset $\{x_1, \ldots, x_n\}$, $x_i \neq x_j$ for $i \neq j$ is called linearly independent if, and only if, $\sum_{i=1}^{n} \lambda_i x_i = 0$, $\lambda_i \in \mathbb{K}$, implies $\lambda_1 = \lambda_2 = \cdots = \lambda_n = 0$, i.e., the only way to write the null vector 0 of V as a linear combination of the vectors $x_1, \ldots, x_n$ is the trivial one with $\lambda_i = 0 \in \mathbb{K}$ for $i = 1, \ldots, n$.

The set of all vectors in V which can be written as some linear combination of elements in the given nonempty subset A is called the linear hull lin A of A (see Definition 15.1.2), i.e.,

$$\lin A = \left\{ x \in V : x = \sum_{i=1}^{n} \lambda_i x_i,\ x_i \in A,\ \lambda_i \in \mathbb{K},\ n \in \mathbb{N} \right\}.$$

It is the smallest linear subspace which contains A. A linearly independent subset $A \subset V$ which generates V, i.e., $\lin A = V$, is called a *basis of the vector space V*.

A linearly independent set $A \subset V$ is called maximal if, and only if, for any linearly independent subset A' the relation $A \subset A'$ implies $A = A'$. In this sense a basis is a maximal linearly independent subset. This means: If one adds an element x of V to a basis B, then the resulting subset $B \cup \{x\}$ is no longer linearly independent. With the help of Zorn's Lemma (or the axiom of choice) one can prove that every vector space has a basis. Such a basis is a purely algebraic concept and is often called a *Hamel basis*.

In 1927, J. Schauder introduced the concept of Hilbert space basis or a basis of a Hilbert space which takes the topological structure of a Hilbert space into account as expressed in the following definition:

Definition 16.1.1 *Let $\mathcal{H}$ be a Hilbert space over the field $\mathbb{K}$ and B a subset of $\mathcal{H}$.*

1. *B is called a **Hilbert space basis** of $\mathcal{H}$ if, and only if, B is linearly independent in the vector space $\mathcal{H}$ and B generates $\mathcal{H}$ in the sense that $[B] = \overline{\lin B} = \mathcal{H}$.*

2. *The Hilbert space $\mathcal{H}$ is called **separable** if, and only if, it has a countable Hilbert space basis $B = \{x_n \in \mathcal{H} : n \in \mathbb{N}\}$ (or a finite basis $B = \{x_1, \ldots, x_N\}$ for some $N \in \mathbb{N}$).*

3. *An orthonormal system $B = \{x_\alpha \in \mathcal{H} : \alpha \in A\}$ in $\mathcal{H}$ which is a Hilbert space basis is called an **orthonormal basis** or **ONB** of $\mathcal{H}$.*

It is important to realize that in general a Hilbert space basis is not an algebraic basis! For instance in the case of a separable Hilbert space a general element in $\mathcal{H}$ is known to have a representation as a series $\sum_{n=1}^{\infty} \lambda_n x_n$ in the elements x_n of the basis but not as a linear combination.

Often a separable Hilbert space is defined as a Hilbert space which has a countable dense subset. Sometimes this definition is more convenient. The equivalence of both definitions is shown in the Exercises.

In the original definition of a Hilbert space the condition of separability was included. However in 1934 F. Rellich and F. Riesz pointed out that for most parts of the theory the separability assumption is not needed.

Nevertheless most Hilbert spaces which one encounters in applications are separable. In the Exercises we discuss an example of a Hilbert space which is *not separable*. This is the *space of almost-periodic functions* on the real line $\mathbb{R}$.

As we know from the Euclidean spaces $\mathbb{R}^n$ it is in general a great advantage in many problems to work with an orthonormal basis $\{\underline{e}_1, \ldots, \underline{e}_n\}$ instead of an arbitrary basis. Here $\underline{e}_i$ is the standard unit vector along coordinate axis i. In a separable infinite dimensional Hilbert space the corresponding basis is an orthonormal Hilbert space basis, or ONB. The proof of the following result describes in detail how to construct an ONB given any Hilbert space basis. Only the case of a separable Hilbert space is considered since this is the case which is needed in most applications. Using the axiom of choice one can also prove the existence of an orthonormal basis in the case of a nonseparable Hilbert space. In the second section of this chapter we use this construction to generate explicitly ONB's for concrete Hilbert spaces of square integrable functions.

Theorem 16.1.1 (Gram–Schmidt orthonormalization) *Every separable Hilbert space $\mathcal{H}$ has an orthonormal basis B.*

Proof. By definition of a separable Hilbert space there is a countable Hilbert space basis $B = \{y_n : n \in \mathbb{N}\} \subset \mathcal{H}$ (or a finite basis; we consider explicitly the first case). Define $z_1 = y_1$; since B is a basis we know $\|y_1\| > 0$ and hence the vector $z_2 = y_2 - \frac{\langle z_1, y_2 \rangle}{\langle z_1, z_1 \rangle} z_1$ is well defined in $\mathcal{H}$. One has $z_1 \perp z_2$ since $\langle z_1, z_2 \rangle = 0$. As elements of the basis B the vectors y_1 and y_2 are linearly independent, therefore the vector z_2 is not the null vector, and certainly the set of vectors $\{z_1, z_2\}$ generates the same linear subspace as the set of vectors $\{y_1, y_2\}$: $[\{z_1, z_2\}] = [\{y_1, y_2\}]$.

We proceed by induction and assume that for some $N \in \mathbb{N}$, $N \geq 2$ the set of vectors $\{z_1, \ldots, z_N\}$ is well defined and has the following properties:

a) $\|z_j\| > 0$ for all $j = 1, \ldots, N$;

b) $\langle z_i, z_j \rangle = 0$ for all $i, j \in \{1, \ldots, N\}, i \neq j$;

c) The set $\{z_1, \ldots, z_N\}$ generates the same linear subspace as the set $\{y_1, \ldots, y_N\}$, i.e., $[\{z_1, \ldots, z_N\}] = [\{y_1, \ldots, y_N\}]$.

This allows us to define
$$z_{N+1} = y_{N+1} - \sum_{i=1}^{N} \frac{\langle z_i, y_{N+1} \rangle}{\langle z_i, z_i \rangle} z_i.$$

The orthogonality condition b) easily implies $\langle z_j, z_{N+1} \rangle = 0$ for $j = 1, \ldots, N$. Hence the set of vectors $\{z_1, \ldots, z_N, z_{N+1}\}$ is pairwise orthogonal too. From the definition of the vector z_{N+1} it is clear that $[\{z_1, \ldots, z_{N+1}\}] = [\{y_1, \ldots, y_{N+1}\}]$ holds. Finally, since the vector y_{N+1} is not a linear combination of the vectors $y_1, \ldots, y_N$ the vector z_{N+1} is not zero. This shows that the set of vectors $\{z_1, \ldots, z_N, z_{N+1}\}$ too has the properties a), b), and c). By the principle of induction we conclude: There is a set of vectors $\{z_k \in \mathcal{H} : k \in \mathbb{N}\}$ such that $\langle z_j, z_k \rangle = 0$ for all $j, k \in \mathbb{N}, j \neq k$ and $[\{z_k : k \in \mathbb{N}\}] = [\{y_k : k \in \mathbb{N}\}] = \mathcal{H}$. Finally we normalize the vectors z_k to obtain an orthonormal basis $B = \{e_k : k \in \mathbb{N}\}$, $e_k = \frac{1}{\|z_k\|} z_k$. □

Theorem 16.1.2 (Characterization of ONB's) *Let $B = \{x_n : n \in \mathbb{N}\}$ be an orthonormal system in a separable Hilbert space $\mathcal{H}$. The following statements are equivalent.*

a) B is maximal (or **complete**), i.e., an ONB.

b) For any $x \in \mathcal{H}$ the condition "$\langle x_n, x \rangle = 0$ for all $n \in \mathbb{N}$" implies $x = 0$.

c) Every $x \in \mathcal{H}$ has the **Fourier expansion**

$$x = \sum_{n=1}^{\infty} \langle x_n, x \rangle x_n.$$

d) For all vectors $x, y \in \mathcal{H}$ the **completeness relation**

$$\langle x, y \rangle = \sum_{n=1}^{\infty} \langle x, x_n \rangle \langle x_n, y \rangle$$

holds.

e) For every $x \in \mathcal{H}$ the **Parseval relation**

$$\|x\|^2 = \sum_{n=1}^{\infty} |\langle x_n, x \rangle|^2$$

holds.

Proof. a) $\Rightarrow$ b): Suppose that there is a $z \in \mathcal{H}$, $z \neq 0$, with the property $\langle x_n, z \rangle = 0$ for all $n \in \mathbb{N}$. Then $B' = \left\{ \frac{z}{\|z\|}, x_1, x_2, \ldots \right\}$ is an orthonormal system in $\mathcal{H}$ in which B is properly contained, contradicting the maximality of B. Hence there is no such vector $z \in \mathcal{H}$ and statement b) follows.

b) $\Rightarrow$ c): Given $x \in \mathcal{H}$ introduce the sequence $x^{(N)} = \sum_{n=1}^{N} \langle x_n, x \rangle x_n$. Bessel's inequality (Corollary 14.1.1) shows that $\|x^{(N)}\|^2 = \sum_{n=1}^{N} |\langle x_n, x \rangle|^2 \leq \|x\|^2$ for all $N \in \mathbb{N}$. Hence the infinite series $\sum_{n=1}^{\infty} |\langle x_n, x \rangle|^2$ converges and its value is less than or equal to $\|x\|^2$. For all $M < N$ we have $\|x^{(N)} - x^{(M)}\|^2 = \sum_{n=M+1}^{N} |\langle x_n, x \rangle|^2$ and the convergence of the series $\sum_{n=1}^{\infty} |\langle x_n, x \rangle|^2$ implies that $(x^{(N)})_{N \in \mathbb{N}}$ is a Cauchy sequence. Hence this sequence converges to a unique point $y \in \mathcal{H}$,

$$y = \lim_{N \to \infty} x^{(N)} = \sum_{n=1}^{\infty} \langle x_n, x \rangle x_n.$$

Since the inner product is continuous we deduce that $\langle x_n, y \rangle = \lim_{N \to \infty} \langle x_n, x^{(N)} \rangle = \langle x_n, x \rangle$ for all $n \in \mathbb{N}$. Therefore $\langle x_n, x - y \rangle = 0$ for all $n \in \mathbb{N}$ and hypothesis b) implies $x - y = 0$, hence statement c) follows.

c) $\Rightarrow$ d): According to statement c) any vector $x \in \mathcal{H}$ has a Fourier expansion, $x = \sum_{n=1}^{\infty} \langle x_n, x \rangle x_n$, similarly for $y \in \mathcal{H}$: $y = \sum_{n=1}^{\infty} \langle x_n, y \rangle x_n$. Continuity of the inner product and orthonormality of $\{x_n : n \in \mathbb{N}\}$ imply the completeness relation:

$$\langle x, y \rangle = \sum_{n=1}^{\infty} \overline{\langle x_n, x \rangle} \langle x_n, y \rangle = \sum_{n=1}^{\infty} \langle x, x_n \rangle \langle x_n, y \rangle.$$

d) $\Rightarrow$ e): Obviously, statement e) is just the special case $x = y$ of statement d).

e) $\Rightarrow$ a): Suppose that the system B is not maximal. Then we can add one unit vector $z \in \mathcal{H}$ to it which is orthogonal to B. Now Parseval's relation e) gives the contradiction

$$1 = \|z\|^2 = \sum_{n=1}^{\infty} |\langle x_n, z \rangle|^2 = \sum_{n=1}^{\infty} 0 = 0.$$

Therefore, when Parseval's relation holds for every $x \in \mathcal{H}$, the system B is maximal. $\square$

As a first application of the characterization of an orthonormal basis we determine explicitly the closed linear hull of an orthonormal system (ONS). As a simple consequence one obtains a characterization of separable Hilbert spaces.

Corollary 16.1.1 *Let $\{x_n : n \in \mathbb{N}\}$ be an orthonormal system in a Hilbert space over the field $\mathbb{K}$. Denote the closed linear hull of this system by M, i.e., $M = [\{x_n : n \in \mathbb{N}\}]$. Then, the following holds:*

1. *$M = \left\{ x_c \in \mathcal{H} : x_c = \sum_{n=1}^{\infty} c_n x_n, \ c = (c_n)_{n \in \mathbb{N}} \in \ell^2(\mathbb{K}) \right\}.$*

2. *The mapping $U : \ell^2(\mathbb{K}) \to M$, defined by $Uc = x_c = \sum_{n=1}^{\infty} c_n x_n$, is an isomorphism and one has*

$$\langle Uc, Uc' \rangle_{\mathcal{H}} = \langle c, c' \rangle_{\ell^2(\mathbb{K})} \quad \forall c, c' \in \ell^2(\mathbb{K}).$$

Proof. For $c = (c_n)_{n \in \mathbb{N}} \in \ell^2(\mathbb{K})$ define a sequence $x^{(N)} = \sum_{n=1}^{N} c_n x_n \in \text{lin } \{x_n : n \in \mathbb{N}\}$ in the Hilbert space $\mathcal{H}$. Since $\{x_n : n \in \mathbb{N}\}$ is an orthonormal system one has for all $N, M \in \mathbb{N}$, $M < N$, $\|x^{(N)} - x^{(M)}\|^2 = \sum_{n=M+1}^{N} |c_n|^2$. It follows that $(x^{(N)})_{N \in \mathbb{N}}$ is a Cauchy sequence in the Hilbert space $\mathcal{H}$ and thus it converges to

$$x_c = \lim_{N \to \infty} x^{(N)} = \sum_{n=1}^{\infty} c_n x_n \in \mathcal{H}.$$

Obviously, x_c belongs to the closure M of the linear hull of the given orthonormal system. Hence $\ell^2(\mathbb{K}) \ni c \mapsto x_c$ defines a map U from $\ell^2(\mathbb{K})$ into M. This map is linear as one easily shows. Under the restriction of the inner product of $\mathcal{H}$ the closed linear subspace M is itself a Hilbert space which has, by definition, the given ONS as a Hilbert space basis. Therefore, by Theorem 16.1.2, for every $x \in M$ one has $x = \sum_{n=1}^{\infty} \langle x_n, x \rangle_{\mathcal{H}} x_n$ and $\|x\|_{\mathcal{H}}^2 = \sum_{n=1}^{\infty} |\langle x_n, x \rangle_{\mathcal{H}}|^2$. Hence every $x \in M$ is the image of the sequence $c = (\langle x_n, x \rangle_{\mathcal{H}})_{n \in \mathbb{N}} \in \ell^2(\mathbb{K})$ under the map U, i.e., U is a linear map from $\ell^2(\mathbb{K})$ onto M, and the inverse map of U is the map $M \ni x \mapsto U^{-1}x = (\langle x_n, x \rangle_{\mathcal{H}})_{n \in \mathbb{N}} \in \ell^2(\mathbb{K})$.

For $c = (c_n)_{n \in \mathbb{N}} \in \ell^2(\mathbb{K})$ we calculate $\langle x_n, Uc \rangle_{\mathcal{H}} = c_n$ for all $n \in \mathbb{N}$ and thus by the completeness relation of Theorem 16.1.2

$$\langle Uc, Uc' \rangle_{\mathcal{H}} = \sum_{n=1}^{\infty} \overline{c_n} c'_n = \langle c, c' \rangle_{\ell^2(\mathbb{K})} \quad \forall c, c' \in \ell^2(\mathbb{K}).$$

In particular one has $\|Uc\|_{\mathcal{H}} = \|c\|_{\ell^2(\mathbb{K})}$ for all $c \in \ell^2(\mathbb{K})$. Thus U is a bijective continuous linear map with continuous inverse which does not change the values of the inner products, i.e., an isomorphism of Hilbert spaces. □

Corollary 16.1.2 *Every infinite dimensional separable Hilbert space $\mathcal{H}$ over the field $\mathbb{K}$ is isomorphic to the sequence space $\ell^2(\mathbb{K})$.*

Proof. If $\{x_n : n \in \mathbb{N}\}$ is an orthonormal basis we know that the closed linear subspace M generated by this basis is equal to the Hilbert space $\mathcal{H}$. Hence, by the previous corollary we conclude. □

Later we will learn that, for instance, the Lebesgue space $L^2(\mathbb{R}^n, dx)$ is a separable Hilbert space. According to Corollary 16.1.2 this Lebesgue space is isomorphic to the sequence space $\ell^2(\mathbb{K})$. Why then is it important to study other separable Hilbert spaces than the sequence space $\ell^2(\mathbb{K})$? These other separable Hilbert spaces have, just as the Lebesgue space, an additional structure which is lost if they are realized as sequence spaces. While linear partial differential operators, for instance Schrödinger operators, can be studied conveniently in the Lebesgue space, this is in general not the case in the sequence space. In the second section of this chapter we will construct explicitly an orthonormal basis for Hilbert spaces $L^2(I)$ of square integrable functions over some interval I. It turns out that the elements of the ONB's constructed there, are 'eigenfunctions' of important differential operators.

The results on the characterization of an orthonormal basis are quite powerful. We illustrate this with the example of the theory of **Fourier expansions** in the Hilbert space $L^2([0, 2\pi], dx)$.

We begin by recalling some classical results. For integrable functions on the interval $[0, 2\pi]$ the integrals

$$c_n = c_n(f) = \frac{1}{\sqrt{2\pi}} \int_0^{2\pi} e^{-inx} f(x) dx = \langle e_n, f \rangle_2$$

are well defined. In the Exercises one shows that the system of functions $e_n, n \in \mathbb{Z}$, $e_n(x) = \frac{e^{inx}}{\sqrt{2\pi}}$, is an orthonormal system in the Hilbert space $L^2([0, \pi], dx)$. With the above numbers c_n one forms the *Fourier series*

$$\sum_{n=-\infty}^{+\infty} c_n(f) e_n$$

of the function f. A classical result from the theory of Fourier series reads (see [Edw79]): If f is continuously differentiable on the interval $[0, 2\pi]$, then the Fourier series converges uniformly to f, i.e., the sequence of partial sums of the Fourier series converges uniformly to f. This implies in particular

$$\lim_{N \to \infty} \| f - \sum_{n=-N}^{N} c_n(f) e_n \|_2 = 0$$

for all $f \in C^1([0, \pi])$.

We claim that the system $\{e_n : n \in \mathbb{Z}\}$ is actually an orthonormal basis of $L^2([0, 2\pi], dx)$. For the proof take any $g \in L^2([0, 2\pi], dx)$ with the property $\langle e_n, g \rangle_2 = 0$ for all $n \in \mathbb{Z}$. From the above convergence result we deduce, for all $f \in C^1([0, 2\pi])$,

$$\langle f, g \rangle_2 = \lim_{N \to \infty} \langle \sum_{n=-N}^{N} c_n(f) e_n, g \rangle_2 = 0.$$

Since $C^1([0, 2\pi])$ is known to be dense in $L^2([0, 2\pi])$ it follows that $g = 0$, by Corollary 15.1.2, hence by Theorem 16.1.2, this system is an orthonormal basis of $L^2([0, 2\pi], dx)$. Therefore, every $f \in L^2([0, 2\pi], dx)$ has a Fourier expansion which converges (in the sense of the L^2-topology). Thus, convergence of the Fourier series in the L^2-topology is 'natural', from the point of view of having convergence of this series for the largest class of functions.

16.2 Weight functions and orthogonal polynomials

Not only for the interval $I = [0, 2\pi]$ are the Hilbert spaces $L^2(I, dx)$ separable, but for any interval $I = [a, b]$, $-\infty \leq a < b \leq +\infty$, as the results of this section will show. Furthermore an orthonormal basis will be constructed explicitly and some interesting properties of the elements of such a basis will be investigated.

The starting point is a *weight function* $\rho : I \to \mathbb{R}$ *on the interval* I which is assumed to have the following properties:

1. On the interval I, the function ρ is strictly positive: $\rho(x) > 0$ for all $x \in I$;

2. if the interval I is not bounded, there are two positive constants α and C such that $\rho(x) e^{\alpha |x|} \leq C$ for all $x \in I$.

The strategy to prove that the Hilbert space $L^2(I, dx)$ is separable is quite simple. A first step shows that the countable set of functions $\rho_n(x) = x^n \rho(x), n = 0, 1, 2, \ldots$ is total in this Hilbert space. The Gram–Schmidt orthonormalization then produces easily an orthonormal basis.

Lemma 16.2.1 *The system of functions* $\{\rho_n : n = 0, 1, 2, \ldots\}$ *is total in the Hilbert space* $L^2(I, dx)$, *for any interval* I.

Proof. For the proof we have to show: If an element $h \in L^2(I, dx)$ satisfies $\langle \rho_n, h \rangle_2 = 0$ for all n, then $h = 0$.

In the case $I \neq \mathbb{R}$ we consider h to be be extended by 0 to $\mathbb{R} \setminus I$ and thus get a function $h \in L^2(\mathbb{R}, dx)$. On the strip $S_\alpha = \{p = u + iv \in \mathbb{C} : u, v \in \mathbb{R}, |v| < \alpha\}$, introduce the auxiliary function

$$F(p) = \int_\mathbb{R} \rho(x) h(x) e^{ipx} dx.$$

The growth restriction on the weight function implies that F is a well defined holomorphic function on S_α (see Exercises). Differentiation of F generates the functions ρ_n in this integral:

$$F^{(n)}(p) = \frac{d^n F}{dp^n}(p) = i^n \int_\mathbb{R} h(x) \rho(x) x^n e^{ipx} dx$$

for $n = 0, 1, 2, \ldots$, and we deduce $F^{(n)}(0) = i^n \langle \rho_n, h \rangle_2 = 0$ for all n. Since F is holomorphic in the strip S_α it follows that $F(p) = 0$ for all $p \in S_\alpha$ (see Theorem 9.3.1) and thus in particular $F(p) = 0$ for all $p \in \mathbb{R}$. But $F(p) = \sqrt{2\pi} \mathcal{L}(\rho h)(p)$ where $\mathcal{L}$ is the inverse Fourier transform (see Theorem 10.1.2), and we know $\langle \mathcal{L}f, \mathcal{L}g \rangle_2 = \langle f, g \rangle_2$ for all $f, g \in L^2(\mathbb{R}, dx)$ (Theorem 10.3.5). It follows that $\langle \rho h, \rho h \rangle_2 = \langle \mathcal{L}(\rho h), \mathcal{L}(\rho h) \rangle_2 = 0$ and thus $\rho h = 0 \in L^2(\mathbb{R}, dx)$. Since $\rho(x) > 0$ for $x \in I$ this implies $h = 0$ and we conclude. □

Technically it is simpler to do the orthonormalization of the system of functions $\{\rho_n : n \in \mathbb{N}\}$ not in the Hilbert space $L^2(I, dx)$ directly but in the Hilbert space $L^2(I, \rho dx)$ which is defined as the space of all equivalence classes of measurable functions $f : I \to \mathbb{K}$ such that $\int_I |f(x)|^2 \rho(x) dx < \infty$ equipped with the inner product $\langle f, g \rangle_\rho = \int_I \overline{f(x)} g(x) \rho(x) dx$. Note that the relation $\langle f, g \rangle_\rho = \langle \sqrt{\rho} f, \sqrt{\rho} g \rangle_2$ holds for all $f, g \in L^2(I, \rho dx)$. It implies that the Hilbert spaces $L^2(I, \rho dx)$ and $L^2(I, dx)$ are (isometrically) isomorphic under the map

$$L^2(I, \rho dx) \ni f \mapsto \sqrt{\rho} f \in L^2(I, dx).$$

This is shown in the Exercises. Using this isomorphism, Lemma 16.2.1 can be restated as saying that the system of powers of x, $\{x^n : n = 0, 1, 2, \ldots\}$ is total in the Hilbert space $L^2(I, \rho dx)$.

We proceed by applying the Gram–Schmidt orthonormalization to the system of powers $\{x^n : n = 0, 1, 2, \ldots\}$ in the Hilbert space $L^2(I, \rho dx)$. This gives a sequence of polynomials P_k of degree k such that $\langle P_k, P_m \rangle_\rho = \delta_{km}$. These polynomials are defined recursively in the following way: $Q_0(x) = x^0 = 1$, and when for $k \geq 1$ the polynomials $Q_0, \ldots, Q_{k-1}$ are defined, we define the polynomial Q_k by

$$Q_k(x) = x_k - \sum_{n=0}^{k-1} \frac{\langle Q_n, x^k \rangle_\rho}{\langle Q_n, Q_n \rangle_\rho} Q_n.$$

Finally the polynomials Q_k are normalized and we arrive at an orthonormal system of polynomials P_k:

$$P_k = \frac{1}{\|Q_k\|_\rho} Q_k, \qquad k = 0, 1, 2, \ldots.$$

Note that according to this construction P_k is a polynomial of degree k with positive coefficient for the power x^k. Theorem 16.1.1 and Lemma 16.2.1 imply that the system of polynomials $\{P_k : k = 0, 1, 2, \ldots\}$ is an orthonormal basis of the Hilbert space $L^2(I, \rho dx)$. If we now introduce the functions

$$e_k(x) = P_k(x)\sqrt{\rho(x)}, \qquad x \in I$$

we obtain an orthonormal basis of the Hilbert space $L^2(I, dx)$. This shows:

Theorem 16.2.1 *For any interval $I = (a, b)$, $-\infty \leq a < b \leq +\infty$ the Hilbert space $L^2(I, dx)$ is separable, and the above system $\{e_k : k = 0, 1, 2, \ldots\}$ is an orthonormal basis.*

Proof. Only the existence of a weight function for the interval I has to be shown. Then by the preceding discussion we conclude. A simple choice of a weight function for any of these intervals is for instance the exponential function $\rho(x) = e^{-\alpha x^2}$, $x \in \mathbb{R}$, for some $\alpha > 0$. □

Naturally, the orthonormal polynomials P_k depend on the interval and the weight function. After some general properties of these polynomials have been studied we will determine the orthonormal polynomials for some intervals and weight functions explicitly.

Lemma 16.2.2 *If Q_m is a polynomial of degree m, then $\langle Q_m, P_k \rangle_\rho = 0$ for all $k > m$.*

Proof. Since $\{P_k : k = 0, 1, 2, \ldots\}$ is an ONB of the Hilbert space $L^2(I, \rho dx)$ the polynomial Q_m has a Fourier expansion with respect to this ONB: $Q_m = \sum_{n=0}^\infty c_n P_n$, $c_n = \langle P_n, Q_m \rangle_\rho$. Since the powers x^k, $k = 0, 1, 2, \ldots$ are linearly independent functions on the interval I and since the degree of Q_m is m and that of P_n is n, the coefficients c_n in this expansion must vanish for $n > m$, i.e., $Q_m = \sum_{n=0}^m c_n P_n$ and thus $\langle P_k, Q_m \rangle_\rho = 0$ for all $k > m$. □

16.2 Weight functions and orthogonal polynomials

Since the orthonormal system $\{P_k : k = 0, 1, 2, \ldots\}$ is obtained by the Gram–Schmidt orthonormalization from the system of powers x^k for $k = 0, 1, 2, \ldots$ with respect to the inner product $\langle \cdot, \cdot \rangle_\rho$, the polynomial P_{n+1} is generated by multiplying the polynomial P_n with x and adding some lower order polynomial as correction. Indeed one has

Proposition 16.2.2 *Let ρ be a weight for the interval $I = (a, b)$ and denote by $\{P_k : k = 0, 1, 2, \ldots\}$ the complete system of orthonormal polynomials for this weight and this interval. Then, for every $n \geq 1$, there are constants A_n, B_n, C_n such that*

$$P_{n+1}(x) = (A_n x + B_n) P_n(x) + C_n P_{n-1}(x) \qquad \forall x \in I.$$

Proof. We know $P_k(x) = a_k x^k + Q_{k-1}(x)$ with some constant $a_k > 0$ and some polynomial Q_{k-1} of degree smaller than or equal to $k - 1$. Thus, if we define $A_n = \frac{a_{n+1}}{a_n}$, it follows that $P_{n+1} - A_n x P_n$ is a polynomial of degree smaller than or equal to n, hence there are constants $c_{n,k}$ such that

$$P_{n+1} - A_n x P_n = \sum_{k=0}^{n} c_{n,k} P_k.$$

Now calculate the inner product with P_j, $j \leq n$:

$$\langle P_j, P_{n+1} - A_n x P_n \rangle_\rho = \sum_{k=0}^{n} c_{n,k} \langle P_j, P_k \rangle_\rho = c_{n,j}.$$

Since the polynomial P_k is orthogonal to all polynomials Q_j of degree $j \leq k - 1$ we deduce that $c_{n,j} = 0$ for all $j < n-1$, $c_{n,n-1} = -A_n \langle x P_{n-1}, P_n \rangle_\rho$, and $c_{n,n} = -A_n \langle x P_n, P_n \rangle_\rho$. The statement follows by choosing $B_n = c_{n,n}$ and $C_n = c_{n,n-1}$. □

Proposition 16.2.3 *For any weight function ρ on the interval I, the kth orthonormal polynomial P_k has exactly k simple real zeroes.*

Proof. Per construction the orthonormal polynomials P_k have real coefficients, have the degree k, and the coefficient c_k is positive. The fundamental theorem of algebra (Theorem 9.3.2) implies: The polynomial P_k has a certain number $m \leq k$ of simple real roots $x_1, \ldots, x_m$ and the roots which are not real occur in pairs of complex conjugate numbers, $(z_j, \overline{z_j})$, $j = m + 1, \ldots, M$ with the same multiplicity n_j, $m + 2\sum_{j=m+1}^{M} n_j = k$. Therefore the polynomial P_k can be written as

$$P_k(x) = c_k \prod_{j=1}^{m}(x - x_j) \prod_{j=m+1}^{M}(x - z_j)^{n_j}(x - \overline{z_j})^{n_j}.$$

Consider the polynomial $Q_m(x) = c_k \prod_{j=1}^{m}(x - x_j)$. It has the degree m and exactly m real simple roots. Since $P_k(x) = Q_m(x) \prod_{j=m+1}^{M} |x - z_j|^{2n_j}$, it follows that $P_k(x) Q_m(x) \geq 0$ for all $x \in I$ and $P_k Q_m \not\equiv 0$, hence $\langle P_k, Q_m \rangle_\rho > 0$. If the degree m of the polynomial Q_m would be smaller than k, we would arrive at a contradiction to the result of the previous lemma, hence $m = k$ and the pairs of complex conjugate roots cannot occur. Thus we conclude. □

In the Exercises, with the same argument, we prove the following extension of this proposition.

Lemma 16.2.3 *The polynomial $Q_k(x, \lambda) = P_k(x) + \lambda P_{k-1}(x)$ has k simple real roots, for any $\lambda \in \mathbb{R}$.*

Lemma 16.2.4 *There are no points $x_0 \in I$ and no integer $k \geq 0$ such that $P_k(x_0) = P_{k-1}(x_0) = 0$.*

Proof. Suppose that for some $k \geq 0$ the orthonormal polynomials P_k and P_{k-1} have a common root $x_0 \in I$: $P_k(x_0) = P_{k-1}(x_0) = 0$. Since we know that these orthonormal polynomials have simple real roots, we know in particular $P'_{k-1}(x_0) \neq 0$ and thus we can take the real number $\lambda_0 = \frac{-P'_k(x_0)}{P'_{k-1}(x_0)}$ to form the polynomial $Q_k(x, \lambda_0) = P_k(x) + \lambda_0 P_{k-1}(x)$. It follows that $Q(x_0, \lambda_0) = 0$ and $Q'_k(x_0) = 0$, i.e., x_0 is a root of $Q_k(\cdot, \lambda)$ with multiplicity at least two. But this contradicts the previous lemma. Hence there is no common root of the polynomials P_k and P_{k-1}. □

Theorem 16.2.4 (Knotensatz) *Let $\{P_k : k = 0, 1, 2, \ldots\}$ be the orthonormal basis for some interval I and some weight function ρ. Then the roots of P_{k-1} separate the roots of P_k, i.e., between two successive roots of P_k there is exactly one root of P_{k-1}.*

Proof. Suppose that $\alpha < \beta$ are two successive roots of the polynomial P_k so that $P_k(x) \neq 0$ for all $x \in (\alpha, \beta)$. Assume furthermore that P_{k-1} has no root in the open interval (α, β). The previous lemma implies that P_{k-1} does not vanish in the closed interval $[\alpha, \beta]$. Since the polynomials P_{k-1} and $-P_{k-1}$ have the same system of roots, we can assume that P_{k-1} is positive in $[\alpha, \beta]$ and P_k is negative in (α, β). Define the function $f(x) = \frac{-P_k(x)}{P_{k-1}(x)}$. It is continuous on $[\alpha, \beta]$ and satisfies $f(\alpha) = f(\beta) = 0$ and $f(x) > 0$ for all $x \in (\alpha, \beta)$. It follows that $\lambda_0 = \sup\{f(x) : x \in [\alpha, \beta]\} = f(x_0)$ for some $x_0 \in (\alpha, \beta)$. Now consider the family of polynomials $Q_k(x, \lambda) = P_k(x) + \lambda P_{k-1}(x) = P_{k-1}(x)(\lambda - f(x))$. Therefore, for all $\lambda \geq \lambda_0$, the polynomials $Q_k(\cdot, \lambda)$ are nonnegative on $[\alpha, \beta]$, in particular $Q_k(x, \lambda_0) \geq 0$ for all $x \in [\alpha, \beta]$. Since $\lambda_0 = f(x_0)$, it follows that $Q_k(x_0, \lambda_0) = 0$, thus $Q_k(\cdot, \lambda_0)$ has a root $x_0 \in (\alpha, \beta)$. Since f has a maximum at x_0 we know $0 = f'(x_0)$. The derivative of f is easily calculated:

$$f'(x) = -\frac{P'_k(x)P_{k-1}(x) - P_k(x)P'_{k-1}(x)}{P_{k-1}(x)^2}.$$

Thus $f'(x_0) = 0$ implies $P'_k(x_0)P_{k-1}(x_0) - P_k(x_0)P'_{k-1}(x_0) = 0$ and therefore $Q'_k(x_0) = P'_k(x_0) + f(x_0)P'_{k-1}(x_0) = 0$. Hence the polynomial $Q_k(\cdot, \lambda_0)$ has a root of multiplicity 2 at x_0. This contradicts Lemma 16.2.3 and therefore the polynomial P_{k-1} has at least one root in the interval (α, β). Since P_{k-1} has exactly $k-1$ simple real roots according to Proposition 16.2.3, we conclude that P_{k-1} has exactly one simple root in (α, β) which proves the theorem. □

Remark 16.2.1 *Consider the function*

$$F(Q) = \int_I Q(x)^2 \rho(x) dx, \qquad Q(x) = \sum_{k=0}^n a_k x^k.$$

Since we can expand Q in terms of the orthonormal basis $\{P_k : k = 0, 1, 2, \ldots\}$, $Q = \sum_{k=0}^n c_k P_k$, $c_k = \langle P_k, Q \rangle_\rho$ the value of the function F can be expressed in terms of the coefficients c_k as $F(Q) = \sum_{k=0}^n c_k^2$ and it follows that the orthonormal polynomials P_k minimize the function $Q \mapsto F(Q)$ under obvious constraints (see Exercises).

16.3 Examples of complete orthonormal systems for $L^2(I, \rho dx)$

For the intervals $I = \mathbb{R}$, $I = \mathbb{R}_+ = [0, \infty)$, and $I = [-1, 1]$ we are going to construct explicitly an orthonormal basis by choosing a suitable weight function and applying the construction explained above. Certainly, the above general results apply to these concrete examples, in particular the 'Knotensatz'.

$I = \mathbb{R}$, $\rho(x) = e^{-x^2}$: Hermite polynomials

Evidently, the function $\rho(x) = e^{-x^2}$ is a weight function for the real line. Therefore, by Lemma 16.2.1, the system of functions $\rho_n(x) = x^n e^{-\frac{x^2}{2}}$ generates the Hilbert space $L^2(\mathbb{R}, dx)$. Finally the Gram–Schmidt orthonormalization produces an orthonormal basis $\{h_n : n = 0, 1, 2, \ldots\}$. The elements of this basis have the form (Rodrigues' formula)

$$h_n(x) = (-1)^n c_n e^{\frac{x^2}{2}} (\frac{d}{dx})^n (e^{-x^2}) = c_n H_n(x) e^{-\frac{x^2}{2}} \quad (16.1)$$

with normalization constants

$$c_n = (2^n n! \sqrt{\pi})^{1/2} \quad n = 0, 1, 2, \ldots.$$

Here the functions H_n are polynomials of degree n, called *Hermite polynomials* and the functions h_n are the *Hermite functions of order n*.

Theorem 16.3.1 *The system of Hermite functions $\{h_n : n = 0, 1, 2, \ldots\}$ is an orthonormal basis of the Hilbert space $L^2(\mathbb{R}, dx)$. The statements of Theorem 16.2.4 apply to the Hermite polynomials.*

Using equation (16.1) one deduces in the Exercises that the Hermite polynomials satisfy the recursion relation

$$H_{n+1}(x) - 2x H_n(x) + 2n H_{n-1}(x) = 0 \quad (16.2)$$

and the differential equation ($y = H_n(x)$)

$$y'' - 2xy' + 2ny = 0. \quad (16.3)$$

These relations show that the Hermite functions are the eigenfunctions of the linear differential operator $-\frac{d^2}{dx^2} + \omega^2$ which is known to describe the linear harmonic oscillator (see [Amr81, GP90, Thi92]). In these references one also finds other methods to prove that the Hermite functions form an orthonormal basis.

$I = \mathbb{R}_+$, $\rho(x) = e^{-x}$: Laguerre polynomials

On the positive real line the exponential function $\rho(x) = e^{-x}$ certainly is a weight function. Hence our general results apply here and we obtain

Theorem 16.3.2 *The system of* **Laguerre functions** $\{\ell_n : n = 0, 1, 2, \ldots\}$ *which is constructed by orthonormalization of the system* $\left\{x^n e^{-\frac{x}{2}} : n = 0, 1, 2, \ldots\right\}$ *in $L^2(\mathbb{R}_+, dx)$ is an orthonormal basis. These Laguerre functions have the following form (Rodrigues' formula):*

$$\ell_n(x) = \frac{1}{n!} L_n(x) e^{-\frac{x}{2}}, \quad L_n(x) = e^x (\frac{d}{dx})^n (x^n e^{-x}), \quad n = 0, 1, 2, \ldots. \quad (16.4)$$

For the system $\{L_n : n = 0, 1, 2, \ldots\}$ of **Laguerre polynomials** *Theorem 16.2.4 applies.*

In the Exercises we show that the Laguerre polynomials of different order are related according to the identity

$$(n+1)L_{n+1}(x) + (x - 2n - 1)L_n(x) + nL_{n-1}(x) = 0, \quad (16.5)$$

and are solutions of the second order differential equation $(y = L_n(x))$

$$xy'' + (1-x)y' + ny = 0. \quad (16.6)$$

In quantum mechanics this differential equation is related to the radial Schrödinger equation for the hydrogen atom.

I = [−1, +1], $\rho(x) = 1$: Legendre polynomials

For any finite interval $I = [a, b]$, $-\infty < a < b < \infty$ one can take any positive constant as a weight function. Thus, Lemma 16.2.1 says that the system of powers $\{x^n : n = 0, 1, 2, \ldots\}$ is a total system of functions in the Hilbert space $L^2([a, b], dx)$. It follows that every element $f \in L^2([a, b], dx)$ is the limit of a sequence of polynomials, in the L^2-norm. Compare this with the Theorem of Stone–Weierstrass which says that every continuous function on $[a, b]$ is the uniform limit of a sequence of polynomials.

For the special case of the interval $I = [-1, 1]$ the Gram–Schmidt orthonormalization of the system of powers leads to a well-known system of polynomials.

Theorem 16.3.3 *The system of* **Legendre polynomials**

$$P_n(x) = \frac{1}{2^n n!} (\frac{d}{dx})^n (x^2 - 1)^n, \quad x \in [-1, 1], \quad n = 0, 1, 2, \ldots \quad (16.7)$$

is an orthogonal basis of the Hilbert space $L^2([-1, 1], dx)$. The Legendre polynomials are normalized according to the relation

$$\langle P_n, P_m \rangle_2 = \frac{2}{2n+1} \delta_{nm}.$$

Again one can show that these polynomials satisfy a recursion relation and a second order differential equation (see Exercises):

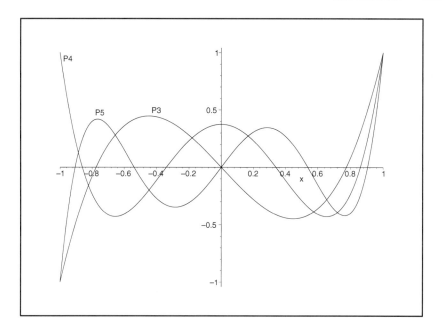

Legendre polynomials P_3, P_4, P_5

$$(n+1)P_{n+1}(x) - (2n+1)xP_n(x) + nP_{n-1}(x) = 0, \tag{16.8}$$

$$(1-x^2)y'' - 2xy' + n(n+1)y = 0, \tag{16.9}$$

where $y = P_n(x)$.

Without further details we mention the weight functions for some other systems of orthogonal polynomials on the interval $[-1, 1]$:

Jacobi $P_n^{\nu,\mu}$ $\rho(x) = (1-x)^\mu$, $\nu, \mu > -1$,

Gegenbauer C_n^λ $\rho(x) = (1-x^2)^{\lambda-\frac{1}{2}}$, $\lambda > -1/2$,

Tschebyschew $1st$ kind $\rho(x) = (1-x)^{-1/2}$,

Tschebyschew $2nd$ kind $\rho(x) = (1-x^2)^{1/2}$.

We conclude this section by an illustration of the Knotensatz for some Legendre polynomials of low order. This graph clearly shows that the zeros of the polynomial P_k are separated by the zeros of the polynomial P_{k+1}, $k = 3, 4$. In addition the orthonormal polynomials are listed explicitly up to order $n = 6$.

16.4 Exercises

1. Prove: A Hilbert space $\mathcal{H}$ is separable if, and only if, $\mathcal{H}$ contains a countable dense subset.

2. **The space of almost-periodic functions:** In the space of complex-valued measurable functions on $\mathbb{R}$ consider the vector space F which is generated by the exponential functions e_λ, $\lambda \in \mathbb{R}$; here $e_\lambda : \mathbb{R} \to \mathbb{C}$ is defined by $e_\lambda(x) = e^{ix\lambda}$ for all $x \in \mathbb{R}$. Thus elements g in F are of the form $g = \sum_{k=1}^{N} a_k e_{\lambda_k}$ for some choice of $N \in \mathbb{N}$, $a_k \in \mathbb{C}$, and $\lambda_k \in \mathbb{R}$. On F we define
$$\langle g, f \rangle = \lim_{T \to \infty} \frac{1}{2T} \int_{-T}^{+T} \overline{g(x)} f(x) dx.$$

a) Show that $\langle \cdot, \cdot \rangle$ defines an inner product on F.
b) Complete the inner product space $(F, \langle \cdot, \cdot \rangle)$ to get a Hilbert space $\mathcal{H}_{ap}$, called the space of almost periodic functions on $\mathbb{R}$.
c) Show that $\mathcal{H}_{ap}$ is not separable.

Hints: Show that $\{e_\lambda : \lambda \in \mathbb{R}\}$ is an orthonormal system in $\mathcal{H}_{ap}$ which is not countable.

3. Consider the functions e_n, $n \in \mathbb{Z}$, defined on the interval $[0, 2\pi]$ by $e_n(x) = \frac{1}{\sqrt{2\pi}} e^{inx}$. Prove: This system is orthonormal basis of the Hilbert space $L^2([0, 2\pi], dx)$.

4. Prove that the function F in Lemma 16.2.1 is well defined and holomorphic in the strip S_α.

Hints: For $p = u + iv \in S_\alpha$ write $ipx = \alpha|x| + ixu - |x|(\alpha + v \operatorname{sign} x)$, group terms appropriately and estimate.

5. Let ρ be a weight function on the interval I. Show: The Hilbert spaces $L^2(I, \rho dx)$ and $L^2(I, dx)$ are isomorphic under the map $f \mapsto \sqrt{\rho} f$.

6. Let P_k, $k = 0, 1, 2, \ldots$, be the system of orthonormal polynomials for the interval I and the weight function ρ. Then the polynomial $Q_k(x, \lambda) = P_k(x) + \lambda P_{k-1}(x)$ has k simple real roots, for any $\lambda \in \mathbb{R}$.

7. Under the assumptions of the previous problem show: The functional
$$f(u) = \int_a^b (u(x) - \sum_{k=0}^{n} a_k P_k(x))^2 \rho(x) dx$$
is minimized by the choice $a_k = \langle P_k, u \rangle_\rho$, $k = 0, 1, \ldots, n$. Here u is a given continuous function.

8. For $n = 0, 1, 2, 3, 4$ calculate the Hermite functions h_n, the Laguerre functions ℓ_n, and the Legendre polynomials P_n explicitly in two ways, first by going through the Gram–Schmidt orthonormalization and then by using the representation of these functions in terms of differentiation of the generating functions given in the last section.

16.4 Exercises

9. Prove the recursion relations (16.2), (16.5), and (16.8).

10. Prove the differential equations (16.3), (16.6), and (16.9) by using the representation of these functions in terms of differentiation of the generating functions.

n	$H_n(x)$	$L_n(x)$	$P_n(x)$
1	$2x$	$1-x$	x
2	$4x^2 - 2$	$1 - 2x + \frac{1}{2}x^2$	$\frac{3}{2}x^2 - \frac{1}{2}$
3	$8x^3 - 12x$	$1 - 3x + \frac{3}{2}x^2 - \frac{1}{6}x^3$	$\frac{5}{2}x^3 - \frac{3}{2}x$
4	$16x^4 - 48x^2 + 12$	$1 - 4x + 3x^2 - \frac{2}{3}x^3 + \frac{1}{24}x^4$	$\frac{35}{8}x^4 - \frac{15}{4}x^2 + \frac{3}{8}$
5	$32x^5 - 160x^3 + 120x$	$1 - 5x + 5x^2 - \frac{5}{3}x^3 + \frac{5}{24}x^4 - \frac{1}{120}x^5$	$\frac{63}{8}x^5 - \frac{35}{4}x^3 + \frac{15}{8}x$
6	$64x^6 - 480x^4 + 720x^2 - 120$	$1 - 6x + \frac{15}{2}x^2 - \frac{10}{3}x^3 + \frac{5}{8}x^4 - \frac{1}{20}x^5 + \frac{1}{720}x^6$	$\frac{231}{16}x^6 - \frac{315}{16}x^4 + \frac{105}{16}x^2 - \frac{5}{16}$

Table 16.1: Orthogonal Polynomials of order ≤ 6

17
Direct Sums and Tensor Products

There are two often used constructions of forming new Hilbert spaces out of a finite or infinite set of given Hilbert spaces. Both constructions are quite important in quantum mechanics and in quantum field theory. This brief chapter introduces these constructions and discusses some examples from physics.

17.1 Direct sums of Hilbert spaces

Recall the construction of the first Hilbert space by D. Hilbert, the space of square summable sequences $\ell^2(\mathbb{K})$ over the field $\mathbb{K}$. Here we take infinitely many copies of the Hilbert space $\mathbb{K}$ and take from each copy an element to form a sequence of elements and define this space as the space of all those sequences for which the square of the norm of these elements form a summable sequence of real numbers. This construction will be generalized by replacing the infinitely many copies of the Hilbert space $\mathbb{K}$ by a countable set of given Hilbert spaces and do the same construction.

Let us first explain the construction of the *direct sum* of a finite number of Hilbert spaces. Suppose we are given two Hilbert spaces $\mathcal{H}_1$ and $\mathcal{H}_2$ over the same field $\mathbb{K}$. Consider the set $\mathcal{H}_1 \times \mathcal{H}_2$ of ordered pairs (x_1, x_2), $x_i \in \mathcal{H}_i$ of elements in these spaces and equip this set in a natural way with the structure of a vector space over the field $\mathbb{K}$. To this end one defines addition and scalar multiplication on $\mathcal{H}_1 \times \mathcal{H}_2$ as follows:

$$(x_1, x_2) + (y_1, y_2) = (x_1 + y_1, x_2 + y_2) \quad \forall x_i, y_i \in \mathcal{H}_i, \ i = 1, 2,$$
$$\lambda \cdot (x_1, x_2) = (\lambda x_1, \lambda x_2) \quad \forall \, x_i \in \mathcal{H}_i, \ \forall \lambda \in \mathbb{K}.$$

It is straightforward to show that with this addition and scalar multiplication the set $\mathcal{H}_1 \times \mathcal{H}_2$ is a vector space over the field $\mathbb{K}$. Next we define a scalar product on this vector space. If $\langle \cdot, \cdot \rangle_i$ denotes the inner product of the Hilbert space $\mathcal{H}_i$, $i = 1, 2$, one defines an inner product $\langle \cdot, \cdot \rangle$ on the vector space $\mathcal{H}_1 \times \mathcal{H}_2$ by

$$\langle (x_1, x_2), (y_1, y_2) \rangle = \langle x_1, y_1 \rangle_1 + \langle x_2, y_2 \rangle_2 \quad \forall x_i, y_i \in \mathcal{H}_i, \ i = 1, 2.$$

In the Exercises one is asked to verify that this expression defines indeed an inner product on $\mathcal{H}_1 \times \mathcal{H}_2$. In another exercise it is shown that the resulting inner product space is complete and thus a Hilbert space. This Hilbert space is denoted by $\mathcal{H}_1 \oplus \mathcal{H}_2$ and is called the *direct sum of the Hilbert spaces $\mathcal{H}_1$ and $\mathcal{H}_2$*.

Now assume that a countable set $\mathcal{H}_i$, $i \in \mathbb{N}$, of Hilbert spaces over the same field $\mathbb{K}$ is given. Consider the set $\mathcal{H}$ of all sequences $\underline{x} = (x_i)_{i \in \mathbb{N}}$ with $x_i \in \mathcal{H}_i$ for all $i \in \mathbb{N}$ such that

$$\sum_{i=1}^{\infty} \|x_i\|_i^2 < \infty \tag{17.1}$$

where $\|\cdot\|_i$ denotes the norm of the Hilbert space $\mathcal{H}_i$. On this set of all such sequences the structure of a vector space over the field $\mathbb{K}$ is introduced in a natural way by defining addition and scalar multiplication as follows:

$$(x_i)_{i \in \mathbb{N}} + (y_i)_{i \in \mathbb{N}} = (x_i + y_i)_{i \in \mathbb{N}} \quad \forall \, x_i, y_i \in \mathcal{H}_i, \ i \in \mathbb{N}, \tag{17.2}$$
$$\lambda \cdot (x_i)_{i \in \mathbb{N}} = (\lambda x_i)_{i \in \mathbb{N}} \quad \forall \, x_i \in \mathcal{H}_i, \ i \in \mathbb{N}. \tag{17.3}$$

It is again an easy exercise to show that with this addition and scalar multiplication the set $\mathcal{H}$ is indeed a vector space over the field $\mathbb{K}$. If $\langle \cdot, \cdot \rangle_i$ denotes the inner product of the Hilbert spaces $\mathcal{H}_i$, $i \in \mathbb{N}$, an inner product on the vector space $\mathcal{H}$ is defined by

$$\langle (x_i)_{i \in \mathbb{N}}, (y_i)_{i \in \mathbb{N}} \rangle = \sum_{i=1}^{\infty} \langle x_i, y_i \rangle_i \quad \forall \, (x_i)_{i \in \mathbb{N}}, (y_i)_{i \in \mathbb{N}} \in \mathcal{H}. \tag{17.4}$$

The proof is left as an exercise. Equipped with this inner product, $\mathcal{H}$ is an inner product space. The following theorem states that $\mathcal{H}$ is complete, and thus a Hilbert space.

Theorem 17.1.1 *Suppose that a countable set of Hilbert spaces $\mathcal{H}_i$, $i \in \mathbb{N}$, over the field $\mathbb{K}$ is given. On the set $\mathcal{H}$ of all sequences $\underline{x} = (x_i)_{i \in \mathbb{N}}$ satisfying condition (17.1), define a vector space structure by relations (17.2), (17.3) and an inner product by relation (17.4). Then $\mathcal{H}$ is a Hilbert space over $\mathbb{K}$, called the* **Hilbert sum** *or* **direct sum** *of the Hilbert spaces $\mathcal{H}_i$, $i \in \mathbb{N}$, and is denoted by*

$$\mathcal{H} = \oplus_{i=1}^{\infty} \mathcal{H}_i. \tag{17.5}$$

If all the Hilbert spaces $\mathcal{H}_i, i \in \mathbb{N}$, are separable, then the direct sum $\mathcal{H}$ is separable too.

Proof. Only the proofs of completeness and of separability of the inner product space are left. In its main steps the proof of completeness is the same as the proof of completeness of the sequence space $\ell^2(\mathbb{K})$ given earlier.

Given a Cauchy sequence $(\underline{x}^{(n)})_{n \in \mathbb{N}}$ in $\mathcal{H}$ and any $\epsilon > 0$, there is an $n_0 \in \mathbb{N}$ such that

$$\|\underline{x}^{(n)} - \underline{x}^{(m)}\| < \epsilon \qquad \forall n, m \geq n_0.$$

Each element $\underline{x}^{(n)}$ of this sequence is itself a sequence $(x_i^{(n)})_{i \in \mathbb{N}}$. Thus, in terms of the inner product (17.4), this Cauchy condition means

$$\sum_{i=1}^{\infty} \|x_i^{(n)} - x_i^{(m)}\|_i^2 < \epsilon^2 \qquad \forall n, m \geq n_0. \tag{17.6}$$

It follows that for every $i \in \mathbb{N}$ the sequence $(x_i^{(n)})_{n \in \mathbb{N}}$ is actually a Cauchy sequence in the Hilbert space $\mathcal{H}_i$ and thus converges to a unique element x_i in this space:

$$x_i = \lim_{n \to \infty} x_i^{(n)} \qquad \forall i \in \mathbb{N}.$$

Condition (17.6) implies, for every $L \in \mathbb{N}$,

$$\sum_{i=1}^{L} \|x_i^{(n)} - x_i^{(m)}\|_i^2 < \epsilon^2 \qquad \forall n, m \geq n_0, \tag{17.7}$$

and thus, by taking the limit $n \to \infty$ in this estimate, it follows that

$$\sum_{i=1}^{L} \|x_i - x_i^{(m)}\|_i^2 \leq \epsilon^2 \qquad \forall m \geq n_0. \tag{17.8}$$

This estimate holds for all $L \in \mathbb{N}$ and the bound is independent of L. Therefore it also holds in the limit $L \to \infty$ (which obviously exists)

$$\sum_{i=1}^{\infty} \|x_i - x_i^{(m)}\|_i^2 \leq \epsilon^2 \qquad \forall m \geq n_0. \tag{17.9}$$

Introducing the sequence $\underline{x} = (x_i)_{i \in \mathbb{N}}$ of limit elements x_i of the sequence $(x_i^{(n)})_{n \in \mathbb{N}}$ estimate (17.9) reads

$$\|\underline{x} - \underline{x}^{(m)}\| \leq \epsilon \qquad \forall m \geq n_0.$$

Therefore, for any fixed $m \geq n_0$, $\|\underline{x}\| \leq \|\underline{x} - \underline{x}^{(m)}\| + \|\underline{x}^{(m)}\| \leq \epsilon + \|\underline{x}^{(m)}\|$, and it follows that the sequence $\underline{x}$ is square summable, i.e., $\underline{x} \in \mathcal{H}$, and that the given Cauchy sequence $(\underline{x}^{(n)})_{n \in \mathbb{N}}$ converges in $\mathcal{H}$ to $\underline{x}$. Thus the inner product space $\mathcal{H}$ is complete.

The proof of separability is left as an exercise. □

17.2 Tensor products

Tensor products of Hilbert spaces are an essential tool in the description of multi-particle systems in quantum mechanics and in relativistic quantum field theory. There are several other areas in physics where tensor products, not only of Hilbert spaces but of vector spaces in general, play a prominent role. Certainly, in various areas of mathematics, the concept of tensor product is essential. Accordingly we begin this section with a brief reminder of the tensor product of vector spaces and then discuss the special aspects of the tensor product of Hilbert spaces.

Given two vector spaces E and F over the same field $\mathbb{K}$, introduce the vector space $\Lambda = \Lambda(E, F)$ of all linear combinations

$$\sum_{j=1}^{N} a_j(x_j, y_j), \qquad a_j \in \mathbb{K}, \quad x_j \in E, \quad y_j \in F, \quad j = 1, \ldots, N \in \mathbb{N}$$

of ordered pairs $(x, y) \in E \times F$. Consider the following four types of elements of a special form in Λ:

$$
\begin{array}{ll}
(x, y_1 + y_2) - (x, y_1) - (x, y_2) & x \in E, \ y_1, y_2 \in F \\
(x_1 + x_2, y) - (x_1, y) - (x_2, y) & x_1, x_2 \in E, \ y \in F \\
(\lambda x, y) - \lambda(x, y) & x \in E, \ y \in F, \ \lambda \in \mathbb{K} \\
(x, \lambda y) - \lambda(x, y) & x \in E, \ y \in F, \ \lambda \in \mathbb{K}.
\end{array}
$$

These special elements generate a linear subspace $\Lambda_0 \subset \Lambda$. The quotient space of Λ with respect to this subspace Λ_0 is called the *tensor product of E and F* and is denoted by $E \otimes F$:

$$E \otimes F = \Lambda(E, F)/\Lambda_0. \tag{17.10}$$

By construction, $E \times F$ is a subspace of $\Lambda(E, F)$; the restriction of the quotient map $Q : \Lambda(E, F) \to \Lambda(E, F)/\Lambda_0$ to this subspace (E, F) is denoted by χ and the image of an element $(x, y) \in (E, F)$ under χ is accordingly called the *tensor product of x and y*,

$$\chi(x, y) = x \otimes y.$$

The *calculation rules of the tensor product* are

$$
\begin{array}{lll}
x \otimes (y_1 + y_2) = x \otimes y_1 + x \otimes y_2 & x \in E, \ y_1, y_2 \in F & (17.11) \\
(x_1 + x_2) \otimes y = x_1 \otimes y + x_2 \otimes y & x_1, x_2 \in E, \ y \in F & (17.12) \\
(\lambda x) \otimes y = \lambda(x \otimes y) & x \in E, \ y \in F, \ \lambda \in \mathbb{K} & (17.13) \\
x \otimes (\lambda y) = \lambda(x \otimes y) & x \in E, \ y \in F, \ \lambda \in \mathbb{K}. & (17.14)
\end{array}
$$

The proof of these rules is left as an Exercise.

The important role of the tensor product in analysis comes from the following (universal) property which roughly says that through the tensor product one can 'linearize' bilinear maps.

Theorem 17.2.1 *Let E, F, G be vector spaces over the field $\mathbb{K}$. Then, for every bi-linear map $b : E \times F \to G$ there is a linear map $\ell : E \otimes F \to G$ such that*

$$b(x, y) = \ell \circ \chi(x, y) = \ell(x \otimes y) \qquad \forall x \in E, y \in F.$$

Proof. The bilinear map $b : E \times F \to G$ has a natural extension $B : \Lambda(E, F) \to G$ defined by $B(\sum_{i=1}^{N} a_i(x_i, y_i)) = \sum_{i=1}^{N} a_i b(x_i, y_i)$. By definition B is linear. It is a small exercise to show that bilinearity of b implies $B(t) = 0$ for all $t \in \Lambda_0$. This allows us to define a linear map $\ell : \Lambda(E, F)/\Lambda_0 \to G$ by $\ell \circ Q(t) = B(t)$ for all $t \in \Lambda(E, F)$. (Q denotes again the quotient map). Thus, for all $(x, y) \in E \times F$, one has $\ell \circ \chi(x, y) = B(x, y) = b(x, y)$. □

17.2 Tensor products

In the first part on distribution theory we introduced the tensor product of test function spaces and of distributions, for instance the tensor product $\mathcal{D}(\Omega_1) \otimes \mathcal{D}(\Omega_2)$ for $\Omega_i \subseteq \mathbb{R}^{n_i}$, $i = 1, 2$, open and nonempty, in a direct way by defining, for all $f_i \in \mathcal{D}(\Omega_i)$, the tensor product $f_1 \otimes f_2$ as a function $\Omega_1 \times \Omega_2 \to \mathbb{K}$ with values $f_1 \otimes f_2(x_1, x_2) = f_1(x_1) f_2(x_2)$ for all $(x_1, x_2) \in \Omega_1 \times \Omega_2$. That this is a special case of the general construction given above is shown in the Exercises.

Now, given two Hilbert spaces $\mathcal{H}_i$, $i = 1, 2$, we know what the algebraic tensor product $\mathcal{H}_1 \otimes \mathcal{H}_2$ of the two vector spaces $\mathcal{H}_1$ and $\mathcal{H}_2$ is. If $\langle \cdot, \cdot \rangle_i$ denotes the inner product of the Hilbert space $\mathcal{H}_i$, we introduce on the vector space $\mathcal{H}_1 \otimes \mathcal{H}_2$, the inner product

$$\langle x_1 \otimes x_2, y_1 \otimes y_2 \rangle = \langle x_1, y_1 \rangle_1 \langle x_2, y_2 \rangle_2 \qquad \forall x_i, y_i \in \mathcal{H}_i, \ i = 1, 2. \quad (17.15)$$

Using the calculation rules of tensor products, this definition is extended to generic elements of the vector space $\mathcal{H}_1 \otimes \mathcal{H}_2$, and in the Exercises we show that this defines indeed an inner product.

In general the inner product space $(\mathcal{H}_1 \otimes \mathcal{H}_2, \langle \cdot, \cdot \rangle)$ is not complete. However, according to the Corollary A.0.1, the completion of an inner product space is a Hilbert space. This completion $\mathcal{H}_1 \tilde{\otimes} \mathcal{H}_2$ is called the *tensor product of the Hilbert spaces* $\mathcal{H}_1$ and $\mathcal{H}_2$ and is usually denoted as

$$\mathcal{H}_1 \otimes \mathcal{H}_2.$$

Note that in this notation the symbol ˜ for the completion has been omitted.

For separable Hilbert spaces there is a direct construction of the tensor product in terms of an orthonormal basis. Suppose that $\{u_i : i \in \mathbb{N}\}$ is an orthonormal basis of the Hilbert space $\mathcal{H}_1$ and $\{v_i : i \in \mathbb{N}\}$ an orthonormal basis of $\mathcal{H}_2$. Now consider the system $S = \{(u_i, v_j) : i, j \in \mathbb{N}\} \subset \mathcal{H}_1 \times \mathcal{H}_2$. This system is orthonormal with respect to the inner product (17.15):

$$\langle (u_i, v_j), (u_p, v_q) \rangle = \langle u_i, u_p \rangle_1 \langle v_j, v_q \rangle_2 = \delta_{ip} \delta_{jq} \qquad \forall i, j, p, q \in \mathbb{N}.$$

The idea now is to define the tensor product $\mathcal{H}_1 \otimes \mathcal{H}_2$ as the Hilbert space in which the system S is an orthonormal basis, i.e.,

$$\mathcal{H}_1 \otimes \mathcal{H}_2 = \left\{ T = \sum_{i,j=1}^{\infty} a_{ij} (u_i, v_j) : a_{ij} \in \mathbb{K}, \ \sum_{i,j=1}^{\infty} |a_{ij}|^2 < \infty \right\}. \quad (17.16)$$

For two elements $T_1, T_2 \in \mathcal{H}_1 \otimes \mathcal{H}_2$ with coefficients a_{ij} respectively b_{ij} it follows easily that

$$\langle T_1, T_2 \rangle = \sum_{i,j=1}^{\infty} \overline{a_{ij}} b_{ij}$$

as one would expect. According to this construction the tensor product of two separable Hilbert spaces is separable.

For every $x \in \mathcal{H}_1$ and $y \in \mathcal{H}_2$ one has $x = \sum_{i=1}^{\infty} a_i u_i$ with $a_i = \langle u_i, x \rangle_1$ and $y = \sum_{j=1}^{\infty} b_j v_j$ with $b_j = \langle v_j, y \rangle_2$ and thus $\langle (u_i, v_j), (x, y) \rangle = \langle u_i, x \rangle_1 \langle v_j, y \rangle_2 = a_i b_j$. Therefore the standard factorization follows:

$$\sum_{i,j=1}^{\infty} |\langle (u_i, v_j), (x, y) \rangle|^2 = \sum_{i,j=1}^{\infty} |a_i|^2 |b_j|^2 = \|x\|_1^2 \|y\|_2^2.$$

By identifying the elements (u_i, v_j) with $u_i \otimes v_j$ one can show that this construction leads to the same result as the general construction of the tensor product of two Hilbert spaces.

Without much additional effort the construction of the tensor product generalizes to more than two factors. Thus, given a finite number of vector spaces $E_1, \ldots, E_n$ over the field $\mathbb{K}$, the n-fold tensor product

$$E_1 \otimes \cdots \otimes E_n$$

is well defined and has similar properties as the tensor product of two vector spaces. In particular, to any n-linear map $b : E_1 \times \cdots \times E_n \to G$ into some vector space over the same field there is a linear map $\ell : E_1 \otimes \cdots \otimes E_n \to G$ such that

$$b(x_1, \ldots, x_n) = \ell(x_1 \otimes \cdots \otimes x_n) \qquad \forall\, x_i \in E_i,\ i = 1, \ldots, n.$$

This applies in particular to the n-fold tensor product

$$\mathcal{H}_1 \otimes \cdots \otimes \mathcal{H}_n$$

of Hilbert spaces $\mathcal{H}_i,\ i = 1, \ldots, n$.

17.3 Some applications of tensor products and direct sums

17.3.1 State space of particles with spin

Originally, in quantum physics the state space (more precisely the space of wave functions) $\mathcal{H}$ for an elementary localizable particle was considered to be the Hilbert space of complex valued square integrable functions in configuration space $\mathbb{R}^3$, i.e., $\mathcal{H} = L^2(\mathbb{R}^3)$. Initially this state space was also used for the quantum mechanical description of an electron. Later through several experiments (Stern–Gerlach, Zeeman) one learned that the electron has an additional internal degree of freedom with two possible values. This internal degree of freedom is called *spin*. Hence the state space for the electron had to be extended by these two additional degrees of freedom and accordingly the state space of the electron is taken to be

$$\mathcal{H}_e = L^2(\mathbb{R}^3) \otimes \mathbb{C}^2 = L^2(\mathbb{R}^3, \mathbb{C}^2). \tag{17.17}$$

17.3 Some applications of tensor products and direct sums

Note that $L^2(\mathbb{R}^3, \mathbb{C}^2)$ is the Hilbert space of all square integrable functions $\psi : \mathbb{R}^3 \to \mathbb{C}^2$ with inner product $\langle \psi, \phi \rangle = \sum_{j=1}^{2} \int_{\mathbb{R}^3} \overline{\psi_j(x)} \phi_j(x) dx$ for all $\psi, \phi \in L^2(\mathbb{R}^3, \mathbb{C}^2)$.

Later other elementary particles were discovered with $p > 2$ internal degrees of freedom. Accordingly their state space was taken to be

$$L^2(\mathbb{R}^3) \otimes \mathbb{C}^p = L^2(\mathbb{R}^3, \mathbb{C}^p).$$

The validity of this identity is shown in the Exercises.

Actually the theory of these internal degrees of freedom or spins is closely related to the representation theory of the group $SU(2)$ (see [Thi02]). $\mathbb{C}^2$ is the representation space of the irreducible representation $D_{1/2}$ of $SU(2)$ and similarly, $\mathbb{C}^{2s+1}$ is the representation space of the irreducible representation D_s of $SU(2)$, $s = n/2, n = 0, 1, 2, \ldots$.

17.3.2 State space of multi-particle systems

In the quantum mechanical description of multi-particle systems the question naturally arises of how the states of the multi-particle system are related to the single particle states of the particles which constitute the multi-particle system. The answer is given by the tensor product of Hilbert spaces. According to the principles of quantum mechanics the state space $\mathcal{H}_n$ of an n-particle system of n identical particles with state space $\mathcal{H}_1$ is

$$\mathcal{H}_n = \mathcal{H}_1 \otimes \cdots \otimes \mathcal{H}_1 \quad n \text{ factors}, \tag{17.18}$$

or a certain subspace thereof depending on the type of particle.

Empirically one found that there are two types of particles, bosons and fermions. The spin of bosons has an integer value $s = 0, 1, 2, \ldots$ while fermions have a spin with half-integer values, i.e., $s = \frac{1}{2}, \frac{3}{2}, \frac{5}{2}, \ldots$. The n-particle state space of n identical bosons is the *totally symmetric n-fold tensor product* of the one particle state space, i.e.,

$$\mathcal{H}_{n,b} = \mathcal{H}_1 \otimes_s \cdots \otimes_s \mathcal{H}_1 \quad n \text{ factors}, \tag{17.19}$$

and the n-particle state space of n identical fermions is the *totally anti-symmetric tensor product* of the one particle state space, i.e.,

$$\mathcal{H}_{n,f} = \mathcal{H}_1 \otimes_a \cdots \otimes_a \mathcal{H}_1 \quad n \text{ factors}. \tag{17.20}$$

Here we use the following notation: $\phi \otimes_s \psi = \frac{1}{2}(\phi \otimes \psi + \psi \otimes \phi)$, respectively $\phi \otimes_a \psi = \frac{1}{2}(\phi \otimes \psi - \psi \otimes \phi)$. In the Exercises some concrete examples of multi-particle state spaces are studied.

In relativistic quantum field theory one considers systems in which elementary particles can be created and annihilated. Thus one needs a state space which allows the description of any number of particles and which allows a change of particle numbers.

Suppose we consider such a system composed of bosons with one particle state space $\mathcal{H}_1$. Then the *Boson Fock space* over $\mathcal{H}_1$

$$\mathcal{H}_B = \oplus_{n=0}^{\infty} \mathcal{H}_{n,b}$$

where $\mathcal{H}_{0,b} = \mathbb{C}$ and $\mathcal{H}_{n,b}$ is given in (17.19) is a Hilbert space which allows the description of a varying number of bosons.

Similarly, the *Fermion Fock space* over the one particle state space $\mathcal{H}_1$

$$\mathcal{H}_F = \oplus_{n=0}^{\infty} \mathcal{H}_{n,f}$$

where again $\mathcal{H}_{0,f} = \mathbb{C}$ and $\mathcal{H}_{n,f}$ is given in (17.20), is a Hilbert space which allows the description of a varying number of fermions.

We conclude this chapter with the remark that in relativistic quantum field theory one can explain, on the basis of well established physical principles, why the n-particle space of bosons has to be a totally symmetric and that of fermions a totally anti-symmetric tensor product of their one particle state space (for a theorem on the connection between spin and statistics, see [Thi02, RS75, Jos65, SW64]).

17.4 Exercises

1. Prove: Through formula (17.15) a scalar product is well defined on the tensor product $\mathcal{H}_1 \otimes \mathcal{H}_2$ of two Hilbert spaces $\mathcal{H}_i$, $i = 1, 2$.

2. Complete the proof of Theorem 17.1.1, i.e., show: If all the Hilbert spaces $\mathcal{H}_i$, $i \in \mathbb{N}$, are separable, so is the direct sum $\mathcal{H} = \oplus_{i=1}^{\infty} \mathcal{H}_i$.

3. Prove the calculation rules for tensor products.

4. Show that the definition of the tensor product $\mathcal{D}(\Omega_1) \otimes \mathcal{D}(\Omega_2)$ of test function spaces $\mathcal{D}(\Omega_i)$ is a special case of the tensor product of vector spaces.

5. Prove the statements in the text about the n-fold tensor product for $n > 2$.

6. On the Hilbert space $\mathbb{C}^2$ consider the matrices

$$\sigma_x = \begin{pmatrix} 0 & 1 \\ 1 & 0 \end{pmatrix}, \quad \sigma_y = \begin{pmatrix} 0 & -i \\ i & 0 \end{pmatrix}, \quad \sigma_z = \begin{pmatrix} 1 & 0 \\ 0 & -1 \end{pmatrix}. \quad (17.21)$$

Show that these matrices are self-adjoint on $\mathbb{C}^2$, i.e., $\sigma^* = \sigma$ (for the definition of the adjoint σ^* see the beginning of Section 19.2) and satisfy the relations

$$\sigma_x \sigma_y = -\sigma_y \sigma_x = i\sigma_z \quad \sigma_y \sigma_z = -\sigma_z \sigma_y = i\sigma_x \quad \sigma_z \sigma_x = -\sigma_x \sigma_z = i\sigma_y. \quad (17.22)$$

The matrices $\sigma_x, \sigma_y, \sigma_z$ are called the *Pauli matrices*. In quantum physics they are used for the description of the *spin* of a particle.

18
Topological Aspects

In our introduction we stressed the analogy between Euclidean spaces and Hilbert spaces. This analogy works well as long as only the vector space and the geometric structures of a Hilbert space are concerned. But in the case of infinite dimensional Hilbert spaces there are essential differences when we look at topological structures on these spaces. It turns out that in an infinite dimensional Hilbert space the unit ball is not compact (with respect to the natural or norm topology) with the consequence that in such a case there are very few compact sets of interest for analysis. Accordingly a weaker topology in which the closed unit ball is compact is of great importance. This topology, called the *weak topology*, is studied in the second section to the extent needed in later chapters.

18.1 Compactness

We begin by recalling some basic concepts related to compact sets. If M is a subset of a normed space X, a system $\mathcal{G}$ of subsets G of X is called a *covering of M* if, and only if, $M \subset \cup_{G \in \mathcal{G}} G$. If all the sets in $\mathcal{G}$ are open such a covering is called an *open covering of M*. A subset K of X is called *compact* if, and only if, every open covering of K contains a finite sub-covering, i.e., there are $G_1, \ldots, G_N \in \mathcal{G}$ such that $K \subset \cup_{i=1}^{N} G_i$.

It is important to be aware of the following basic facts about compact sets. A compact set $K \subset X$ is closed and bounded in the normed space $(X, \|\cdot\|)$. A closed subset of a compact set is compact.

Every infinite sequence $(x_n)_{n \in \mathbb{N}}$ in a compact set K contains a subsequence which converges to a point in K (*Theorem of Bolzano–Weierstrass*). If K is a set such that every infinite sequence in K has a convergent subsequence, then K is called *sequentially compact*. One shows (see Exercises) that in a normed space a set is compact if, and only if, it is sequentially compact. This is very convenient in applications and is used frequently. B. Bolzano was the first to point out the significance of this property for a rigorous introduction to analysis.

A continuous real valued function is bounded on a compact set, attains its minimal and maximal values (*Theorem of Weierstrass*) and is equi-continuous (*Theorem of Heine*).

The *covering theorem of Heine–Borel* states that a subset $K \subset \mathbb{K}^n$ is compact if, and only if, it is closed and bounded. In infinite dimensional normed spaces this equivalence is not true as the following important theorem shows:

Theorem 18.1.1 (Theorem of F. Riesz) *Suppose $(X, \|\cdot\|)$ is a normed space and $B = \overline{B_1(0)}$ denotes the closed unit ball with centre 0. Then B is compact if, and only if, X is finite dimensional.*

Proof. If X is finite dimensional, then B is compact because of the Heine–Borel covering theorem.

Conversely assume that B is compact in the normed space $(X, \|\cdot\|)$. Denote by $B(a, r)$ the open ball with centre $a \in X$ and radius $r > 0$. Then $\mathcal{G} = \{B(a, r) : a \in B\}$ is an open covering of B for any $r > 0$. Compactness of B implies that there is a finite sub-cover, i.e., there are points $a_1, \ldots, a_N \in B$ such that

$$B \subseteq \cup_{i=1}^N B(a_i, r). \tag{18.1}$$

Now observe $B(a_i, r) = a_i + rB(0, 1)$ and denote by V the linear subspace of X generated by the vectors $a_1, \ldots, a_N$. Certainly, V has a dimension smaller than or equal to N and is thus closed in X. Relation (18.1) implies

$$B \subseteq \cup_{i=1}^N (a_i + rB(0, 1)) \subseteq V + rB(0, 1) \subseteq V + rB. \tag{18.2}$$

By iterating this relation we obtain, for $n = 1, 2, \ldots$

$$B \subseteq V + r^n B. \tag{18.3}$$

Choose $0 < r < 1$. It follows that

$$B \subseteq \cap_{n \in \mathbb{N}} (V + r^n B) = \overline{V} = V.$$

Since B is the closed unit ball of X we know $X = \cup_{n=1}^\infty nB$ and thus

$$X \subseteq \cup_{n=1}^\infty nV = V.$$

Therefore X has a finite dimension smaller than or equal to N. □

For an infinite dimensional Hilbert space there is another proof of the fact that its closed unit ball is not compact. For such a Hilbert space one can find an orthonormal system with infinitely many elements: $\{e_n : n \in \mathbb{N}\} \subset B$. For $n, m \in \mathbb{N}, n \neq m$ one has $\|e_n - e_m\| = \sqrt{2}$. Thus no subsequence of the sequence $(e_n)_{n \in \mathbb{N}}$ is a Cauchy sequence; therefore no subsequence converges and hence B is not sequentially compact.

Remark 18.1.1 *An obvious consequence of Theorem 18.1.1 is that in an infinite dimensional normed space X, compact sets have an empty interior. Hence in such* a case the only continuous function $f : X \to \mathbb{K}$ with compact support is the null function.

Recall that a space is called **locally compact** if, and only if, every point has a compact neighborhood. Hence a locally compact normed space is finite dimensional.

18.2 The weak topology

As the Theorem of F. Riesz shows, the closed unit ball in an infinite dimensional Hilbert space $\mathcal{H}$ is not (sequentially) compact. We are going to introduce a weaker topology on $\mathcal{H}$ with respect to which the convenient characterization of compact sets as we know it from the Euclidean spaces $\mathbb{K}^n$ is available. In particular the theorem of Bolzano–Weierstrass is valid for this weak topology. Though we introduced the weak topology in the part on distributions we repeat it for the present particular case.

Definition 18.2.1 *Let X be a normed space and X' its topological dual. The **weak topology on X**, $\sigma(X, X')$, is the coarsest locally convex topology on X such that all $f \in X'$ are continuous. A basis of neighborhoods of a point $x_0 \in X$ for the topology $\sigma(X, X')$ is given by the following system of sets:*

$$U(x_0; f_1, \ldots, f_n; r), \qquad f_1, \ldots, f_n \in X', \quad r > 0, \quad n \in \mathbb{N},$$

$$U(x_0; f_1, \ldots, f_n; r) = \{x \in X : |f_i(x - x_0)| < r, \ i = 1, \ldots, n\}.$$

In particular, for a Hilbert space $\mathcal{H}$, a basis of neighborhoods for the weak topology is

$$U(x_0; y_1, \ldots, y_n; r), \qquad y_1, \ldots, y_n \in \mathcal{H}, \quad r > 0, \quad n \in \mathbb{N},$$

$$U(x_0; y_1, \ldots, y_n; r) = \{x \in \mathcal{H} : |\langle y_i, x - x_0\rangle| < r, \ i = 1, \ldots, n\}.$$

Certainly, Corollary 15.3.1 has been used in the description of the elements of a neighborhood basis for the weak topology of a Hilbert space. It is important to be aware of the following elementary facts about the topology $\sigma = \sigma(X, X')$ of a normed space X. It has fewer open and thus fewer closed sets than the strong or norm topology. Hence, if a subset $A \subset X$ is closed for σ it is also closed for the strong topology. But the converse does not hold in general. However for convex sets we will learn later in this section that such a set is closed for σ if, and only if, it is closed for the strong topology.

In case of a finite dimensional normed space X, the weak and the strong topology coincide. One can actually show that this property characterizes finite dimensional normed spaces. This is discussed in the Exercises.

Though it should be clear from the above definition we formulate the concepts of convergence for the weak topology explicitly.

Definition 18.2.2 *Let $\mathcal{H}$ be a Hilbert space with inner product $\langle \cdot, \cdot \rangle$ and $(x_n)_{n \in \mathbb{N}}$ a sequence in $\mathcal{H}$.*

1. The sequence $(x_n)_{n\in\mathbb{N}}$ **converges weakly** to $x \in \mathcal{H}$ if, and only if, for every $u \in \mathcal{H}$ the numerical sequence $(\langle u, x_n\rangle)_{n\in\mathbb{N}}$ converges to the number $\langle u, x\rangle$. x is called the **weak limit** of the sequence $(x_n)_{n\in\mathbb{N}}$.

2. The sequence $(x_n)_{n\in\mathbb{N}}$ is a **weak Cauchy sequence**, i.e., a Cauchy sequence for the weak topology, if, and only if, for every $u \in \mathcal{H}$ the numerical sequence $(\langle u, x_n\rangle)_{n\in\mathbb{N}}$ is a Cauchy sequence.

Some immediate consequences of these definitions are:

Lemma 18.2.1 *Suppose $\mathcal{H}$ is a Hilbert space with inner product $\langle \cdot, \cdot \rangle$.*

a) A weakly convergent sequence is a weak Cauchy sequence.

b) A sequence has at most one weak limit.

c) Every infinite orthonormal system converges weakly to zero.

Proof. Part a) is obvious from the definition. For Part b) assume that a sequence $(x_n)_{n\in\mathbb{N}} \subset \mathcal{H}$ has the points $x, y \in \mathcal{H}$ as weak limits. For every $u \in \mathcal{H}$ it follows that
$$\langle u, x - y\rangle = \lim_{n\to\infty} \langle u, x_n - x_n\rangle = 0,$$
and hence $x - y \in \mathcal{H}^\perp = \{0\}$, thus $x = y$.

Suppose $\{x_n : n \in \mathbb{N}\}$ is an infinite orthonormal system in $\mathcal{H}$. For every $u \in \mathcal{H}$ Bessel's inequality (see Corollary 14.1.1) implies that
$$\sum_{n=1}^{\infty} |\langle x_n, u\rangle|^2 \le \|u\|^2 < \infty$$
and therefore $\langle x_n, u\rangle \to 0$. Since $u \in \mathcal{H}$ is arbitrary we conclude. □

Before we continue with some deeper results about the weak topology on a Hilbert space we would like to pause a little for a heuristic discussion of the intuitive meaning of the concept of weak convergence.

Consider the wave equation in one dimension
$$\partial_t^2 u - \partial_x^2 u = 0$$
where $\partial_t = \frac{\partial}{\partial t}$ and similarly $\partial_x = \frac{\partial}{\partial x}$ and look for solutions u which are in the Hilbert space $\mathcal{H} = L^2(\mathbb{R})$ with respect to the space variable x for each time t, i.e., $u(\cdot, t) \in L^2(\mathbb{R})$ for each $t \ge 0$, given a smooth initial condition $u_0 \in C^2(\mathbb{R})$ with support in the interval $[-1, 1]$ which is symmetric, $u_0(-x) = u_0(x)$:
$$u(\cdot, 0) = u_0, \qquad \partial_t u(\cdot, 0) = 0.$$

The solution is easily found to be $u(x, t) = \frac{1}{2}(u_0(x - t) + u_0(x + t))$. Obviously, the support of $u(\cdot, t)$ is contained in the set $S_t = [-1-t, +1-t] \cup [-1+t, +1+t]$. For $t > 1$ the two functions $x \mapsto u_0(x - t)$ and $x \mapsto u_0(x + t)$ have a disjoint support and thus for all $t > 1$,
$$\|u(\cdot, t)\|_2^2 = \int_{\mathbb{R}} |u(x, t)|^2 dx = \frac{1}{2}\|u_0\|_2^2.$$

The support S_t of $u(\cdot, t)$ moves to "infinity" as $t \to +\infty$. This implies that $u(\cdot, t)$ converges weakly to 0 as $t \to \infty$: For every $v \in L^2(\mathbb{R})$ one finds

$$|\langle v, u(\cdot, t)\rangle_2| = |\int_{S_t} \overline{v(x)} u(x,t) dx| \leq \sqrt{\int_{S_t} |v(x)|^2 dx} \sqrt{\int_{S_t} |u(x,t)|^2 dx}$$

$$\leq \frac{\|u_0\|}{\sqrt{2}} \sqrt{\int_{S_t} |v(x)|^2 dx}.$$

Since $v \in L^2(\mathbb{R})$, given $\epsilon > 0$ there is $R > 0$ such that $\int_{|x| \geq R} |v(x)|^2 dx \leq \epsilon^2$. For $|t|$ sufficiently large the support S_t is contained in $\{x \in \mathbb{R} : |x| \geq R\}$. Hence for such t we can continue the above estimate by

$$\leq \|u_0\|_2 \epsilon / \sqrt{2}$$

and we conclude that $\langle v, u(\cdot, t)\rangle_2 \to 0$ as $|t| \to \infty$.

The way in which weak convergence is achieved in this example is not atypical for weak convergence in $L^2(\mathbb{R}^n)$! Later in our discussion of quantum mechanical scattering theory we will encounter a similar phenomenon. There, scattering states in $L^2(\mathbb{R}^n)$ will be defined as those functions $t \mapsto \phi(\cdot, t) \in L^2(\mathbb{R}^n)$ for which

$$\lim_{|t| \to \infty} \int_{|x| \leq R} |\phi(x,t)|^2 dx = 0$$

for every $R \in (0, \infty)$.

How are strong and weak convergence related? Certainly, if a sequence $(x_n)_{n \in \mathbb{N}}$ converges strongly to $x \in \mathcal{H}$, then it also converges weakly and has the same limit. This follows easily from Schwarz' inequality: $|\langle u, x - x_n\rangle| \leq \|u\| \|x - x_n\|$, for any $u \in \mathcal{H}$. The relation between both concepts of convergence is fully understood as the following theorem shows.

Theorem 18.2.1 *Let $\mathcal{H}$ be a Hilbert space with inner product $\langle \cdot, \cdot \rangle$ and $(x_n)_{n \in \mathbb{N}}$ a sequence in $\mathcal{H}$. This sequence converges strongly to $x \in \mathcal{H}$ if, and only if, it converges weakly to x and $\lim_{n \to \infty} \|x_n\| = \|x\|$.*

Proof. That weak convergence is necessary for strong convergence has been shown above. The basic estimate for norms

$$| \|x\| - \|x_n\| | \leq \|x - x_n\|$$

(see Corollary 14.1.2) implies that $\lim_{n \to \infty} \|x_n\| = \|x\|$.

In order to see that they are sufficient, consider a sequence which converges weakly to $x \in \mathcal{H}$ and for which the sequence of norms converges to the norm of x. Since the norm is defined in terms of the inner product one has

$$\|x - x_n\|^2 = \langle x - x_n, x - x_n \rangle = \|x\|^2 + \|x_n\|^2 - \langle x, x_n \rangle - \langle x_n, x \rangle \qquad \forall n \in \mathbb{N}.$$

Weak convergence implies that

$$\lim_{n \to \infty} \langle x, x_n \rangle = \lim_{n \to \infty} \langle x_n, x \rangle = \|x\|^2.$$

Since also $\lim_{n \to \infty} \|x_n\| = \|x\|$ is assumed, we deduce $\|x - x_n\|^2 \to 0$ as $n \to \infty$ and strong convergence follows. □

There are some simple but important facts implied by the these results.

- The open unit ball $B_1 = \{x \in \mathcal{H} : \|x\| < 1\}$ of an infinite dimensional Hilbert space $\mathcal{H}$ is not open for the weak topology. Since otherwise every set which is open for the strong topology would be open for the weak topology and thus both topologies would be identical.

- The unit sphere $S_1 = \{x \in \mathcal{H} : \|x\| = 1\}$ of an infinite dimensional Hilbert space $\mathcal{H}$ is closed for the strong but not for the weak topology. The weak closure of S_1, i.e., the closure of S_1 with respect to the weak topology is equal to the closed unit ball $\overline{B_1} = \{x \in \mathcal{H} : \|x\| \leq 1\}$. (See Exercises)

A first important step towards showing that the closed unit ball of a Hilbert space is compact for the weak topology is to show that strongly bounded sequences have weakly convergent subsequences.

Theorem 18.2.2 *Every sequence $(x_n)_{n \in \mathbb{N}}$ in a Hilbert space $\mathcal{H}$ which is strongly bounded, i.e., there is an $M < \infty$ such that $\|x_n\| \leq M$ for all $n \in \mathbb{N}$, has a weakly convergent subsequence.*

Proof. The given sequence generates a closed linear subspace $\mathcal{H}_0 = [\{x_n : n \in \mathbb{N}\}]$ in $\mathcal{H}$.

Consider the numerical sequence $A_n^1 = \langle x_1, x_n \rangle, n = 1, 2, \ldots$. By Schwarz' inequality it is bounded: $|A_n^1| \leq \|x_1\| \|x_n\| \leq M^2$. The Bolzano–Weierstrass theorem ensures the existence of a convergent subsequence $A_{n^1(j)}^1 = \langle x_1, x_{n^1(j)} \rangle, j \in \mathbb{N}$. Next consider the numerical sequence $A_{n^1(j)}^2 = \langle x_2, x_{n^1(j)} \rangle$, $j \in \mathbb{N}$. It too is bounded by M^2 and again by Bolzano–Weierstrass we can find a convergent subsequence $A_{n^2(j)}^2 = \langle x_2, x_{n^2(j)} \rangle$, $j \in \mathbb{N}$.

This argument can be iterated and thus generates a sequence $x_{n^i(j)}, i = 1, 2, \ldots$ of subsequences of our original sequence with the property that $(x_{n^{i+1}(j)})_{j \in \mathbb{N}}$ is a subsequence of $(x_{n^i(j)})_{j \in \mathbb{N}}$. Finally we consider the diagonal sequence $(x_{m(j)})_{j \in \mathbb{N}}$ where we use $m(j) = n^j(j)$. Then all numerical sequences $\langle x_k, x_{m(j)} \rangle, j \in \mathbb{N}$, converge since for $j > k$ this sequence is a subsequence of the convergent sequence $(A_{n^k(j)}^k)_{j \in \mathbb{N}}$. It follows that $\lim_{j \to \infty} \langle x, x_{m(j)} \rangle$ exists for all $x \in V = \text{lin}\{x_n : n \in \mathbb{N}\}$. Hence $\lim_{j \to \infty} \langle x_{m(j)}, x \rangle$ exists for all $x \in V$. We call this limit $T(x)$. Basic rules of calculation imply that $T : V \to \mathbb{K}$ is linear. The estimate $|\langle x_{m(j)}, x \rangle| \leq \|x\| \|x_{m(j)}\| \leq M\|x\|$ implies $|T(x)| \leq M\|x\|$ and thus T is a continuous linear functional on the subspace V. The Extension Theorem 15.3.2 implies that there is a unique continuous linear functional $\hat{T}$ on $\mathcal{H}$ such that $\|\hat{T}\| = \|T\|$. Furthermore, by Theorem 15.3.1, there is a unique vector $y \in \mathcal{H}_0$ such that $\hat{T}(x) = \langle y, x \rangle$ for all $x \in \mathcal{H}$, and we deduce that y is the weak limit of the sequence $(x_{m(j)})_{j \in \mathbb{N}}$ (first we have $\langle y, x \rangle = \lim_{j \to \infty} \langle x_{m(j)}, x \rangle$ for all $x \in V$, then by continuous extension for all $x \in \mathcal{H}$; details are considered in the Exercises). □

One of the fundamental principles of functional analysis is the *uniform boundedness principle*. It is also widely used in the theory of Hilbert spaces. In Appendix 34.4 we prove this principle in the generality which is needed in the theory of generalized functions. In this section we give a direct proof for Banach spaces. This version obviously is sufficient for the theory of Hilbert spaces.

Definition 18.2.3 *Let X be a Banach space with norm $\|\cdot\|$ and $\{T_\alpha : \alpha \in A\}$ a family of continuous linear functionals on X (A an arbitrary index set). One says that this family is*

1. **pointwise bounded** *if, and only if, for every $x \in X$ there is a real constant $C_x < \infty$ such that*
$$\sup_{\alpha \in A} |T_\alpha(x)| \leq C_x;$$

2. **uniformly bounded** *or* **norm bounded** *if, and only if*
$$\sup_{\alpha \in A} \sup \{|T_\alpha(x)| : x \in X, \ \|x\| \leq 1\} = C < \infty.$$

Clearly, every uniformly bounded family of continuous linear functionals is pointwise bounded. For a certain class of spaces (see Appendix 34.4) the converse is also true and is called the *principle of uniform boundedness* or the *uniform boundedness principle*. It was first proven by Banach and Steinhaus for Banach spaces.

We prepare for the proof of this fundamental result by an elementary lemma.

Lemma 18.2.2 *A family $\{T_\alpha : \alpha \in A\}$ of continuous linear functionals on a Banach space X is uniformly bounded if, and only if, this family is uniformly bounded on some ball $B_r(x_0) = \{x \in X : \|x - x_0\| < r\}$, i.e.,*
$$\sup_{\alpha \in A} \sup_{x \in B_r(x_0)} |T_\alpha(x)| = C < \infty.$$

Proof. If the given family is uniformly bounded we know that there is some positive constant C_0 such that $|T_\alpha(x)| \leq C_0$ for all $x \in B = B_1(0)$ and all $\alpha \in A$. A ball $B_r(x_0)$ with centre x_0 and radius $r > 0$ is obtained from the unit ball B by translation and scaling: $B_r(x_0) = x_0 + rB$. Thus every $x \in B_r(x_0)$ can be written as $x = x_0 + ry$ with $y \in B$ and therefore
$$|T_\alpha(x)| = |T_\alpha(x_0 + ry)| = |T_\alpha(x_0) + rT_\alpha(y)|$$
$$\leq |T_\alpha(x_0)| + r|T_\alpha(y)| \leq C_0\|x_0\| + rC_0.$$

Hence the family $\{T_\alpha : \alpha \in A\}$ is uniformly bounded on the ball $B_r(x_0)$ by $(r + \|x_0\|)C_0$.

Conversely, assume that the family $\{T_\alpha : \alpha \in A\}$ is uniformly bounded on some ball $B_r(x_0)$ with bound C. The points y in the unit ball B have the representation $y = (x - x_0)/r$ in terms of the points $x \in B_r(x_0)$. It follows, for all $y \in B$ and all $\alpha \in A$:
$$|T_\alpha(y)| = \frac{1}{r}|T_\alpha(x - x_0)| \leq \frac{1}{r}(|T_\alpha(x)| + |T_\alpha(x_0)|) \leq \frac{2C}{r} < \infty,$$
and we conclude. $\square$

Theorem 18.2.3 (Banach–Steinhaus) *A family $\{T_\alpha : \alpha \in A\}$ of continuous linear functionals on a Banach space X is uniformly bounded if, and only if, it is pointwise bounded.*

Proof. Let $\mathcal{T} = \{T_\alpha : \alpha \in A\}$ be a pointwise bounded family of continuous linear functionals on X. We prove the uniform bound
$$\sup_{\alpha \in A} \|T_\alpha\| < \infty$$
indirectly.

Assume that $\mathcal{T}$ is not uniformly bounded. Lemma 18.2.2 implies that $\mathcal{T}$ is not uniformly bounded on any of the balls $B_r(x_0)$, $x_0 \in X$, $r > 0$. It follows that for every $p \in \mathbb{N}$ there are an index $\alpha_p \in A$ and a point $x_p \in B = B_1(0)$ such that $|T_{\alpha_p}(x_p)| > p$.

Begin with $p = 1$. Since T_{α_1} is continuous there is an $\epsilon_1 > 0$ such that $|T_{\alpha_1}(x)| > 1$ for all $x \in B_{\epsilon_1}(x_1)$. By choosing ϵ_1 small enough we can ensure $B_{\epsilon_1}(x_1) \subset B$. Again by Lemma 18.2.2 we know that the family $\mathcal{T}$ is not uniformly bounded on the ball $B_{\epsilon_1}(x_1)$. Hence there are a point $x_2 \in B_{\epsilon_1}(x_1)$ and an index $\alpha_2 \in A$ such that $|T_{\alpha_2}(x_2)| > 2$. Continuity of T_{α_2} implies the existence of $\epsilon_2 \in (0, \epsilon_1/2)$ such that $|T_{\alpha_2}(x)| > 2$ for all $x \in B_{\epsilon_2}(x_2) \subset B_{\epsilon_1}(x_1)$.

On the basis of Lemma 18.2.2 these arguments can be iterated. Thus we obtain a sequence of points $(x_p)_{p \in \mathbb{N}} \subset B$, a decreasing sequence of positive numbers ϵ_p and a sequence of indices $\alpha_p \in A$ such that

a) $|T_{\alpha_p}(x)| > p$ for all $x \in B_{\epsilon_p}(x_p)$;

b) $B_{\alpha_{p+1}}(x_{p+1}) \subset B_{\alpha_p}(x_p)$ for all $p \in \mathbb{N}$;

c) $0 < \epsilon_{p+1} < \frac{\epsilon_p}{2} < \frac{\epsilon_1}{2^p}$.

Property b) implies $\|x_{p+1} - x_p\| < \epsilon_p$ and thus by c), for all $m \in \mathbb{N}$:

$$\|x_{p+m} - x_p\| = \left\|\sum_{i=0}^{m-1} (x_{p+i+1} - x_{p+i})\right\| \leq \sum_{i=0}^{m-1} \|x_{p+i+1} - x_{p+i}\|$$
$$< \sum_{i=0}^{m-1} \epsilon_{p+i} < \sum_{i=0}^{m-1} \frac{\epsilon_1}{2^{p+i}} \to 0 \quad \text{as } p \to \infty.$$

This shows that $(x_p)_{p \in \mathbb{N}}$ is a Cauchy sequence in the Banach space X, hence it converges to a point $x \in X$. This point belongs to all the balls $B_{\epsilon_p}(x_p)$ because of b). At this point x the family $\mathcal{T}$ is bounded by assumption. This is a contradiction to the construction according to property a). We conclude that the family $\mathcal{T}$ is uniformly bounded. □

Remark 18.2.1 1. *The statement of the Banach–Steinhaus theorem can be rephrased as follows: If a family $\{T_\alpha : \alpha \in A\}$ of continuous linear functionals on a Banach space X is not uniformly bounded, then there is a point $x_0 \in X$ such that $\sup_{\alpha \in A} |T_\alpha(x_0)| = +\infty$.*

2. *One can also prove the principle of uniform boundedness by using the fact that a Banach space X is a Baire space, i.e., if X is represented as the countable union of closed sets X_n, $X = \cup_{n \in \mathbb{N}} X_n$, then at least one of the sets X_n must contain an open nonempty ball (see Appendix C). Given a pointwise bounded family $\{T_\alpha : \alpha \in A\}$ of continuous linear functionals of X we apply this to the sets*

$$X_n = \{x \in X : |T_\alpha(x)| \leq n \; \forall \alpha \in A\} \qquad n \in \mathbb{N}.$$

The pointwise bounds ensure that the union of these sets X_n represents X. It thus follows that the family is bounded on some open ball and by Lemma 18.2.2 we conclude.

3. *The theorem of Riesz–Fréchet (Theorem 15.3.1) states that the continuous linear functionals T on a Hilbert space $\mathcal{H}$ can be identified with the points $u \in \mathcal{H}$: $T = T_u$, $u \in \mathcal{H}$, $T_u(x) = \langle u, x \rangle$ for all $x \in \mathcal{H}$. Theorem 18.2.3 implies: If a set $A \subset \mathcal{H}$ is weakly bounded, then it is uniformly bounded, i.e., bounded in norm. (See the Exercises for details.)*

18.2 The weak topology   243

4. In order to verify whether a set A is bounded (i.e., whether A is contained in some finite ball) it suffices, because of Theorem 18.2.3, to verify that it is weakly bounded. As in the case of a finite dimensional Hilbert space, this amounts to verifying that A is 'bounded in every coordinate direction' and this is typically much easier.

A weakly convergent sequence $(x_n)_{n \in \mathbb{N}}$ in a Hilbert space $\mathcal{H}$ is obviously pointwise bounded and thus bounded in norm. This proves

Lemma 18.2.3 *Every weakly convergent sequence in a Hilbert space is bounded in norm.*

Now we are well prepared to prove the second major result of this section.

Theorem 18.2.4 *Every Hilbert space $\mathcal{H}$ is sequentially complete with respect to the weak topology.*

Proof. Suppose we are given a weak Cauchy sequence $(x_n)_{n \in \mathbb{N}} \subset \mathcal{H}$. For every $u \in \mathcal{H}$ the numerical sequence $(\langle x_n, u \rangle)_{n \in \mathbb{N}}$ then is a Cauchy sequence and thus converges to some number in the field $\mathbb{K}$. Call this number $T(u)$. It follows that this sequence is pointwise bounded. Hence it is norm bounded, i.e., there is some constant $C \in [0, \infty)$ such that $\|x_n\| \leq C$ for all $n \in \mathbb{N}$. Since $T(u) = \lim_{n \to \infty} \langle x_n, u \rangle$ it follows by Schwarz' inequality $|T(u)| \leq C\|u\|$. Basic rules of calculation for limits imply that the function $T : \mathcal{H} \to \mathbb{K}$ is linear. Thus T is a continuous linear functional on $\mathcal{H}$, and we know that such functionals are of the form $T = T_x$ for a unique vector $x \in \mathcal{H}$, $T_x(u) = \langle x, u \rangle$ for all $u \in \mathcal{H}$. We conclude that $\langle x, u \rangle = \lim_{n \to \infty} \langle x_n, u \rangle$ for $u \in \mathcal{H}$. Hence the Cauchy sequence $(x_n)_{n \in \mathbb{N}}$ converges weakly to the point $x \in \mathcal{H}$. The Hilbert space $\mathcal{H}$ is weakly sequentially complete. □

Theorem 18.2.5 (Banach–Saks) *Suppose that $(x_n)_{n \in \mathbb{N}}$ is a weakly convergent sequence with limit x. Then there exists a subsequence $(x_{n(j)})_{j \in \mathbb{N}}$ such that the sequence of arithmetic means of this subsequence converges strongly to x, i.e.,*

$$\lim_{m \to \infty} \frac{1}{m} \sum_{j=1}^{m} x_{n(j)} = x.$$

Proof. Since weakly convergent sequences are norm bounded, there is a constant M such that $\|x - x_n\| \leq M$ for all $n \in \mathbb{N}$. We define the subsequence successively and start with $n(1) = 1$. Because of weak convergence of the given sequence there is an $n(2) \in \mathbb{N}$ such that $|\langle x_{n(2)} - x, x_{n(1)} - x \rangle| < 1$. Suppose that $n(1), \ldots, n(k)$ have been constructed. Since the given sequence converges weakly to x there is an $n(k+1) \in \mathbb{N}$ such that

$$|\langle x_{n(k+1)} - x, x_{n(i)} - x \rangle| < \frac{1}{k}, \quad i = 1, \ldots, k.$$

Now we estimate $\frac{1}{m} \sum_{j=1}^{m} x_{n(j)} - x$ in norm, taking the choice of the subsequence into account in the last step:

$$\|\frac{1}{k}\sum_{i=1}^{k}(x_{n(i)} - x)\|^2 = \frac{1}{k^2}\sum_{i,j=1}^{k}\langle x_{n(j)} - x, x_{n(i)} - x\rangle$$

$$= \frac{1}{k^2}(\sum_{i=1}^{k}\langle x_{n(i)} - x, x_{n(i)} - x\rangle + 2\sum_{1\le i<j\le k}\langle x_{n(j)} - x, x_{n(i)} - x\rangle)$$

$$\le \frac{1}{k^2}(kM^2 + 2\sum_{j=2}^{k}\sum_{i=1}^{j-1}|\langle x_{n(j)} - x, x_{n(i)} - x\rangle|)$$

$$\le \frac{1}{k^2}(kM^2 + 2\sum_{j=2}^{k}\sum_{i=1}^{j-1}\frac{1}{j-1}) = \frac{1}{k^2}(kM^2 + 2k).$$

Clearly, the upper bound in this estimate converges to zero as $k \to \infty$ and thus proves strong convergence of the sequence of arithmetic means. □

An immediate and very important consequence of the Theorem of Banach and Saks is the conclusion that the weak and the strong closure of convex sets coincide.

Corollary 18.2.1 *For convex sets A of a Hilbert space $\mathcal{H}$, the strong closure $\overline{A}^s$ and the weak closure $\overline{A}^w$ are the same:*

$$A \subset \mathcal{H}, \; A \text{ convex} \Rightarrow \overline{A}^s = \overline{A}^w. \tag{18.4}$$

Proof. Since the strong topology is finer than the weak topology the closure with respect to the strong topology is always contained in the closure with respect to the weak topology: $\overline{A}^s \subseteq \overline{A}^w$. Convexity implies that both closures actually agree.

Recall that a subset A of a vector space is called convex if, and only if, $tx + (1-t)y \in A$ whenever $x, y \in A$ and $0 < t < 1$. Now take any point $x \in \overline{A}^w$. Then there is a sequence $(x_n)_{n \in \mathbb{N}} \subset A$ which converges weakly to x. We can find a subsequence $(x_{n(i)})_{i \in \mathbb{N}}$ such that the sequence $(\xi_m)_{m \in \mathbb{N}}$ of arithmetic means

$$\xi_m = \frac{1}{m}\sum_{i=1}^{m} x_{n(i)}$$

converges strongly to x. Since A is convex we know $\xi_m \in A$ for all $m \in \mathbb{N}$. Hence x is also the limit of a strongly convergent sequence of points in A, thus $x \in \overline{A}^s$ and therefore $\overline{A}^w \subseteq \overline{A}^s$. This proves equality of both sets. □

The weak topology σ on a Hilbert space $\mathcal{H}$ is not metrizable. However, when restricted to the closed unit ball, it is metrizable according to the following lemma which we mention without proof.

Lemma 18.2.4 *On closed balls $\overline{B}_r(x_0)$ in a Hilbert space, the weak topology σ induces the topology of a metric space.*

Gathering all the results of this section, the announced compactness of the closed unit ball with respect to the weak topology follows easily.

Theorem 18.2.6 *The closed balls $\overline{B}_r(x_0)$, $r > 0$, $x_0 \in \mathcal{H}$, in a Hilbert space $\mathcal{H}$ are weakly compact.*

Proof. Since the weak topology is metrizable on these balls it suffices to show sequential compactness. Given any sequence $(x_n)_{n \in \mathbb{N}} \subset \overline{B}_r(x_0)$ there is a weakly convergent subsequence by Theorem 18.2.2 and the weak limit of this subsequence belongs to the ball $\overline{B}_r(x_0)$ because of Corollary 18.2.1. This proves sequential compactness. □

Actually closed balls in any reflexive Banach space, and not only in Hilbert spaces, are weakly compact. This fact plays a very fundamental rôle in optimization problems and will be discussed in more detail in Part C.

18.3 Exercises

1. Prove: On a finite dimensional normed space the weak and the norm topology coincide. And conversely, if the norm and the weak topology of a normed space coincide then this space has a finite dimension.

2. Fill in the details of the last step in the proof of Theorem 18.2.2.

    Hints: It suffices to show $\lim_{j \to \infty} \langle x_{m(j)}, x \rangle = \langle y, x \rangle$ for $x \in \mathcal{H}_0$. For $x \in \mathcal{H}_0$ write $\langle y - x_{m(j)}, x \rangle = \langle y - x_{m(j)}, x_\epsilon \rangle + \langle y - x_{m(j)}, x - x_\epsilon \rangle$ with a suitable choice of $x_\epsilon \in V$, given any $\epsilon > 0$.

3. For a subset A of a Hilbert space $\mathcal{H}$ prove: If for every $x \in \mathcal{H}$ there is a finite constant C_x such that $|\langle u, x \rangle| \leq C_x$ for all $u \in A$, then there is a constant $C < \infty$ such that $\|u\| \leq C$ for all $u \in A$.

19
Linear Operators

For a Hilbert space one can distinguish three structures, namely the linear, the geometric and the topological structure. This chapter begins with the study of mappings which are compatible with these structures. In this first chapter on linear operators the topological structure is not taken into account and accordingly the operators studied in this chapter are not considered to be continuous. Certainly, this will be relevant only in the case of infinite dimensional Hilbert spaces, since on a finite dimensional vector space every linear function is continuous.

Mappings which are compatible with the linear structure are called *linear operators*. The topics of the first section are the basic definitions and facts about linear operators. The next section takes the geometrical structure into account insofar as consequences of the existence of an inner product are considered. The following section builds on the results of the second section and develops the basic theory of a special class of operators which play a fundamental role in quantum physics. These studies will be continued in later chapters. Finally the fourth section discusses some first examples from quantum mechanics.

19.1 Basic facts

Recall that any mapping is specified by giving the following data: A domain, a target space, and a rule which tells us how to assign to an element in the domain an element in the target space. When the domain and the target space carry a linear structure, one can consider those mappings which respect these structures. Such

248   19. Linear Operators

mappings are called *linear*. Accordingly one defines linear operators in Hilbert spaces.

Definition 19.1.1 *Let $\mathcal{H}$ and $\mathcal{K}$ be two Hilbert spaces over the field $\mathbb{K}$. A* **linear operator from $\mathcal{H}$ into $\mathcal{K}$** *is a mapping $A : D(A) \to \mathcal{K}$ where $D(A)$ is a linear subspace of $\mathcal{H}$ such that*

$$A(\alpha_1 x_1 + \alpha_2 x_2) = \alpha_1 A(x_1) + \alpha_2 A(x_2) \quad \forall x_i \in D(A), \ \forall \alpha_i \in \mathbb{K}, \ i = 1, 2.$$

The linear subspace $D(A)$ is called the **domain of A**.
  If $\mathcal{K} = \mathcal{H}$, a linear operator A from $\mathcal{H}$ into $\mathcal{K}$ is called a **linear operator in $\mathcal{H}$**.

Following tradition we write Ax instead of $A(x)$ for $x \in D(A)$ for a linear operator A.

In many studies of linear operators A from $\mathcal{H}$ into $\mathcal{K}$, the following two subspaces play a distinguished role: The *kernel or nullspace* $N(A)$ and the *range* ran A of A:

$$N(A) = \{x \in D(A) : Ax = 0\},$$

$$\text{ran } A = \{y \in \mathcal{K} : y = Ax \text{ for some } x \in D(A)\}.$$

It is very easy to show that $N(A)$ is a linear subspace of $D(A)$ and ran A is a linear subspace of $\mathcal{K}$.

Recall that a mapping $f : D(f) \to \mathcal{K}$, $D(f) \subset \mathcal{H}$, is called *injective* if, and only if, $f(x_1) = f(x_2)$, $x_1, x_2 \in D(f)$ implies $x_1 = x_2$. Thus, a linear operator A from $\mathcal{H}$ into $\mathcal{K}$ is injective if, and only if, its nullspace $N(A)$ is trivial, i.e., $N(A) = \{0\}$. Similarly, a linear operator A is *surjective* if, and only if, its range equals the target space, i.e., ran $A = \mathcal{K}$.

Suppose that A is an injective operator from $\mathcal{H}$ into $\mathcal{K}$. Then there is a linear operator B from $\mathcal{K}$ into $\mathcal{H}$ with domain $D(B) = \text{ran } A$ and ran $B = D(A)$ such that $BAx = x$ for all $x \in D(A)$. B is called the **inverse operator of** A and is usually written as A^{-1}.

Let us consider some simple examples of operators in the Hilbert space $\mathcal{H} = L^2([0, 1], dx)$: First we specify several linear subspaces of $\mathcal{H}$:

$$D_0 = \{f : (0, 1) \to \mathbb{C} : f \text{ continuous, supp } f \subset (0, 1)\},$$
$$D_\alpha = \{\psi \in \mathcal{H} : \psi = x^\alpha \phi, \ \phi \in \mathcal{H}\},$$
$$D_\infty = \left\{\psi \in \mathcal{H} : \frac{1}{x}\psi \in \mathcal{H}\right\}.$$

Here α is some number ≥ 1. It is clear that these three sets are actually linear subspaces of $\mathcal{H} = L^2([0, 1], dx)$ and that they are all different:

$$D_0 \subsetneq D_\alpha \subsetneq D_\infty.$$

As the rule which assigns to an element ψ in any of these subspaces an element in $L^2([0, 1], dx)$, we take the multiplication with $\frac{1}{x}$. One checks that indeed $\frac{1}{x}\psi \in$

$L^2([0, 1], dx)$ in all three cases and that this assignment is linear. Thus we get three different linear operators:

$$A_0 : \quad D(A_0) = D_0, \quad D_0 \ni \psi \mapsto \frac{1}{x}\psi,$$

$$A_\alpha : \quad D(A_\alpha) = D_\alpha, \quad D_\alpha \ni \psi \mapsto \frac{1}{x}\psi,$$

$$A_\infty : \quad D(A_\infty) = D_\infty, \quad D_\infty \ni \psi \mapsto \frac{1}{x}\psi.$$

Note that for the multiplication with $\frac{1}{x}$ one cannot have, as a domain, the whole Hilbert space $L^2([0, 1], dx)$, since the function $\psi = 1$ is square-integrable but $\frac{1}{x} \cdot 1$ is not square integrable on the interval $[0, 1]$.

For a situation as in this example an appropriate terminology is introduced in the following definition.

Definition 19.1.2 *Let $\mathcal{H}$ and $\mathcal{K}$ be two Hilbert spaces over the field $\mathbb{K}$ and A, B two linear operators from $\mathcal{H}$ into $\mathcal{K}$. B is called an **extension of A**, in symbols, $A \subseteq B$ if, and only if, $D(A) \subseteq D(B)$ and $Ax = Bx$ for all $x \in D(A)$. Then A is also called a **restriction of B**, namely to the subspace $D(A)$ of $D(B)$: $A = B|D(A)$.*

Using this terminology we have for our example:

$$A_0 \subset A_\alpha \subset A_\infty.$$

In the Exercises further examples of linear operators are discussed. These examples and the examples discussed above show a number of features one has to be aware of:

1. Linear operators from a Hilbert space $\mathcal{H}$ into another Hilbert space $\mathcal{K}$ are not necessarily defined on all of $\mathcal{H}$. The domain as a linear subspace of $\mathcal{H}$ is an essential part of the definition of a linear operator.

2. Even if the assignment $\mathcal{H} \ni \psi \mapsto A\psi$ makes sense mathematically the vector ψ might not be in the domain of A. Consider for example the case $\mathcal{H} = \mathcal{K} = L^2(\mathbb{R})$ and the function $\psi \in L^2(\mathbb{R})$, $\psi(x) = \frac{1}{1+x^2}$ and let A stand for the multiplication with the function $1+x^2$. Then the multiplication of ψ with this function makes good mathematical sense and the result is the function $f = 1$, but ψ is not in the domain of this multiplication operator since $1 \notin L^2(\mathbb{R})$. Thus there are linear operators which are only defined on a proper subspace of the Hilbert space.

3. Whether or not a linear operator A can be defined on all of the Hilbert space $\mathcal{H}$ can be decided by investigating the set

$$\left\{ \frac{\|A\psi\|}{\|\psi\|} : \psi \in D(A), \psi \neq 0 \right\}.$$

If this set is not bounded, the operator A is called an *unbounded linear operator*. These operators are not continuous and in dealing with them special care has to be taken. (See later sections). If the above set is bounded the operator A is called a *bounded linear operator*. They respect the topological structure too, since they are continuous.

The fact that the domain of a linear operator is not necessarily equal to the whole space causes a number of complications. We mention two. Suppose that A, B are two linear operators from the Hilbert space $\mathcal{H}$ into the Hilbert space $\mathcal{K}$. The *addition* of A and B can naturally only be defined on the domain $D(A+B) = D(A) \cap D(B)$ by $(A+B)\psi = A\psi + B\psi$. However even if both domains are dense in $\mathcal{H}$ their intersection might be trivial, i.e., $D(A) \cap D(B) = \{0\}$ and then the resulting definition is not of interest.

Similarly, the natural definition of a *product* or the *composition* of two linear operators can lead to a trivial result. Suppose A, B are two linear operators in the Hilbert space $\mathcal{H}$. Then their *product* $A \cdot B$ is naturally defined as the *composition* on the domain $D(A \cdot B) = \{\psi \in D(B) : B\psi \in D(A)\}$ by $(A \cdot B)\psi = A(B(\psi))$ for all $\psi \in D(A \cdot B)$. But again it can happen that $D(A \cdot B)$ is trivial though $D(A)$ and $D(B)$ are dense in $\mathcal{H}$.

In the next chapter on quadratic forms we will learn how one can improve on some of these difficulties.

We conclude this section with a remark on the importance of the domain of a linear operator in a Hilbert space. As we have seen, a linear operator usually can be defined on many different domains. Which domain is relevant? This depends on the kind of problem in which the linear operator occurs. Large parts of the theory of linear operators in Hilbert spaces have been developed in connection with quantum mechanics where the linear operators are supposed to represent observables of a quantum mechanical system and as such they should be self-adjoint (a property to be addressed later). It turns out that typically linear operators are self-adjoint on precisely one domain. Also the spectrum of a linear operator depends in a very sensitive way on the domain. These statements will become obvious when we have developed the corresponding parts of the theory.

19.2 Adjoints, closed and closable operators

In the complex Hilbert space $\mathcal{H} = \mathbb{C}^n$ with inner product $\langle x, y \rangle = \sum_{i=1}^{n} \overline{x_i} y_i$ for all $x, y \in \mathbb{C}^n$, consider a matrix $A = (a_{ij})_{i,j=1,\ldots,n}$ with complex coefficients. Let us calculate, for $x, y \in \mathbb{C}^n$,

$$\langle x, Ay \rangle = \sum_{i,j=1}^{n} \overline{x_i} a_{ij} y_j = \sum_{i,j=1}^{n} \overline{x_i \overline{a_{ij}}} y_j = \langle A^* x, y \rangle$$

where we define the *adjoint matrix* A^* by $(A^*)_{ji} = \overline{a_{ij}}$, i.e., $A^* = \overline{A}^t$ is the transposed complex conjugate matrix. This shows that for any $n \times n$ matrix A

19.2 Adjoints, closed and closable operators

there exists an adjoint matrix A^* such that for all $x, y \in \mathbb{C}^n$ one has

$$\langle x, Ay \rangle = \langle A^*x, y \rangle.$$

Certainly, in case of a linear operator in an infinite dimensional Hilbert space $\mathcal{H}$ the elementary calculation in terms of components of the vector is not available. Nevertheless we are going to show that, for any densely defined linear operator A in a Hilbert space $\mathcal{H}$, there is a unique *adjoint operator* A^* such that

$$\langle x, Ay \rangle = \langle A^*x, y \rangle \qquad \forall x \in D(A^*), \ \forall y \in D(A) \tag{19.1}$$

holds.

Theorem 19.2.1 (Existence and uniqueness of the adjoint operator) *For every densely defined linear operator A in a Hilbert space $\mathcal{H}$ there is a unique adjoint operator A^* such that relation (19.1) holds. The domain of the adjoint is defined as*

$$D(A^*) = \{x \in \mathcal{H} : \exists\, C_x < \infty, \ |\langle x, Ay \rangle| \leq C_x \|y\| \ \forall y \in D(A)\}. \tag{19.2}$$

The adjoint is maximal among all linear operators B which satisfy $\langle Bx, y \rangle = \langle x, Ay \rangle$ for all $x \in D(B)$ and all $y \in D(A)$.

Proof. In the Exercises it is shown that the set $D(A^*)$ is indeed a linear subspace of $\mathcal{H}$ which contains at least the zero element of $\mathcal{H}$. Take any $x \in D(A^*)$ and define a function $T_x : D(A) \to \mathbb{C}$ by $T_x(y) = \langle x, Ay \rangle$ for all $y \in D(A)$. Linearity of A implies that T_x is a linear function and this linear functional is bounded by $|T_x(y)| \leq C_x \|y\|$, since $x \in D(A^*)$. Thus Theorems 15.3.1 and 15.3.2 apply and we get a unique element $x^* \in \mathcal{H}$ such that $T_x(y) = \langle x^*, y \rangle$ for all $y \in D(A)$. This defines an assignment $D(A^*) \ni x \mapsto x^* \in \mathcal{H}$ which we denote by A^*, i.e., $A^*x = x^*$ for all $x \in D(A^*)$.

By definition of T_x, the mapping $D(A^*) \ni x \mapsto T_x \in \mathcal{H}'$ is antilinear; by our convention the inner product is antilinear in the first argument. We conclude that $A^* : D(A^*) \to \mathcal{H}$ is linear. By construction this linear operator A^* satisfies relation (19.1) and is called the adjoint of the operator A.

Suppose that B is a linear operator in $\mathcal{H}$ which satisfies

$$\langle Bx, y \rangle = \langle x, Ay \rangle \qquad \forall x \in D(B), \ \forall y \in D(A).$$

It follows immediately that $D(B) \subset D(A^*)$. If $x \in D(B)$ is given, take $C_x = \|Bx\|$ for the constant in the definition of $D(A^*)$. Therefore, for $x \in D(B)$ we have $\langle Bx, y \rangle = \langle A^*x, y \rangle$ for all $y \in D(A)$, or $Bx - A^*x \in D(A)^\perp = \{0\}$ since $D(A)$ is dense. We conclude that B is a restriction of the adjoint operator A^*, $B \subseteq A^*$. Therefore the adjoint operator A^* is the 'maximal' operator which satisfies relation (19.1). $\square$

Remark 19.2.1

The assumption in Theorem 19.2.1 that the operator is densely defined is essential for uniqueness of the adjoint. In case this assumption is not satisfied one can still define an adjoint, but in many ways. Some details are discussed in the Exercises.

Sometimes it is more convenient to use the equivalent definition (19.6) in the Exercises for the domain of the adjoint of a densely defined linear operator.

As equations (19.1) and (19.2) clearly show, the adjoint depends in an essential way on the inner product of the Hilbert space. Two different but topologically equivalent inner products (i.e., both inner products define the same topology) give rise to two different adjoints for a densely defined linear operator. Again, some details are discussed in the Exercises.

There is a simple but in many applications quite useful relation between the range of a densely defined linear operator and the null space of its adjoint. The relation reads as follows:

Lemma 19.2.1 *For a densely defined linear operator A in a Hilbert space $\mathcal{H}$ the orthogonal complement of the range of A is equal to the nullspace of the adjoint A^*:*

$$(\operatorname{ran} A)^\perp = N(A^*). \tag{19.3}$$

Proof. The proof is simple. $y \in (\operatorname{ran} A)^\perp$ if, and only if, $0 = \langle y, Ax \rangle$ for all $x \in D(A)$. This identity implies first that $y \in D(A^*)$ and then $0 = \langle A^*y, x \rangle$ for all $x \in D(A)$. Since $D(A)$ is dense, we deduce $y \in (\operatorname{ran} A)^\perp$ if, and only if, $A^*y = 0$, i.e., $y \in N(A^*)$. □

As a first straightforward application we state:

Lemma 19.2.2 *Let A be a densely defined injective linear operator in the Hilbert space $\mathcal{H}$ with dense range $\operatorname{ran} A$. Then the adjoint of A has an inverse which is given by the adjoint of the inverse of A:*

$$(A^*)^{-1} = (A^{-1})^*.$$

Proof. See Exercises! □

In the definition of the adjoint of a densely defined linear operator A the explicit definition (19.2) of the domain plays an important role. Even in concrete examples it is not always straightforward to translate this explicit definition into a concrete description of the domain of the adjoint, and this in turn has the consequence that it is not a simple task to decide when a linear operator is equal to its adjoint, i.e., whether the linear operator is self-adjoint. We discuss a relatively simple example where we can obtain an explicit description of the domain of the adjoint.

Example 19.2.1 *In the Hilbert space $\mathcal{H} = L^2([0, 1])$ consider the linear operator of multiplication with the function $x^{-\alpha}$ for some $\alpha > 1/2$ on the domain $D(A) = \{f \in L^2([0, 1]) : f = \chi_n \cdot g, \ g \in L^2([0, 1]), \ n \in \mathbb{N}\}$ where χ_n is the characteristic function of the subinterval $[\frac{1}{n}, 1]$, i.e., $D(A)$ consists of those elements in $L^2([0, 1])$ which vanish in some neighborhood of zero. It is easy to verify that $\lim_{n \to \infty} \chi_n \cdot g = g$ in $L^2([0, 1])$. Hence A is densely defined and thus has a unique adjoint. If $g \in D(A^*)$, then there is some constant such that for all $n \in \mathbb{N}$ and all $f \in L^2([0, 1])$*

$$|\langle g, A\chi_n \cdot f \rangle_2| \leq C \|\chi_n \cdot f\|_2.$$

Now we use the fact that the multiplication of functions is commutative and obtain

$$\langle g, A\chi_n \cdot f \rangle_2 = \int_0^1 \overline{g(x)} x^{-\alpha} f(x) dx = \langle \chi_n \cdot Ag, f \rangle_2.$$

Since obviously $\|\chi_n \cdot f\|_2 \leq \|f\|_2$ the estimate

$$|\langle \chi_n \cdot Ag, f \rangle_2| \leq C \|f\|_2$$

results for all $n \in \mathbb{N}$ and all $f \in L^2([0,1])$. It follows that $\|\chi_n \cdot Ag\|_2 \leq C$ for all $n \in \mathbb{N}$, i.e.,

$$\int_{\frac{1}{n}}^1 |x^{-\alpha} g(x)|^2 dx \leq C^2 \quad \forall n \in \mathbb{N}$$

and thus in the limit $n \to \infty$ we deduce $x^{-\alpha} g \in L^2([0,1])$ and therefore the explicit characterization of the domain of the adjoint reads

$$D(A^*) = \left\{ g \in L^2([0,1]) : x^{-\alpha} g \in L^2([0,1]) \right\}.$$

In this example the domain of the operator is properly contained in the domain of the adjoint.

Other examples are studied in the Exercises.

As is well known from analysis, the graph of a function often reveals important details. This applies in particular to linear operators and their graphs. Let us recall the definition of the *graph* $\Gamma(A)$ of a linear operator from a Hilbert space $\mathcal{H}$ into a Hilbert space $\mathcal{K}$, with domain $D(A)$:

$$\Gamma(A) = \{(x, y) \in \mathcal{H} \times \mathcal{K} : x \in D(A), \ y = Ax\}. \tag{19.4}$$

Clearly, $\Gamma(A)$ is a linear subspace of $\mathcal{H} \times \mathcal{K}$. It is an important property of the operator A whether the graph is closed or not. Accordingly these operators are singled out in the following definition.

Definition 19.2.1 *A linear operator A from $\mathcal{H}$ into $\mathcal{K}$ is called* **closed** *if, and only if, its graph $\Gamma(A)$ is a closed subspace of $\mathcal{H} \times \mathcal{K}$.*

When one has to use the concept of a closed linear operator the following characterization is very helpful.

Theorem 19.2.2 *Let A be a linear operator from a Hilbert space $\mathcal{H}$ into a Hilbert space $\mathcal{K}$. The following statements are equivalent.*

a) *A is closed.*

b) *For every sequence $(x_n)_{n \in \mathbb{N}} \subset D(A)$ which converges to some $x \in \mathcal{H}$ and for which the sequence of images $(Ax_n)_{n \in \mathbb{N}}$ converges to some $y \in \mathcal{K}$, it follows that $x \in D(A)$ and $y = Ax$.*

c) For every sequence $(x_n)_{n \in \mathbb{N}} \subset D(A)$ which converges weakly to some $x \in \mathcal{H}$ and for which the sequence of images $(Ax_n)_{n \in \mathbb{N}}$ converges weakly to some $y \in \mathcal{K}$, it follows that $x \in D(A)$ and $y = Ax$.

d) Equipped with the inner product,

$$\langle (x, Ax), (y, Ay) \rangle = \langle x, y \rangle_{\mathcal{H}} + \langle Ax, Ay \rangle_{\mathcal{K}} \equiv \langle x, y \rangle_A \qquad (19.5)$$

for $(x, Ax), (y, Ay) \in \Gamma(A)$, the graph $\Gamma(A)$ is a Hilbert space.

Proof. The graph $\Gamma(A)$ is closed if, and only if, every point $(x, y) \in \mathcal{H} \times \mathcal{K}$ in the closure of $\Gamma(A)$ actually belongs to this graph. And a point (x, y) belongs to closure of $\Gamma(A)$ if, and only if, there is a sequence of points $(x_n, Ax_n) \in \Gamma(A)$ which converges to (x, y) in the Hilbert space $\mathcal{H} \times \mathcal{K}$. The hypothesis in statement a) says that we consider a sequence in the graph of A which converges to the point $(x, y) \in \mathcal{H} \times \mathcal{K}$. The conclusion in this statement expresses the fact that this limit point is a point in the graph of A. Hence statements a) and b) are equivalent.

Since a linear subspace of the Hilbert space $\mathcal{H} \times \mathcal{K}$ is closed if, and only if, it is weakly closed, the same reasoning proves the equivalence of statements a) and c).

Finally consider the graph $\Gamma(A)$ as an inner product space with the inner product $\langle \cdot, \cdot \rangle_A$. A Cauchy sequence in this space is a sequence $((x_n, Ax_n))_{n \in \mathbb{N}} \subset \Gamma(A)$ such that for every $\epsilon > 0$ there is an $n_0 \in \mathbb{N}$ such that

$$\|(x_n, Ax_n) - (x_m, Ax_m)\|_A = \sqrt{\|x_n - x_m\|_{\mathcal{H}}^2 + \|Ax_n - Ax_m\|_{\mathcal{K}}^2} \leq \epsilon \quad \forall n, m \geq n_0.$$

Completeness of this inner product space expresses the fact that such a sequence converges to a point $(x, Ax) \in \Gamma(A)$.

According to the above Cauchy condition the sequence $(x_n)_{n \in \mathbb{N}} \subset D(A)$ is a Cauchy sequence in the Hilbert space $\mathcal{H}$ and the sequence $(Ax_n)_{n \in \mathbb{N}}$ is a Cauchy sequence in the Hilbert space $\mathcal{K}$. Thus these sequences converge to a point $x \in \mathcal{H}$, respectively to a point $y \in \mathcal{K}$. Now A is closed if, and only if, this limit point (x, y) belongs to the graph $\Gamma(A)$, i.e., if and only if, this inner product space is complete. □

It is instructive to compare the concept of a closed linear operator with that of a continuous linear operator. One can think of a closed operator A from $\mathcal{H}$ into $\mathcal{K}$ as a 'quasi-continuous' operator in the following sense: If a sequence $(x_n)_{n \in \mathbb{N}} \subset D(A)$ converges in $\mathcal{H}$ and if the sequence of images $(Ax_n)_{n \in \mathbb{N}}$ converges in $\mathcal{K}$, then

$$\lim_{n \to \infty} Ax_n = A(\lim_{n \to \infty} x_n).$$

In contrast continuity of A means: Whenever the sequence $(x_n)_{n \in \mathbb{N}} \subset D(A)$ converges in $\mathcal{H}$, the sequence of images $(Ax_n)_{n \in \mathbb{N}}$ converges in $\mathcal{K}$ and the above relation between both limits is satisfied.

As one would expect, not all linear operators are closed. A simple example of such an operator which is not closed is the operator of multiplication with $x^{-\alpha}$ in $L^2([0, 1])$ discussed earlier in this section. To see that this operator is not closed take the following sequence $(f_n)_{n \in \mathbb{N}} \subset D(A)$ defined by

$$f_n(x) = \begin{cases} x^\alpha, & \frac{1}{n} < x \leq 1, \\ 0, & 0 \leq x \leq \frac{1}{n}. \end{cases}$$

The sequence of images then is

$$g_n(x) = Af_n(x) = \begin{cases} 1, & \frac{1}{n} < x \le 1, \\ 0, & 0 \le x \le \frac{1}{n}. \end{cases}$$

Clearly, both sequences have a limit f, respectively g, in $L^2([0, 1])$, $f(x) = x^\alpha$, $g(x) = 1$, for all $x \in [0, 1]$. Obviously $g = Af$, but $f \notin D(A)$; hence this operator is not closed.

Thus there are linear operators which are not closed. Some of these linear operators might have extensions which are closed. This is addressed in the following definition.

Definition 19.2.2 *A linear operator A from $\mathcal{H}$ into $\mathcal{K}$ is called* **closable** *if, and only if, A has an extension B which is a closed linear operator from $\mathcal{H}$ into $\mathcal{K}$.*

The **closure** *of a linear operator A, denoted by $\overline{A}$, is the smallest closed extension of A, if it exists.*

Naturally, in the definition of closure the natural ordering among linear operators is used. This means: If B is a closed extension of A, then $\overline{A} \subseteq B$.

For densely defined linear operators one has a convenient characterization of those operators which are closable as we learn in the following

Theorem 19.2.3 (Closability of densely defined operators) *Suppose A is a densely defined linear operator in the Hilbert space $\mathcal{H}$. It follows that*

a) *If B is an extension of A, then the adjoint A^* of A is an extension of the adjoint B^* of B: $A \subseteq B \Rightarrow B^* \subseteq A^*$.*

b) *The adjoint A^* of A is closed: $\overline{A^*} = A^*$.*

c) *A is closable if, and only if, its adjoint A^* is densely defined, and in this case the closure of A is equal to the bi-adjoint $A^{**} = (A^*)^*$ of A: $\overline{A} = A^{**}$.*

Proof. Suppose $A \subseteq B$ and $y \in D(B^*)$, i.e., there is a constant C such that $|\langle y, Bx \rangle| \le C\|x\|$ for all $x \in D(B)$. Since $D(A) \subseteq D(B)$ and $Ax = Bx$ for all $x \in D(A)$, we deduce $y \in D(A^*)$ (one can use the same constant C) and $\langle B^*y, x \rangle = \langle y, Bx \rangle = \langle y, Ax \rangle = \langle A^*y, x \rangle$ for all $x \in D(A)$. Hence $B^*y - A^*y \in D(A)^\perp = \{0\}$ since $D(A)$ is dense. Therefore $D(B^*) \subseteq D(A^*)$ and $B^*y = A^*y$ for all $y \in D(B^*)$, i.e., $B^* \subseteq A^*$.

In order to prove part b) take any sequence $((y_n, A^*y_n))_{n \in \mathbb{N}} \subset \Gamma(A^*)$ which converges in $\mathcal{H} \times \mathcal{H}$ to a point (y, z). It follows that, for all $x \in D(A)$,

$$\langle y, Ax \rangle = \lim_{n \to \infty} \langle y_n, Ax \rangle = \lim_{n \to \infty} \langle A^*y_n, x \rangle = \langle z, x \rangle,$$

and we deduce $|\langle y, Ax \rangle| \le \|z\|\|x\|$ for all $x \in D(A)$, hence $y \in D(A^*)$ and $\langle z, x \rangle = \langle y, Ax \rangle = \langle A^*y, x \rangle$ for all $x \in D(A)$ and thus $z = A^*y$, i.e., $(y, z) = (y, A^*y) \in \Gamma(A^*)$. Therefore A^* is closed.

Finally, for the proof of part c), observe that a linear operator A is closable if, and only if, the closure $\overline{\Gamma(A)}$ of its graph $\Gamma(A)$ is the graph of a linear operator. Furthermore, an easy exercise shows that a linear subspace $M \subset \mathcal{H} \times \mathcal{H}$ is the graph of a linear operator in $\mathcal{H}$ if, and only if,

$$(0, y) \in M \Rightarrow y = 0.$$

We know (Corollary 15.1.1): $\overline{\Gamma(A)} = (\Gamma(A)^\perp)^\perp$. Now

$$(x, y) \in \Gamma(A)^\perp \Leftrightarrow 0 = \langle (x, y), (z, Az) \rangle_{\mathcal{H} \times \mathcal{H}} = \langle x, z \rangle + \langle y, Az \rangle \ \forall z \in D(A),$$

i.e., $\Leftrightarrow y \in D(A^*)$, $x = -A^*y$. Similarly,

$$(u, v) \in (\Gamma(A)^\perp)^\perp \Leftrightarrow 0 = \langle (u, v), (x, y) \rangle_{\mathcal{H} \times \mathcal{H}} = \langle u, -A^*y \rangle + \langle v, y \rangle \, \forall y \in D(A^*),$$

i.e., $\Leftrightarrow \langle u, A^*y \rangle = \langle v, y \rangle \, \forall y \in D(A^*)$. This shows: $(0, v) \in \overline{\Gamma(A)} \Leftrightarrow v \in D(A^*)^\perp$. Therefore $\overline{\Gamma(A)}$ is the graph of a linear operator if, and only if, $D(A^*)^\perp = \{0\}$ and thus we conclude by Corollary 15.1.2 that $\overline{\Gamma(A)}$ is the graph of a linear operator if, and only if, $D(A^*)$ is dense.

Now suppose that A^* is densely defined. Then we know that its adjoint $(A^*)^* = A^{**}$ is well defined. The above calculations show

$$(u, v) \in \overline{\Gamma(A)} \Leftrightarrow \langle u, A^*y \rangle = \langle v, y \rangle \qquad \forall y \in D(A^*),$$

i.e., $\Leftrightarrow u \in D(A^{**})$, $v = A^{**}u$, and therefore

$$\overline{\Gamma(A)} = \Gamma(A^{**}).$$

Since the closure $\overline{A}$ is defined through the relation $\Gamma(\overline{A}) = \overline{\Gamma(A)}$ this proves

$$\Gamma(\overline{A}) = \Gamma(A^{**})$$

and the proof is complete. □

19.3 Symmetric and self-adjoint operators

In the previous section, for densely defined linear operators in a Hilbert space the adjoints were defined. In general one can not compare a densely defined linear operator A with its adjoint A^*. However there are important classes of such operators where such a comparison is possible. If we can compare the operators A and A^*, there are two prominent cases to which we direct our attention in this section: 1) The adjoint A^* is equal to A. 2) The adjoint is an extension of A. These two classes of operators are distinguished by proper names according to the

Definition 19.3.1 *A densely defined linear operator A in a Hilbert space $\mathcal{H}$ is called*

   a) **symmetric** *if, and only if, $A \subseteq A^*$;*

   b) **self-adjoint** *if, and only if, $A = A^*$;*

   c) **essentially self-adjoint** *if, and only if, A is symmetric and its closure $\overline{A}$ is self-adjoint.*

In the definition of an essentially self-adjoint operator we obviously rely on the following result.

Corollary 19.3.1 *A symmetric operator A is always closable. Its closure is the bi-adjoint of A: $\overline{A} = A^{**}$ and the closure is symmetric too.*

Proof. For a symmetric operator the adjoint is densely defined so that Theorem 19.2.3 applies. Since the adjoint is always closed, the relation $A \subseteq A^*$ implies $\overline{A} \subseteq A^*$ and hence the closure is symmetric: $\overline{A} = A^{**} \subseteq (\overline{A})^*$. □

Another simple but useful observation about the relation between closure and adjoint is

19.3 Symmetric and self-adjoint operators

Corollary 19.3.2 *Let A be a densely defined linear operator with closure $\overline{A}$. Then the adjoint of the closure is equal to the adjoint:*
$$(\overline{A})^* = A^*.$$

Proof. The simple proof is left as an exercise. □

Thus for symmetric operators we can assume that they are closed. By definition, the closure of an essentially self-adjoint operator is self-adjoint. From the discussion above we deduce for such an operator that
$$A^* = A^{**} = \overline{A}$$
and conversely, if this relation holds for a symmetric operator it is essentially self-adjoint.

Certainly, a self-adjoint operator is essentially self-adjoint ($A^* = A$ implies $A^{**} = A^*$). However in general an essentially self-adjoint operator is not self-adjoint, but such an operator has a unique self-adjoint extension, namely its closure. The proof is easy: Suppose that B is a self-adjoint extension of A. $A \subseteq B$ implies first $B^* \subseteq A^*$ and then $A^{**} \subseteq B^{**}$. Since $B^* = B$ and $A^* = A^{**}$ we get $B = B^* \subseteq A^* = A^{**} \subseteq B^{**} = B$, i.e., $B = A^* = \overline{A}$.

The importance of the concept of an essentially self-adjoint operator is based on the fact that an operator can be essentially self-adjoint on many different domains while it is self-adjoint on precisely one domain. The flexibility in the domain of an essentially self-adjoint operator is used often in the construction of self-adjoint operators, for instance in quantum mechanics. For differential operators such as Schrödinger operators it is not very difficult to find a dense domain D_0 on which this operator is symmetric. If one succeeds in showing that the operator is essentially self-adjoint on D_0, one knows that it has a unique self-adjoint extension, namely its closure. This requires that the domain D_0 is large enough. If one only knows that the operator is symmetric on this domain, the problem of constructing all different self-adjoint extensions arises. Even more flexibility in the initial choice of the domain is assured through the use of the concept of *core* of a closed operator.

Definition 19.3.2 *Suppose that A is a closed linear operator. A subset D of the domain $D(A)$ of A is called a **core of A** if, and only if, the closure of the restriction of A to the linear subspace $\mathrm{lin}\, D$ is equal to A: $\overline{A|\mathrm{lin}\, D} = A$.*

It is important to be aware of the fine differences of the various classes of linear operators we have introduced. The following table gives a useful survey for a densely defined linear operator A.

operator A	properties
symmetric	$A \subseteq \overline{A} = A^{**} \subseteq A^*$
closed and symmetric	$A = \overline{A} = A^{**} \subseteq A^*$
essentially self-adjoint	$A \subseteq \overline{A} = A^{**} = A^*$
self-adjoint	$A = \overline{A} = A^{**} = A^*.$

The direct proof of self-adjointness of a given linear operator is usually impossible. Fortunately there are several quite general criteria available. Below the two basic characterizations of self-adjointness are proven.

Theorem 19.3.1 (Self-adjointness) *For a symmetric operator A in a Hilbert space $\mathcal{H}$, the following statements are equivalent.*

   a) A is self-adjoint: $A^* = A$;

   b) A is closed and $N(A^* \pm iI) = \{0\}$;

   c) $\operatorname{ran}(A \pm iI) = \mathcal{H}$.

Proof. We proceed with the equivalence proof in the following order: a) $\Rightarrow$ b) $\Rightarrow$ c) $\Rightarrow$ a).
Suppose A is self-adjoint. Then A is certainly closed. Consider $\phi_\pm \in N(A^* \pm iI)$. $A^*\phi_\pm = \mp i\phi_\pm$ implies
$$\mp i\langle \phi_\pm, \phi_\pm \rangle = \langle \phi_\pm, A^*\phi_\pm \rangle = \langle A\phi_\pm, \phi_\pm \rangle = \langle A^*\phi_\pm, \phi_\pm \rangle = \pm i\langle \phi_\pm, \phi_\pm \rangle,$$
and thus $\langle \phi_\pm, \phi_\pm \rangle = \|\phi_\pm\|^2 = 0$, i.e., $\phi_\pm = 0$.

Next assume that A is a closed symmetric operator such that $N(A^* \pm iI) = \{0\}$. Relation (19.3) gives $N(A^* + \bar{z}I) = (\operatorname{ran}(A+zI))^\perp$ and therefore $\overline{\operatorname{ran}(A \pm iI)} = (N(A^* \pm iI))^\perp = \mathcal{H}$. Hence for the proof of c) it suffices to show that $\operatorname{ran}(A \pm iI)$ is closed. Suppose $x = \lim_{n\to\infty} x_n$, $x_n = (A \pm iI)y_n$, $y_n \in D(A)$, $n \in \mathbb{N}$. It is straightforward to calculate, for all $n, m \in \mathbb{N}$,
$$\|x_n - x_m\|^2 = \|(A \pm iI)(y_n - y_m)\|^2 = \|Ay_n - Ay_m\|^2 + \|y_n - y_m\|^2.$$

Therefore with $(x_n)_{n\in\mathbb{N}}$ also the two sequences $(y_n)_{n\in\mathbb{N}}$ and $(Ay_n)_{n\in\mathbb{N}}$ are Cauchy sequences in the Hilbert space $\mathcal{H}$ and thus they converge too, to y, respectively z. Since A is closed, $y \in D(A)$ and $z = Ay$, hence $x = (A \pm iI)y \in \operatorname{ran}(A \pm iI)$, and this range is closed. Statement c) follows.

Finally assume c). Since A is symmetric it suffices to show that the domain of the adjoint is contained in the domain of A. Consider any $y \in D(A^*)$, then $(A^* - iI)y \in \mathcal{H}$. Hypothesis c) implies that there is some $\xi \in D(A)$ such that $(A^* - iI)y = (A - iI)\xi = (A^* - iI)\xi$, hence $(A^* - iI)(y - \xi) = 0$ or $y - \xi \in N(A^* - iI) = (\operatorname{ran}(A+iI))^\perp = \{0\}$. This proves $y = \xi \in D(A)$ and finally $D(A^*) = D(A)$, i.e., $A^* = A$. □

The proof of this theorem has also established the following relation between the closure of the range of a symmetric operator and the range of the closure of the operator:
$$\overline{\operatorname{ran}(A \pm iI)} = \operatorname{ran}(\overline{A} \pm iI).$$

Together with Corollary 19.3.2 this observation implies

Corollary 19.3.3 *For a symmetric operator A in a Hilbert space $\mathcal{H}$ the following statements are equivalent:*

   a) A is essentially self-adjoint;

   b) $N(A^* \pm iI) = \{0\}$;

   c) $\overline{\operatorname{ran}(A \pm iI)} = \mathcal{H}$.

In particular one knows for a closed symmetric operator that

$$\operatorname{ran}(A + iI) \qquad \text{and} \qquad \operatorname{ran}(A - iI)$$

are closed linear subspaces of $\mathcal{H}$. Without proof we mention that a closed symmetric operator has self-adjoint extensions if, and only if, the orthogonal complements of these subspaces have the same dimension:

$$\dim (\operatorname{ran}(A+iI))^\perp = \dim (\operatorname{ran}(A-iI))^\perp$$

or

$$\dim N(A^* + iI) = \dim N(A^* - iI).$$

The main difficulty in applying these criteria for self-adjointness is that one usually does not know the explicit form of the adjoint so that it is not obvious at all to check whether $N(A^* \pm iI)$ is trivial. Later, in connection with our study of Schrödinger operators we will learn how in special cases one can master this difficulty.

19.4 Examples

The concepts and the results of the previous three sections are illustrated by several examples which are discussed in some detail.

19.4.1 Operator of multiplication

Suppose that $g : \mathbb{R}^n \to \mathbb{C}$ is a continuous (but not necessarily bounded) function. We want to define the multiplication with g as a linear operator in the Hilbert space $\mathcal{H} = L^2(\mathbb{R}^n)$. To this end the natural domain

$$D_g = \left\{ f \in L^2(\mathbb{R}^n) : g \cdot f \in L^2(\mathbb{R}^n) \right\}$$

is introduced. With this domain we denote the operator of multiplication with g by M_g, $(M_g f)(x) = g(x) f(x)$ for almost all $x \in \mathbb{R}^n$ and all $f \in D_g$. This operator is densely defined since it contains the dense subspace

$$D_0 = \left\{ \chi_r \cdot f : f \in L^2(\mathbb{R}^n), r > 0 \right\} \subset L^2(\mathbb{R}^n).$$

Here χ_r denotes the characteristic function of the closed ball of radius r and centre 0. The reader is asked to prove this statement as an exercise. As a continuous function, g is bounded on the closed ball with radius r, by a constant C_r let us say. Thus the elementary estimate

$$\|g \cdot \chi_r f\|^2 = \int_{|x| \leq r} |g(x)|^2 |f(x)|^2 dx \leq C_r^2 \|f\|_2^2 \qquad \forall f \in L^2(\mathbb{R}^n), \forall r > 0$$

proves $D_0 \subseteq D_g$ and the operator M_g is densely defined. In order to determine the adjoint of M_g, take any $h \in D(M_g^*)$; then $h^* = M_g^* h \in L^2(\mathbb{R}^n)$ and for all $f \in D_g$ one has $\langle h, M_g f \rangle = \langle h^*, f \rangle$, in particular for all $\chi_r f$, $f \in L^2(\mathbb{R}^n)$,

$r > 0$, $\langle h^*, \chi_r f \rangle = \langle h, M_g \chi_r f \rangle$. Naturally, the multiplication with χ_r commutes with the multiplication with g, thus

$$\langle h, M_g \chi_r f \rangle = \langle h, \chi_r M_g f \rangle = \langle \chi_r h, M_g f \rangle = \langle \chi_r M_{\bar{g}} h, f \rangle,$$

or $\langle \chi_r h^*, f \rangle = \langle \chi_r M_{\bar{g}} h, f \rangle$ for all $f \in L^2(\mathbb{R}^n)$ and all $r > 0$. It follows that $\chi_r h^* = \chi_r \bar{g} h$ for all $r > 0$ and therefore

$$\int_{|x| \leq r} |(\bar{g} h)(x)|^2 dx = \int_{|x| \leq r} |h^*(x)|^2 dx \leq \|h^*\|_2^2$$

for all $r > 0$. We deduce $\bar{g} \cdot h \in L^2(\mathbb{R}^n)$ and $h^* = \bar{g} \cdot h = M_{\bar{g}} h$, hence $h \in D_{\bar{g}} = D_g$ and $M_g^* = M_{\bar{g}}$. This shows that the adjoint of the operator of multiplication with the continuous function g is the multiplication with the complex conjugate function $\bar{g}$, on the same domain. Therefore this multiplication operator is always closed. In particular the operator of multiplication with a real valued continuous function is self-adjoint.

Our arguments are valid not only for continuous functions but for all measurable functions g which are bounded on all compact subsets of $\mathbb{R}^n$. In this case the operator of multiplication with g is the prototype of a self-adjoint operator, as we will learn in later chapters.

19.4.2 Momentum operator

As a simple model of the momentum operator in a one dimensional quantum mechanical system we discuss the operator

$$P = i \frac{d}{dx}$$

in the Hilbert space $\mathcal{H} = L^2([0, 1])$. Recall that a function $f \in L^2([0, 1])$ is called absolutely continuous if, and only if, there is a function $g \in L^1([0, 1])$ such that for all $0 \leq x_0 < x \leq 1$ one has $f(x) - f(x_0) = \int_{x_0}^{x} g(y) dy$. It follows that f has a derivative $f' = g$ almost everywhere. Initially we are going to use as a domain for P the subspace

$$D = \left\{ f \in L^2([0, 1]) : f \text{ is absolutely continuous}, \ f' \in L^2([0, 1]) \right\}.$$

This subspace is dense in $L^2([0, 1])$ and clearly P is well defined by

$$(Pf)(x) = if'(x) \quad \text{for almost all } x \in [0, 1], \ \forall f \in D.$$

For arbitrary $f, g \in D$ one has

$$\langle f, Pg \rangle_2 = \int_0^1 \overline{f(x)} i g'(x) dx = i \overline{f(x)} g(x) \Big|_0^1 - i \int_0^1 \overline{f'(x)} g(x) dx$$
$$= i[\overline{f(1)} g(1) - \overline{f(0)} g(0)] + \langle Pf, g \rangle_2.$$

Hence P will be symmetric on all domains D' for which

$$\overline{f(1)}g(1) - \overline{f(0)}g(0) = 0 \qquad \forall\, f, g \in D'$$

holds. These are the subspaces

$$D_\gamma = \left\{ f \in D : f(1) = e^{i\gamma} f(0) \right\}, \qquad \gamma \in \mathbb{R}$$

as one sees easily. In this way we have obtained a one parameter family of symmetric operators

$$P_\gamma = P|_{D_\gamma}, \quad D(P_\gamma) = D_\gamma, \qquad \gamma \in \mathbb{R}.$$

These operators are all extensions of the symmetric operator $P_\infty = P|D_\infty$, $D_\infty = \{f \in D : f(1) = f(0) = 0\}$.

Lemma 19.4.1 *For all $\gamma \in \mathbb{R}$ the symmetric operator P_γ is self-adjoint.*

Proof. For $f \in D(P_\gamma^*)$ we know $f^* = P_\gamma^* f \in L^2([0,1]) \subset L^1([0,1])$, hence $h_c(x) = \int_0^x f^*(y)dy + c$ is absolutely continuous and satisfies $h'_c(x) = f^*(x)$ almost everywhere. Clearly h_c belongs to $L^2([0,1])$, thus $h_c \in D$. Now calculate, for all $g \in D_\gamma$:

$$\langle f, P_\gamma g\rangle_2 = \langle f^*, g\rangle_2 = \int_0^1 \overline{h'_c(x)} g(x)dx = \overline{h_c(x)}g(x)\big|_0^1 - \int_0^1 \overline{h_c(x)}g'(x)dx$$
$$= [\overline{h_c(1)}e^{i\gamma} - \overline{h_c(0)}]g(0) - \langle ih_c, P_\gamma g\rangle_2, \quad \text{or}$$
$$\langle f + ih_c, P_\gamma g\rangle_2 = [\overline{h_c(1)}e^{i\gamma} - \overline{h_c(0)}]g(0) \qquad \forall\, g \in D_\gamma.$$

Observe that the subspace $\left\{ u \in L^2([0,1]) : u = g', g \in D_\gamma, g(0) = 0 \right\}$ is dense in $L^2([0,1])$. This implies $f(x) + ih_c(x) = 0$ almost everywhere and thus $f \in D$ and $if' = h'_c = f^*$. From the above identity we now deduce $[\overline{h_c(1)}e^{i\gamma} - \overline{h_c(0)}]g(0) = 0$ for all $g \in D_\gamma$, and it follows that $h_c(1)e^{-i\gamma} - h_c(0) = 0$, hence $f \in D_\gamma$. Since $f \in D(P_\gamma^*)$ was arbitrary, this shows that $D(P_\gamma^*) = D_\gamma$ and $P_\gamma^* f = P_\gamma f$ for all $f \in D_\gamma$. Hence, for every $\gamma \in \mathbb{R}$, the operator P_γ is self-adjoint. □

We conclude that the operator P_∞ has a one parameter family of self-adjoint extensions P_γ, $\gamma \in \mathbb{R}$. Our argument shows moreover that every self-adjoint extension of P_∞ is of this form.

19.4.3 Free Hamilton operator

In suitable units the Hamilton operator of a free quantum mechanical particle in Euclidean space $\mathbb{R}^3$ is

$$H_0 = -\Delta_3$$

on a suitable domain $D(H_0) \subset L^2(\mathbb{R}^3)$. Recall Plancherel's Theorem 10.3.5. It says that the Fourier transform $\mathcal{F}_2$ is a 'unitary' mapping of the Hilbert space $L^2(\mathbb{R}^3)$. Theorem 10.3.1 implies that for all f in

$$D(H_0) = \left\{ f = \mathcal{F}_2 \tilde{f} \in L^2(\mathbb{R}^3) : p^2 \tilde{f} \in L^2(\mathbb{R}^3) \right\}$$

one has
$$H_0 f = \mathcal{F}_2(M_{p^2} \tilde{f}).$$

Here $\tilde{f}$ denotes the inverse Fourier transform of f. Since we know from our first example that the operator of multiplication with the real valued function $g(p) = p^2$ is self-adjoint on the domain $\{g \in L^2(\mathbb{R}^3) : p^2 g \in L^2(\mathbb{R}^3)\}$, unitarity of $\mathcal{F}_2$ implies that H_0 is self-adjoint on the domain $D(H_0)$ specified above. This will be evident when we have studied unitary operators in some details later (Section 22.2).

19.5 Exercises

1. Let $g : \mathbb{R} \to \mathbb{C}$ be a bounded continuous function. Denote by M_g the multiplication of a function $f : \mathbb{R} \to \mathbb{C}$ with the function g. Show: M_g defines a linear operator in the Hilbert space $\mathcal{H} = L^2(\mathbb{R})$ with domain $D(M_g) = L^2(\mathbb{R})$.

2. Denote by D_n the space of all continuous functions $f : \mathbb{R} \to \mathbb{C}$ for which
$$p_{0,n}(f) = \sup_{x \in \mathbb{R}}(1 + x^2)^{n/2}|f(x)|$$
is finite. Show: $D_{n+1} \subsetneq D_n$ for $n = 0, 1, 2, \ldots$ and for $n = 1, 2, \ldots$ D_n is a dense linear subspace of the Hilbert space $\mathcal{H} = L^2(\mathbb{R})$. Denote by Q multiplication with the variable x, i.e., $(Qf)(x) = xf(x)$ for all $x \in \mathbb{R}$. Show that Q defines a linear operator Q_n in $L^2(\mathbb{R})$ with domain $D(Q_n) = D_n$ and $Q_{n+1} \subseteq Q_n$ for $n = 2, 3, \ldots$.

3. Denote by $\mathcal{C}^k(\mathbb{R})$ the space of all functions $f : \mathbb{R} \to \mathbb{C}$ which have continuous derivatives up to order k. Define $(Pf)(x) = i\frac{df}{dx}(x)$ for all $x \in \mathbb{R}$ and all $f \in \mathcal{C}^k(\mathbb{R})$ for $k \geq 1$. Next define the following subset of $L^2(\mathbb{R})$: $D_k = \left\{ f \in L^2(\mathbb{R}) \cap \mathcal{C}^k(\mathbb{R}) : \frac{df}{dx} \in L^2(\mathbb{R}) \right\}$. Show that D_k is a dense linear subspace of $L^2(\mathbb{R})$. Then show that P defines a linear operator P_k in $L^2(\mathbb{R})$ with domain $D(P_k) = D_k$ and that $P_{k+1} \subseteq P_k$ for $k = 1, 2, \ldots$.

4. Show that the set (19.2) is a linear subspace of the Hilbert space.

5. Prove: The domain $D(A^*)$ of the adjoint of a densely defined linear operator A (see 19.2) is a linear subspace of $\mathcal{H}$ which can also be defined as
$$D(A^*) = \left\{ x \in \mathcal{H} : \sup_{\substack{y \in D(A) \\ \|y\|=1}} |\langle x, Ay \rangle| < \infty \right\}. \tag{19.6}$$

6. Let A be a linear operator in a Hilbert space $\mathcal{H}$ whose domain $D(A)$ is not dense in $\mathcal{H}$. Characterize the nonuniqueness in the definition of an adjoint.

7. Let $\mathcal{H}$ be a Hilbert space with inner product $\langle \cdot, \cdot \rangle$. Suppose that there is another inner product $\langle \cdot, \cdot \rangle_1$ on the vector space $\mathcal{H}$ and there are two positive numbers α, β such that

$$\alpha \langle x, x \rangle \leq \langle x, x \rangle_1 \leq \beta \langle x, x \rangle \qquad \forall x \in \mathcal{H}.$$

Consider the anti-linear canonical isomorphisms J (respectively J_1) between $\mathcal{H}$ and its topological dual $\mathcal{H}'$, defined by $J(x)(y) = \langle x, y \rangle$ (respectively $J_1(x)(y) = \langle x, y \rangle_1$) for all $x, y \in \mathcal{H}$. Prove:

(a) Both inner products define the same topology on $\mathcal{H}$.

(b) Both $(\mathcal{H}, \langle \cdot, \cdot \rangle)$ and $(\mathcal{H}, \langle \cdot, \cdot \rangle_1)$ are Hilbert spaces.

(c) Let A be a linear operator in $\mathcal{H}$ with domain $D(A) = \mathcal{H}$ and $\|Ax\| \leq C\|x\|$ for all $x \in \mathcal{H}$, for some constant $0 \leq C < \infty$. This operator then defines an operator $A' : \mathcal{H}' \to \mathcal{H}'$ by $\ell \mapsto A'\ell$, $A'\ell(x) = \ell(Ax)$ for all $\ell \in \mathcal{H}'$ and all $x \in \mathcal{H}$. Use the maps J and J_1 to relate the adjoints A^* (respectively A_1^*) of A with respect to the inner product $\langle \cdot, \cdot \rangle$ (respectively with respect to $\langle \cdot, \cdot \rangle_1$) to the operator A'.

(d) Show the relation between A^* and A_1^*.

8. Prove Lemma 19.2.2.

9. Prove Corollary 19.3.2.

20
Quadratic Forms

Quadratic forms are a powerful tool for the construction of self-adjoint operators, in particular in situations when the natural strategy fails (for instance for the addition of linear operators). For this reason we give a brief introduction into the theory of quadratic forms. After the basic concepts have been introduced and have been explained by some examples we give the main results of the representation theory of quadratic forms including detailed proofs. The power of these representation theorems is illustrated through several important applications (Friedrichs extensions, form sum of operators).

20.1 Basic concepts. Examples

We begin by collecting the basic concepts of the theory of quadratic forms on a Hilbert space.

Definition 20.1.1 *Let $\mathcal{H}$ be a complex Hilbert space with inner product $\langle \cdot, \cdot \rangle$. A* **quadratic form** *E with domain $D(E) = D$ where D is a linear subspace of $\mathcal{H}$ is a mapping $E : D \times D \to \mathbb{C}$ which is anti-linear in the first and linear in the second argument. A quadratic form E in $\mathcal{H}$ is called*

 a) **symmetric** *if, and only if, $\overline{E(\phi, \psi)} = E(\psi, \phi)$ for all $\phi, \psi \in D(E)$;*

 b) **densely defined** *if, and only if, its domain $D(E)$ is dense in $\mathcal{H}$;*

 c) **semi-bounded** *(from below) if, and only if, there is a $\lambda \in \mathbb{R}$ such that for all $\psi \in D(E)$,*

$$E(\psi, \psi) \geq -\lambda \|\psi\|^2;$$

this number λ is called a **lower bound** of E.

d) **positive** if, and only if, E is semi-bounded with lower bound $\lambda = 0$.

e) **continuous** if, and only if, there is a constant C such that

$$|E(\phi, \psi)| \leq C \|\phi\| \|\psi\| \qquad \forall \phi, \psi \in D(E).$$

Based on these definitions one introduces several other important concepts.

Definition 20.1.2 a) A semi-bounded quadratic form E with lower bound λ is called **closed** if, and only if, the **form domain** $D(E)$ is complete when it is equipped with the **form norm**

$$\|\psi\|_E = \sqrt{E(\psi, \psi) + (\lambda + 1)\|\psi\|^2}.$$

b) A quadratic form F with domain $D(F)$ is called an **extension** of a quadratic form E with domain $D(E)$ if, and only if, $D(E) \subset D(F)$ and $F(\phi, \psi) = E(\phi, \psi)$ for all $\phi, \psi \in D(E)$.

c) A quadratic form is called **closable** if, and only if, it has a closed extension.

d) A subset $D' \subset D(E)$ of the domain of a closed quadratic form E is called a **core** if, and only if, D' is dense in the form domain $D(E)$ equipped with the form norm $\|\cdot\|_E$.

These definitions are illustrated by several not atypical examples.

1. The inner product $\langle \cdot, \cdot \rangle$ of a complex Hilbert space $\mathcal{H}$ is a positive closed quadratic form with domain $\mathcal{H}$.

2. Suppose that A is a linear operator in the complex Hilbert space $\mathcal{H}$ with domain $D(A)$. In a natural way we can associate with A two quadratic forms with form domain $D(A)$:

$$E_1(\phi, \psi) = \langle \phi, A\psi \rangle, \qquad E_2(\phi, \psi) = \langle A\phi, A\psi \rangle.$$

We now relate properties of these quadratic forms to properties of the linear operator A. The form E_2 is always positive and symmetric. The quadratic forms are densely defined if, and only if, the operator A is. If the operator A is symmetric, the quadratic form E_1 is densely defined and symmetric. The form E_1 is semi-bounded if, and only if, the operator A is bounded from below, i.e., if, and only if, there is some $\lambda \in \mathbb{R}$ such that $\langle \psi, A\psi \rangle \geq \lambda \langle \psi, \psi \rangle$ for all $\psi \in D(A)$.

Since the form norm $\|\cdot\|_{E_2}$ is equal to the graph norm of the operator A, the quadratic form E_2 is closed if, and only if, the operator A is closed.

20.1 Basic concepts. Examples   267

It is important to note that even for a closed operator A the quadratic form E_1 is not necessarily closed.

Both quadratic forms E_1 and E_2 are continuous if A is continuous, i.e., if there is some constant C such that $\|A\psi\| \leq C\|\psi\|$ for all $\psi \in D(A)$.

The proof of all these statements is left as an exercise.

3. Suppose $\Omega \subset \mathbb{R}^n$ is an open nonempty set. We know that $D = C_0^\infty(\Omega)$ is a dense subspace in the complex Hilbert space $\mathcal{H} = L^2(\Omega)$. On D a quadratic form E is well defined by

$$E(\phi, \psi) = \int_\Omega \overline{\nabla \phi} \cdot \nabla \psi \, dx = \int_\Omega \sum_{i=1}^n \overline{\frac{\partial \phi}{\partial x_i}}(x) \frac{\partial \psi}{\partial x_i}(x) dx \quad \forall \phi, \psi \in D.$$

This quadratic form is called the *Dirichlet form on* Ω. It is densely defined and positive, but not closed. However the Dirichlet form has a closed extension. The completion of the domain $C_0^\infty(\Omega)$ with respect to the form norm $\|\cdot\|_E$,

$$\|\psi\|_E^2 = \int_\Omega [|\psi(x)|^2 + |\nabla \psi(x)|^2] dx$$

is just the Sobolev space $H_0^1(\Omega)$ which is a Hilbert space with the inner product

$$\langle \phi, \psi \rangle_{1,2} = \langle \phi, \psi \rangle_2 + \sum_{i=1}^n \langle \partial_i \phi, \partial_i \psi \rangle_2.$$

Here we use the abbreviation $\partial_i = \frac{\partial}{\partial x_i}$ and $\langle \cdot, \cdot \rangle_2$ is the inner product of the Hilbert space $L^2(\Omega)$. (Basic facts about completions are given in the Appendix 34.1).

4. As in the previous example we are going to define a quadratic form E with domain $D(E) = C_0^\infty(\Omega)$ in the Hilbert space $\mathcal{H} = L^2(\Omega)$. Suppose we are given real valued functions $A_j \in L_{loc}^2(\Omega)$, $j = 1, \ldots, n$. On $D(E)$ we define

$$E(\phi, \psi) = \int_\Omega \sum_{j=1}^n \overline{(-i\partial_j \phi + A_j \phi)}(-i\partial_j \psi + A_j \psi) dx.$$

(Note that the assumption $A_j \in L_{loc}^2(\Omega)$ ensures $A_j \phi \in L^2(\Omega)$ for all $\phi \in D(E)$.) This quadratic form is densely defined and positive, but not closed. Later we will come back to this example.

5. In the Hilbert space $\mathcal{H} = L^2(\mathbb{R})$ the subspace $D = C_0^\infty(\mathbb{R})$ is dense. On this subspace we define a quadratic form E_δ,

$$E_\delta(f, g) = \overline{f(0)} g(0).$$

It is trivial to see that E_δ is a positive quadratic form. It is also trivial to see that E_δ is not closed. We show now that E_δ does not have any closed extension. To this

end consider a sequence of functions $f_n \in D$ which have the following properties:
i) $0 \leq f_n(x) \leq 1$ for all $x \in \mathbb{R}$; ii) $f_n(0) = 1$; iii) supp $f_n \subseteq [-\frac{1}{n}, \frac{1}{n}]$, for all $n \in \mathbb{N}$. It follows that, as $n \to \infty$,

$$\|f_n\|_2^2 \leq \frac{2}{n} \to 0, \qquad \|f_n\|_{E_\delta}^2 = |f_n(0)|^2 + \|f_n\|_2^2 \leq 1 + \frac{2}{n} \to 1.$$

Property i) implies that $(f_n)_{n \in \mathbb{N}}$ is a Cauchy sequence with respect to the form norm $\|\cdot\|_{E_\delta}$. Suppose F is a closed extension of E_δ. Then we have the contradiction

$$0 = F(0, 0) = \lim_{n \to \infty} F(f_n, f_n) = \lim_{n \to \infty} E_\delta(f_n, f_n) = 1,$$

hence E_δ has no closed extension.

These examples show:

1. There are closed, closable, nonclosable, symmetric, semi-bounded, densely defined, and continuous quadratic forms.

2. Even positive and symmetric quadratic forms are not necessarily closable (see Example 5), in contrast to the situation for symmetric operators.

3. Positive quadratic forms are closable in special cases when they are defined in terms of linear operators.

4. There are positive quadratic forms which can not be defined in terms of a linear operator in the sense of Example 2 (see the Exercises).

20.2 Representation of quadratic forms

We have learned in the previous section that linear operators can be used to define quadratic forms (Example 2) and we have mentioned an example of a densely defined positive quadratic form which cannot be represented by a linear operator in the sense of this example. Naturally the question arises which quadratic forms can be represented in terms of a linear operator. The main result of this section will be that densely defined, semi-bounded, closed quadratic forms can be represented by self-adjoint operators bounded from below, and this correspondence is one-to-one.

We begin with the simplest case.

Theorem 20.2.1 *Let E be a densely defined continuous quadratic form in the Hilbert space $\mathcal{H}$. Then there is a unique continuous linear operator $A : \mathcal{H} \to \mathcal{H}$ such that*

$$E(x, y) = \langle x, Ay \rangle \qquad \forall\, x, y \in D(E).$$

In particular this quadratic form can be extended to the quadratic form $F(x, y) = \langle x, Ay \rangle$ with domain $\mathcal{H}$.

Proof. Since E is supposed to be continuous the estimate $|E(x, y)| \leq C\|x\| \|y\|$ is available for all $x, y \in D(E)$. Since $D(E)$ is dense in $\mathcal{H}$, every $x \in \mathcal{H}$ is the limit of a sequence $(x_n)_{n \in \mathbb{N}} \subset D(E)$. Thus, given $x, y \in \mathcal{H}$, there are sequences $(x_n)_{n \in \mathbb{N}}, (y_n)_{n \in \mathbb{N}} \subset D(E)$ such that $x = \lim_{n \to \infty} x_n$ and $y = \lim_{n \to \infty} y_n$. As convergent sequences in the Hilbert space $\mathcal{H}$ these sequences are bounded, by some M_1, respectively M_2. For all $n, m \in \mathbb{N}$ we estimate the quadratic form as follows:

$$\begin{aligned}|E(x_n, y_n) - E(x_m, y_m)| &= |E(x_n - x_m, y_n) + E(x_m, y_n - y_m)| \\ &\leq |E(x_n - x_m, y_n)| + |E(x_m, y_n - y_m)| \\ &\leq C\|x_n - x_m\| \|y_n\| + C\|x_m\| \|y_n - y_m\| \\ &\leq C M_2 \|x_n - x_m\| + C M_1 \|y_n - y_m\|.\end{aligned}$$

This shows that $(E(x_n, y_n))_{n \in \mathbb{N}}$ is a Cauchy sequence in the field $\mathbb{C}$. We denote its limit by $F(x, y)$. As above one shows that this limit does not depend on the sequences $(x_n)_{n \in \mathbb{N}}$ and $(y_n)_{n \in \mathbb{N}}$ but only on their limits x, respectively y. Thus $F: \mathcal{H} \times \mathcal{H} \to \mathbb{C}$ is well defined.

Basic rules of calculation for limits imply that F too is a quadratic form, i.e., anti-linear in the first and linear in the second argument. Furthermore, F satisfies the same estimate $|F(x, y)| \leq C\|x\| \|y\|$ on $\mathcal{H} \times \mathcal{H}$. Hence, for every $x \in \mathcal{H}$ the mapping $\mathcal{H} \ni y \mapsto F(x, y) \in \mathbb{C}$ is a continuous linear functional. The Riesz–Fréchet theorem (Theorem 15.3.1) implies that there is a unique $x^* \in \mathcal{H}$ such that $F(x, y) = \langle x^*, y \rangle$ for all $y \in \mathcal{H}$. Since F is anti-linear in the first argument as the inner product, the mapping $\mathcal{H} \ni x \mapsto x^*$ is linear and thus defines a linear operator $B: \mathcal{H} \to \mathcal{H}$ by $Bx = x^*$ for all $x \in \mathcal{H}$. This shows that $F(x, y) = \langle Bx, y \rangle$ for all $x, y \in \mathcal{H}$ and thus, defining $A = B^*$, we get

$$F(x, y) = \langle x, Ay \rangle \quad \forall x, y \in \mathcal{H}.$$

Since F is continuous, the bound $|\langle x, Ay \rangle| \leq C\|x\| \|y\|$ is available for all $x, y \in \mathcal{H}$. We deduce easily that $\|Ax\| \leq C\|x\|$ for all $x \in \mathcal{H}$ and hence the operator A is continuous. □

Considerably deeper are the following two results which represent the core of the representation theory for quadratic forms.

Theorem 20.2.2 (First representation theorem) *Let $\mathcal{H}$ be a complex Hilbert space and E a densely defined, closed, and semi-bounded quadratic form in $\mathcal{H}$. Then there is a self-adjoint operator A in $\mathcal{H}$ which is bounded from below and which defines the quadratic form in the following sense:*

a)
$$E(x, y) = \langle x, Ay \rangle \quad \forall x \in D(E), \ \forall y \in D(A) \subset D(E). \tag{20.1}$$

b) *The domain $D(A)$ of the operator is a core of the quadratic form E.*

c) *If for $y \in D(E)$ there exists $y^* \in \mathcal{H}$ such that $E(x, y) = \langle x, y^* \rangle$ for all elements x of a core of E, then it follows: $y \in D(A)$ and $y^* = Ay$, i.e.,*

  i) *the operator A is uniquely determined by Equation 20.1;*

  ii) *if D' is a core of the quadratic form E, then the domain of the operator A is characterized by*

$$D(A) = \left\{ y \in D(E) : \exists C_y \in \mathbb{R}_+, \ |E(x, y)| \leq C_y \|x\| \ \forall x \in D' \right\}.$$

Proof. If $\lambda \geq 0$ is a bound of the quadratic form E, the form norm $\|\cdot\|_E$ of E comes from the inner product

$$\langle x, y \rangle_E = E(x, y) + (\lambda + 1)\langle x, y \rangle \tag{20.2}$$

on $D(E)$. Since E is closed, the form domain $D(E)$ equipped with the inner product $\langle \cdot, \cdot \rangle_E$ is a complex Hilbert space which we call $\mathcal{H}_E$. $E(x, x) + \lambda \|x\|^2 \geq 0$ implies that

$$\|x\| \leq \|x\|_E \qquad \forall x \in D(E). \tag{20.3}$$

Thus, for fixed $x \in \mathcal{H}$, the mapping $\mathcal{H}_E \ni y \mapsto \langle x, y \rangle \in \mathbb{C}$ defines a continuous linear functional on the Hilbert space $\mathcal{H}$. Apply the theorem of Riesz–Fréchet to get a unique $x^* \in \mathcal{H}_E$ such that $\langle x, y \rangle = \langle x^*, y \rangle_E$ for all $y \in \mathcal{H}$. As in the proof of the representation theorem for continuous quadratic forms, it follows that the map $x \mapsto x^*$ defines a linear operator $J : \mathcal{H} \to \mathcal{H}_E$. Hence J is characterized by the identity

$$\langle x, y \rangle = \langle Jx, y \rangle_E \qquad \forall x \in \mathcal{H}, \forall y \in \mathcal{H}_E. \tag{20.4}$$

Since E is densely defined, the domain $D(E)$ is dense in $\mathcal{H}$. Suppose $Jx = 0$, then $\langle x, y \rangle = 0$ for all $y \in D(E)$ and therefore $x = 0$, i.e., the operator J is injective. Per construction we have

$$J\mathcal{H} \subseteq \mathcal{H}_E \subseteq \mathcal{H}.$$

This allows us to calculate, for all $x, y \in \mathcal{H}$, using equation (20.4),

$$\langle x, Jy \rangle = \langle Jx, Jy \rangle_E = \overline{\langle Jy, Jx \rangle_E} = \overline{\langle y, Jx \rangle} = \langle Jx, y \rangle,$$

i.e., the operator J is symmetric. It is also bounded since

$$\|Jy\| \leq \|Jy\|_E = \sup \{|\langle x, Jy \rangle_E| : x \in \mathcal{H}_E, \|x\|_E \leq 1\}$$
$$= \sup \{|\langle x, y \rangle| : x \in \mathcal{H}_E, \|x\|_E \leq 1\} \leq \|y\|.$$

Hence J is a self-adjoint continuous operator with trivial null space $N(J) = \{0\}$, $\operatorname{ran} J \subseteq \mathcal{H}_E$, and $\|J\|' = \sup \{\|Jy\| : y \in \mathcal{H}, \|y\| \leq 1\} \leq 1$. The range of J is dense in $\mathcal{H}$ since its orthogonal complement is trivial: $(\operatorname{ran} J)^\perp = N(J^*) = N(J) = \{0\}$. Here Lemma 19.2.1 is used.

Now we can define a linear operator A as a simple modification of the inverse of J:

$$Ay = J^{-1}y - (\lambda + 1)y \qquad \forall y \in D(A) \equiv \operatorname{ran} J. \tag{20.5}$$

By Lemma 19.2.2 or Theorem 19.3.1 the operator J^{-1} is self-adjoint, hence A is self-adjoint. This operator A indeed represents the quadratic form E as claimed in equation (20.1). To see this take any $x \in D(E)$ and any $y \in D(A) \subset D(E)$ and calculate

$$E(x, y) = \langle x, y \rangle_E - (\lambda + 1)\langle x, y \rangle = \langle x, J^{-1}y \rangle - (\lambda + 1)\langle x, y \rangle = \langle x, Ay \rangle.$$

Since λ is a bound of the quadratic form E, the operator A is bounded from below: For all $y \in D(A)$ the estimate $\langle y, Ay \rangle + \lambda \langle y, y \rangle = E(y, y) + \lambda \langle y, y \rangle \geq 0$ is available.

Next we show that the domain of the operator A is a core of the quadratic form E by showing that $D(A) = \operatorname{ran} J$ is dense in the Hilbert space $\mathcal{H}_E$. Take any $x \in (\operatorname{ran} J)^\perp \subseteq \mathcal{H}_E$. Equation (20.4) implies that

$$\langle x, y \rangle = \langle x, Jy \rangle_E = 0 \qquad \forall y \in \mathcal{H},$$

and thus $x = 0$ and accordingly $D(A)$ is dense in $\mathcal{H}_E$.

Finally we prove Part c). Suppose D' is a core of E and suppose that for some $y \in D(E)$ there is a $y^* \in \mathcal{H}$ such that

$$E(x, y) = \langle x, y^* \rangle \qquad \forall x \in D'.$$

Since $\|x\| \leq \|x\|_E$, both sides of this identity are continuous with respect to the form norm and thus this identity has a unique $\| \cdot \|_E$-continuous extension to all of the form domain $D(E)$. In particular, for all $x \in D(A) \subset D(E)$, we know $E(x, y) = \langle x, y^* \rangle$. But for $x \in D(A)$ and $y \in D(E)$ the representation $E(x, y) = \langle Ax, y \rangle$ holds according to equation (20.1). This shows that

$$\langle Ax, y \rangle = E(x, y) = \langle x, y^* \rangle \qquad \forall x \in D(A),$$

and it follows that $y \in D(A^*)$ and $y^* = A^* y$. But A is self-adjoint. The characterization of the domain $D(A)$ then is obvious from the above considerations. Thus we conclude. □

Theorem 20.2.3 (Second representation theorem) *Under the same assumption as in the first representation theorem, let λ be a bound of E and let A be the self-adjoint operator determined by E according to Theorem 20.2.2. Then it follows:*

a) $A + \lambda I \geq 0$;

b) $D(\sqrt{A + \lambda I}) = D(E)$ *and for all* $x, y \in D(E)$ *the identity*

$$E(x, y) + \lambda \langle x, y \rangle = \langle \sqrt{A + \lambda I} x, \sqrt{A + \lambda I} y \rangle \qquad (20.6)$$

holds;

c) a subset $D' \subset D(E)$ is a core of the quadratic form E if, and only if, it is a core of the operator $\sqrt{A + \lambda I}$.

Proof. The fact that a positive self-adjoint operator B has a unique square root $\sqrt{B}$, which is a self-adjoint operator with domain $D(\sqrt{B}) \supseteq D(B)$ and characteristic identity $(\sqrt{B})^2 = B$, will be shown in the chapter on spectral theory. $D(B)$ is a core of the square root $\sqrt{B}$. Here we simply use these results for an interesting and important extension of the first representation theorem of quadratic forms.

Thus, the positive operator $A + \lambda I$ has a unique self-adjoint square root $\sqrt{A + \lambda I}$ on a domain $D = D(\sqrt{A + \lambda I}) \supseteq D(A)$. As in Example 2 of the previous section, define a quadratic form E' on this domain by $E'(x, y) = \langle \sqrt{A + \lambda I} x, \sqrt{A + \lambda I} y \rangle$. Since $\sqrt{A + \lambda I}$ is self-adjoint we know from Example 2 that E' is a positive, closed, and densely defined quadratic form. On $D(A) \subseteq D(E')$ we can relate this form to the operator A itself and thus to the original quadratic form E:

$$E'(x, y) = \langle \sqrt{A + \lambda I} x, \sqrt{A + \lambda I} y \rangle = \langle x, \sqrt{A + \lambda I} \sqrt{A + \lambda I} y \rangle$$
$$= \langle x, (A + \lambda I) y \rangle = E(x, y) + \langle x, y \rangle.$$

According to the results from spectral theory the domain of A is a core of the operator $\sqrt{A + \lambda I}$. Hence $D(A)$ is a core of the quadratic form E'. According to Part b) of Theorem 20.2.2, $D(A)$ is also a core of the quadratic form E. Hence the quadratic forms E' and $E + \lambda \langle \cdot, \cdot \rangle$ agree. This proves Part b). Part c) follows immediately from Part b). □

20.3 Some applications

Given two densely defined operators we will construct, under certain assumptions about these operators for three important cases, self-adjoint operators using the representation theorems of quadratic forms. The results which we obtain in this way have many applications, in particular in quantum mechanics, but not only there.

Theorem 20.3.1 *Suppose that B is a densely defined closed linear operator in the complex Hilbert space $\mathcal{H}$. Then, on the domain*

$$D(B^*B) = \{x \in D(B) : Bx \in D(B^*)\},$$

the operator B^*B is positive and self-adjoint. The domain $D(B^*B)$ is a core of the operator B.

Proof. On the domain of the operator define a quadratic form $E(x, y) = \langle Bx, By \rangle$ for all $x, y \in D(B)$. One proves (see Example 2 above) that this is a densely defined, positive, and closed quadratic form. So the first representation theorem applies: There is a unique self-adjoint operator A with domain $D(A) \subseteq D(B)$ such that $\langle Bx, By \rangle = \langle x, Ay \rangle$ for all $x \in D(B)$ and all $y \in D(A)$. This implies first that $By \in D(B^*)$ for $y \in D(A)$ and then that B^*B is an extension of A: $A \subseteq B^*B$. Hence B^*B is a densely defined linear operator. Now it follows easily that B^*B is symmetric and thus $A \subseteq B^*B \subset (B^*B)^* \subseteq A^* = A$, i.e., $A = B^*B$. The second part of the first representation finally proves that $D(B^*B) = D(A)$ is a core of the operator B. □

As we had argued earlier a symmetric operator can have, in some cases, many self-adjoint extensions. For positive symmetric operators one can construct a 'smallest' self-adjoint extension, using again the representation results for quadratic forms.

Theorem 20.3.2 (Friedrichs extension) *Let A be a positive (or lower bounded) symmetric linear operator in a complex Hilbert space $\mathcal{H}$. Then A has a positive self-adjoint extension A_F which is the smallest among all positive self-adjoint extensions in the sense that it has the smallest form domain. This extension A_F is called the* **Friedrichs extension** *of A.*

Proof. We give the proof for the case of a positive symmetric operator. The necessary modifications for the case of a lower bounded symmetric operator are obvious (compare the proofs of the representation theorems).

On the domain of the operator define a quadratic form $E(x, y) = \langle x, Ay \rangle$ for all $x, y \in D(E) = D(A)$. E is a densely defined positive quadratic form and $\langle x, y \rangle_E = E(x, y) + \langle x, y \rangle$ defines an inner product on $D(E)$. This inner product space has a completion $\mathcal{H}_E$ which is a Hilbert space and in which the space $D(E)$ is sequentially dense. The quadratic form E has an extension E_1 to this Hilbert space which is defined as $E_1(x, y) = \lim_{n \to \infty} E(x_n, y_n)$ whenever $x = \lim_{n \to \infty} x_n$, $y = \lim_{n \to \infty} y_n$, $x_n, y_n \in D(E)$ for all $n \in \mathbb{N}$. The resulting quadratic form E_1 is a closed densely defined positive quadratic form. It is called the **closure of the quadratic form** E.

The first representation theorem, applied to the quadratic form E_1, gives a unique positive self-adjoint operator A_F such that

$$E_1(x, y) = \langle x, A_F y \rangle \qquad \forall x \in D(E_1), \ \forall y \in D(A_F) \subseteq D(E_1).$$

For $x, y \in D(A)$ one has $\langle x, Ay \rangle = E(x, y) = E_1(x, y)$ and hence A_F is an extension of A.

Finally we prove that A_F is the smallest self-adjoint positive extension of A. Suppose $B \geq 0$ is a self-adjoint extension of A. The associated quadratic form $E_B(x, y) = \langle x, By \rangle$ on $D(B)$ then is an extension of the form E. Hence the closure $\tilde{E}_B$ of the quadratic form E_B is an extension of the closure E_1 of the quadratic form E. The second representation theorem implies: The form domain of $\tilde{E}_B$ is the domain $D(\sqrt{B})$ and the form domain of E_1 is the domain $D(\sqrt{A_F})$, hence $D(\sqrt{A_F}) \subseteq D(\sqrt{B})$ and thus we conclude. □

Note that in this proof we have used the following facts about positive self-adjoint operators B which are of interest on their own. Recall first that the domain $D(B)$ is contained in the domain of the square root $\sqrt{B}$ of B and that $D(B)$ is a core for the operator $\sqrt{B}$. With B we can associate two densely defined positive quadratic forms: $E_1(x, y) = \langle x, By \rangle$ with domain $D(E_1) = D(B)$ and $E_2(x, y) = \langle \sqrt{B}x, \sqrt{B}y \rangle$ with domain $D(E_2) = D(\sqrt{B})$. E_2 is a closed extension of E_1 and actually the closure of E_1 (see Exercises). Thus $D(\sqrt{B})$ is called the *form domain* of the positive self-adjoint operator B.

Our last application of the representation theorems for quadratic forms is concerned with the sum of two positive self-adjoint operators. There are examples of such operators for which the intersection of their domains is trivial and thus the natural way to define their sum gives an uninteresting result. In some cases quadratic forms and their representation can help to define the *form sum* of such operators.

Suppose A and B are two positive self-adjoint operators in the Hilbert space $\mathcal{H}$ such that $D = D(\sqrt{A}) \cap D(\sqrt{B})$ is dense in $\mathcal{H}$. Then a densely defined positive quadratic form E is naturally defined on D by

$$E(x, y) = \langle \sqrt{A}x, \sqrt{A}y \rangle + \langle \sqrt{B}x, \sqrt{B}y \rangle.$$

The closure E_1 of this quadratic form is then a closed positive densely defined quadratic form to which the first representation theorem can be applied. Hence there is a unique positive self-adjoint operator C with domain $D(C) \subset D(E_1)$ such that for all $x \in D(E_1)$ and all $y \in D(C)$ the standard representation $E_1(x, y) = \langle x, Cy \rangle$ holds. This self-adjoint operator C is called the **form sum of A and B**. One writes

$$C = A \dotplus B. \tag{20.7}$$

Typically the construction of the form sum is used in the theory of Schrödinger operators in those cases where the potential V has a too strong local singularity which prevents V to be locally square integrable. A simple case for this construction is considered below.

Let $H_0 = \frac{1}{2m} P^2$ be the free Hamilton operator in the Hilbert space $\mathcal{H} = L^2(\mathbb{R}^n)$. On $D_0 = \mathcal{C}_0^\infty(\mathbb{R}^n)$ the momentum operator P is given by $-i\nabla$. (Some details of the construction of the free Hamilton operator as a self-adjoint operator in $L^2(\mathbb{R}^n)$ are considered in the exercise using Theorem 20.3.1.) Suppose that the potential V is a nonnegative function in $L^1_{loc}(\mathbb{R}^n)$ which does not belong to $L^2_{loc}(\mathbb{R}^n)$. Then $V \cdot \phi$ is not necessarily square-integrable for $\phi \in D_0$ so that we cannot define the interacting Schrödinger operator $H_0 + V$ on D_0 by $(H_0 + V)\phi = H_0\phi + V\phi$. However, the assumption $0 \leq V \in L^1_{loc}(\mathbb{R}^n)$ ensures that the interacting Schrödinger operator can be constructed as a self-adjoint operator as the form sum of the free Schrödinger operator H_0 and the interaction V.

On D_0 a positive quadratic form E_0 is well defined by

$$E_0(\phi, \psi) = \frac{1}{2m} \sum_{j=1}^n \langle \partial_j \phi, \partial_j \psi \rangle_2 + \langle \sqrt{V}\phi, \sqrt{V}\psi \rangle_2.$$

Here, as usual, we use the notation $\partial_j = \frac{\partial}{\partial x_j}$ and $\langle \cdot, \cdot \rangle_2$ is the inner product of the Hilbert space $L^2(\mathbb{R}^n)$. This quadratic form is closable. Applying the first representation theorem to the closure E of this quadratic form E_0 defines the form sum $H_0 \dotplus V$ of H_0 and V as a self-adjoint positive operator. Thus we get for all

$\phi, \psi \in D_0$,

$$\langle \phi, (H_0 \dot{+} V)\psi \rangle_2 = \frac{1}{2m} \sum_{j=1}^{n} \langle \partial_j \phi, \partial \psi \rangle_2 + \langle \sqrt{V}\phi, \sqrt{V}\psi \rangle_2. \tag{20.8}$$

20.4 Exercises

1. Prove: A semi-bounded quadratic form is not necessarily symmetric.

2. Let A be a linear operator in the complex Hilbert space $\mathcal{H}$ and associate to it the quadratic forms E_1 and E_2. Prove all the statements of the second example of the first section about the relation between the operator A and these quadratic forms.

3. Prove: There is no linear operator A in $L^2(\mathbb{R})$ with domain $D(A) \supseteq C_0^\infty(\mathbb{R})$ such that

$$\langle f, Ag \rangle_2 = \overline{f(0)} g(0) \quad \text{or} \quad \langle Af, Ag \rangle_2 = \overline{f(0)} g(0) \quad \forall f, g \in C_0^\infty(\mathbb{R}).$$

4. On the subspace $D_0 = C_0^\infty(\mathbb{R}) \subset L^2(\mathbb{R})$ the momentum operator P_0 is defined by $P_0 \phi = -i \frac{d\phi}{dx}$. Show: P_0 is symmetric. Determine the domain $D(P_0^*)$ of the adjoint of P_0 and the adjoint P_0^* itself. Finally show that P_0^* is self-adjoint.

   Hints: Use the Fourier transform on $L^2(\mathbb{R})$ and recall the example of the free Hamilton operator in the previous chapter.

5. Using the results of the previous problem and Theorem 20.3.1 determine the domain on which the free Hamilton operator $H_0 = \frac{1}{2m} P^2$ is self-adjoint.

6. Give the details of the proof of the fact that a densely defined positive quadratic form E_0 is closable and characterize its closure E, i.e., characterize the elements of the domain of the closure and the values of E at elements of its domain $D(E)$, in terms of certain limits.

7. Find the closure of the quadratic form of Example 4 in the first section of this chapter. Which self-adjoint operator does this closed quadratic form represent?

21
Bounded Linear Operators

Linear operators from a Hilbert space $\mathcal{H}$ into a Hilbert space $\mathcal{K}$ are those mappings $\mathcal{H} \to \mathcal{K}$ which are compatible with the vector space structure on both spaces. Similarly, the bounded or continuous linear operators are those which are compatible with both the vector space and the topological structures on both spaces. The fact that a linear map $\mathcal{H} \to \mathcal{K}$ is continuous if, and only if, it is bounded follows easily from Corollary 2.1.1. (A linear map between topological vector spaces is continuous if, and only if, it is continuous at the origin which in turn is equivalent to the linear map being bounded).

This chapter studies the fundamental properties of single bounded linear operators and of the set of all bounded linear operators $\mathfrak{B}(\mathcal{H})$ on a Hilbert space $\mathcal{H}$. In particular, a product and various important topologies will be introduced in $\mathfrak{B}(\mathcal{H})$. Also examples of bounded operators which are important in quantum physics will be presented.

21.1 Preliminaries

Let $\mathcal{H}$ and $\mathcal{K}$ be two Hilbert spaces over the same field $\mathbb{K}$ and A a linear operator from $\mathcal{H}$ into $\mathcal{K}$. A is called **bounded** if, and only if, the set

$$\{\|Ax\|_{\mathcal{K}} : x \in D(A),\ \|x\|_{\mathcal{H}} \leq 1\} \tag{21.1}$$

is bounded. If A is bounded its **norm** is defined as the least upper bound of this set:

$$\|A\| = \sup \{\|Ax\|_{\mathcal{K}} : x \in D(A),\ \|x\|_{\mathcal{H}} \leq 1\}. \tag{21.2}$$

21. Bounded Linear Operators

We will show later that $A \mapsto \|A\|$ is indeed a norm on the vector space of all bounded linear operators from $\mathcal{H}$ into $\mathcal{K}$.

Linear operators which are not bounded are called **unbounded**.

There are several different ways to express that a linear operator is bounded.

Lemma 21.1.1 *Let A be a linear operator from a Hilbert space $\mathcal{H}$ into a Hilbert space $\mathcal{K}$. The following statements are equivalent:*

a) *A is bounded;*

b) *The set $\{\|Ax\|_{\mathcal{K}} : x \in D(A), \|x\|_{\mathcal{H}} = 1\}$ is bounded and the norm of A is $\|A\| = \sup\{\|Ax\|_{\mathcal{K}} : x \in D(A), \|x\|_{\mathcal{H}} = 1\}$;*

c) *The set $\left\{\frac{\|Ax\|_{\mathcal{K}}}{\|x\|_{\mathcal{H}}} : x \in D(A), x \neq 0\right\}$ is bounded and the norm of A is $\|A\| = \sup\left\{\frac{\|Ax\|_{\mathcal{K}}}{\|x\|_{\mathcal{H}}} : x \in D(A), x \neq 0\right\}$;*

d) *There is a $C \in \mathbb{R}_+$ such that $\|Ax\|_{\mathcal{K}} \leq C\|x\|_{\mathcal{H}}$ for all $x \in D(A)$ and the norm is $\|A\| = \inf\{C \in \mathbb{R}_+ : \|Ax\|_{\mathcal{K}} \leq C\|x\|_{\mathcal{H}} \,\forall\, x \in D(A)\}$.*

Proof. This is a straightforward exercise. □

Corollary 21.1.1 *If A is a bounded linear operator from a Hilbert space $\mathcal{H}$ into a Hilbert space $\mathcal{K}$, then*

$$\|Ax\|_{\mathcal{K}} \leq \|A\|\,\|x\|_{\mathcal{H}} \qquad \forall\, x \in D(A). \tag{21.3}$$

Thus A has always a unique continuous extension to the closure $\overline{D(A)}$ of its domain. In particular, if A is densely defined, this extension is unique on all of $\mathcal{H}$; if $D(A)$ is not dense, then one can extend A on $D(A)^\perp$ for instance by 0. Hence in all cases, a bounded linear operator A can be considered to have the domain $D(A) = \mathcal{H}$.

Proof. For $x \in D(A)$, $x \neq 0$, estimate (21.3) is evident from Part c) of Lemma 21.1.1. For $x = 0$ we have $Ax = 0$ and thus (21.3) is satisfied.

Concerning the extension observe that the closure of a linear subspace is again a linear subspace. If $\overline{D(A)} \ni x = \lim_{n\to\infty} x_n$, $(x_n)_{n\in\mathbb{N}} \subset D(A)$, then estimate (21.3) implies immediately that $(Ax_n)_{n\in\mathbb{N}}$ is a Cauchy sequence in $\mathcal{K}$ and thus has a unique limit which is called $\overline{A}x$, i.e.,

$$\overline{A}(\lim_{n\to\infty} x_n) = \lim_{n\to\infty} Ax_n.$$

Finally one shows that the limit $\lim_{n\to\infty} Ax_n$ does not depend on the approximating sequence $(x_n)_{n\in\mathbb{N}}$, but only on its limit x. □

However there are linear operators in an infinite dimensional Hilbert space which are defined on all of the space but which are not bounded. Thus the converse of the above corollary does not hold.

Proposition 21.1.1 *In infinite dimensional Hilbert spaces $\mathcal{H}$ there are linear operators A with domain $D(A) = \mathcal{H}$ which are not bounded.*

Proof. Since we will not use this result we only give a sketch of the proof. The axiom of choice (or Zorn's Lemma) implies that there exists a maximal set H of linearly independent vectors in $\mathcal{H}$, i.e., a Hamel basis. This means that every $x \in \mathcal{H}$ has a unique representation as a linear combination of elements h_j of the Hamel basis H:

$$x = \sum_{j=1}^{n} a_j h_j \qquad a_j \in \mathbb{K}, \ j = 1, \ldots, n \in \mathbb{N}.$$

Choose a sequence $(h_n)_{n \in \mathbb{N}} \subset H$ and define $Ah_n = nh_n$ for all $n \in \mathbb{N}$ and extend A by linearity to all of $\mathcal{H}$:

$$Ax = \sum_{j=1}^{n} a_j Ah_j.$$

If in the linear combination an element h_j occurs which does not belong to the sequence chosen above define $Ah_j = h_j$ or $= 0$. Then the domain of A is $\mathcal{H}$ and A is not bounded. □

In practice these everywhere defined but unbounded linear operators are not important. Usually one has some more information about the linear operator than just the fact that it is defined everywhere. And indeed if such a linear operator is symmetric, then it follows that it is bounded.

Theorem 21.1.2 (Hellinger–Toeplitz Theorem) *Suppose A is a linear operator in the Hilbert space $\mathcal{H}$ with domain $D(A) = \mathcal{H}$. If A is symmetric, i.e., if $\langle x, Ay \rangle = \langle Ax, y \rangle$ for all $x, y \in \mathcal{H}$, then A is bounded.*

Proof. For the indirect proof assume that A is unbounded. Then there is a sequence $(y_n)_{n \in \mathbb{N}} \subset \mathcal{H}$, $\|y_n\| = 1$ for all $n \in \mathbb{N}$ such that $\|Ay_n\| \to \infty$ as $n \to \infty$. Now define a sequence of linear functionals $T_n : \mathcal{H} \to \mathbb{K}$ by $T_n(x) = \langle y_n, Ax \rangle = \langle Ay_n, x \rangle$ for all $x \in \mathcal{H}$. The second representation of T_n implies by Schwarz' inequality that every functional T_n is continuous. For fixed $x \in \mathcal{H}$ we can use the first representation of T_n to show that the sequence $(T_n(x))_{n \in \mathbb{N}}$ is bounded: $|\langle y_n, Ax \rangle| \leq \|y_n\| \|Ax\| \leq \|Ax\|$ for all $n \in \mathbb{N}$. Thus the uniform boundedness principle (Theorem 18.2.5) implies that there is a $C \in \mathbb{R}_+$ such that $\|T_n\| \leq C$ for all $n \in \mathbb{N}$. But this gives a contradiction to the construction of the y_n: $\|Ay_n\|^2 = T_n(Ay_n) \leq \|T_n\| \|Ay_n\| \leq C \|Ay_n\|$ implies $\|Ay_n\| \leq C$. □

21.2 Examples

In order to gain some insight into the various ways in which a linear operator in a Hilbert space is bounded, respectively unbounded, we study several examples in concrete Hilbert spaces of square integrable functions.

1. **Linear operators of differentiation** such as the momentum operator are unbounded in Hilbert spaces of square integrable functions. Consider for example the momentum operator $P = -i \frac{d}{dx}$ in the Hilbert space $\mathcal{H} = L^2([0,1])$. The functions $e_n(x) = e^{inx}$ obviously have the norm 1, $\|e_n\|_2^2 = \int_0^1 |e^{inx}|^2 dx = 1$ and for Pe_n we find $\|Pe_n\|_2^2 = \int_0^1 |-ie_n'(x)|^2 dx = n^2$, hence $\frac{\|Pe_n\|_2}{\|e_n\|_2} = n$ and the linear operator P is not bounded (on any domain which contains these exponential functions).

2. **Bounded multiplication operators.** Suppose g is an essentially bounded measurable function on $\mathbb{R}^n$. Then the operator of multiplication M_g with g is a bounded

operator in the Hilbert space $L^2(\mathbb{R}^n)$ since in this case, for almost all $x \in \mathbb{R}^n$, $|g(x)| \leq \|g\|_\infty$, and thus

$$\|M_g f\|_2^2 = \int_{\mathbb{R}^n} |g(x)f(x)|^2 dx \leq \int_{\mathbb{R}^n} \|g\|_\infty^2 |f(x)|^2 dx = \|g\|_\infty^2 \|f\|_2^2$$

for all $f \in L^2(\mathbb{R}^n)$.

3. **Unbounded operators of multiplication**. Consider the operator of multiplication with a function which has a sufficiently strong local sigularity, for instance the function $g(x) = x^{-\alpha}$ for $2\alpha > 1$ in the Hilbert space $L^2([0, 1])$. In the exercises we show that this operator is unbounded.

Another way that a multiplication operator M_g in $L^2(\mathbb{R})$ is not bounded is that the function g is not bounded at 'infinity'. A very simple example is $g(x) = x$ for all $x \in \mathbb{R}$ on the domain

$$D_n = \left\{ f \in \mathcal{C}(\mathbb{R}) : p_{0,n}(f) = \sup_{x \in \mathbb{R}} (1 + x^2)^{n/2} |f(x)| < \infty \right\}.$$

Consider the sequence of functions

$$f_j(x) = \begin{cases} 1 & \text{for } x \in [-j, j], \\ 0 & \text{for } x \notin [-j-1, j+1], \\ \text{linear and continuous} & \text{otherwise.} \end{cases}$$

Certainly, for every $j \in \mathbb{N}$, $f_j \in D_n$ ($n \in \mathbb{N}$ fixed). A straightforward calculation shows

$$\|f_j\|_2^2 \leq 2(j+1) \quad \text{and} \quad \|M_g\|_2^2 \geq \frac{2}{3} j^3,$$

hence $\|M_g f_j\|_2 \geq \frac{j}{\sqrt{6}} \|f_j\|_2$. We conclude that this multiplication operator is not bounded.

4. **Integral operators of Hilbert–Schmidt**. Let $k \in L^2(\mathbb{R}^n \times \mathbb{R}^n)$ be given. Then $\|k\|_2^2 = \int_{\mathbb{R}^n} \int_{\mathbb{R}^n} |k(x,y)|^2 dx dy$ is finite and thus, for almost all $x \in \mathbb{R}^n$ the integral $\int_{\mathbb{R}^n} |k(x,y)|^2 dy$ is finite and thus allows us to define a linear map $K : L^2(\mathbb{R}^n) \to L^2(\mathbb{R}^n)$ by

$$(K\psi)(x) = \int_{\mathbb{R}^n} k(x,y)\psi(y) dy \quad \text{for almost all } x \in \mathbb{R}^n.$$

Again for almost all $x \in \mathbb{R}^n$ this image is bounded by

$$|(K\psi)(x)|^2 \leq \int_{\mathbb{R}^n} |k(x,y)|^2 dy \int_{\mathbb{R}^n} |\psi(y)|^2 dy = \int_{\mathbb{R}^n} |k(x,y)|^2 dy \|\psi\|_2^2$$

where Schwarz' inequality is used. We deduce $\|K\psi\|_2 \leq \|k\|_2 \|\psi\|_2$ for all $\psi \in L^2(\mathbb{R}^n)$ and the integral operator K with kernel k is bounded. Such integral operators are called *Hilbert–Schmidt operators*. They played a very important role in the initial stage of the theory of Hilbert spaces. We indicate briefly some basic aspects.

If $\{e_j : j \in \mathbb{N}\}$ is an orthonormal basis of the Hilbert space $L^2(\mathbb{R}^n)$, every $\psi \in L^2(\mathbb{R}^n)$ has a Fourier expansion with respect to this basis: $\psi = \sum_{j=1}^{\infty} \langle e_j, \psi \rangle_2 e_j$, $\sum_{j=1}^{\infty} |\langle e_j, \psi \rangle_2|^2 = \|\psi\|_2^2$. Similarly, $K\psi = \sum_{i=1}^{\infty} \langle e_i, K\psi \rangle_2 e_i$ and $\|K\psi\|_2^2 = \sum_{i=1}^{\infty} |\langle e_i, K\psi \rangle_2|^2$. Continuity of the operator K and of the inner product imply

$$\langle e_i, K\psi \rangle_2 = \langle e_i, K(\sum_{j=1}^{\infty} \langle e_j, \psi \rangle_2 e_j) \rangle_2 = \sum_{j=1}^{\infty} \langle e_i, Ke_j \rangle_2 \langle e_j, \psi \rangle_2.$$

Hence the action of the integral operator K on $\psi \in L^2(\mathbb{R}^n)$ can be represented as the action of the infinite matrix $(K_{ij})_{i,j\in\mathbb{N}}$ on the sequence $\underline{\psi} = (\psi_j)_{j\in\mathbb{N}} \in \ell^2(\mathbb{K})$ of expansion coefficients of ψ, where $K_{ij} = \overline{\langle e_i, Ke_j \rangle_2}$ and $\psi_j = \langle e_j, \psi \rangle_2$. Because of Parseval's relation, since $e_{ij} = e_i \otimes e_j$, $i, j \in \mathbb{N}$, is an orthonormal basis of the Hilbert space $L^2(\mathbb{R}^n \times \mathbb{R}^n)$, the matrix elements are square summable, $\sum_{i,j=1}^{\infty} |K_{ij}|^2 = \|K\|_2^2$.

Now this matrix representation for the integral operator K allows us to rewrite the integral equation as infinite linear system over the space $\ell^2(\mathbb{K})$ of square summable numerical sequences.

Given $f \in L^2(\mathbb{R}^n)$, consider for instance the integral equation

$$u + Ku = f$$

for an unknown function $u \in L^2(\mathbb{R}^n)$. As a linear system over $\ell^2(\mathbb{K})$ this integral equation reads

$$u_i + \sum_{j=1}^{\infty} K_{ij} u_j = f_i, \qquad i = 1, 2, \ldots$$

where naturally $u_i = \langle e_i, u \rangle_2$ and $f_i = \langle e_i, f \rangle_2$ for all $i \in \mathbb{N}$ are square summable sequences.

5. **Spin operators**. In quantum physics the spin as an internal degree of freedom plays a very important role. In mathematical terms it is described by a bounded operator, more precisely by a triple $S = (S_1, S_2, S_3)$ of bounded operators. These operators will be discussed briefly.

We had mentioned before that the state space of an elementary localizable particle with spin $s = \frac{j}{2}$, $j = 0, 1, 2, \ldots$, is the Hilbert space

$$\mathcal{H}_s = L^2(\mathbb{R}^3) \otimes \mathbb{C}^{2s+1} = L^2(\mathbb{R}^3, \mathbb{C}^{2s+1}).$$

The elements of $\mathcal{H}_s$ are $2s + 1$-tuples of complex valued functions f_m, $m = -s, -s+1, \ldots, s-1, s$, in $L^2(\mathbb{R}^3)$. The inner product of $\mathcal{H}_s$ is

$$\langle f, g \rangle = \sum_{m=-s}^{s} \int_{\mathbb{R}^3} \overline{f_m(x)} g_m(x) dx \qquad \forall\, f, g \in \mathcal{H}_s.$$

21. Bounded Linear Operators

The spin operators S_j act on this space according to the following rules.

$$S_1 = \frac{1}{2}(S_+ + S_-), \quad S_2 = -\frac{i}{2}(S_+ + S_-),$$
$$(S_3 f)_m(x) = m f_m(x), \quad m = -s, -s+1, \ldots, s-1, s,$$
$$(S_+ f)_m(x) = \sqrt{(s+m)(s-m+1)} f_{m-1}(x),$$
$$(S_- f)_m(x) = \sqrt{(s-m)(s+m+1)} f_+(x).$$

Clearly these operators are linear and bounded in $\mathcal{H}_s$. In the Exercises we show that they are self-adjoint: $S_j^* = S_j$ for $j = 1, 2, 3$. Introducing the commutator notation $[A, B] = AB - BA$ for two bounded linear operators one finds interesting commutation relations for these spin operators:

$$[S_1, S_2] = i S_3, \quad [S_2, S_3] = i S_1, \quad [S_3, S_1] = i S_2.$$

Furthermore, the operator $S^2 = S_1^2 + S_2^2 + S_3^2 = S_+ S_- - S_3^2 + S_3$ turns out to be proportional to the identity operator $I_{\mathcal{H}_s}$ on $\mathcal{H}_s$:

$$(S^2 f)_m(x) = s(s+1) f_m(x), \quad \text{i.e., } S^2 = s(s+1) I_{\mathcal{H}_s}.$$

Without going into further details we mention that the operators given above are a realization or 'representation' of the commutation relations for the S_j.

6. **Wiener–Hopf operators**. For a given function $g \in L^1(\mathbb{R})$ define a map $K_g : L^2(\mathbb{R}_+) \to L^2(\mathbb{R}_+)$ by

$$(K_g f)(x) = \int_0^\infty g(x-y) f(y) dy \quad \text{for almost all } x \in \mathbb{R}_+, \; \forall f \in L^2(\mathbb{R}_+).$$

It is not quite trivial to show that this operator is indeed a bounded linear operator. It is done in the Exercises. These Wiener–Hopf operators have a wide range of applications. They are used for instance in the analysis of boundary value problems, in filtering problems in information technology and metereology, and time series analysis in statistics.

We conclude this section with a discussion of the famous **Heisenberg commutation relations**

$$[Q, P] \subseteq i I$$

for the position operator Q and momentum operator P in quantum mechanics. The standard realization of these commutation relations in the Hilbert space $L^2(\mathbb{R})$ we had mentioned before: Q is realized as the multiplication operator with the coordinate variable x while the momentum operator then is $P = -i\frac{d}{dx}$, both on suitable domains which have been studied in detail earlier. Recall that both operators are unbounded. It is an elementary calculation to verify these commutation relations for this case, for instance on the dense subspace $C_0^\infty(\mathbb{R})$.

Now we ask the question whether there are other realizations of these commutation relations in terms of bounded operators. A clear answer is given in the following lemma.

Lemma 21.2.1 (Lemma of Wielandt) *There are no bounded linear operators Q and P in a Hilbert space $\mathcal{H}$ which satisfy the commutation relations $[Q, P] = QP - PQ = iI$ where I is the identity operator in $\mathcal{H}$.*

Proof. We are going to derive a contradiction from the assumption that two bounded linear operators satisfy these commutation relations.
Observe first that $P^{n+1}Q - QP^{n+1} = P^n[PQ - QP] + [P^nQ - QP^n]P = -iP + [P^n, Q]P$. A proof of induction with respect to n gives $[P^{n+1}, Q] = -i(n+1)P^n$ and thus

$$(n+1)\|P^n\| = \|[P^{n+1}, Q]\| \leq \|P^{n+1}Q\| + \|QP^{n+1}\|.$$

In the following section one learns that $\|AB\| \leq \|A\|\|B\|$ holds for bounded linear operators A, B. Thus we continue the above estimate

$$(n+1)\|P^n\| \leq \|P^n\|\|PQ\| + \|QP\|\|P^n\| \leq 2\|P^n\|\|Q\|\|P\|.$$

According to the commutation relation we know $\|P\| > 0$. The relation $[P^2, Q] = -iP$ implies $\|P^2\| > 0$ and per induction, $\|P^n\| > 0$ for all $n \in \mathbb{N}$, hence we can divide our estimate by $\|P^n\|$ to get $n + 1 \leq 2\|Q\|\|P\|$ for all $n \in \mathbb{N}$, a contradiction. □

21.3 The space $\mathfrak{L}(\mathcal{H}, \mathcal{K})$ of bounded linear operators

Given two Hilbert spaces $\mathcal{H}$ and $\mathcal{K}$ over the field $\mathbb{K}$, the set of all bounded linear operators $A : \mathcal{H} \to \mathcal{K}$ is denoted by $\mathfrak{L}(\mathcal{H}, \mathcal{K})$. This section studies the basic properties of this set.

First of all, on this set $\mathfrak{L}(\mathcal{H}, \mathcal{K})$ the structure of a $\mathbb{K}$-vector space can naturally be introduced by defining an addition and a scalar multiplication according to the following rules. For $A, B \in \mathfrak{L}(\mathcal{H}, \mathcal{K})$ define a map $A + B : \mathcal{H} \to \mathcal{K}$ by

$$(A + B)x = Ax + Bx \qquad \forall x \in \mathcal{H},$$

i.e., we add two bounded operators by adding, at each point $x \in \mathcal{H}$, the images Ax and Bx. It is straightforward to show that $A + B$, defined in this way, is again a bounded linear operator. The verification is left as an exercise. Similarly, one multiplies a bounded linear operator $A \in \mathfrak{L}(\mathcal{H}, \mathcal{K})$ with a number $\lambda \in \mathbb{K}$ by multiplying, at every point $x \in \mathcal{H}$, the value Ax with λ,

$$(\lambda \cdot A)x = \lambda \cdot (Ax) \qquad \forall x \in \mathcal{H}.$$

In future we will follow the tradition and write this scalar multiplication $\lambda \cdot A$ simply as λA. Since the target space $\mathcal{K}$ is a vector space it is clear that with this addition and scalar multiplication the set $\mathfrak{L}(\mathcal{H}, \mathcal{K})$ becomes a vector space over the field $\mathbb{K}$. The details are filled in as an exercise.

Proposition 21.3.1 *For two Hilbert spaces $\mathcal{H}$ and $\mathcal{K}$ over the field $\mathbb{K}$ the set $\mathfrak{L}(\mathcal{H}, \mathcal{K})$ of all bounded linear operators $A : \mathcal{H} \to \mathcal{K}$ is a vector space over the field $\mathbb{K}$. The function $A \mapsto \|A\|$ defined by*

$$\|A\| = \sup\{\|Ax\|_{\mathcal{K}} : x \in \mathcal{H}, \|x\|_{\mathcal{H}} = 1\}$$

is a norm on $\mathfrak{L}(\mathcal{H}, \mathcal{K})$.

Proof. The first part of the proof has been given above. In order to prove that the function $A \mapsto \|A\|$ actually is a norm on the vector space $\mathfrak{L}(\mathcal{H}, \mathcal{K})$, recall that for any $A, B \in \mathfrak{L}(\mathcal{H}, \mathcal{K})$ and any $x \in \mathcal{H}$ one knows $\|Ax\|_\mathcal{K} \leq \|A\| \|x\|_\mathcal{H}$ and $\|Bx\|_\mathcal{K} \leq \|B\| \|x\|_\mathcal{H}$ and it follows that

$$\|(A+B)x\|_\mathcal{K} = \|Ax + Bx\|_\mathcal{K} \leq \|Ax\|_\mathcal{K} + \|Bx\|_\mathcal{K} \leq \|A\| \|x\|_\mathcal{H} + \|B\| \|x\|_\mathcal{H}.$$

Hence

$$\|A + B\| = \sup\left\{\|(A+B)x\|_\mathcal{K} : x \in \mathcal{H}, \|x\|_\mathcal{H} = 1\right\} \leq \|A\| + \|B\|$$

is immediate. The rule $\|\lambda A\| = |\lambda| \|A\|$ for all $\lambda \in \mathbb{K}$ and all $A \in \mathfrak{L}(\mathcal{H}, \mathcal{K})$ is obvious from the definition. Finally, if $\|A\| = 0$ for $A \in \mathfrak{L}(\mathcal{H}, \mathcal{K})$ then $\|Ax\|_\mathcal{K} = 0$ for all $x \in \mathcal{H}$ and hence $Ax = 0$ for all $x \in \mathcal{H}$, i.e., $A = 0$. We conclude that $\|\cdot\|$ is a norm on $\mathfrak{L}(\mathcal{H}, \mathcal{K})$. □

Proposition 21.3.2 *Let $\mathcal{H}$ and $\mathcal{K}$ be two Hilbert spaces over the field $\mathbb{K}$. Every operator $A \in \mathfrak{L}(\mathcal{H}, \mathcal{K})$ has an adjoint A^* which is a bounded linear operator $\mathcal{K} \to \mathcal{H}$. The map $A \mapsto A^*$ has the following properties:*

a) $A^{**} = A$ for all $A \in \mathfrak{L}(\mathcal{H}, \mathcal{K})$;

b) $(A + B)^* = A^* + B^*$ for all $A, B \in \mathfrak{L}(\mathcal{H}, \mathcal{K})$;

c) $(\lambda A)^* = \overline{\lambda} A^*$ for all $A \in \mathfrak{L}(\mathcal{H}, \mathcal{K})$ and all $\lambda \in \mathbb{K}$;

d) $\|A^*\| = \|A\|$.

Proof. Take any $A \in \mathfrak{L}(\mathcal{H}, \mathcal{K})$. For all $x \in \mathcal{H}$ and all $y \in \mathcal{K}$ the estimate

$$|\langle y, Ax\rangle_\mathcal{K}| \leq \|A\| \|x\|_\mathcal{H} \|y\|_\mathcal{K}$$

holds. Fix $y \in \mathcal{K}$. Then this estimate says that $x \mapsto \langle y, Ax\rangle_\mathcal{K}$ is a continuous linear functional on $\mathcal{H}$, hence by the Theorem of Riesz – Fréchet, there is a unique $y^* \in \mathcal{H}$ such that this functional is of the form $x \mapsto \langle y^*, x\rangle_\mathcal{H}$, i.e.,

$$\langle y, Ax\rangle_\mathcal{K} = \langle y^*, x\rangle_\mathcal{H} \qquad \forall x \in \mathcal{H}.$$

In this way we get a map $y \mapsto y^*$ from $\mathcal{K}$ into $\mathcal{H}$ which is called the adjoint A^* of A, i.e., $A^* y = y^*$. This gives, for all $x \in \mathcal{H}$ and all $y \in \mathcal{K}$ the identity

$$\langle y, Ax\rangle_\mathcal{K} = \langle A^* y, x\rangle_\mathcal{H}.$$

Linearity of A^* is evident from this identity. For the norm of A^* one finds

$$\|A^*\| = \sup\left\{\|A^* y\|_\mathcal{H} : y \in \mathcal{K}, \|y\|_\mathcal{K} = 1\right\}$$
$$= \sup\left\{|\langle A^* y, x\rangle_\mathcal{H}| : x \in \mathcal{H}, \|x\|_\mathcal{H} = 1, y \in \mathcal{K}, \|y\|_\mathcal{K} = 1\right\}$$
$$= \sup\left\{|\langle y, Ax\rangle_\mathcal{K}| : x \in \mathcal{H}, \|x\|_\mathcal{H} = 1, y \in \mathcal{K}, \|y\|_\mathcal{K} = 1\right\}$$
$$= \|A\|.$$

Hence A^* is bounded and Part d) follows.

The bi-adjoint $A^{**} = (A^*)^*$ is defined in the same way as a bounded linear operator $\mathcal{H} \to \mathcal{K}$ through the identity

$$\langle y, A^{**} x\rangle_\mathcal{K} = \langle A^* y, x\rangle_\mathcal{H}$$

for all $y \in \mathcal{K}$ and all $x \in \mathcal{H}$. But by definition of the adjoint A^* both terms are equal to $\langle y, Ax\rangle_\mathcal{K}$. We deduce $A^{**} = A$. Parts b) and c) are easy calculations and are left as an exercise. □

In Proposition 21.3.1 we learned that the space of all bounded linear operators from a Hilbert space $\mathcal{H}$ into a Hilbert space $\mathcal{K}$ is a normed space. This is actually true under considerably weaker assumptions when the Hilbert spaces are replaced by normed spaces X and Y over the same field. In this case a linear map $A : X \to Y$

is bounded if, and only if, there is a $C \in \mathbb{R}_+$ such that $\|Ax\|_Y \leq C\|x\|_X$ for all $x \in X$. Then the norm of A is defined as in the case of Hilbert spaces: $\|A\| = \sup\{\|Ax\|_Y : x \in X, \|x\|_X = 1\}$. Thus we arrive at the normed space $\mathcal{L}(X, Y)$ of bounded linear operators $X \to Y$. If the target space Y is complete, then this space is complete too, a very widely used result. Certainly, this applies also to the case $\mathcal{L}(\mathcal{H}, \mathcal{K})$ of Hilbert spaces.

Theorem 21.3.3 *Let X and Y be normed spaces over the field $\mathbb{K}$. If Y is complete, then the normed space $\mathcal{L}(X, Y)$ is also complete.*

Proof. The proof that $\mathcal{L}(X, Y)$ is a normed space is the same as for the case of Hilbert spaces. Therefore we prove here completeness of this space.

If $(A_n)_{n \in \mathbb{N}} \subset \mathcal{L}(X, Y)$ is a Cauchy sequence, then for every $\epsilon > 0$ there is an $n_0 \in \mathbb{N}$ such that $\|A_n - A_m\| \leq \epsilon$ for all $n, m \geq n_0$. Now take any $x \in X$ and consider the sequence $(A_n x)_{n \in \mathbb{N}} \subset Y$. Since $\|A_n x - A_m x\|_Y = \|(A_n - A_m) x\|_Y \leq \|A_n - A_m\| \|x\|_X$ this sequence is a Cauchy sequence in Y and thus converges to a unique element $y = y(x) \in Y$, $y(x) = \lim_{n \to \infty} A_n x$. The rules of calculation for limits imply that $x \mapsto y(x)$ is a linear function $A : X \to Y$, $Ax = \lim_{n \to \infty} A_n x$. A is bounded too: Since $\|A_n x - A_m x\|_Y \leq \epsilon \|x\|_X$ for all $n, m \geq n_0$ it follows that, by taking the limit $n \to \infty$, $\|Ax - A_m x\|_Y \leq \epsilon \|x\|_X$ and thus for fixed $m \geq n_0$

$$\|Ax\|_Y \leq \|Ax - A_m x\|_Y \|x\|_X + \|A_m x\|_Y \leq (\epsilon + \|A_m\|) \|x\|_X,$$

i.e., A is bounded and the proof is complete. □

Corollary 21.3.1 *Let X be a normed space over the field $\mathbb{K}$. Then the topological dual $X' = \mathcal{L}(X, \mathbb{K})$ is complete.*

Proof. The field $\mathbb{K} = \mathbb{R}, \mathbb{C}$ is complete so that the previous theorem applies. □

21.4 The C*-algebra $\mathfrak{B}(\mathcal{H})$

The case of the Banach space $\mathcal{L}(\mathcal{H}, \mathcal{K})$ of bounded linear operators from a Hilbert space $\mathcal{H}$ into a Hilbert space $\mathcal{K}$ in which $\mathcal{K} = \mathcal{H}$ deserves special attention since there some additional important structure is available, namely one can naturally define a product through the composition. Following the tradition, the Banach space $\mathcal{L}(\mathcal{H}, \mathcal{H})$ is denoted by $\mathfrak{B}(\mathcal{H})$. For A, B the composition $A \circ B : \mathcal{H} \to \mathcal{H}$ is again a bounded linear operator from $\mathcal{H}$ into itself since for all $x \in \mathcal{H}$ we have $\|A \circ Bx\|_{\mathcal{H}} = \|A(Bx)\|_{\mathcal{H}} \leq \|A\| \|Bx\|_{\mathcal{H}} \leq \|A\| \|B\| \|x\|_{\mathcal{H}}$. This composition is used to define a product on $\mathfrak{B}(\mathcal{H})$:

$$A \cdot B = A \circ B \qquad \forall A, B \in \mathfrak{B}(\mathcal{H}).$$

The standard rules of composition of functions and the fact that the functions involved are linear imply that this product satisfies the following relations, for all $A, B, C \in \mathfrak{B}(\mathcal{H})$:

$$(A \cdot B) \cdot C = A \cdot (B \cdot C), \quad (A+B) \cdot C = A \cdot C + B \cdot C, \quad A \cdot (B+C) = A \cdot B + A \cdot C,$$

i.e., this product is *associative* and *distributive* but not commutative. One also has $A \cdot (\lambda B) = \lambda (A \cdot B)$. Equipped with this product the Banach space $\mathfrak{B}(\mathcal{H})$ is a *normed algebra*.

According to Proposition 21.3.2 every $A \in \mathfrak{B}(\mathcal{H})$ has an adjoint $A^* \in \mathfrak{B}(\mathcal{H})$. Products in $\mathfrak{B}(\mathcal{H})$ are transformed according to the following rule which is shown in the Exercises:

$$(A \cdot B)^* = B^* \cdot A^* \qquad \forall A, B \in \mathfrak{B}(\mathcal{H}).$$

As a matter of convenience we omit the '·' for this product and write accordingly $AB \equiv A \cdot B$.

Theorem 21.4.1 *Let $\mathcal{H}$ be a Hilbert space. Then the space $\mathfrak{B}(\mathcal{H})$ of all bounded linear operators $A : \mathcal{H} \to \mathcal{H}$ is a **C*-algebra**, i.e., a complete normed algebra with involution *. For all $A, B \in \mathfrak{B}(\mathcal{H})$ one has*

a) $\|AB\| \leq \|A\| \|B\|$;

b) $\|A^*\| = \|A\|$;

c) $\|AA^*\| = \|A^*A\| = \|A\|^2$;

d) $\|I_\mathcal{H}\| = 1$.

If the dimension of $\mathcal{H}$ is larger than 1, then the algebra $\mathfrak{B}(\mathcal{H})$ is non-Abelian.

Proof. Parts a) and b) have been shown above. Part d) is trivial.
By a) and b) we know $\|AA^*\| \leq \|A\| \|A^*\| = \|A\|^2$. The estimate

$$\|Ax\|^2 = \langle Ax, Ax \rangle = \langle x, A^*Ax \rangle \leq \|x\| \|A^*Ax\| \leq \|A^*A\| \|x\|^2$$

implies $\|A\|^2 \leq \|A^*A\|$ and thus $\|A\|^2 = \|A^*A\|$. Because of b) we can exchange A^* and A and Part c) holds. Multiplication of 2×2-matrices is already not commutative. □

Theorem 21.4.1 states that $\mathfrak{B}(\mathcal{H})$ is a complete normed algebra with involution. In this statement it is the *norm* or *uniform topology* to which we refer. However there are important problems when weaker topologies on $\mathfrak{B}(\mathcal{H})$ are needed. Accordingly we discuss briefly weaker topologies on this space.

In order to put these topologies into perspective we recall the definition of neighborhoods for the norm topology. Neighborhoods of a point $A \in \mathfrak{B}(\mathcal{H})$ for the *norm topology* are all sets which contain a set of the form

$$U_r(A) = \{B \in \mathfrak{B}(\mathcal{H}) : \|B - A\| < r\}$$

for some $r > 0$. A basis of neighborhoods at the point $A \in \mathfrak{B}(\mathcal{H})$ for the *strong topology* on $\mathfrak{B}(\mathcal{H})$ are the sets

$$U_{r, y_1, \ldots, y_n}(A) = \{B \in \mathfrak{B}(\mathcal{H}) : \|(B - A)y_j\|_\mathcal{H} < r, \ j = 1, \ldots, n\}$$

with $r > 0$ and any finite collection of points $y_1, \ldots, y_n \in \mathcal{H}$. Finally a basis of neighborhoods at $A \in \mathfrak{B}(\mathcal{H})$ for the *weak topology* on $\mathfrak{B}(\mathcal{H})$ are the sets

$$U_{r, y_1, \ldots, y_n, x_1, \ldots, x_n}(A) = \{B \in \mathfrak{B}(\mathcal{H}) : |\langle x_j, (B - A)y_j \rangle| < r, \ j = 1, \ldots, n\}$$

for any finite collection of points $x_j, y_j \in \mathcal{H}$, $j = 1, \ldots, n$.

In practice we will not be using the definitions of these topologies in terms of a neighborhood basis but the notions of convergence which these definitions imply. Therefore we state these explicitly.

Definition 21.4.1 *A sequence $(A_n)_{n \in \mathbb{N}} \subset \mathcal{B}(\mathcal{H})$ converges to $A \in \mathcal{B}(\mathcal{H})$ with respect to the*

a) *norm topology if, and only if, $\lim_{n \to \infty} \|A - A_n\| = 0$;*

b) *strong topology if, and only if, $\lim_{n \to \infty} \|Ax - A_n x\|_{\mathcal{H}} = 0$ for every $x \in \mathcal{H}$;*

c) *weak topology if, and only if, $\lim_{n \to \infty} |\langle y, Ax \rangle - \langle y, A_n x \rangle| = 0$ for every pair of points $x, y \in \mathcal{H}$.*

The estimate $\|Ax - A_n x\|_{\mathcal{H}} \leq \|A - A_n\| \|x\|_{\mathcal{H}}$ shows that norm convergence always implies strong convergence and similarly, according to the estimate $|\langle y, Ax \rangle - \langle y, A_n x \rangle| \leq \|y\|_{\mathcal{H}} \|Ax - A_n x\|_{\mathcal{H}}$, strong convergence always implies weak convergence. The converses of these statements do not hold. The norm topology is really stronger than the strong topology which in turn is stronger than the weak topology. The terminology is thus consistent.

Some examples will help to explain the differences between these topologies. On the Hilbert space $\mathcal{H} = \ell^2(\mathbb{C})$ consider the operator S_n which replaces the first n elements of the sequence $\underline{x} = (x_1, \ldots, x_n, x_{n+1}, \ldots)$ by 0,

$$S_n \underline{x} = (0, \ldots, 0, x_{n+1}, x_{n+2}, \ldots).$$

The norm of S_n is easily calculated: $\|S_n\| = 1$ for all $n \in \mathbb{N}$. Thus $(S_n)_{n \in \mathbb{N}}$ does not converge to 0 in norm. But this sequence converges to 0 in the strong topology since for any $\underline{x} \in \ell^2(\mathbb{C})$ we find $\|S_n \underline{x}\|_2^2 = \sum_{j=n+1}^{\infty} |x_j|^2 \to 0$ as $n \to \infty$.

Next define a bounded operator $W_n : \ell^2(\mathbb{C}) \to \ell^2(\mathbb{C})$ by

$$W_n \underline{x} = (0, \ldots, 0, x_1, x_2, \ldots),$$

i.e., W_n shifts $\underline{x} = (x_1, x_2, \ldots)$ by n places to ∞. Clearly $\|W_n \underline{x}\|_2 = \|\underline{x}\|_2$ for all $\underline{x} \in \ell^2(\mathbb{C})$. Now take any $\underline{y} \in \ell^2(\mathbb{C})$ and calculate $\langle \underline{y}, W_n \underline{x} \rangle_2 = \sum_{j=n+1}^{\infty} \overline{y_j} x_{j-n}$, hence $|\langle \underline{y}, W_n \underline{x} \rangle|^2 \leq \sum_{j=n+1}^{\infty} |y_j|^2 \|\underline{x}\|_2^2 \to 0$ as $n \to \infty$. This implies that the sequence $(W_n)_{n \in \mathbb{N}}$ converges to 0 in the weak but not in the strong topology.

Finally we address the question whether these three topologies we have introduced on the C^*-algebra $\mathcal{B}(\mathcal{H})$ are compatible with the algebra operations. The answer is given in

Proposition 21.4.2 *Let $\mathcal{B}(\mathcal{H})$ be the C^*-algebra of bounded linear operators on a Hilbert space $\mathcal{H}$. Then the following holds:*

a) *Addition and scalar multiplication are continuous with respect to the norm, the strong and the weak topology on $\mathcal{B}(\mathcal{H})$;*

b) *the product* $(A, B) \mapsto AB$ *is continuous with respect to the norm topology;*

c) *the involution* $A \mapsto A^*$ *is continuous with respect to the weak topology.*

Continuity with respect to a topology not mentioned in statements a) – c) is in general not given.

Proof. All three topologies we have introduced on $\mathfrak{B}(\mathcal{H})$ are locally convex topologies on a vector space. Thus Part a) is trivial. The estimate $\|AB\| \leq \|A\|\|B\|$ for all $A, B \in \mathfrak{B}(\mathcal{H})$ implies continuity of the product with respect to the norm topology. Suppose a sequence $(A_n)_{n \in \mathbb{N}} \subset \mathfrak{B}(\mathcal{H})$ converges weakly to $A \in \mathfrak{B}(\mathcal{H})$. Then the sequence of adjoints $(A_n^*)_{n \in \mathbb{N}}$ converges to A^* since for every pair $x, y \in \mathcal{H}$ we have, as $n \to \infty$,

$$\langle A_n^* x, y \rangle = \langle x, A_n y \rangle \to \langle x, Ay \rangle = \langle A^* x, y \rangle.$$

Explicit examples in infinite dimensional Hilbert spaces show that the involution $A \mapsto A^*$ is not continuous with respect to the strong and the norm topology and that the multiplication is not continuous with respect to the strong and the weak topology. These counterexamples are done as exercises. □

The fundamental role which C^*-algebras play in local quantum physics is explained in full detail in [Haa96].

21.5 Calculus in the C^*-algebra $\mathfrak{B}(\mathcal{H})$

21.5.1 Preliminaries

On the C^*-algebra $\mathfrak{B}(\mathcal{H})$ one can do calculus since we can add and multiply elements and one can take limits. With these operations one can calculate certain functions $f(A)$ of elements $A \in \mathfrak{B}(\mathcal{H})$. Suppose that f is analytic in the disk $|z| < R$ for some $R > 0$. Then f has a power series expansion $\sum_{n=0}^{\infty} a_n z^n$ which converges for $|z| < R$, i.e., $f(z) = \lim_{N \to \infty} f_N(z)$ where $f_N(z) = \sum_{n=0}^{N} a_n z^n$ is a partial sum. For any $A \in \mathfrak{B}(\mathcal{H})$ the polynomial

$$f_N(A) = \sum_{n=0}^{N} a_n A^n$$

is certainly a well defined element in $\mathfrak{B}(\mathcal{H})$. And so is the limit in the norm topology of $\mathfrak{B}(\mathcal{H})$ if it exists. We claim: For $A \in \mathfrak{B}(\mathcal{H})$, $\|A\| < R$, this sequence of partial sums has a limit in $\mathfrak{B}(\mathcal{H})$. It suffices to show that this sequence is a Cauchy sequence. Since the power series converges, given $\epsilon > 0$ and $\|A\| \leq r < R$ there is $n_0 \in \mathbb{N}$ such that $\sum_{j=n}^{m} |a_j||z|^j < \epsilon$ for all $m > n \geq n_0$ and all $|z| \leq r$. Therefore $\|f_m(A) - f_n(A)\| = \|\sum_{j=n}^{m} a_j A^j\| \leq \sum_{j=n}^{m} |a_j|\|A\|^j < \epsilon$ for all $m > n \geq n_0$, and this sequence is indeed a Cauchy sequence and thus converges to a unique element $f(A) \in \mathfrak{B}(\mathcal{H})$, usually written as

$$f(A) = \sum_{n=0}^{\infty} a_n A^n.$$

21.5 Calculus in the C^*-algebra $\mathfrak{B}(\mathcal{H})$

Let us consider two wellknown examples. The geometric series $\sum_{n=0}^{\infty} z^n$ is known to converge for $|z| < 1$ to the function $(1 - z)^{-1}$. Hence, for every $A \in \mathfrak{B}(\mathcal{H})$, $\|A\| < 1$, we get ($I = I_{\mathcal{H}}$, the identity operator on $\mathcal{H}$)

$$(I - A)^{-1} = \sum_{n=0}^{\infty} A^n. \tag{21.4}$$

The operator series $\sum_{n=0}^{\infty} A^n$ is often called the *Neumann series*. It was first introduced in the study of integral equations to calculate the inverse of $I - A$.

Another important series is the exponential series $\sum_{n=0}^{\infty} \frac{1}{n!} z^n$ which is known to have a radius of convergence $R = \infty$. Hence for every $A \in \mathfrak{B}(\mathcal{H})$

$$e^A = \sum_{n=0}^{\infty} \frac{1}{n!} A^n \tag{21.5}$$

is a well defined element in $\mathfrak{B}(\mathcal{H})$. If $A, B \in \mathfrak{B}(\mathcal{H})$ commute, i.e., $AB = BA$, then one can show, as for complex numbers, $e^{A+B} = e^A e^B$. As a special case consider $U(t) = e^{tA}$ for $t \in \mathbb{C}$ for some fixed $A \in \mathfrak{B}(\mathcal{H})$. One finds

$$U(t + s) = U(t)U(s) \quad \forall t, s \in \mathbb{C}, \qquad U(0) = I.$$

This family of operators $U(t) \in \mathfrak{B}(\mathcal{H})$, $t \in \mathbb{C}$ has interesting applications for the solution of differential equations in $\mathcal{H}$. Take some $x_0 \in \mathcal{H}$ and consider the function $x : \mathbb{C} \to \mathcal{H}$,

$$x(t) = U(t)x_0 = e^{tA}x_0 \qquad t \in \mathbb{C}.$$

We have $x(0) = x_0$ and for $t, s \in \mathbb{C}$

$$x(t) - x(s) = e^{sA}[e^{(t-s)A}x_0 - x_0].$$

In the Exercises one proves, as an identity in $\mathcal{H}$,

$$\lim_{t \to s} \frac{x(t) - x(s)}{t - s} = Ax(s),$$

i.e., the function $x(t)$ is differentiable (actually it is analytic) and satisfies the differential equation

$$x'(t) = Ax(t), \quad t \in \mathbb{C}, \qquad x(0) = x_0.$$

Therefore $x(t) = e^{tA}x_0$ is a solution of the initial value problem $x'(t) = Ax(t)$, $x(0) = x_0$.

Such differential equations are used often for the description of the *time evolution* of physical systems. Compared to the time evolution of systems in classical mechanics the exponential bound $\|x(t)\| \leq e^{|t|\|A\|} \|x_0\|_{\mathcal{H}}$ for all $t \in \mathbb{R}$ corresponds to the case of bounded vector fields governing the time evolution.

21.5.2 Polar decomposition of operators

Recall the polar representation of a complex number $z = e^{i \arg z}|z|$ where the modulus of z is the positive square root of the product of the complex number and its complex conjugate: $|z| = \sqrt{\bar{z}z}$. In this section we will present an analogue for bounded linear operators on a Hilbert space, called the polar decomposition. In a first step the square root of a positive operator is defined using the power series representation of the square root, a result which is of great interest on its own. Thus one can define the modulus $|A|$ of a bounded linear operator A as the positive square root of A^*A. The phase factor in the polar decomposition of complex numbers will be replaced in the case of operators by a partial isometry, i.e., an operator which is isometric on the orthogonal complement of its null space.

It is a wellknown fact (see also the Exercises) that the Taylor expansion at $z = 0$ of the function $z \mapsto \sqrt{1-z}$ converges absolutely for $|z| \leq 1$:

$$\sqrt{1-z} = 1 - \sum_{j=1}^{\infty} a_j z^j \quad \forall\, |z| \leq 1. \tag{21.6}$$

The coefficients a_j of this expansion are all positive and known explicitly.

Similarly to the previous two examples this power series will be used to define the square root of a positive linear operator.

Theorem 21.5.1 (Square root lemma) *Let $A \in \mathcal{B}(\mathcal{H})$ be positive, i.e., $0 \leq \langle x, Ax \rangle$ for all $x \in \mathcal{H}$. Then there is a unique positive operator $B \in \mathcal{B}(\mathcal{H})$ such that $B^2 = A$. This operator B commutes with every bounded linear operator which commutes with A. One calls B the positive square root of A and writes $B = \sqrt{A}$.*

If $\|A\| \leq 1$, then $\sqrt{A}$ has the norm convergent power series expansion

$$\sqrt{A} = \sqrt{I - (I - A)} = I - \sum_{j=1}^{\infty} a_j (I - A)^j \tag{21.7}$$

where the coefficients are those of equation (21.6). The general case is easily reduced to this one.

Proof. For a positive operator A of norm ≤ 1 one has $\|I - A\| = \sup_{\|x\|=1} |\langle x, (I-A)x \rangle| \leq 1$. Hence we know that the series in equation (21.7) converges in norm to some bounded linear operator B. Since the square of the series (21.6) is known to be $1 - z$, the square of the series (21.7) is $I - (I - A) = A$, thus $B^2 = A$.

In order to show positivity of B observe that $0 \leq I - A \leq I$ implies $0 \leq \langle x, (I - A)^n x \rangle \leq 1$ for all $x \in \mathcal{H}$, $\|x\| = 1$. The series (21.7) for B implies that

$$\langle x, Bx \rangle = \langle x, x \rangle - \sum_{j=1}^{\infty} a_j \langle x, (I - A)^j x \rangle \geq 1 - \sum_{j=1}^{\infty} a_j \geq 0$$

where in the last step the estimate $\sum_{j=1}^{\infty} a_j \leq 1$ is used (see Exercises). Therefore $B \geq 0$.

The partial sums of the series (21.7) commute obviously with every bounded operator which commutes with A. Thus the norm limit B does the same.

Suppose $0 \le C \in \mathcal{B}(\mathcal{H})$ satisfies $C^2 = A$. Then $CA = CC^2 = AC$, thus C commutes with A and hence with B. Calculate $(B - C)B(B - C) + (B - C)C(B - C) = (B^2 - C^2)(B - C) = 0$ and note that the two summands are positive operators, hence both of them vanish and so does their difference $(B - C)B(B - C) - (B - C)C(B - C) = (B - C)^3 = 0$. It follows that $\|B - C\|^4 = \|(B - C)^4\| = 0$ since $B - C$ is self-adjoint. We conclude $B - C = 0$. □

Definition 21.5.1 *The function* $|\cdot| : \mathcal{B}(\mathcal{H}) \to \mathcal{B}(\mathcal{H})$ *defined by* $|A| = \sqrt{A^*A}$ *for all* $A \in \mathcal{B}(\mathcal{H})$ *is called the* **modulus**. *Its values are positive bounded operators.*

Theorem 21.5.2 (Polar decomposition) *For every bounded linear operator A on the Hilbert space $\mathcal{H}$ the polar decomposition*

$$A = U|A| \tag{21.8}$$

holds. Here $|A|$ is the modulus of A and U is a partial isometry with null space $N(U) = N(A)$. U is uniquely determined by this condition and its range is $\overline{\operatorname{ran} A}$.

Proof. The definition of the modulus implies for all $x \in \mathcal{H}$

$$\||A|x\|^2 = \langle x, |A|^2 x \rangle = \langle x, A^*Ax \rangle = \|Ax\|^2,$$

hence $N(A) = N(|A|) = (\operatorname{ran}|A|)^\perp$, and we have the orthogonal decomposition $\mathcal{H} = N(|A|) \oplus \overline{\operatorname{ran}|A|}$ of the Hilbert space. Now define a map $U : \mathcal{H} \to \mathcal{H}$ with $N(U) = N(|A|)$ by continuous extension of $U(|A|x) = Ax$ for all $x \in \mathcal{H}$. Because of the identity given above U is a well defined linear operator which is isometric on $\overline{\operatorname{ran}|A|}$. Its range is $\overline{\operatorname{ran} A}$. On the basis of equation (21.8) and the condition $N(U) = N(A)$ the proof of uniqueness is straightforward. □

21.6 Exercises

1. Prove Lemma 21.1.1.

2. Prove that the operator of multiplication with the function $g(x) = x^{-\alpha}$, $2\alpha > 1$, is unbounded in the Hilbert space $L^2([0, 1])$.

   Hints: Consider the functions

   $$f_n(x) = \begin{cases} 1 & \frac{1}{n} \le x \le 1, \\ 0 & 0 \le x < \frac{1}{n}. \end{cases}$$

   For these functions one can calculate the relevant norms easily.

3. Prove all the statements about the spin operators in the section on examples of bounded linear operators.

4. Prove that the Wiener–Hopf operators are well-defined bounded linear operators in $L^2(\mathbb{R}_+)$.

   Hints: Consider the space $L^2(\mathbb{R}_+)$ as a subspace of $L^2(\mathbb{R})$ and use the results on the relations between multiplication and convolution under Fourier transformation given in Part A, Chapter 10.

5. Prove parts b) and c) of Proposition 21.3.2.

6. For $A, B \in \mathfrak{B}(\mathcal{H})$ prove: $(A + B)^* = A^* + B^*$ and $(A \cdot B)^* = B^* \cdot A^*$.

7. For $A \in \mathfrak{B}(\mathcal{H})$ and $x_0 \in \mathcal{H}$, define $x(t) = e^{tA}x_0$ for $t \in \mathbb{C}$ and show that this function $\mathbb{C} \to \mathcal{H}$ is differentiable on $\mathbb{C}$. Calculate its (complex) derivative.

8. In the Hilbert space $\mathcal{H} = \ell^2(\mathbb{C})$, denote by e_j $j \in \mathbb{N}$ the standard basis vectors (the sequence e_j has a 1 at position j, otherwise all elements are 0). Then every $\underline{x} \in \ell^2(\mathbb{C})$ has the Fourier expansion $\underline{x} = \sum_{j=1}^{\infty} x_j e_j$ with $(x_j)_{j \in \mathbb{N}}$ a square summable sequence of numbers. Define a bounded linear operator $A \in \mathfrak{B}(\mathcal{H})$ by

$$A \sum_{j=1}^{\infty} x_j e_j = \sum_{j=2}^{\infty} x_j e_{j-1}$$

and show:

a) $A^* \sum_{j=1}^{\infty} x_j e_j = \sum_{j=1}^{\infty} x_j e_{j+1}$;

b) The sequence $A_n = A^n$ converges to 0 in the strong topology;

c) $A_n^* = (A^*)^n$ does not converge strongly to 0;

d) $A_n A_n^* = I$ for all $n \in \mathbb{N}$;

e) deduce that the product is continuous neither with respect to the strong nor with respect to the weak topology.

9. Though in general the involution is not strongly continuous on $\mathfrak{B}(\mathcal{H})$ it is strongly continuous on a linear subspace $\mathcal{N}$ of *normal operators* in $\mathfrak{B}(\mathcal{H})$, i.e., bounded operators with the property

$$A^*A = AA^*.$$

Prove: If $(A_n)_{n \in \mathbb{N}} \subset \mathcal{N}$ converges strongly to $A \in \mathcal{N}$ then the sequence of adjoints $(A_n^*)_{n \in \mathbb{N}}$ converges strongly to A^*.

Hints: Show first that $\|(A^* - A_n^*)x\|_{\mathcal{H}}^2 = \|(A - A_n)x\|_{\mathcal{H}}^2$ for $x \in \mathcal{H}$.

10. Show: The algebra $Q = M_2(\mathbb{C})$ of complex 2×2 matrices is not a C^*-algebra when it is equipped with the norm

$$\|A\| = \sqrt{\operatorname{Tr}(AA^*)} = \sqrt{\sum_{k,j=1}^{2} |A_{kj}|^2}.$$

Hints: Take the matrix $A = \begin{pmatrix} 1 & i \\ 0 & 1 \end{pmatrix}$ and calculate $\|A^*\|^2$ and $\|AA^*\|^2$.

11. Determine the Taylor series of the function $f(z) = \sqrt{1-z}$ at $z = 0$. It is of the form
$$\sqrt{1-z} = 1 - \sum_{j=1}^{\infty} a_j z^j \qquad |z| < 1$$
where the coefficients a_j are positive (they are known explicitly). Prove: $\sum_{j=1}^{\infty} a_j \leq 1$. Deduce that the above power series for $\sqrt{1-z}$ converges for $|z| \leq 1$.

Hints: For any $N \in \mathbb{N}$ write $\sum_{j=1}^{N} a_j = \lim_{x \to 1} \sum_{j=1}^{N} a_j x^j$ with $0 < x < 1$. Since the coefficients are positive one has $\sum_{j=1}^{N} a_j x^j < \sum_{j=1}^{\infty} a_j x^j = 1 - \sqrt{1-x}$.

22
Special Classes of Bounded Operators

22.1 Projection operators

Let e be a unit vector in a Hilbert space $\mathcal{H}$ over the field $\mathbb{K}$ with inner product $\langle \cdot, \cdot \rangle$. Define $P_e : \mathcal{H} \to \mathcal{H}$ by $P_e x = \langle e, x \rangle e$ for all $x \in \mathcal{H}$. Evidently, P_e is a bounded linear operator with null space $N(P_e) = \{e\}^\perp$ and range ran $P_e = \mathbb{K}e$. In addition P_e satisfies $P_e^* = P_e$ and $P_e^2 = P_e$ which is also elementary to prove. The operator P_e is the simplest example of the class of projection operators or projectors to be studied in this section.

Definition 22.1.1 *A bounded linear operator P on a Hilbert space $\mathcal{H}$ which is symmetric, $P^* = P$, and idempotent, $P^2 = P$, is called a* **projector** *or* **projection operator**.

The set of all projection operators on a Hilbert space $\mathcal{H}$ is denoted by $\mathfrak{P}(\mathcal{H})$, i.e.,

$$\mathfrak{P}(\mathcal{H}) = \left\{ P \in \mathfrak{B}(\mathcal{H}) : P^* = P = P^2 \right\}.$$

With the help of the following proposition one can easily construct many examples of projectors explicitly.

Proposition 22.1.1 *Let $\mathcal{H}$ be a Hilbert space of the field $\mathbb{K}$. Projectors on $\mathcal{H}$ have the following properties:*

a) *For every $P \in \mathfrak{P}(\mathcal{H})$, $P \neq 0$, $\|P\| = 1$;*

b) *a bounded operator $P \in \mathfrak{B}(\mathcal{H})$ is a projector if, and only if, $P^\perp = I - P$ is a projector;*

c) if $P \in \mathfrak{P}(\mathcal{H})$, then

$$\mathcal{H} = \operatorname{ran} P \oplus \operatorname{ran} P^\perp, \quad P|_{\operatorname{ran} P} = I_{\operatorname{ran} P}, \quad P|_{\operatorname{ran} P^\perp} = 0;$$

d) there is a one-to-one correspondence between projection operators P on $\mathcal{H}$ and closed linear subspaces M of $\mathcal{H}$, i.e., the range $\operatorname{ran} P$ of a projector P is a closed linear subspace of $\mathcal{H}$, and conversely to every closed linear subspace $M \subset \mathcal{H}$ there is exactly one $P \in \mathfrak{P}(\mathcal{H})$ such that the range of this projector is M;

e) Suppose that $\{e_n : n = 1, \ldots, N\}$, $N \in \mathbb{N}$ or $N = \infty$ is an orthonormal system in $\mathcal{H}$, then the projection operator onto the closed linear subspace $M = [\{e_n : n = 1, \ldots, N\}]$ generated by the orthonormal system is

$$P_N x = \sum_{n=1}^{N} \langle e_n, x \rangle e_n \quad \forall x \in \mathcal{H}. \tag{22.1}$$

Proof. By definition any projector satisfies $P = P^*P$, thus by Theorem 21.4.1 $\|P\| = \|P^*P\| = \|P\|^2$, and therefore $\|P\| \in \{0, 1\}$ and Part a) follows.

To prove b) we show that the operator $I - P$ satisfies the defining relations of a projector ($Q = Q^* = Q^2$) if, and only if, P does: $I - P = (I - P)^* = I - P^* \Leftrightarrow P = P^*$ and $I - P = (I - P)^2 = I - P - P + P^2 \Leftrightarrow -P + P^2 = 0$.

For the proof of Part c) observe that the relation $I = P + P^\perp$ implies immediately that every $x \in \mathcal{H}$ is the sum of an element in the range of P and an element in the range of $P^\perp$. In Part d) we prove that the range of a projector is a closed linear subspace. Thus $\operatorname{ran} P \oplus \operatorname{ran} P^\perp$ gives indeed a decomposition of $\mathcal{H}$ into closed orthogonal subspaces. The image of $Px \in \operatorname{ran} P$ under P is $PPx = Px$, since $P^2 = P$ and similarly, the image of $P^\perp x \in \operatorname{ran} P^\perp$ under P is $PP^\perp x = P(I - P)x = 0$, thus the second and third statement in Part c) follow.

Let P be a projector on $\mathcal{H}$ and y an element in the closure of the range of P, i.e., there is a sequence $(x_n)_{n \in \mathbb{N}} \subset \mathcal{H}$ such that $y = \lim_{n \to \infty} Px_n$. Since a projector is continuous we deduce $Py = \lim_{n \to \infty} PPx_n = \lim_{n \to \infty} Px_n = y$, thus $y \in \operatorname{ran} P$ and the range of a projector is a closed linear subspace.

Now given a closed linear subspace $M \subset \mathcal{H}$, we apply to each $x \in \mathcal{H}$ the Projection Theorem 15.1.1 to get a unique decomposition of x into $u_x \in M$ and $v_x \in M^\perp$, $x = u_x + v_x$. The uniqueness condition allows us to conclude that the mapping $x \mapsto u_x$ is linear. Since $\|x\|^2 = \|u_x\|^2 + \|v_x\|^2 \geq \|u_x\|^2$ the linear map $P_M : \mathcal{H} \to M$ defined by $P_M x = u_x$ is bounded. Next apply the projection theorem to $x, y \in \mathcal{H}$ to get

$$\langle x, P_M y \rangle = \langle u_x + v_x, u_y \rangle = \langle u_x, u_y \rangle = \langle P_M x, P_M y \rangle = \langle P_M x, y \rangle,$$

hence $P_M = P_M^* = P_M^* P_M = P_M^2$ and thus P_M is a projector. Per construction its range is the given closed subspace M. This proves Part d).

The proof of Part e) is done explicitly for the case $N = \infty$. Then the closed linear hull M of the linear subspace generated by the given orthonormal system is described in Corollary 16.1.1 as

$$M = \left\{ x \in \mathcal{H} : x = \sum_{n=1}^{\infty} c_n e_n, \ c_n \in \mathbb{K}, \ \sum_{n=1}^{\infty} |c_n|^2 < \infty \right\}.$$

Given $x \in \mathcal{H}$, Bessels' inequality (Corollary 14.1.1) states that $\sum_{n=1}^{\infty} |\langle e_n, x \rangle|^2 \leq \|x\|^2$, hence $P_M x = \sum_{n=1}^{\infty} \langle e_n, x \rangle e_n \in M$ and $\|P_M x\| \leq \|x\|$. It follows that P_M is a bounded linear operator into M. By definition $P_M e_n = e_n$ and thus $P_M^2 x = P_M x$ for all $x \in \mathcal{H}$ and we conclude $P_M^2 = P_M$. Next we prove symmetry of the operator P_M. For all $x, y \in \mathcal{H}$ the following chain of identities holds using continuity of the inner product:

$$\langle x, P_M y \rangle = \langle x, \sum_{n=1}^{\infty} \langle e_n, y \rangle e_n \rangle = \sum_{n=1}^{\infty} \langle x, e_n \rangle \langle e_n, y \rangle$$
$$= \sum_{n=1}^{\infty} \langle \langle e_n, x \rangle e_n, y \rangle = \langle P_M x, y \rangle,$$

hence $P_M^* = P_M$ and the operator P_M is a projector. Finally from the characterization of M repeated above it is clear that P_M maps onto M. □

This proposition allows us, for instance, to construct projection operators $P_j = P_{M_j}$ such that

$$P_1 + P_2 + \cdots + P_N = I$$

for any given family $M_1, \ldots, M_N$ of pair-wise orthogonal closed linear subspaces M_j of a Hilbert space $\mathcal{H}$ such that $\mathcal{H} = M_1 \oplus M_2 \oplus \cdots \oplus M_N$. Such a family of projection operators is called a *resolution of the identity*. Later in connection with the spectral theorem for self-adjoint operators we will learn about a continuous analogue. Thus, intuitively, projectors are the basic building blocks of self-adjoint operators.

Recall that a bounded monotone increasing sequence of real numbers converges. The same is true for sequences of projectors if the appropriate notion of monotonicity is used.

Definition 22.1.2 *Let $\mathcal{H}$ be a Hilbert space with inner product $\langle \cdot, \cdot \rangle$. We say that a bounded linear operator A on $\mathcal{H}$ is smaller than or equal to a bounded linear operator B on $\mathcal{H}$, in symbols $A \leq B$, if, and only if, for all $x \in \mathcal{H}$ one has $\langle x, Ax \rangle \leq \langle x, Bx \rangle$.*

We prepare the proof of the convergence of a monotone increasing sequence of projectors by

Lemma 22.1.1 *For two projectors P, Q on a Hilbert space $\mathcal{H}$ the following statements are equivalent:*

a) $P \leq Q$;

b) $\|Px\| \leq \|Qx\|$ for all $x \in \mathcal{H}$;

c) $\operatorname{ran} P \subseteq \operatorname{ran} Q$;

d) $P = PQ = QP$.

Proof. Since any projector satisfies $P = P^*P$, the inequality $\langle x, Px \rangle \leq \langle x, Qx \rangle$ holds if, and only if, $\langle Px, Px \rangle \leq \langle Qx, Qx \rangle$ holds, for all $x \in \mathcal{H}$, thus a) and b) are equivalent.

Assume $\operatorname{ran} P \subseteq \operatorname{ran} Q$ and recall that $y \in \mathcal{H}$ is an element of the range of the projector Q if, and only if, $Qy = y$. The range of P is $P\mathcal{H}$ which by assumption is contained in $\operatorname{ran} Q$, hence $QPx = Px$ for all $x \in \mathcal{H}$ which says $QP = P$; and conversely, if $QP = P$ holds, then clearly $\operatorname{ran} P \subseteq \operatorname{ran} Q$. Since projectors are self-adjoint we know that $P = QP = (QP)^* = P^*Q^* = PQ$, therefore statements c) and d) are equivalent.

If d) holds, then $Px = PQx$ and thus $\|Px\| = \|PQx\| \leq \|Qx\|$ for all $x \in \mathcal{H}$, and conversely if $\|Px\| \leq \|Qx\|$ for all $x \in \mathcal{H}$, then $Px = QPx + Q^\perp Px$ implies $\|Px\|^2 = \|QPx\|^2 + \|Q^\perp Px\|^2$ and hence $Q^\perp Px = 0$ for all $x \in \mathcal{H}$, therefore $P = QP$ and b) and d) are equivalent. $\square$

Theorem 22.1.2 *A monotone increasing sequence $(P_j)_{j \in \mathbb{N}}$ of projectors on a Hilbert space $\mathcal{H}$ converges strongly to a projector P on $\mathcal{H}$.*
The null space of the limit is $N(P) = \bigcap_{j=1}^\infty N(P_j)$ and its range is $\operatorname{ran} P = \overline{\bigcup_{j=1}^\infty \operatorname{ran} P_j}$.

Proof. $P_j \leq P_{j+1}$ means according to Lemma 22.1.1 that $\|P_j x\| \leq \|P_{j+1} x\| \leq \|x\|$ for all $x \in \mathcal{H}$. Thus the monotone increasing and bounded sequence $(\|P_j x\|)_{j \in \mathbb{N}}$ of numbers converges. Lemma 22.1.1 implies also that $P_k = P_k P_j = P_j P_k$ for all $k \leq j$ and therefore

$$\|P_j x - P_k x\|^2 = \langle P_j x, P_j x\rangle - \langle P_j x, P_k x\rangle - \langle P_k x, P_j x\rangle + \langle P_k x, P_k x\rangle$$
$$= \langle x, P_j x\rangle - \langle x, P_k x\rangle = \|P_j x\|^2 - \|P_k x\|^2$$

for all $j \geq k$. Since the numerical sequence $(\|P_j x\|)_{j \in \mathbb{N}}$ converges, we deduce that the sequence of vectors $(P_j x)_{j \in \mathbb{N}}$ is a Cauchy sequence in $\mathcal{H}$ and thus converges to some vector in $\mathcal{H}$ which we denote by Px,

$$Px = \lim_{j \to \infty} P_j x.$$

Since this applies to every $x \in \mathcal{H}$, a map $\mathcal{H} \ni x \to Px \in \mathcal{H}$ is well defined. Standard rules of calculation for limits imply that this map P is linear. The bound $\|P_j x\| \leq \|x\|$ for all $j \in \mathbb{N}$ implies that $\|Px\| \leq \|x\|$ holds for every $x \in \mathcal{H}$, i.e., P is a bounded linear operator on $\mathcal{H}$.

Next we show that this operator is symmetric and idempotent. Continuity of the inner product implies for all $x, y \in \mathcal{H}$,

$$\langle x, Py\rangle = \lim_{j \to \infty} \langle x, P_j y\rangle = \lim_{j \to \infty} \langle P_j x, y\rangle = \langle Px, y\rangle$$

and P is symmetric.

Our starting point of the proof of the relation $P = P^2$ is the observation that $\lim_{j \to \infty} \langle P_j x, P_j y\rangle = \langle Px, Py\rangle$ which follows from the estimate

$$|\langle Px, Py\rangle - \langle P_j x, P_j y\rangle| = |\langle Px - P_j x, P_j y\rangle + \langle Px, Py - P_j y\rangle|$$
$$\leq |\langle Px - P_j x, P_j y\rangle| + |\langle Px, Py - P_j y\rangle|$$
$$\leq \|y\| \|Px - P_j x\| + \|Px\| \|Py - P_j y\|$$

and the strong convergence of the sequence $(P_j)_{j \in \mathbb{N}}$. With this result the identity $P = P^2$ is immediate: For all $x, y \in \mathcal{H}$ it implies that $\langle x, Py\rangle = \lim_{j \to \infty} \langle x, P_j y\rangle = \lim_{j \to \infty} \langle P_j x, P_j y\rangle = \langle Px, Py\rangle$ and thus $P = P^2$.

If a vector x belongs to the kernel of all the projectors P_j, then $Px = \lim_{j \to \infty} P_j x = 0$ implies $x \in N(P)$. Conversely $P_j \leq P$ implies $P_j x = 0$ for all $j \in \mathbb{N}$ if $Px = 0$.

By monotonicity we know $\|P_j x\| \leq \lim_{k \to \infty} \|P_k x\| = \|Px\|$ for all $j \in \mathbb{N}$, hence by Lemma 22.1.1 $\operatorname{ran} P_j \subseteq \operatorname{ran} P$ for all $j \in \mathbb{N}$, and therefore the closure of union of the ranges of the projectors P_j is contained in the range of P. Since $Px = \lim_{j \to \infty} P_j x$ it is obvious that the range of the limit P is contained in the closure of the union of the ranges $\operatorname{ran} P_j$. $\square$

22.2 Unitary operators

22.2.1 Isometries

The subject of this subsection is the linear maps between two Hilbert spaces which do not change the length or norm of vectors. These bounded operators are called isometries.

Definition 22.2.1 *For two Hilbert spaces $\mathcal{H}$ and $\mathcal{K}$ over the same field $\mathbb{K}$ any linear map $A : \mathcal{H} \to \mathcal{K}$ with the property*

$$\|Ax\|_{\mathcal{K}} = \|x\|_{\mathcal{H}} \quad \forall x \in \mathcal{H}$$

is called an **isometry** *(between $\mathcal{H}$ and $\mathcal{K}$).*

Since the norm of a Hilbert space is defined in terms of an inner product the following convenient characterization of isometries is easily available.

Proposition 22.2.1 *Given two Hilbert spaces $\mathcal{H}$, $\mathcal{K}$ over the field $\mathbb{K}$ and a bounded linear operator $A : \mathcal{H} \to \mathcal{K}$, the following statements hold.*

a) *A is an isometry $\Leftrightarrow A^*A = I_{\mathcal{H}}$;*

b) *Every isometry A has an inverse operator $A^{-1} : \operatorname{ran} A \to \mathcal{H}$ and this inverse is $A^{-1} = A^*|_{\operatorname{ran} A}$;*

c) *If A is an isometry, then $AA^* = P_{\operatorname{ran} A}$ is the projector onto the range of A.*

Proof. The adjoint $A^* : \mathcal{K} \to \mathcal{H}$ of A is defined by the identity $\langle A^*y, x\rangle_{\mathcal{H}} = \langle y, Ax\rangle_{\mathcal{K}}$ for all $x \in \mathcal{H}$ and all $y \in \mathcal{K}$. Thus using the definition of an isometry we get $\langle x, A^*Ax\rangle_{\mathcal{H}} = \langle Ax, Ax\rangle_{\mathcal{K}}$ for all $x \in \mathcal{H}$. The polarization identity implies that $\langle x_1, x_2\rangle_{\mathcal{H}} = \langle x_1, A^*Ax_2\rangle_{\mathcal{H}}$ for all $x_1, x_2 \in \mathcal{H}$ and therefore $A^*A = I_{\mathcal{H}}$. The converse is obvious.
    Certainly, an isometry is injective and thus on its range it has an inverse $A^{-1} : \operatorname{ran} A \to \mathcal{H}$. The characterization $A^*A = I_{\mathcal{H}}$ of Part a) allows us to identify the inverse as $A^*|_{\operatorname{ran} A}$.
    For Part c) we use the orthogonal decomposition $\mathcal{K} = \overline{\operatorname{ran} A} \oplus (\operatorname{ran} A)^{\perp}$ and determine AA^* on both subspaces. For $y \in (\operatorname{ran} A)^{\perp}$ the equation $0 = \langle y, Ax\rangle_{\mathcal{K}} = \langle A^*y, x\rangle_{\mathcal{H}}$ for all $x \in \mathcal{H}$ implies $A^*y = 0$ and thus $AA^*y = 0$. For $y \in \operatorname{ran} A$, Part b) gives $AA^*y = AA^{-1}y = y$ and we conclude. □

22.2.2 Unitary operators

According to Proposition 22.2.1 the range of an isometric operator $A : \mathcal{H} \to \mathcal{K}$ contains characteristic information about the operator. In general the range is a proper subspace of the target space $\mathcal{K}$. The case where this range is equal to the target space deserves special attention. These operators are discussed in this subsection.

Definition 22.2.2 *A surjective isometry $U : \mathcal{H} \to \mathcal{K}$ is called a* **unitary operator**.

22. Special Classes of Bounded Operators

On the basis of Proposition 22.2.1 unitary operators can be characterized as follows.

Proposition 22.2.2 *For a bounded linear operator $U : \mathcal{H} \to \mathcal{K}$ these statements are equivalent:*

a) *U is unitary;*

b) *$U^*U = I_\mathcal{H}$ and $UU^* = I_\mathcal{K}$;*

c) *$U\mathcal{H} = \mathcal{K}$ and $\langle Ux, Uy \rangle_\mathcal{K} = \langle x, y \rangle_\mathcal{H}$ for all $x, y \in \mathcal{H}$.*

Note that Part c) of this proposition identifies unitary operators as those surjective bounded linear operators which do not change the value of the inner product. Thus unitary operators respect the full structure of Hilbert spaces (linear, topological, metric and geometric structure). Accordingly unitary operators are the *isomorphisms of Hilbert spaces*. In the chapter on separable Hilbert spaces (Chapter 16) we had constructed an important example of such an isomorphism of Hilbert spaces: There we constructed a unitary map from a separable Hilbert space over the field $\mathbb{K}$ onto the sequence space $\ell^2(\mathbb{K})$.

Note also that in the case of finite dimensional spaces every isometry is a unitary operator. The proof is done as an exercise. In the case of infinite dimensions there are many isometric operators which are not unitary. A simple example is discussed in the Exercises.

For unitary operators of a Hilbert space $\mathcal{H}$ onto itself the composition of mappings is well defined. The composition of two unitary operators U, V on the Hilbert space $\mathcal{H}$ is again a unitary operator since by Part b) of Proposition 22.2.2 $(UV)^*(UV) = V^*U^*UV = V^*V = I_\mathcal{H}$ and $(UV)(UV)^* = UVV^*U^* = UU^* = I_\mathcal{H}$. Thus the unitary operators of a Hilbert space $\mathcal{H}$ form a group, denoted by $\mathfrak{U}(\mathcal{H})$.

This group $\mathfrak{U}(\mathcal{H})$ contains many important and interesting sub-groups. For quantum mechanics the *one-parameter groups of unitary operators* play a prominent rôle.

Definition 22.2.3 *A family of unitary operators $\{U(t) : t \in \mathbb{R}\} \subset \mathfrak{U}(\mathcal{H})$ is called a* **one-parameter group of unitary operators in** *$\mathcal{H}$ if, and only if, $U(0) = I_\mathcal{H}$ and $U(s)U(t) = U(s+t)$ for all $s, t \in \mathbb{R}$.*

Naturally one can view a one-parameter group of unitary operators on $\mathcal{H}$ as a representation of the additive group $\mathbb{R}$ by unitary operators on $\mathcal{H}$. The importance of these groups for quantum mechanics comes from the fact that the time evolution of quantum systems is typically described by such a group.

Under a weak continuity hypothesis the general form of these groups is known.

Theorem 22.2.3 (Stone) *Let $\{U(t) : t \in \mathbb{R}\}$ be a one-parameter group of unitary operators on the complex Hilbert space $\mathcal{H}$ which is strongly continuous, i.e., for every $x \in \mathcal{H}$ the function $\mathbb{R} \ni t \to U(t)x \in \mathcal{H}$ is continuous. Then the set*

$$D = \left\{ x \in \mathcal{H} : \lim_{t \to 0} \frac{1}{t}[U(t)x - x] \text{ exists} \right\}$$

22.2 Unitary operators

is a dense linear subspace of $\mathcal{H}$ and on D a linear operator A is well defined by

$$iAx = \lim_{t \to 0} \frac{1}{t}[U(t)x - x] \qquad \forall x \in D.$$

This operator A is self-adjoint (on D). It is called the **infinitesimal generator** of the group which often is expressed in the notation

$$U(t) = e^{itA}, \qquad t \in \mathbb{R}.$$

Proof. Since the group U is strongly continuous, the function $\mathbb{R} \ni t \to U(t)x \in \mathcal{H}$ is continuous and bounded (by $\|x\|$) for every $x \in \mathcal{H}$. For every function $f \in \mathcal{D}(\mathbb{R})$, the function $t \mapsto f(t)U(t)x$ is thus a continuous function of compact support for which the existence of the Riemann integral and some basic estimates are shown in the Exercises. This allows us to define a map $J : \mathcal{D}(\mathbb{R}) \times \mathcal{H} \to \mathcal{H}$ by this integral:

$$J(f, x) = \int_{\mathbb{R}} f(t)U(t)x dt.$$

Since U is strongly continuous, given $\epsilon > 0$, there is $r > 0$ such that $\sup_{-r \leq t \leq r} \|U(t)x - x\| \leq \epsilon$. Choose a nonnegative function $\rho_r \in \mathcal{D}(\mathbb{R})$ with the properties $\int_{\mathbb{R}} \rho_r(t)dt = 1$ and $\operatorname{supp} \rho_r \subseteq [-r, r]$ (such functions exist according to the chapter on test functions) and estimate

$$\|J(\rho_r, x) - x\| = \|\int_{\mathbb{R}} \rho_r(t)[U(t)x - x]dt\| \leq \int_{\mathbb{R}} \|\rho_r(t)[U(t)x - x]\|dt$$

$$\leq \|\rho_r\|_1 \sup_{-r \leq t \leq r} \|U(t)x - x\| \leq \epsilon.$$

Therefore the set $D_0 = \{J(f, x) : f \in \mathcal{D}(\mathbb{R}), x \in \mathcal{H}\}$ is dense in the Hilbert space $\mathcal{H}$. By changing the integration variables we find the transformation law of the vectors $J(f, x)$ under the group U:

$$U(t)J(f, x) = J(f_{-t}, x), \qquad f_a(x) = f(x - a) \ \forall x \in \mathbb{R}.$$

This transformation law and the linearity of J with respect to the first argument imply that the group U is differentiable on D_0: The relation $U(s)J(f, x) - J(f, x) = J(f_s - f, x)$ gives

$$\lim_{s \to 0} \frac{1}{s}[U(s)J(f, x) - J(f, x)] = \lim_{s \to 0} J(\frac{f_{-s} - f}{s}, x) = J(-f', x)$$

where we have used that $(f_{-s} - f)/s$ converges uniformly to $-f'$ and that uniform limits and Riemann integration commute.

Define a function $A : D_0 \to D_0$ by $AJ(f, x) = -iJ(-f', x)$ and extend this definition by linearity to the linear hull D of D_0 to get a densely defined linear operator $A : D \to D$. Certainly, the linear subspace D is also left invariant by the action of the group, $U(t)D \subset D$ for all $t \in \mathbb{R}$ and a straightforward calculation shows

$$AU(t)\psi = U(t)A\psi \qquad \forall t \in \mathbb{R}, \ \forall \psi \in D.$$

The symmetry of the operator A is the result of this operator being defined as the derivative of a unitary group (modulo the constant $-i$): For all $f, g \in \mathcal{D}$ and all $x, y \in \mathcal{H}$ the following chain of equations holds:

$$\langle AJ(f, x), J(g, y) \rangle = \langle \lim_{s \to 0} \frac{U(s) - I}{is} J(f, x), J(g, y) \rangle$$

$$= \lim_{s \to 0} \langle \frac{U(s) - I}{is} J(f, x), J(g, y) \rangle$$

$$= \lim_{s \to 0} \langle J(f, x), (\frac{U(s) - I}{is})^* J(g, y) \rangle$$

$$= \lim_{s \to 0} \langle J(f, x), \frac{U(-s) - I}{-is} J(g, y) \rangle$$

$$= \langle J(f, x), AJ(g, y) \rangle.$$

Certainly by linearity this symmetry relation extends to all of D.

Next, using Corollary 19.3.3 we show that A is actually essentially self-adjoint on D. This is done by proving $N(A^* \pm iI) = \{0\}$. Suppose $\phi \in D(A^*)$ satisfies $A^*\phi = i\phi$. Then, for all $\psi \in D$,

$$\frac{d}{dt}\langle U(t)\psi, \phi\rangle = \langle iAU(t)\psi, \phi\rangle = \langle U(t)\psi, -iA^*\phi\rangle = \langle U(t)\psi, \phi\rangle,$$

i.e., the function $h(t) = \langle U(t)\psi, \phi\rangle$ satisfies the differential equation $h'(t) = h(t)$ for all $t \in \mathbb{R}$ and it follows that $h(t) = h(0)e^t$. Since the group U is unitary the function h is bounded and this is the case only if $h(0) = \langle \psi, \phi\rangle = 0$. This argument applies to all $\psi \in D$, hence $\phi \in D^\perp = \{0\}$ (D is dense). Similarly one shows that $A^*\phi = -i\phi$ is satisfied only for $\phi = 0$. Hence Corollary 19.3.3 proves A to be essentially self-adjoint, thus the closure $\overline{A}$ of A is self-adjoint.

When spectral calculus has been developed we will be able to define the exponential function $e^{it\overline{A}}$ of an unbounded self-adjoint operator and then we can show that this exponential function indeed is equal to the given unitary group. □

The continuity hypothesis in Stone's theorem can be relaxed. It suffices to assume that the group is weakly continuous, i.e., that $\mathbb{R} \ni t \mapsto \langle x, U(t)y\rangle$ is continuous for every choice of $x, y \in \mathcal{H}$. This is so since on the class $\mathfrak{U}(\mathcal{H})$ the weak and strong topology coincide (see Exercises). In separable Hilbert spaces the continuity hypothesis can be relaxed even further to weak measurability, i.e., the map $\mathbb{R} \ni t \mapsto \langle x, U(t)y\rangle \in \mathbb{K}$ is measurable, for every $x, y \in \mathcal{H}$.

22.2.3 Examples of unitary operators

In the section on Fourier transformation for tempered distributions we learned that the Fourier transform $\mathcal{F}_2$ on the Hilbert space $L^2(\mathbb{R}^n)$ is a unitary operator. In the same Hilbert space we consider several other examples of unitary operators, respectively groups of such operators.

For $f \in L^2(\mathbb{R}^n)$ and $a \in \mathbb{R}^n$, define $f_a(x) = f(x - a)$ for all $x \in \mathbb{R}^n$ and then define $U_a : L^2(\mathbb{R}^n) \to L^2(\mathbb{R}^n)$ by $U_a f = f_a$ for all $f \in L^2(\mathbb{R}^n)$. For all $f, g \in L^2(\mathbb{R}^n)$ one has

$$\langle U(a)f, U(a)g\rangle_2 = \int_{\mathbb{R}^n} \overline{f(x-a)}g(x-a)dx = \int_{\mathbb{R}^n} \overline{f(y)}g(y)dy = \langle f, g\rangle_2$$

and $U(a)$ is an isometry. Given $f \in L^2(\mathbb{R}^n)$, define $g = f_{-a}$ and calculate $U(a)g = g_a = f$, hence $U(a)$ is surjective and thus a unitary operator, i.e., $U(a) \in \mathfrak{U}(L^2(\mathbb{R}^n))$ for all $a \in \mathbb{R}^n$. In addition we find

$$U(0) = I_{L^2(\mathbb{R}^n)}, \qquad U(a)U(b) = U(a+b), \quad \forall a, b \in \mathbb{R}^n,$$

i.e., $\{U(a) : a \in \mathbb{R}^n\}$ is an *n-parameter group of unitary operators* on $L^2(\mathbb{R}^n)$. Naturally, $U(a)$ has the interpretation of the operator of translation by a.

22.3 Compact operators

In the introduction to the theory of Hilbert spaces we mentioned that substantial a part of this theory has its origin in D. Hilbert's research on the problem to extend the well-known theory of eigenvalues of matrices to the case of 'infinite dimensional matrices' or linear operators in an infinite dimensional space. A certain limit (to be specified later) of finite dimensional matrices gives a class of operators which are called compact. Accordingly the early results in the theory of bounded operators on infinite dimensional Hilbert spaces were mainly concerned with this class of compact operators which were typically investigated in separable spaces. We take a slightly more general approach.

Definition 22.3.1 *Let $\mathcal{H}$ and $\mathcal{K}$ be two Hilbert spaces over the field $\mathbb{K}$. A bounded linear operator $K : \mathcal{H} \to \mathcal{K}$ is called* **compact** *(or* **completely continuous**) *if, and only if, it maps every bounded set in $\mathcal{H}$ onto a precompact set of $\mathcal{K}$ (this means that the closure of the image of a bounded set is compact), i.e., if, and only if, for every bounded sequence $(e_n)_{n\in\mathbb{N}} \subset \mathcal{H}$ the sequence of images $(Ke_n)_{n\in\mathbb{N}} \subset \mathcal{K}$ contains a convergent subsequence.*

In particular in concrete problems the following characterization of compact operators is very helpful.

Theorem 22.3.1 (Characterization of compact operators) *Let $\mathcal{H}$ and $\mathcal{K}$ be two Hilbert spaces over the field $\mathbb{K}$ and $A : \mathcal{H} \to \mathcal{K}$ a bounded linear operator. A is compact if it satisfies one (and thus all) of the following equivalent conditions.*

a) *The image of the open unit ball $B_1(0) \subset \mathcal{H}$ under A is precompact in $\mathcal{K}$;*

b) *The image of every bounded set $B \subset \mathcal{H}$ under A is precompact in $\mathcal{K}$;*

c) *For every bounded sequence $(x_n)_{n\in\mathbb{N}} \subset \mathcal{H}$ the sequence of images $(Ax_n)_{n\in\mathbb{N}}$ in $\mathcal{K}$ contains a convergent subsequence;*

d) *The operator A maps weakly convergent sequences in $\mathcal{H}$ into norm convergent sequences in $\mathcal{K}$.*

Proof. The proof proceeds according the following steps: $a) \Rightarrow b) \Rightarrow c) \Rightarrow d) \Rightarrow a)$.

Assume a), that is assume $A(B_1(0))$ is precompact in $\mathcal{K}$ and consider any bounded set $B \subset \mathcal{H}$. It follows that B is contained in some ball $B_r(0) = rB_1(0)$ for suitable $r > 0$, hence $A(B) \subseteq A(rB_1(0)) = rA(B_1(0))$, thus $A(B)$ is precompact and b) follows.

Next assume b) and recall the Bolzano–Weierstrass result that a metric space is compact if, and only if, every bounded sequence contains a convergent subsequence. Hence statement c) holds.

Now assume c) and consider a sequence $(x_n)_{n\in\mathbb{N}} \subset \mathcal{H}$ which converges weakly to $x \in \mathcal{H}$. For any $z \in \mathcal{K}$ we find $\langle z, Ax_n \rangle_\mathcal{K} = \langle A^*z, x_n \rangle_\mathcal{H} \to \langle A*z, x \rangle_\mathcal{H} = \langle z, Az \rangle_\mathcal{K}$ as $n \to \infty$, i.e., the sequence of images converges weakly to the image of the weak limit.

Suppose that the sequence $(y_n = Ax_n)_{n\in\mathbb{N}}$ does not converge in norm to $y = Ax$. Then there are $\epsilon > 0$ and a subsequence $(Ax_{n(j)})_{j\in\mathbb{N}}$ such that $\|y - y_{n(j)}\|_\mathcal{K} \geq \epsilon$ for all $j \in \mathbb{N}$. Since $(x_{n(j)})_{j\in\mathbb{N}}$ is a bounded sequence there is a subsequence $(x_{n(j_i)})_{i\in\mathbb{N}}$ for which $(y_{n(j_i)} = Ax_{n(j_i)})_{i\in\mathbb{N}}$ converges in norm, because c) is assumed. The limit of this sequence is $y = Ax$ since the weak limit of this sequence is y, but this is a contradiction to the construction of the subsequence $(y_{n(j)})_{j\in\mathbb{N}}$ and thus $\lim_{n\to\infty} Ax_n = Ax$ in norm. This proves Part d).

Finally assume d). Take any sequence $(x_n)_{n\in\mathbb{N}} \subset B_1(0)$. By Theorem 18.2.2 there is a weakly convergent subsequence $(x_{n(j)})_{j\in\mathbb{N}}$. According to assumption d) the sequence $(Ax_{n(j)})_{j\in\mathbb{N}} \subset A(B_1(0))$ of images converges in norm. The Theorem of Bolzano–Weierstrass implies that $A(B_1(0))$ is precompact. Thus we conclude. □

Definition 22.3.2 *A bounded linear operator $A : \mathcal{H} \to \mathcal{K}$ with finite dimensional range is called an operator of **finite rank**.*

The general form of an operator A of finite rank is easily determined. The result is (see Exercise)

$$Ax = \sum_{j=1}^{N} \langle f_j, x \rangle_{\mathcal{H}} e_j \qquad \forall x \in \mathcal{H}$$

where $\{e_1, \ldots, e_N\}$ is some finite orthonormal system in $\mathcal{K}$ and $f_1, \ldots, f_N$ are some vectors in $\mathcal{H}$. If now a sequence $(x_n)_{n \in \mathbb{N}} \subset \mathcal{H}$ converges weakly to $x \in \mathcal{H}$ then, for $j = 1, \ldots, N$, $\langle f_j, x_n \rangle_{\mathcal{H}} \to \langle f_j, x \rangle_{\mathcal{H}}$ and thus, as $n \to \infty$,

$$Ax_n = \sum_{j=1}^{N} \langle f_j, x_n \rangle_{\mathcal{H}} e_j \to \sum_{j=1}^{N} \langle f_j, x \rangle_{\mathcal{H}} e_j = Ax.$$

We conclude that operators of finite rank are compact.

The announced approximation of compact operators by matrices take the following precise form.

Theorem 22.3.2 *In a separable Hilbert space $\mathcal{H}$ every compact operator A is the norm limit of a sequence of operators of finite rank.*

Proof. Let $\{e_j : j \in \mathbb{N}\}$ be an orthonormal basis of $\mathcal{H}$ and introduce the projectors P_n onto the subspace $[e_1, \ldots, e_n]$ spanned by the first n basis vectors. Proposition 22.1.1 implies that the sequence of projectors P_n converges strongly to the identity I. Define

$$d_n = \sup\left\{\|AP_n^\perp x\| : \|x\| = 1\right\} = \sup\{\|Ay\| : y \in S_n\}$$

where $S_n = \left\{y \in [e_1, \ldots, e_n]^\perp : \|y\| = 1\right\}$. Clearly $(d_n)_{n \in \mathbb{N}}$ is a monotone decreasing sequence of positive numbers. Thus this sequence has a limit $d \geq 0$. For every $n \in \mathbb{N}$ there is $y_n \in S_n$ such that $\|Ay_n\| \geq \frac{d_n}{2}$. $y_n \in S_n$ means: $\|y_n\| = 1$ and $y_n = P_n^\perp y_n$, hence $\langle x, y_n \rangle = \langle P_n^\perp x, y_n \rangle \to 0$ as $n \to \infty$ since $P_n^\perp x \to 0$ in $\mathcal{H}$, for every $x \in \mathcal{H}$, i.e., the sequence $(y_n)_{n \in \mathbb{N}}$ converges weakly to 0. Compactness of A implies that $\|Ay_n\| \to 0$ and thus $d = 0$. Finally observe

$$d_n = \|A - AP_n\| \qquad \forall n \in \mathbb{N},$$

hence the compact operator A is the norm limit of the sequence of operators AP_n,

$$AP_n x = \sum_{j=1}^{n} \langle e_j, x \rangle A e_j \qquad \forall x \in \mathcal{H},$$

which are of finite rank. $\square$

The set of compact operators is stable under uniform limits:

Theorem 22.3.3 *Suppose $\mathcal{H}$ and $\mathcal{K}$ are Hilbert spaces over the field $\mathbb{K}$ and $A_n : \mathcal{H} \to \mathcal{K}$, $n \in \mathbb{N}$, are compact operators, and suppose that A is the norm limit of this sequence. Then A is compact.*

Proof. Take a sequence $(x_n)_{n\in\mathbb{N}}$ in the unit ball of $\mathcal{H}$. We are going to construct a subsequence for which the sequence of images under A converges in $\mathcal{K}$. Then Theorem 22.3.1 implies compactness of A.

Since A_1 is compact there is a subsequence $(x_{n_1(j)})_{j\in\mathbb{N}}$ such that the sequence $(A_1 x_{n_1(j)})_{j\in\mathbb{N}}$ converges in $\mathcal{K}$, to y_1 let us say. Since A_2 is compact there is a subsequence $(x_{n_2(j)})_{j\in\mathbb{N}}$ of the first subsequence such that $(A_2 x_{n_2(j)})_{j\in\mathbb{N}}$ converges in $\mathcal{K}$ with limit y_2. Iterating this argument produces a sequence of subsequences

$$(x_{n_{k+1}(j)})_{j\in\mathbb{N}} \subset (x_{n_k(j)})_{j\in\mathbb{N}} \qquad k \in \mathbb{N}$$

such that for $k = 1, 2, \ldots$

$$\lim_{j\to\infty} A_k x_{n_k(j)} = y_k \in \mathcal{K}.$$

Finally form the diagonal sequence $z_k = x_{n_k(k)}$ to obtain a subsequence of the original sequence such that $A_n z_k \to y_n$ as $k \to \infty$, for all $n \in \mathbb{N}$.

Now we show convergence of the sequence $(A z_k)_{k\in\mathbb{N}}$ by showing that it is a Cauchy sequence in $\mathcal{K}$. Given $\epsilon > 0$ there is $j_0 \in \mathbb{N}$ such that $\|A - A_j\| < \epsilon/4$ for all $j \geq j_0$. Since the sequence $(A_{j_0} z_k)_{k\in\mathbb{N}}$ converges in $\mathcal{K}$ there is $k_0 \in \mathbb{N}$ such that $\|A_{j_0} z_k - A_{j_0} z_m\|_\mathcal{K} < \epsilon/2$ for all $k, m \geq k_0$. For all $k, m \geq k_0$ we thus find

$$\|A z_k - A z_m\|_\mathcal{K} \leq \|A - A_{j_0}\| \|z_k - z_m\|_\mathcal{K} + \|A_{j_0} z_k - A_{j_0} z_m\|_\mathcal{K} < \frac{\epsilon}{4} 2 + \frac{\epsilon}{2} = \epsilon,$$

and this sequence is indeed a Cauchy sequence. Thus we conclude. □

Some important properties of compact operators are collected in

Theorem 22.3.4 *For a Hilbert space $\mathcal{H}$ denote the set of all compact operators $A : \mathcal{H} \to \mathcal{H}$ by $\mathfrak{K}(\mathcal{H})$. Then*

a) *$\mathfrak{K}(\mathcal{H})$ is a linear subspace of $\mathfrak{B}(\mathcal{H})$;*

b) *$A \in \mathfrak{B}(\mathcal{H})$ is compact if, and only if, its adjoint A^* is compact;*

c) *$A \in \mathfrak{K}(\mathcal{H})$, $\lambda \in \mathbb{K}$, $\lambda \neq 0$, $\Rightarrow \dim N(A - \lambda I) < \infty$, i.e., the eigenspaces of compact operators for eigenvalues different from zero are finite dimensional;*

d) *$\mathfrak{K}(\mathcal{H})$ is a closed subalgebra of the C^*-algebra $\mathfrak{B}(\mathcal{H})$ and thus itself a C^*-algebra;*

e) *$\mathfrak{K}(\mathcal{H})$ is a closed ideal in $\mathfrak{B}(\mathcal{H})$, i.e.,*

$$\mathfrak{B}(\mathcal{H}) \cdot \mathfrak{K}(\mathcal{H}) \subset \mathfrak{K}(\mathcal{H}) \qquad \text{and} \qquad \mathfrak{K}(\mathcal{H}) \cdot \mathfrak{B}(\mathcal{H}) \subset \mathfrak{K}(\mathcal{H}).$$

Proof. With the exception of Part b) the proofs are relatively simple. Here we only prove Part c), the other parts are done as an exercise.

Suppose $\lambda \neq 0$ is an eigenvalue of A. Then we have $I|_{N(A-\lambda I)} = \frac{1}{\lambda} A|_{N(A-\lambda I)}$. Thus the identity operator on the subspace $N(A - \lambda I)$ is compact, hence this subspace must have a finite dimension. □

As a conclusion of this section a simple example of a compact operator in an infinite dimensional Hilbert space is discussed.

Suppose that $\{e_j : j \in \mathbb{N}\}$ is an orthonormal basis of the Hilbert space $\mathcal{H}$. Define a linear operator by continuous linear extension of $A e_j = \frac{1}{j} e_j$, $j \in \mathbb{N}$, i.e., define

$$Ax = \sum_{j=1}^\infty \frac{a_j}{j} e_j, \quad x = \sum_{j=1}^\infty a_j e_j, \quad \sum_{j=1}^\infty |a_j|^2 = \|x\|^2.$$

It follows that the open unit ball $B_1(0)\mathcal{H}$ is mapped by A onto a subset which is isomorphic to a subset of the *Hilbert cube* W (under the standard isomorphism between a separable Hilbert space over $\mathbb{K}$ and $\ell^2(\mathbb{K})$),

$$W = \left\{ (v_n)_{n\in\mathbb{N}} \in \ell^2(\mathbb{K}) : |v_n| \leq \frac{1}{n}, \ \forall n \in \mathbb{N} \right\}.$$

The following lemma shows compactness of the Hilbert cube, hence A is a compact operator.

Lemma 22.3.1 *The Hilbert cube W is a compact subset of the Hilbert space $\ell^2(\mathbb{K})$.*

Proof. In an infinite dimensional Hilbert space the closed bounded sets are not compact (in the norm topology) but always weakly compact (see Theorem 18.2.6). Compactness of W follows from the observation that the strong and the weak topology coincide on W. For this it suffices to show that every weakly convergent sequence $(x_n)_{n\in\mathbb{N}} \subset W$ with weak limit $x \in W$ also converges strongly to x. Given $\epsilon > 0$ there is $p \in \mathbb{N}$ such that $\sum_{j=p+1}^{\infty} \frac{1}{j^2} < \epsilon$. Because of weak convergence of the sequence we can now find $N \in \mathbb{N}$ such that $\sum_{j=1}^{p} |x_{n,j} - x_j|^2 < \epsilon$ for all $n \geq N$ (we use the notation $x_n = (x_{n,j})_{j\in\mathbb{N}}$). This gives $\sum_{j=1}^{\infty} |x_{n,j} - x_j|^2 < 3\epsilon$ for all $n \geq N$ and thus x is the strong limit of the sequence $(x_n)_{n\in\mathbb{N}} \subset W$. □

22.4 Trace class operators

In separable Hilbert spaces there is a subclass of the compact operators which is frequently used in quantum mechanics. These are the trace class operators. The symmetric, positive and normalized trace class operators are the density matrices or statistical operators of quantum mechanics. We present a short introduction to this class of operators.

Definition 22.4.1 *A bounded linear operator A on a separable Hilbert space $\mathcal{H}$ is called a* **trace class operator** *if, and only if, for some orthonormal basis $\{e_n : n \in \mathbb{N}\}$ the sum*

$$\sum_{n=1}^{\infty} \langle e_n, |A|e_n \rangle$$

is finite where $|A|$ is the modulus of A.

Some basic properties of trace class operators are collected in

Proposition 22.4.1 *Let A be a trace class operator on the separable Hilbert space $\mathcal{H}$. Then*

a) For any two orthonormal bases $\{e_n : n \in \mathbb{N}\}, \{f_n : n \in \mathbb{N}\}$ of $\mathcal{H}$

$$\sum_{n=1}^{\infty} \langle e_n, |A|e_n \rangle = \sum_{n=1}^{\infty} \langle f_n, |A|f_n \rangle < \infty;$$

thus the **trace** $\mathrm{Tr}(|A|)$ of $|A|$ is well defined as the common value of any of these sums;

b) A^2 is a trace class operator;

c) $|A|$ is compact.

Proof. For a) it suffices to show: If $\{e_n : n \in \mathbb{N}\}$ is the orthonormal basis used in the definition and if $\{f_n : n \in \mathbb{N}\}$ is any other orthonormal basis, the two sums agree:.

$$\sum_{n=1}^\infty \langle e_n, |A|e_n\rangle = \sum_{n=1}^\infty \langle \sqrt{|A|}e_n, \sqrt{|A|}e_n\rangle = \sum_{n=1}^\infty \|\sqrt{|A|}e_n\|^2$$

$$= \sum_{n=1}^\infty \sum_{j=1}^\infty |\langle f_j, \sqrt{|A|}e_n\rangle|^2 = \sum_{n=1}^\infty \sum_{j=1}^\infty |\langle \sqrt{|A|}f_j, e_n\rangle|^2$$

$$= \sum_{j=1}^\infty \sum_{n=1}^\infty |\langle \sqrt{|A|}f_j, e_n\rangle|^2 = \sum_{j=1}^\infty \sum_{n=1}^\infty |\langle \sqrt{|A|}f_j, e_n\rangle|^2$$

$$= \sum_{j=1}^\infty \|\sqrt{|A|}f_j\|^2 = \sum_{j=1}^\infty \langle f_j, |A|f_j\rangle.$$

Since in this calculation only sums of positive terms are involved, the interchange of the order of summation is justified.

Since the modulus of an operator commutes with the operator, one has $|A^2| = |A|^2$ and thus

$$\sum_{j=1}^\infty \||A|e_j\|^2 \le \sum_{j=1}^\infty \|\sqrt{|A|}\|^2 \|\sqrt{|A|}e_j\|^2 = \|\sqrt{|A|}\|^2 \sum_{j=1}^\infty \langle e_j, |A|e_j\rangle < \infty,$$

and the operator $|A^2|$ has a finite trace.

Since $|A^2|$ has a finite trace one can use any orthonormal basis to calculate it. Suppose $\{e_n : n \in \mathbb{N}\}$ is an orthonormal basis and for $N \in \mathbb{N}$ take any $y \in [e_1, \ldots, e_N]^\perp$, $\|y\| = 1$; then we have

$$\sum_{j=1}^N \||A|e_j\|^2 + \||A|y\|^2 \le \mathrm{Tr}(|A|^2),$$

and therefore

$$\||A|P_N^\perp\| = \sup\left\{\||A|y\| : y \in [e_1, \ldots, e_N]^\perp, \|y\| = 1\right\} \le \mathrm{Tr}(|A|^2) - \sum_{j=1}^N \||A|e_j\|^2,$$

where P_N is the orthogonal projector onto the subspace $[e_1, \ldots, e_N]$. Since the right-hand side of this estimate converges to 0 as $N \to \infty$, this proves that $\|(|A| - |A|P_N)\| \to 0$ as $N \to \infty$. We conclude that $|A|$ is the norm limit of the finite rank operators $|A|P_N$, $|A|P_N x = \sum_{j=1}^N \langle e_j, x\rangle|A|e_j$. Theorem 22.3.3 implies that the operator $|A|$ is compact. □

Theorem 22.4.2 (Characterization of trace class operators) *A bounded linear operator on a separable Hilbert space $\mathcal{H}$ is a trace class operator if, and only if, there is a sequence $\underline{\lambda} = (\lambda_n)_{n\in\mathbb{N}} \in \ell^1(\mathbb{K})$ (i.e., $\sum_{n=1}^\infty |\lambda_n| < \infty$) and there are orthonormal bases $\{e_n : n \in \mathbb{N}\}$ and $\{x_n : n \in \mathbb{N}\}$ of $\mathcal{H}$ such that*

$$Ax = \sum_{n=1}^\infty \lambda_n \langle e_n, x\rangle x_n \qquad \forall x \in \mathcal{H}. \tag{22.2}$$

*If A has the form (22.2), then the **trace norm** $\|A\|_1 = \mathrm{Tr}(|A|)$ satisfies*

$$\|A\|_1 = \mathrm{Tr}(|A|) = \|\underline{\lambda}\|_1 = \sum_{n=1}^{\infty} |\lambda_n|. \tag{22.3}$$

Proof. Suppose A is defined by equation (22.2). For any $x \in \mathcal{H}$ the estimate

$$\sum_{n=1}^{\infty} |\lambda_n \langle e_n, x \rangle|^2 \le (\sup_{n \in \mathbb{N}} \|\lambda_n\|)^2 \sum_{n=1}^{\infty} |\langle e_n, x \rangle|^2 \le (\sup_{n \in \mathbb{N}} |\lambda_n|)^2 \|x\|^2$$

implies that A is a well-defined and bounded operator. Linearity follows easily.

If the operator A has the form (22.2), its adjoint is easily determined as $A^* y = \sum_{n=1}^{\infty} \overline{\lambda_n} \langle x_n, y \rangle e_n$ and thus $A^* A y = \sum_{n=1}^{\infty} \overline{\lambda_n} \lambda_n \langle e_n, y \rangle e_n$. Since the square root is unique we get

$$|A| = \sum_{n=1}^{\infty} |\lambda_n| P_{e_n},$$

hence $\sum_{n=1}^{\infty} \langle e_n, |A| e_n \rangle = \sum_{n=1}^{\infty} |\lambda_n| = \|\underline{\lambda}\|_1 < \infty$ and A is indeed a trace class operator and relation (22.3) holds.

Conversely assume that A is a trace class operator. Proposition 22.4.1 implies that the modulus of A is a positive self-adjoint operator with the property $\sum_{n=1}^{\infty} \langle e_n, |A| e_n \rangle < \infty$ for any orthonormal basis of $\mathcal{H}$. The spectral theorem for positive compact operators (see Theorem 25.1.1) states that $|A|$ has a decreasing sequence (or a finite number) $\rho_1 \ge \rho_2 \ldots \rho_N \ge \ldots$ of eigenvalues of finite multiplicity such that $\rho_N \to 0$ as $N \to \infty$. The orthonormal system of eigenfunctions e_j, $|A|e_j = \rho_j e_j$, is complete if, and only if, $|A|$ is injective. Thus $|A|$ has the form

$$|A|x = \sum_{n=1}^{\infty} \rho_n \langle e_n, x \rangle e_n \tag{22.4}$$

where $\{e_n : n \in \mathbb{N}\}$ is an orthonormal basis of $\mathcal{H}$ and $\rho_n \ge 0$. The condition $\sum_{n=1}^{\infty} \langle e_n, |A|e_n \rangle < \infty$ implies that

$$\sum_{n=1}^{\infty} \langle e_n, |A|e_n \rangle = \sum_{n=1}^{\infty} \rho_n < \infty,$$

and thus the sequence of eigenvalues of $|A|$ is summable. Finally apply the polar decomposition $A = U|A|$ with some partial isometry U (see Theorem 21.5.2) to get for all $x \in \mathcal{H}$,

$$Ax = U|A|x = \sum_{n=1}^{\infty} \rho_n \langle e_n, x \rangle U e_n.$$

Since U is an isometry on the range of $|A|$ which is spanned by the vectors e_n with $\rho_n > 0$, the image vectors $U e_n$ are orthonomal too (see Proposition 14.1.2). The orthonormal system $\{U e_n : \rho_n > 0\}$ can be extended to give an orthonormal basis $\{x_n : n \in \mathbb{N}\}$ of $\mathcal{H}$. Thus the form (22.2) results for the trace class operator A. □

Theorem 22.4.3 (Trace of trace class operators) *Denote by $\mathfrak{T}(\mathcal{H})$ the set of all trace class operators on a separable Hilbert space $\mathcal{H}$, then*

a) *$\mathfrak{T}(\mathcal{H})$ is a linear subspace of $\mathfrak{K}(\mathcal{H})$ and $A \mapsto \|A\|_1$ defines a norm on $\mathfrak{T}(\mathcal{H})$ and one has $\|A\| \le \|A\|_1$;*

b) *on $\mathfrak{T}(\mathcal{H})$ a linear functional $\mathrm{Tr}(\cdot) : \mathfrak{T}(\mathcal{H}) \to \mathbb{K}$ is well defined by*

$$\mathrm{Tr}(A) = \sum_{n=1}^{\infty} \langle e_n, A e_n \rangle \tag{22.5}$$

where $\{e_n : n \in \mathbb{N}\}$ is any orthonormal basis of $\mathcal{H}$. This functional has the following properties:

i) $|\text{Tr}(A)| \leq \|A\|_1$ for all $A \in \mathfrak{T}(\mathcal{H})$;

ii) if $A \in \mathfrak{T}(\mathcal{H})$, then $A^* \in \mathfrak{T}(\mathcal{H})$ and $\text{Tr}(A^*) = \overline{\text{Tr}(A)}$;

iii) $\text{Tr}(AB) = \text{Tr}(BA)$ for all $A, B \in \mathfrak{T}(\mathcal{H})$.

Proof. Proposition 22.4.1 implies that every trace class operator is compact. It is elementary to show that $A \mapsto \|A\|_1$ is a norm on $\mathfrak{T}(\mathcal{H})$. Because of the representation (22.2) we know for all $x, y \in \mathcal{H}$ with $\|x\| = \|y\| = 1$ that

$$|\langle y, Ax \rangle| \leq \sum_{n=1}^{\infty} |\lambda_n| |\langle e_n, x \rangle| |\langle y, x_n \rangle| \leq \sum_{n=1}^{\infty} |\lambda_n| \leq \|A\|_1,$$

hence $\|A\| \sup\{|\langle y, Ax \rangle| : \|x\| = \|y\| = 1\} \leq \|A\|_1$. This proves the first part.

Since $\sum_{n=1}^{\infty} |\langle e_n, Ae_n \rangle| \leq \|A\|_1$ the functional $\text{Tr}(\cdot)$ is certainly defined on $\mathfrak{T}(\mathcal{H})$ for any choice of the orthonormal basis which is used to evaluate the series. It is an elementary calculation (see Exercises) to show that for any other orthonormal basis $\{f_n : n \in \mathbb{N}\}$ of $\mathcal{H}$ one has

$$\sum_{n=1}^{\infty} \langle f_n, Af_n \rangle = \sum_{n=1}^{\infty} \langle e_n, Ae_n \rangle,$$

hence $\text{Tr}(\cdot)$ is a well-defined functional on the space of trace class operators. The estimate for its values now is obvious and Part i) follows.

If $A \in \mathfrak{T}(\mathcal{H})$ has the form $Ax = \sum_{n=1}^{\infty} \lambda_n \langle e_n, x \rangle x_n$, it is shown in the Exercises that the adjoint of A has the form $A^* y = \sum_{n=1}^{\infty} \overline{\lambda_n} \langle x_n, y \rangle e_n$ and thus is an operator of trace class too. The proof of the identity in Part ii) is a simple calculation.

The simple proof of Part iii) is based on the use of the completeness relation. For any $A, B \in \mathfrak{T}(\mathcal{H})$ the following chain of identities holds:

$$\text{Tr}(AB) = \sum_{n=1}^{\infty} \langle e_n ABe_n \rangle = \sum_{n=1}^{\infty} \langle A^* e_n, Be_n \rangle$$

$$= \sum_{n=1}^{\infty} \sum_{j=1}^{\infty} \langle A^* e_n, e_j \rangle \langle e_j, Be_n \rangle = \sum_{j=1}^{\infty} \sum_{n=1}^{\infty} \langle B^* e_j, e_n \rangle \langle e_n, Ae_j \rangle$$

$$= \sum_{j=1}^{\infty} \langle B^* e_j, Ae_j \rangle = \text{Tr}(BA).$$

Because all the series involved converge absolutely the order of summation may be exchanged. This completes the proof of the second part and thus of the theorem. □

Remark 22.4.1   *1. It should be clear that the representation of a trace class operator A in a Hilbert space $\mathcal{H}$ in terms of an absolutely summable sequence of numbers and two orthonormal bases of $\mathcal{H}$ in equation (22.2) is not unique. Similarly the freedom in the choice of the orthonormal basis in the evalution of the trace is reflected in the identity*

$$\text{Tr}(UAU^{-1}) = \text{Tr}(A) \qquad \forall A \in \mathfrak{T}(\mathcal{H}), \quad \forall U \in \mathfrak{U}(\mathcal{H}) \qquad (22.6)$$

which follows easily from Part b) iii) of Theorem 22.4.3.

2. In the case of concrete Hilbert spaces the trace can often be evaluated explicitly without much effort, usually easier than for instance the operator norm. Consider the Hilbert–Schmidt integral operator K in $L^2(\mathbb{R}^n)$ discussed earlier. It is defined in terms of a kernel $k \in L^2(\mathbb{R}^n \times \mathbb{R}^n)$ by

$$K\psi(x) = \int_{\mathbb{R}^n} k(x,y)\psi(y)dy \quad \forall \psi \in L^2(\mathbb{R}^n).$$

In the Exercises we show that

$$Tr(K^*K) = \int_{\mathbb{R}^n} \int_{\mathbb{R}^n} \overline{k(x,y)}k(x,y)dxdy.$$

A special class of trace class operators is of great importance for quantum mechanics, which we briefly mention.

Definition 22.4.2 A **density matrix** or **statistical operator** W on a separable Hilbert space $\mathcal{H}$ is a trace class operator which is symmetric ($W^* = W$), positive ($\langle x, Wx \rangle \geq 0$ for all $x \in \mathcal{H}$), and normalized (Tr $W = 1$).

Note that in a complex Hilbert space symmetry is implied by positivity. In quantum mechanics density matrices are usually denoted by ρ. Density matrices can be characterized explicitly.

Theorem 22.4.4 A bounded linear operator W on a separable Hilbert space $\mathcal{H}$ is a density matrix if, and only if, there are a sequence of nonnegative numbers $\rho_n \geq 0$ with $\sum_{n=1}^{\infty} \rho_n = 1$ and an orthonormal basis $\{e_n : n \in \mathbb{N}\}$ of $\mathcal{H}$ such that for all $x \in \mathcal{H}$,

$$Wx = \sum_{n=1}^{\infty} \rho_n \langle e_n, x \rangle e_n, \qquad (22.7)$$

i.e., $W = \sum_{n=1}^{\infty} \rho_n P_{e_n}$, P_{e_n} = projector onto the subspace $\mathbb{K} e_n$.

Proof. Using the characterization (22.2) of trace class operators this proof is left as an exercise. □

In the remark above we have mentioned the integral operators of Hilbert–Schmidt. These operators are a special case of the Hilbert–Schmidt operators which are defined as follows:

Definition 22.4.3 A bounded linear operator A on the Hilbert space $\mathcal{H}$ is called a **Hilbert–Schmidt operator** if, and only if, $\text{Tr}(A^*A)$ is finite.

22.5 Some applications in Quantum Mechanics

The results of this chapter have important applications in quantum mechanics, but also in other areas. We mention, respectively sketch, some of these applications briefly.

We begin with a reminder of some of the basic principles of quantum mechanics (see for instance [Jau73, Ish95]).

22.5 Some applications in Quantum Mechanics

1. The *states* of a quantum mechanical system are described in terms of density matrices on a separable complex Hilbert space $\mathcal{H}$.

2. The *observables* of the systems are represented by self-adjoint operators in $\mathcal{H}$.

3. The *mean value* or *expectation value* of an observable a in a state z is equal to the expectation value $E(A, W)$ of the corresponding operators in $\mathcal{H}$; if the self-adjoint operator A represents the observable a and the density matrix W represents the state z, this means that

$$m(a, z) = E(A, W) = \text{Tr}(AW).$$

Naturally, the mean value $m(a, z)$ is considered as the mean value of the results of a measurement procedure. Here we have to assume that AW is a trace class operator, reflecting the fact that not all observables can be measured in all states.

4. Examples of density matrices W are projectors P_e on $\mathcal{H}$, $e \in \mathcal{H}$, $\|e\| = 1$, i.e., $Wx = \langle e, x \rangle e$. Such states are called *vector states* and e the representing vector. Then clearly $E(A, P_e) = \langle e, Ae \rangle = \text{Tr}(P_e A)$.

5. *Convex combinations* of states, i.e., $\sum_{j=1}^{n} \lambda_j W_j$ of states W_j are again states (here $\lambda_j \geq 0$ for all j and $\sum_{j=1}^{n} \lambda_j = 1$). Those states which can not be represented as nontrivial convex combinations of other states are called *extremal* or *pure* states. Under quite general conditions one can prove: There are extremal states and the set of all convex combinations of pure states is dense in the space of all states (Theorem of Krein–Milman, [Rud73], not discussed here).

Thus we learn, that and how, projectors and density matrices enter in quantum mechanics.

Next we discuss a basic application of Stone's Theorem 22.2.3 on groups of unitary operators. As we had argued earlier, the Hilbert space of an elementary localizable particle in one dimension is the separable Hilbert space $L^2(\mathbb{R})$. The translation of elements $f \in L^2(\mathbb{R})$ is described by the unitary operators $U(a)$, $a \in \mathbb{R}$: $(U(a)f)(x) = f_a(x) = f(x - a)$. It is not difficult to show that this one-parameter group of unitary operators acts strongly continuous on $L^2(\mathbb{R})$: One shows $\lim_{a \to 0} \|f_a - f\|_2 = 0$. Now Stone's theorem applies. It says that this group is generated by a self-adjoint operator P which is defined on the domain

$$D = \left\{ f \in L^2(\mathbb{R}) : \lim_{a \to 0} \frac{1}{a}(f_a - f) \text{ exists in } L^2(\mathbb{R}) \right\}$$

by

$$Pf = \frac{1}{i} \lim_{a \to 0} \frac{1}{a}(f_a - f) \qquad \forall f \in D.$$

The domain D is known to be $D = W^1(\mathbb{R}) \equiv \{f \in L^2(\mathbb{R}) : f' \in L^2(\mathbb{R})\}$ and clearly $Pf = -if' = -i\frac{df}{dx}$. This operator P represents the momentum of the particle which is consistent with the fact that P generates the translations:

$$U(a) = e^{-iaP}.$$

As an illustration of the use of trace class operators and the trace functional we discuss a general form of the *Heisenberg uncertainty principle*. Given a density matrix W on a separable Hilbert space $\mathcal{H}$, introduce the set

$$O_W = \{A \in \mathcal{B}(\mathcal{H}) : A^*AW \in \mathfrak{T}(\mathcal{H})\}$$

and a functional on $O_W \times O_W$,

$$(A, B) \mapsto \langle A, B \rangle_W = \text{Tr}(A^*BW).$$

One shows (see Exercises) that this is a sesquilinear form on O_W which is positive semi-definite ($\langle A, A \rangle_W \geq 0$), hence the Cauchy–Schwarz inequality applies, i.e.,

$$|\langle A, B \rangle_W| \leq \sqrt{\langle A, A \rangle_W}\sqrt{\langle B, B \rangle_W} \qquad \forall A, B \in O_W.$$

Now consider two self-adjoint operators such that all the operators AAW, BBW, AW, BW, ABW, BAW are of trace class. Then the following quantities are well defined:

$$\overline{A} = A - \langle A \rangle_W I, \qquad \overline{B} = B - \langle B \rangle_W I$$

and then

$$\Delta_W(A) = \sqrt{\text{Tr}(\overline{A}\overline{A}W)} = \sqrt{\text{Tr}(A^2W) - \langle A \rangle_W^2},$$

$$\Delta_W(B) = \sqrt{\text{Tr}(\overline{B}\overline{B}W)} = \sqrt{\text{Tr}(B^2W) - \langle B \rangle_W^2}.$$

The quantity $\Delta_W(A)$ is called the *uncertainty* of the observable 'A' in the state 'W'. Next calculate the expectation value of the commutator $[A, B] = AB - BA$. One finds

$$\text{Tr}([A, B]W) = \text{Tr}([\overline{A}, \overline{B}]W) = \text{Tr}(\overline{A}\overline{B}W) - \text{Tr}(\overline{B}\overline{A}W) = \langle \overline{A}, \overline{B} \rangle_W - \langle \overline{B}, \overline{A} \rangle_W$$

and by the above inequality this expectation value is bounded by the product of the uncertainties:

$$|\text{Tr}([A, B]W)| \leq |\langle \overline{A}, \overline{B} \rangle_W| + |\langle \overline{B}, \overline{A} \rangle_W| \leq \Delta_W(A)\Delta_W(B) + \Delta_W(B)\Delta_W(A).$$

Usually this estimate of the expectation value of the commutator in terms of the uncertainties is written as

$$\frac{1}{2}|\text{Tr}([A, B]W)| \leq \Delta_W(A)\Delta_W(B)$$

and called the *Heisenberg uncertainty relations* (for the 'observables' A, B).

Actually in quantum mechanics many observables are represented by unbounded self-adjoint operators. Then the above calculations do not apply directly and thus typically they are not done for a general density matrix as above but for pure states only. Originally they were formulated by Heisenberg corresponding to the observables of the position and the momentum, represented by the self-adjoint operators Q and P with the commutator $[Q, P] \subseteq iI$ and thus on suitable pure states ψ the famous version

$$\frac{1}{2} \leq \Delta_\psi(Q)\Delta_\psi(P)$$

of these uncertainty relations follows.

22.6 Exercises

1. Consider the Hilbert space $\mathcal{H} = \mathbb{K}^n$ and an isometric map $A : \mathbb{K}^n \to \mathbb{K}^n$. Prove: A is unitary.

2. In the Hilbert space $\mathcal{H} = \ell^2(\mathbb{K})$ with canonical basis $\{e_n : n \in \mathbb{N}\}$ define a linear operator A by $A(\sum_{n=1}^\infty c_n e_n) = \sum_{n=1}^\infty c_n e_{n+1}, c_n \in \mathbb{K}, \sum_{n=1}^\infty |c_n|^2 < \infty$. Show: A is isometric but not unitary.

3. Show: The weak and strong operator topologies coincide on the space $\mathfrak{U}(\mathcal{H})$ of unitary operators on a Hilbert space $\mathcal{H}$.

4. For a continuous function $x : \mathbb{R} \to \mathcal{H}$ on the real line with values in a Hilbert space $\mathcal{H}$ which has a compact support, prove the existence of the Riemann integral

$$\int_\mathbb{R} x(t)dt$$

and the estimate

$$\| \int_\mathbb{R} x(t)dt \| \leq \int_\mathbb{R} \|x(t)\|dt.$$

Hints: As a continuous real valued function of compact support the function $t \mapsto \|x(t)\|$ is known to be Riemann integrable, hence

$$\int_\mathbb{R} \|x(t)\|dt = \lim_{N \to \infty} \sum_{i=1}^N \|x(t_{N,i})\|\frac{L}{N}$$

where $\{t_{N,i} : i = 1, \ldots, N\}$ is an equidistant partition of the support of the function x of length L. From the existence of this limit deduce that the sequence

$$S_N = \sum_{i=1}^N x(t_{N,i})\frac{L}{N}, \quad N \in \mathbb{N}$$

is a Cauchy sequence in the Hilbert space $\mathcal{H}$ and thus this sequence has a limit in $\mathcal{H}$ which is the Riemann integral of the vector valued function x:

$$\int_{\mathbb{R}} x(t)dt = \lim_{N\to\infty} \sum_{i=1}^{N} x(t_{N,i}) \frac{L}{N}.$$

The estimate for the norm of the Riemann integral follows easily.

Deduce also the standard properties of a Riemann integral, i.e., show that it is linear in the integrand, additive in its domain of integration and that the fundamental theorem of calculus holds also for the vector-valued version.

5. Complete the proof of Theorem 22.3.4.

    Hints: For the proof of Part b) see also [RS80].

6. For a trace class operator A on a separable Hilbert space $\mathcal{H}$ and two orthonormal bases $\{e_n : n \in \mathbb{N}\}$ and $\{f_n : n \in \mathbb{N}\}$, prove

$$\sum_{n=1}^{\infty} \langle e_n, A e_n \rangle = \sum_{n=1}^{\infty} \langle f_n, A f_n \rangle.$$

7. Using Theorem 22.4.2 determine the form of the adjoint of a trace class operator A on $\mathcal{H}$ explicitly.

8. For a Hilbert–Schmidt operator K with kernel $k \in L^2(\mathbb{R}^n \times \mathbb{R}^n)$ show that

$$Tr(K^*K) = \int_{\mathbb{R}^n} \int_{\mathbb{R}^n} \overline{k(x,y)} k(x,y) dx dy.$$

9. Prove the characterization (22.7) of a density matrix W.

    Hints: One can use $W^* = W = |W| = \sqrt{W^*W}$ and the explicit representation of the adjoint of a trace class operator (see the previous problem).

10. Show: A density matrix W on a Hilbert space $\mathcal{H}$ represents a vector state, i.e., can be written as the projector P_ψ onto the subspace generated by a vector $\psi \in \mathcal{H}$ if, and only if, $W^2 = W$.

23
Self-adjoint Hamilton Operators

The time evolution of a classical mechanical system is governed by the Hamilton function. Similarly, the Hamilton operator determines the time evolution of a quantum mechanical system and this operator provides information about the total energy of the system in specific states. In both cases it is important that the Hamilton operator is self-adjoint in the Hilbert space of the quantum mechanical system. Thus we are faced with the mathematical task of constructing a self-adjoint Hamilton operator out of a given classical Hamilton function. The Hamilton function is the sum of the kinetic and the potential energy. For the construction of the Hamilton operator this typically means that we have to add to unbounded self-adjoint operators.

In the chapter on quadratic forms we have explained a strategy which allows to add two unbounded positive operators even if the intersection of their domains of definition is too small for the natural addition of unbounded operators to be meaningful. Now we consider the case where the domain of the potential operator contains the domain of the free Hamilton operator. Then obviously the addition of the two operators is not a problem. But the question of self-adjointness of the sum remains. The key to the solution of this problem is to consider the potential energy as a small perturbation of the free Hamilton operator, in a suitable way. Then indeed self-adjointness of the sum on the domain of the free Hamilton operator follows.

A first section introduces the basic concepts and results of the theory of Kato perturbations (see the book of T. Kato, [Kat66]) which is then applied to the case of Hamilton operators discussed above.

23. Self-adjoint Hamilton Operators

23.1 Kato perturbations

As in most parts of this book related to quantum mechanics, in this section $\mathcal{H}$ is assumed to be a complex Hilbert space. The starting point is

Definition 23.1.1 *Suppose A, B are two densely defined linear operators in $\mathcal{H}$. B is called a **Kato perturbation** of A if, and only if, $D(A) \subset D(B)$ and there are real numbers $0 \leq a < 1$ and b such that*

$$\|Bx\| \leq a\|Ax\| + b\|x\| \qquad \forall x \in D(A). \tag{23.1}$$

This notion of a Kato perturbation is very effective in solving the problem of self-adjointness of the sum, under natural restrictions.

Theorem 23.1.1 (Kato–Rellich Theorem) *Suppose A is a self-adjoint and B is a symmetric operator in $\mathcal{H}$. If B is a Kato perturbation of A, then the sum $A + B$ is self-adjoint on the domain $D(A)$.*

Proof. According to Part c) of Theorem 19.3.1 it suffices to show that for some number $c > 0$ we have ran $(A + B + icI) = \mathcal{H}$. For every $x \in D(A)$ and $c \in \mathbb{R}$ a simple calculation gives

$$\|(A + icI)x\|^2 = \|Ax\|^2 + c^2\|x\|^2. \tag{23.2}$$

Hence, for $c \neq 0$, the operator $A + icI$ is injective and thus has an inverse on its range which is equal to $\mathcal{H}$ by Theorem 19.3.1 and which has values in the domain of A. Therefore the elements $x \in D(A)$ can be represented as $x = (A + icI)^{-1}y$, $y \in \mathcal{H}$ and the above identity can be rewritten as

$$\|y\|^2 = \|A(A + icI)^{-1}y\|^2 + c^2\|(A + icI)^{-1}y\|^2 \qquad \forall y \in \mathcal{H}.$$

And this identity has two implications:

$$\|A(A + icI)^{-1}\| \leq 1, \qquad \|(A + icI)^{-1}\| \leq \frac{1}{|c|}, \quad c \neq 0.$$

Now use the assumption that B is a Kato perturbation of A. For $c > 0$ and $x = (A + icI)^{-1}y \in D(A)$ the following estimate results:

$$\|B(A + icI)^{-1}y\| \leq a\|A(A + icI)^{-1}y\| + b\|(A + icI)^{-1}y\| \leq (a + \frac{b}{c})\|y\|.$$

We deduce $\|B(A+icI)^{-1}\| \leq (a+\frac{b}{c})$. Since $a < 1$ is assumed there is a $c_0 > 0$ such that $(a+\frac{b}{c_0}) < 1$. Thus $C = B(A + icI)^{-1}$ is a bounded operator with $\|C\| < 1$ and therefore the operator $I + C$ is invertible with inverse given by the Neumann series (see equation (21.4)). This means in particular that the operator $I + C$ has the range ran $(I+C) = \mathcal{H}$. Since A is self-adjoint one knows ran $(A \pm ic_0 I) = \mathcal{H}$ and therefore that the range of $A + B \pm ic_0 I = (I + C)(A \pm ic_0 I)$ is the whole Hilbert space. Thus we conclude. □

One can read the Kato–Rellich theorem as saying that self-adjointness of operators is a property which is stable against certain small symmetric perturbations. But naturally in a concrete case it might be quite difficult to establish whether or not a given symmetric operator is a Kato perturbation of a given self-adjoint operator. Thus the core of the following section is to prove that certain classes of potential operators are indeed Kato perturbations of the free Hamilton operator.

23.2 Kato perturbations of the free Hamiltonian

Though it can be stated more generally, we present the case of a three dimensional system explicitly. The Hamilton function of a particle of mass $m > 0$ in the force field associated with a potential V is

$$H(p,q) = \frac{1}{2m}p^2 + V(q)$$

where $q \in \mathbb{R}^3$ is the position variable and $p \in \mathbb{R}^3$ the momentum of the particle.

Recall the realization of the position operator $Q = (Q_1, Q_2, Q_3)$ and of the momentum operator $P = (P_1, P_2, P_3)$ in the Hilbert space $\mathcal{H} = L^2(\mathbb{R}^3)$ of such a system. The domain of Q is

$$D(Q) = \left\{ \psi \in L^2(\mathbb{R}^3) : x_j \psi \in L^2(\mathbb{R}^3), \ j = 1, 2, 3 \right\}$$

and on this domain the component Q_j is defined as the multiplication with the component x_j of the variable $x \in \mathbb{R}^3$. Such multiplication operators have been shown to be self-adjoint. Then the observable of potential energy $V(Q)$ is defined on the domain

$$D(V) = \left\{ \psi \in L^2(\mathbb{R}^3) : V \cdot \psi \in L^2(\mathbb{R}^3) \right\} \quad \text{by} \quad (V(Q)\psi)(x) = V(x)\psi(x)$$

for almost all $x \in \mathbb{R}^3$, $\forall \psi \in D(V)$. We assume V to be a real valued function which is locally square integrable. Then, as we have discussed earlier, V is self-adjoint.

The momentum operator P is the generator of the three parameter group of translations defined by the unitary operators $U(a)$, $a \in \mathbb{R}^3$, $U(a)\psi = \psi_a$, for all $\psi \in L^2(\mathbb{R}^3)$. As in the one dimensional case discussed explicitly, this group is strongly continuous and thus Stone's Theorem 22.2.3 applies and according to this theorem the domain of P is characterized by

$$D(P) = \left\{ \psi \in L^2(\mathbb{R}^3) : \lim_{s \to 0} \frac{1}{s}[\psi_{se_j} - \psi] \in L^2(\mathbb{R}^3), \ j = 1, 2, 3 \right\}$$

where e_j is the unit vector in coordinate direction j. Representing the elements ψ of $L^2(\mathbb{R}^3)$ as images under the Fourier transform $\mathcal{F}_2$, $\psi = \mathcal{F}_2(\tilde{\psi})$, the domain $D(P)$ is conveniently described as $D(P) = \left\{ \psi = \mathcal{F}_2(\tilde{\psi}) : \tilde{\psi} \in D(\tilde{Q}) \right\}$ where $D(\tilde{Q}) = \left\{ \tilde{\psi} \in L^2(\mathbb{R}^3) : q_j \tilde{\psi}(q) \in L^2(\mathbb{R}^3), \ j = 1, 2, 3 \right\}$. Then the action of the momentum operator is $P\psi = \mathcal{F}_2(\tilde{Q}\tilde{\psi})$.

Similarly the domain of the free Hamilton operator

$$H_0 = \frac{1}{2m}P^2$$

is $D(H_0) = \left\{ \psi = \mathcal{F}_2(\tilde{\psi}) : q^2 \tilde{\psi}(q) \in L^2(\mathbb{R}^3) \right\}$. H_0 is self-adjoint on this domain.

The verification that large classes of potential operators V are Kato perturbations of the free Hamiltonian is prepared by

Lemma 23.2.1 *All $\psi \in D(H_0) \subset L^2(\mathbb{R}^3)$ are bounded by*

$$\|\psi\|_\infty \leq 2^{-3/2}\pi^{-1/2}(r^{-1/2}2m\|H_0\psi\|_2 + r^{3/2}\|\psi\|_2), \qquad \text{any } r > 0. \quad (23.3)$$

Proof. For every $\psi \in D(H_0)$ we know $(1+q^2)\tilde\psi(q) \in L^2(\mathbb{R}^3)$ and $(1+q^2)^{-1} \in L^2(\mathbb{R}^3)$ and thus deduce $\tilde\psi(q) = (1+q^2)^{-1}(1+q^2)\tilde\psi(q) \in L^1(\mathbb{R}^3)$. The Cauchy–Schwarz inequality implies

$$\|\tilde\psi\|_1 = \int_{\mathbb{R}^3} (1+q^2)^{-1}(1+q^2)|\tilde\psi(q)|dq$$
$$\leq \|(1+q^2)^{-1}\|_2 \|(1+q^2)\tilde\psi(q)\|_2 \leq \pi(\|q^2\tilde\psi\|_2 + \|\tilde\psi\|_2).$$

Now scale the function with $r > 0$, i.e., consider $\tilde\psi_r(q) = r^3\tilde\psi(rq)$. A simple integration shows

$$\|\tilde\psi_r\|_1 = \|\tilde\psi\|_1, \quad \|\tilde\psi_r\|_2 = r^{3/2}\|\tilde\psi\|_2, \quad \|q^2\tilde\psi_r\|_2 = r^{-1/2}\|q^2\tilde\psi\|_2$$

and thus implies

$$\|\tilde\psi\|_1 = \|\tilde\psi_r\|_1 \leq \pi(r^{-1/2}\|q^2\tilde\psi\|_2 + r^{3/2}\|\tilde\psi\|_2).$$

For the Fourier transformation the estimate $\|\psi\|_\infty \leq \|\tilde\psi\|_1$ is well known and estimate (23.3) follows. □

Theorem 23.2.1 *Any potential of the form $V = V_1 + V_2$ with real valued functions $V_1 \in L^2(\mathbb{R}^3)$ and $V_2 \in L^\infty(\mathbb{R}^3)$ is a Kato perturbation of the free Hamilton operator and thus the Hamilton operator $H = H_0 + V(Q)$ is self-adjoint on the domain $D(H_0)$.*

Proof. For every $\psi \in D(H_0)$ we estimate as follows:

$$\|V\psi\|_2 \leq \|V_1\psi\|_2 + \|V_2\psi\|_2 \leq \|V_1\|_2\|\psi\|_\infty + \|V\|_\infty\|\psi\|_2.$$

Now the term $\|\psi\|_\infty$ is estimated by our lemma and thus

$$\|V\psi\|_2 \leq a(r)\|H_0\psi\|_2 + b(r)\|\psi\|_2$$

with

$$a(r) = (2\pi)^{-1/2}m\|V_1\|_2\, r^{-1/2}, \qquad b(r) = 2^{-3/2}\pi^{-1/2}r^{3/2}\|V_1\|_2 + \|V_2\|_\infty.$$

For sufficiently large r the factor $a(r)$ is smaller than 1 so that Theorem 23.1.1 applies and proves self-adjointness of $H_0 + V(Q)$. □

23.3 Exercises

1. Show that $(1+q^2)^{-1} \in L^2(\mathbb{R}^3)$ and calculate $\|(1+q^2)^{-1}\|_2$.

2. Prove: Potentials of the form $V(x) = \frac{a}{|x|^\rho}$ with some constant a are Kato perturbations of the free Hamilton operator in $L^2(\mathbb{R}^3)$ if $0 < \rho \leq 1$.

    Hints: Denote by χ_R the characteristic function of the ball with radius $R > 0$ and define $V_1 = \chi_R V$ and $V_2 = (1-\chi_R)V$.

24
Elements of Spectral Theory

The *spectrum* $\sigma(A)$ of a linear operator in an infinite dimensional Hilbert space $\mathcal{H}$ is the appropriate generalization of the set of all eigenvalues of a linear operator in a finite dimensional Hilbert space. This statement we intend to establish in this and the following two chapters.

If A is a complex $N \times N$ matrix, i.e., a linear operator in the Hilbert space $\mathbb{C}^N$, one has a fairly simple criterium for eigenvalues: $\lambda \in \mathbb{C}$ is an eigenvalue of A if, and only if, there is a $\psi_\lambda \in \mathbb{C}^N$, $\psi_\lambda \neq 0$, such that $A\psi_\lambda = \lambda \psi_\lambda$ or $(A - \lambda I)\psi_\lambda = 0$. This equation has a nontrivial solution if, and only if, the matrix $A - \lambda I$ is not invertible. In the space of matrices one has a convenient criterium to decide whether or not a matrix is invertible. On this space the determinant is well defined and convenient to use: Thus $A - \lambda I$ is not invertible if, and only if, $\det(A - \lambda I) = 0$. Therefore the set $\sigma(A)$ of eigenvalues of the $N \times N$ matrix A is given by

$$\sigma(A) = \{\lambda \in \mathbb{C} : A - \lambda I \text{ is not invertible}\} \tag{24.1}$$
$$= \{\lambda \in \mathbb{C} : \det(A - \lambda I) = 0\} = \{\lambda_1, \ldots, \lambda_N\}, \tag{24.2}$$

since the polynomial $\det(A - \lambda I)$ of degree N has exactly N roots in $\mathbb{C}$.

In an infinite dimensional Hilbert space one does not have a substitute for the determinant function which is general enough to cover all cases of interest (in special cases one can define such a function, and we will mention it briefly later). Thus in infinite dimensional Hilbert space one can only use the first characterization of $\sigma(A)$ which is *independent of the dimension of the space*. If we proceed with this definition the above identity ensures consistency with the finite dimensional case.

24. Elements of Spectral Theory

24.1 Basic concepts and results

Suppose that $\mathcal{H}$ is a complex Hilbert space and D is a linear subspace of this space. Introduce the set of bounded linear operators on $\mathcal{H}$ which map into D:

$$\mathcal{B}(\mathcal{H}, D) = \{A \in \mathcal{B}(\mathcal{H}) : \operatorname{ran} A \subseteq D\}.$$

Our basis definition now reads:

Definition 24.1.1 *Given a linear operator A with domain D in a complex Hilbert space $\mathcal{H}$, the set*

$$\rho(A) = \left\{z \in \mathbb{C} : A - zI \text{ has an inverse operator } (A - zI)^{-1} \in \mathcal{B}(\mathcal{H}, D)\right\} \tag{24.3}$$

*is called the **resolvent set of** A and its complement*

$$\sigma(A) = \mathbb{C} \setminus \rho(A) \tag{24.4}$$

*the **spectrum of** A. Finally the function*

$$R_A : \sigma(A) \to \mathcal{B}(\mathcal{H}, D), \qquad R_A(z) = (A - zI)^{-1} \tag{24.5}$$

*is the **resolvent of** A.*

Given a point $z \in \mathbb{C}$, it is in general not straightforward to decide when the operator $A - zI$ has an inverse in $\mathcal{B}(\mathcal{H}, D)$. Here the auxiliary concept of a regular point is a good help.

Definition 24.1.2 *Suppose that A is a linear operator in $\mathcal{H}$ with domain D. The **set of regular points of** A is the set*

$$\rho_r(A) = \{z \in \mathbb{C} : \exists \delta(z) > 0 \text{ such that } \|(A - zI)x\| \geq \delta(z)\|x\| \ \forall x \in D\}. \tag{24.6}$$

The relation between regular points and points of the resolvent set is obvious and it is also clear that the set of regular points is open. For a closed operator the resolvent set is open too.

Lemma 24.1.1 *Suppose A is a linear operator in $\mathcal{H}$ with domain D. Then the following holds:*

a) $\rho(A) \subseteq \rho_r(A)$;

b) $\rho_r(A) \subset \mathbb{C}$ *is open;*

c) *if A is closed the resolvent set is open too.*

Proof. If $z \in \rho(A)$ is given, then the resolvent $R_A(z)$ is a bounded linear operator $\mathcal{H} \to D$ such that $x = R_A(z)(A - zI)x$ for all $x \in D$, hence $\|x\| \leq \|R_A(z)\| \|(A - zI)x\|$ for all $x \in D$. For arbitrary $\epsilon > 0$ define $\delta(z) = (\|R_A(z)\| + \epsilon)^{-1}$. With this choice of $\delta(z)$ we easily see that $z \in \rho_r(A)$ and Part a) is proven.

Given $z_0 \in \rho_r(A)$ there is a $\delta(z_0) > 0$ such that $\|(A - z_0 I)x\| \geq \delta(z_0)\|x\|$ for all $x \in D$. For all $z \in \mathbb{C}$ with $|z - z_0| < \frac{1}{2}\delta(z_0)$ we estimate

$$\|(A - zI)x\| = \|(A - z_0)x - (z - z_0)x\| \geq |\|(A - z_0)x\| - \|(z - z_0)x\||$$
$$\geq \|(A - z_0)x\| - \frac{1}{2}\delta(z_0)\|x\| \geq \frac{1}{2}\delta(z_0)\|x\| \quad \forall x \in D,$$

hence with the point z_0 the disk $\{z \in \mathbb{C} : |z - z_0| < \frac{1}{2}\delta(z_0)\}$ is contained in $\rho_r(A)$ too. Thus this set is open.

Now we assume that the operator A is closed and z_0 is a point in the resolvent set of A. Then $R_A(z_0)$ is a bounded linear operator and thus $r = \|R_A(z_0)\|^{-1} > 0$. For all $z \in \mathbb{C}$ with $|z - z_0| < r$ this implies that $C = (z - z_0)R_A(z_0)$ is a bounded operator $\mathcal{H} \to D$ with $\|C\| < 1$. Hence the Neumann series for C converges and it defines the inverse of $I - C$:

$$(I - C)^{-1} = \sum_{n=0}^{\infty} C^n.$$

For $z \in \mathbb{C}$ observe that $A - zI = A - z_0 I - (z - z_0)I = (A - z_0)[I - (z - z_0)R_A(z_0)]$, and it follows that for all points $z \in \mathbb{C}$ with $|z - z_0| < r$ the inverse of $A - zI$ exists and is given by

$$(A - zI)^{-1} = (I - C)^{-1} R_A(z_0) = \sum_{n=0}^{\infty} (z - z_0)^n R_A(z_0)^{n+1}.$$

In order to show that this inverse operator maps into the domain D, consider the partial sum $S_N = \sum_{n=0}^{N} (z - z_0)^n R_A(z_0)^{n+1}$ of this series. As a resolvent the operator $R_A(z_0)$ maps into D, hence all the partial sums S_N map into D. For $x \in \mathcal{H}$ we know that

$$y = (A - zI)^{-1} x = \lim_{N \to \infty} S_N x$$

in the Hilbert space $\mathcal{H}$. We claim $y \in D$. To see this calculate

$$(A - zI)S_N x = [I - (z - z_0)R_A(z_0)](A - z_0 I)\sum_{n=0}^{\infty}(z - z_0)^n R_A(z_0)^{n+1} x$$

$$= (I - C)\sum_{n=0}^{N} C^n x.$$

We deduce $\lim_{N \to \infty}(A - zI)S_N x = x$. Since A is closed, it follows that $y \in D$ and $(A - zI)y = x$. This proves that $(A - zI)^{-1}$ maps into D and thus is equal to the resolvent $R_A(z)$, for all $|z - z_0| < r$. And the resolvent set is therefore open. □

Corollary 24.1.1 *For a closed linear operator A in a complex Hilbert space $\mathcal{H}$ the resolvent is an analytic function $R_A : \rho(A) \to \mathfrak{B}(\mathcal{H})$. For any point $z_0 \in \rho(A)$ one has the power series expansion*

$$R_A(z) = \sum_{n=0}^{\infty} (z - z_0)^n R_A(z_0)^{n+1} \tag{24.7}$$

which converges in $\mathfrak{B}(\mathcal{H})$ for all $z \in \mathbb{C}$ with $|z - z_0| < \|R_A(z_0)\|^{-1}$.
Furthermore the **resolvent identity**

$$R_A(z) - R_A(\zeta) = (z - \zeta)R_A(z)R_A(\zeta) \quad \forall z, \zeta \in \rho(A) \tag{24.8}$$

holds and shows that the resolvents at different points commute.

Proof. The power series expansion has been established in the proof of Lemma 24.1.1. Since the resolvent maps into the domain of the operator A one has

$$R_A(z) - R_A(\zeta) = R_A(z)(A - \zeta I)R_A(\zeta) - R_A(z)(A - zI)R_A(\zeta)$$
$$= R_A(z)[(A - \zeta I) - (A - zI)]R_A(\zeta) = (z - \zeta)R_A(z)R_A(\zeta)$$

which proves the resolvent identity.

Note that a straightforward iteration of the resolvent identity also gives the power series expansion of the resolvent. □

Note that according to our definitions the operator $A - zI$ is injective for a regular point $z \in \mathbb{C}$ and has thus a bounded inverse on its range. For a point z in the resolvent set $\rho(A)$ the operator $A - zI$ is in addition surjective and its inverse maps the Hilbert space $\mathcal{H}$ into the domain D. Since regular points have a simple characterization one would like to know when a regular point belongs to the resolvent set. To this end we introduce the spaces $\mathcal{H}_A(z) = \operatorname{ran}(A - zI) = (A - zI)D \subseteq \mathcal{H}$. If the operator A is closed these subspaces are closed. For a regular point z the operator $A - zI$ has an inverse operator $(A - zI)^{-1} : \mathcal{H}_A(z) \to D$ which is bounded in norm by $\frac{1}{\delta(z)}$. After these preparations we can easily decide when a regular point belongs to the resolvent set. This is the case if, and only if, $\mathcal{H}_A(z) = \mathcal{H}$. In the generality in which we have discussed this problem thus far one cannot say much. However for densely defined closed operators and then for self-adjoint operators we know how to proceed.

Recall that a densely defined operator A has a unique adjoint A^* and that the relation $(\operatorname{ran}(A - zI))^\perp = N(A^* - \bar{z}I)$ holds; therefore

$$\rho_r(A) \subseteq \rho(A) \Leftrightarrow N(A^* - \bar{z}I) = \{0\}.$$

For a self-adjoint operator this criterium is easily verified. Suppose $z \in \rho_r(A)$ and $x \in N(A^* - \bar{z}I)$, i.e., $A^*x = \bar{z}x$. Since A is self-adjoint it follows that $x \in D$ and $Ax = \bar{z}x$ and therefore $\bar{z}\langle x, x \rangle = \langle x, Ax \rangle = \langle Ax, x \rangle = z\langle x, x \rangle$. We conclude that either $x = 0$ or $z = \bar{z}$. The latter case implies $(A - zI)x = 0$ which contradicts the assumption that z is a regular point, hence $x = 0$. This nearly proves

Theorem 24.1.1 *For a self-adjoint operator A in a complex Hilbert space $\mathcal{H}$ the resolvent set $\rho(A)$ and the set $\rho_r(A)$ of regular points coincide and the spectrum $\sigma(A)$ is a nonempty closed subset of $\mathbb{R}$.*

Proof. As the complement of the open resolvent set, the spectrum $\sigma(A)$ is closed. For the proof of $\sigma(A) \subseteq \mathbb{R}$ we use the identity $\rho(A) = \rho_r(A)$. For all points $z = \alpha + i\beta$ one has for all $x \in D$,

$$\|(A - zI)x\|^2 = \|(A - \alpha I)x + i\beta x\|^2 = \|(A - \alpha I)x\|^2 + \|\beta x\|^2 \geq |\beta|^2 \|x\|^2,$$

and this lower bound shows that all points $z = \alpha + i\beta$ with $\beta \neq 0$ are regular points.

Here we prove that the spectrum of a bounded operator is not empty. The general case of an unbounded self-adjoint operator follows easily from the spectral theorem which is discussed in a later chapter (see Theorem 26.3.1).

Suppose that the spectrum $\sigma(A)$ of the bounded self-adjoint operator is empty. Then the resolvent R_A is an entire analytic function with values in $\mathcal{B}(\mathcal{H})$ (see Corollary 24.1.1). For all points $z \in \mathbb{C}$, $|z| > 2\|A\|$, the resolvent is bounded: $R_A(z) = -z^{-1}(I - \frac{1}{z}A)^{-1} = -\frac{1}{z}\sum_{n=0}^{\infty}(\frac{A}{z})^n$ implies the bound

$$\|R_A(z)\| \leq \sum_{n=0}^{\infty} \|A\|^n |z|^{-n-1} = \frac{1}{|z| - \|A\|} \leq \frac{1}{\|A\|}.$$

As an analytic function, R_A is bounded on the compact set $\{z \in \mathbb{C} : |z| \le 2\|A\|\}$ and hence R_A is a bounded entire function. The theorem of Liouville (Corollary 9.3.1) implies that R_A is constant. This contradiction implies that the spectrum is not empty. □

Since for a self-adjoint operator the resolvent set and the set of regular points are the same, a real number belongs to the spectrum if, and only if, it is not a regular point. Taking the definition of a regular point into account, points of the spectrum can be characterized in the following way.

Theorem 24.1.2 (Weyl's criterium) *A real number λ belongs to the spectrum of a self-adjoint operator A in a complex Hilbert space $\mathcal{H}$ if, and only if, there is a sequence $(x_n)_{n\in\mathbb{N}} \subset D(A)$ such that $\|x_n\| = 1$ for all $n \in \mathbb{N}$ and*

$$\lim_{n\to\infty} \|(A - \lambda I)x_n\| = 0.$$

In the following section we study several explicit examples. These examples show that in infinite dimensional Hilbert spaces the spectrum does not only consist of eigenvalues, but contains various other parts which have no analogue in the finite dimensional case. The following definition gives a first division of the spectrum into the set of all eigenvalues and some remainder. Later, with the help of the spectral theorem, a finer division of the spectrum will be introduced and investigated.

Definition 24.1.3 *Let A be a closed operator in a complex Hilbert space $\mathcal{H}$ and $\sigma(A)$ its spectrum. The **point spectrum** $\sigma_p(A)$ of A is the set of all eigenvalues, i.e.,*

$$\sigma_p(A) = \{\lambda \in \sigma(A) : N(A - \lambda I) \ne \{0\}\}.$$

*The complement $\sigma(A)\setminus\sigma_p(A)$ of the point spectrum is the **continuous spectrum** $\sigma_c(A)$. Finally, the **discrete spectrum** $\sigma_d(A)$ of A is the set of all eigenvalues λ of finite multiplicity which are isolated in $\sigma(A)$, i.e.,*

$$\sigma_d(A) = \{\lambda \in \sigma_p(A) : \dim N(A - \lambda I) < \infty, \; \lambda \text{ isolated in } \sigma(A)\}.$$

As in the finite dimensional case the eigenspaces to different eigenvalues of a self-adjoint operator are orthogonal.

Corollary 24.1.2 *Suppose A is a self-adjoint operator in a complex Hilbert space and $\lambda_j \in \sigma_p(A)$, $j = 1, 2$ are two eigenvalues. If $\lambda_1 \ne \lambda_2$, then the corresponding eigenspaces are orthogonal: $N(A - \lambda_1 I) \perp N(A - \lambda_2)$.*

Proof. If $A\psi_j = \lambda_j \psi_j$, then $(\lambda_1 - \lambda_2)\langle\psi_1, \psi_2\rangle = \langle\lambda_1\psi_1, \psi_2\rangle - \langle\psi_1, \lambda_2\psi_2\rangle = \langle A\psi_1, \psi_2\rangle - \langle\psi_1, A\psi_2\rangle = 0$, hence $\langle\psi_1, \psi_2\rangle = 0$ since $\lambda_1 \ne \lambda_2$. □

We conclude this section with the observation that the spectrum of linear operators does not change under unitary transformations, more precisely:

Proposition 24.1.3 *If A_j is a closed operator in the complex Hilbert space $\mathcal{H}_j$, $j = 1, 2$, and if there is a unitary map $U : \mathcal{H}_1 \to \mathcal{H}_2$ such that $D(A_2) = UD(A_1)$ and $A_2 = UA_1U^{-1}$, then both operators have the same spectrum: $\sigma(A_1) = \sigma(A_2)$.*

Proof. See Exercises. □

24.2 The spectrum of special operators

In general it is quite a difficult problem to determine the spectrum of a closed or self-adjoint operator. The best one can do typically is to give some estimate in those cases where more information about the operator is available. In special cases, for instance in cases of self-adjoint realizations of certain differential operators, one can determine the spectrum exactly. We consider a few examples.

Proposition 24.2.1 *The spectrum $\sigma(U)$ of a unitary operator U on a complex Hilbert space $\mathcal{H}$ is contained in the unit circle $\{z \in \mathbb{C} : |z| = 1\}$.*

Proof. If $|z| < 1$, then we write $U - zI = U(I - zU^{-1})$. Since the operator zU^{-1} has a norm smaller than 1, the Neumann series can be used to find the bounded inverse of $U - zI$. Similarly, for $|z| > 1$, we write $U - zI = -z(I - \frac{1}{z}U)$. This time the Neumann series for the operator $\frac{1}{z}U$ allows us to calculate the inverse. Thus all points $z \in \mathbb{C}$ with $|z| < 1$ or $|z| > 1$ belong to the resolvent set and therefore the spectrum is contained in the unit circle. $\square$

It is somewhat surprising that the spectrum of the Fourier Transformation $\mathcal{F}_2$ on the Hilbert space $L^2(\mathbb{R})$ can be calculated.

Proposition 24.2.2 *The spectrum of the Fourier transformation $\mathcal{F}_2$ on the Hilbert space $L^2(\mathbb{R})$ is $\sigma(\mathcal{F}_2) = \{1, i, -1, -i\}$.*

Proof. The system of Hermite functions $\{h_n : n = 0, 1, 2, \ldots\}$ is an orthonormal basis of $L^2(\mathbb{R})$ (see equation (16.1)). In the Exercises we show by induction with respect to the order n that

$$\mathcal{F}_2(h_n) = (-i)^n h_n, \quad n = 0, 1, 2, \ldots$$

holds. Thus we know a complete set of orthonormal eigenfunctions together with the corresponding eigenvalues. Therefore we can represent the Fourier transformation as

$$\mathcal{F}_2 = \sum_{n=0}^{\infty} (-i)^n P_n$$

where P_n is the projector onto the subspace generated by the eigenfunction h_n. In the following example we determine the spectrum of operators which are represented as a series of projectors onto an orthonormal basis with any arbitrary coefficients s_n. One finds that the spectrum of such an operator is the closure of the set of coefficients which in the present case is $\{1, i, -1, -i\}$. $\square$

Example 24.2.1 1. *$e_n : n \in \mathbb{N}$ is an orthonormal basis of the complex Hilbert space $\mathcal{H}$ and $\{s_n : n \in \mathbb{N}\} \subset \mathbb{C}$ is some sequence of complex numbers. Introduce the set*

$$D = \left\{ x \in \mathcal{H} : \sum_{n=1}^{\infty} |s_n|^2 |\langle e_n, x \rangle|^2 < \infty \right\}$$

and for $x \in D$ define

$$Ax = \sum_{n=1}^{\infty} s_n \langle e_n, x \rangle e_n. \tag{24.9}$$

In the Exercises we show that A is a densely defined closed linear operator with adjoint

$$A^*y = \sum_{n=1}^{\infty} \overline{s_n} \langle e_n, y \rangle e_n \qquad \forall\, y \in D.$$

We claim:

$$\sigma(A) = \overline{\{s_n : n \in \mathbb{N}\}}.$$

If $z \in \mathbb{C}$ and $z \notin \overline{\{s_n : n \in \mathbb{N}\}}$, then $\delta(z) = \inf\{|s_n - z| : n \in \mathbb{N}\} > 0$ and thus for all $x \in D$,

$$\|(A-zI)x\|^2 = \sum_{n=1}^{\infty} |s_n - z|^2 |\langle e_n, x\rangle|^2 \geq \delta(z)^2 \sum_{n=1}^{\infty} |\langle e_n, x\rangle|^2 = \delta(z)^2 \|x\|^2.$$

Thus these points are regular points of the operator A. They are actually points of the resolvent set since one shows that the inverse of $A - zI$ is given by

$$(A - zI)^{-1}x = \sum_{n=1}^{\infty} \frac{1}{s_n - z} \langle e_n, x\rangle e_n$$

and this operator maps $\mathcal{H}$ into D. We conclude $\rho(A) = \mathbb{C}\setminus\overline{\{s_n : n \in \mathbb{N}\}}$ and this proves our claim.

2. For a continuous function $g : \mathbb{R}^n \to \mathbb{C}$ define the domain

$$D_g = \left\{ f \in L^2(\mathbb{R}^n) : gf \in L^2(\mathbb{R}^n) \right\}.$$

As we have shown earlier the operator M_g of multiplication with the function g is a densely defined closed linear operator in the Hilbert space $\mathcal{H} = L^2(\mathbb{R}^n)$ and its adjoint is the operator of multiplication with the complex conjugate function $\overline{g}$. We claim that the spectrum of the operator M_g is the closure of the range of the function g,

$$\sigma(M_g) = \overline{\mathrm{ran}\, g}$$

where $\mathrm{ran}\, g = \{\lambda \in \mathbb{C} : \lambda = g(x) \text{ for some } x \in \mathbb{R}^n\}$. Since $\mathbb{C}\setminus\overline{\mathrm{ran}\, g}$ is open, every point $z \in \mathbb{C}\setminus\overline{\mathrm{ran}\, g}$ has a positive distance from $\overline{\mathrm{ran}\, g}$, i.e.,

$$\delta(z) = \inf\{|g(x) - z| : x \in \mathbb{R}^n\} > 0.$$

Therefore $x \mapsto (g(x)-z)^{-1}$ is a continuous function on $\mathbb{R}^n$ which is bounded by $\frac{1}{\delta(z)}$. It follows that $(M_g - zI)^{-1}$, defined by $(M_g - zI)^{-1} f = \frac{f}{g-z}$ is a bounded linear operator on $L^2(\mathbb{R}^n)$. Since the integral

$$\int_{\mathbb{R}^n} |\frac{g(x)}{g(x) - z} f(x)|^2 dx$$

is finite for all $f \in L^2(\mathbb{R}^n)$ the operator $(M_g - zI)^{-1}$ maps $L^2(\mathbb{R}^n)$ into the domain D_g of M_g. This proves that $\rho(M_g) = \mathbb{C}\backslash\overline{\operatorname{ran} g}$ and we conclude.

In the case that the function g is real valued and not constant this is an example of an operator whose spectrum contains open intervals, i.e., the continuous spectrum is not empty in this case. In this case the operator M_g has no eigenvalues (see Exercises).

24.3 Comments on spectral properties of linear operators

In Definition 24.1.3 the complement $\sigma_c(A) = \sigma(A)\backslash\sigma_p(A)$ of the point spectrum has been called the continuous spectrum of A. This terminology is quite unfortunate since it is often rather misleading: The continuous spectrum can be a discrete set. To see this, consider Example 24.2.1 and choose there the sequence $s_n = \frac{1}{n}$. Then the spectrum of the operator A defined through this sequence is $\sigma(A) = \overline{\{\frac{1}{n} : n \in \mathbb{N}\}} = \{1, \frac{1}{2}, \frac{1}{3}, \ldots, 0\}$ while the point spectrum is $\sigma_p(A) = \{1, \frac{1}{2}, \frac{1}{3}, \ldots\}$, hence the continuous spectrum is just one point: $\sigma_c(A) = \sigma(A)\backslash\sigma_p(A) = \{0\}$.

It is very important to be aware of the fact that the spectrum of an operator depends on its domain in a very sensitive way. To illustrate this point we are going to construct two unbounded linear operators which consist of the same rule of assignment but on different domains. The resulting operators have completely different spectra.

In the Hilbert space $\mathcal{H} = L^2([0, 1])$ introduce two dense linear subspaces

$$D_1 = \left\{ f \in L^2([0, 1]) : f \text{ is absolutely continuous, } f' \in L^2([0, 1]) \right\},$$

$$D_2 = \{f \in D_1 : f(0) = 0\}.$$

Denote by P_j the operator of differentiation $i\frac{d}{dx}$ on the domain D_j, $j = 1, 2$. Both operators P_1, P_2 are closed.

For every $\lambda \in \mathbb{C}$ the exponential function e_λ, $e_\lambda(x) = e^{-i\lambda x}$, belongs to the domain D_1 and clearly $(P_1 - \lambda I)e_\lambda = 0$. We conclude that $\sigma(P_1) = \mathbb{C}$.

Elementary calculations show that the operator R_λ defined by

$$(R_\lambda f)(x) = i \int_0^x e^{-i\lambda(x-y)} f(y)dy \qquad \forall f \in L^2([0, 1])$$

has the following properties: R_λ maps $L^2([0, 1])$ into D_2 and

$$(\lambda I - P_2)R_\lambda = I, \qquad R_\lambda(\lambda I - P_2) = I_{|D_2}.$$

Clearly R_λ is a bounded operator on $L^2([0, 1])$, hence $R_\lambda \in \mathcal{B}(L^2([0, 1]), D_2)$. This is true for every $\lambda \in \mathbb{C}$ and we conclude that $\rho(P_2) = \mathbb{C}$, hence $\sigma(P_2) = \emptyset$.

Without proof we mention an interesting result about the spectrum of a closed symmetric operator. The spectrum determines whether such an operator is self-adjoint or not!

Theorem 24.3.1 *A closed symmetric operator A in a complex Hilbert space $\mathcal{H}$ is self-adjoint if, and only if, its spectrum $\sigma(A)$ is a subset of $\mathbb{R}$.*

This result is certainly another strong motivation why in quantum mechanics observables should be represented by self-adjoint operators and not only by symmetric operators, since in quantum mechanics the expectation values of observables have to be real.

One can also show that the spectrum of an essentially self-adjoint operator is contained in $\mathbb{R}$. The converse of these results reads: If the spectrum $\sigma(A)$ of a symmetric operator A in a complex Hilbert space $\mathcal{H}$ is contained in $\mathbb{R}$, then the operator is either self-adjoint or essentially self-adjoint. This implies that the spectrum $\sigma(A)$ of a symmetric but neither self-adjoint nor essentially self-adjoint operator contains complex points.

But clearly there are nonsymmetric operators with purely real spectrum. For instance, in the complex Hilbert space $\mathcal{H} = \mathbb{C}^2$ take the real matrix

$$A = \begin{pmatrix} a & c \\ 0 & b \end{pmatrix} \qquad a, b \in \mathbb{R}, \ c \neq 0.$$

Obviously, $\sigma(A) = \{a, b\}$.

When we stated and proved earlier that the spectrum of a bounded linear operator in a complex Hilbert space is not empty, it was essential that we considered a Hilbert space over the field of complex numbers. A simple example of a bounded operator in a real Hilbert space with empty spectrum is

$$\mathcal{H} = \mathbb{R}^2, \qquad A = \begin{pmatrix} 0 & a^2 \\ -b^2 & 0 \end{pmatrix}, \qquad a, b \in \mathbb{R}, \ ab \neq 0.$$

The proof is obvious.

Finally we comment on the possibility to define a substitute for the determinant for linear operators in an infinite dimensional space. If the self-adjoint bounded linear operator A in a complex Hilbert space $\mathcal{H}$ has suitable spectral properties, then indeed a kind of determinant function det A can be defined. Suppose A can be written as $A = I + R$ with a self-adjoint trace class operator R. Then one defines

$$\det A = e^{\operatorname{Tr} \log A}.$$

The book [RS78] contains a fairly detailed discussion of this problem.

24.4 Exercises

1. Prove Proposition 24.1.3.

2. Consider the Hermite functions $h_n, n = 0, 1, 2, \ldots$. Use the recursion relation (16.2) for the Hermite polynomials to deduce the recursion relation

$$h_{n+1}(x) = (2n+2)^{-1/2}[xh_n(x) - h'_n(x)]$$

for the Hermite functions. Then prove by induction: $\mathcal{F}_2 h_n = (-i)^n h_n$.

3. Prove that the operator defined by equation (24.9) is densely defined and closed and determine its adjoint.

4. Show: The self-adjoint operator of multiplication with a real-valued continuous function which is not constant has no eigenvalues.

5. Prove the details in the examples of Section 24.3 on spectral properties of linear operators.

25
Spectral Theory of Compact Operators

Compact operators are defined as linear operators with very strong continuity requirements. They are those continuous operators which map weakly convergent sequences into strongly convergent ones (see Theorem 22.3.1). As a consequence their generic form is relatively simple and their spectrum consists only of eigenvalues. These results and some applications are discussed in this chapter. Compact operators were studied intensively in the early period of Hilbert space theory (1904–1940).

25.1 The results of Riesz and Schauder

The key to the spectral theory of self-adjoint compact operators A is a lemma which states that either $\|A\|$ or $-\|A\|$ is an eigenvalue of A. This lemma actually solves the extremal problem: *Find the maximum of the function* $x \mapsto \langle x, Ax \rangle$ *on the set* $S_1 = \{x \in \mathcal{H} : \|x\| = 1\}$. In the last part of this book a general theory for such extremal problems under constraints (and many other similar problems) will be presented. Here however we present a direct proof which is independent of these results.

Lemma 25.1.1 *Suppose that A is a compact self-adjoint operator in a complex Hilbert space $\mathcal{H}$. Then at least one of the two numbers $\pm \|A\|$ is an eigenvalue of A.*

Proof. By definition the norm of the operator can be calculated as $\|A\| = \sup\{|\langle x, Ax \rangle| : \|x\| = 1\}$. Thus there is a sequence $(x_n)_{n \in \mathbb{N}}$ in S_1 such that $\|A\| = \lim_{n \to \infty} |\langle x_n, Ax_n \rangle|$. We can assume that $\lim_{n \to \infty} \langle x_n, Ax_n \rangle$ exists, otherwise we would take a subsequence. Call this limit a. Then we know $|a| = \|A\|$. Since A is self-adjoint this limit is real. Since the closed unit ball of $\mathcal{H}$ is weakly compact (Theorem 18.2.6) there is a subsequence $(x_{n(j)})_{j\in\mathbb{N}}$ which converges weakly and for which the sequence of images $(Ax_{n(j)})_{j\in\mathbb{N}}$ converges strongly to x, respectively to y. The estimate

$$0 \leq \|Ax_{n(j)} - ax_{n(j)}\|^2 = \|Ax_{n(j)}\|^2 - 2a\langle x_{n(j)}, Ax_{n(j)}\rangle + a^2 \leq 2a^2 - 2a\langle x_{n(j)}, Ax_{n(j)}\rangle$$

shows that the sequence $(Ax_{n(j)} - ax_{n(j)})_{j\in\mathbb{N}}$ converges strongly to 0. Since we know strong convergence of the sequence $(Ax_{n(j)})_{j\in\mathbb{N}}$ we deduce that the sequence $(ax_{n(j)})_{j\in\mathbb{N}}$ converges not only weakly but strongly to ax, hence $\|x\| = 1$. Continuity of A implies $\lim_{j\to\infty} Ax_{n(j)} = Ax$ and thus $Ax = ax$. Hence a is an eigenvalue of A. □

Repeated application of this lemma determines the spectrum of a compact self-adjoint operator.

Theorem 25.1.1 (Riesz–Schauder theorem) *Suppose A is a self-adjoint compact operator on a complex Hilbert space. Then*

a) A has a sequence of real eigenvalues $\lambda_j \neq 0$ which can be enumerated in such a way that $|\lambda_1| \geq |\lambda_2| \geq |\lambda_3| \geq \cdots$;

b) if there are infinitely many eigenvalues, then $\lim_{j\to\infty} \lambda_j = 0$, and the only accumulation point of the set of eigenvalues is the point 0;

c) the multiplicity of every eigenvalue $\lambda_j \neq 0$ is finite;

d) if e_j is the eigenvector for the eigenvalue λ_j, then every vector in the range of A has the representation

$$Ax = \sum_{j=1}^{\infty} \lambda_j \langle e_j, x\rangle e_j;$$

e) $\sigma(A) = \{\lambda_1, \lambda_2, \ldots, 0\}$ but 0 is not necessarily an eigenvalue of A.

Proof. Lemma 25.1.1 gives the existence of an eigenvalue λ_1 with $|\lambda_1| = \|A\|$ and a normalized eigenvector e_1. Introduce the orthogonal complement $\mathcal{H}_1 = \{e_1\}^\perp$ of this eigenvector. The operator A maps the space $\mathcal{H}_1$ into itself: For $x \in \mathcal{H}_1$ we find $\langle e_1, Ax\rangle = \langle Ae_1, x\rangle = \langle \lambda_1 e_1, x\rangle = 0$, hence $Ax \in \mathcal{H}_1$. The restriction of the inner product of $\mathcal{H}$ to $\mathcal{H}_1$ makes this space a Hilbert space and the restriction $A_1 = A_{|\mathcal{H}_1}$ of A to this Hilbert space is again a self-adjoint compact operator. Clearly, its norm is bounded by that of A: $\|A_1\| \leq \|A\|$.

Now apply Lemma 25.1.1 to the operator A_1 on the Hilbert space $\mathcal{H}_1$ to get an eigenvalue λ_2 and a normalized eigenvector $e_2 \in \mathcal{H}_1$ such that $|\lambda_2| = \|A_1\| \leq \|A\| = |\lambda_1|$.

Next introduce the subspace $\mathcal{H}_2 = \{e_1, e_2\}^\perp$. Again, the operator A leaves this subspace invariant and thus the restriction $A_2 = A_{|\mathcal{H}_2}$ is a self-adjoint compact operator in the Hilbert space $\mathcal{H}_2$.

Since we assume that the Hilbert space $\mathcal{H}$ is infinite dimensional this argument can be iterated infinitely often and thus leads to a sequence of eigenvectors e_j and of eigenvalues λ_j with $|\lambda_{j+1}| \leq |\lambda_j|$. If there is an $r > 0$ such that $r \leq |\lambda_j|$, then the sequence of vectors $y_j = e_j/\lambda_j$ is bounded, and hence there is a weakly convergent subsequence $y_{j(k)}$. Compactness of A implies convergence of the sequence of images $Ay_{j(k)} = e_{j(k)}$, a contradiction since for an orthonormal system one has $\|e_{j(k)} - e_{j(m)}\| = \sqrt{2}$ for $k \neq m$. This proves parts a) and b).

To prove c) observe that on the eigenspace $E_j = N(A - \lambda_j I)$ the identity operator $I_{|E_j}$ is equal to the compact operator $\frac{1}{\lambda_j} A_{|E_j}$ and thus this space has to be finite dimensional.

The projector onto the subspace $[e_1, \ldots, e_n]$ spanned by the first n eigenvectors is $P_n x = \sum_{j=1}^n \langle e_j, x \rangle e_j$. Then $I - P_n$ is the projector onto $[e_1, \ldots, e_n]^\perp = \mathcal{H}_{n+1}$ and hence $\|A(I - P_n)x\| \leq |\lambda_{n+1}| \|(I - P_n)x\| \leq |\lambda_{n+1}| \|x\| \to 0$ as $n \to \infty$. Since $AP_n x = \sum_{j=1}^n \lambda_j \langle e_j, x \rangle e_j$ Part d) follows.

Finally, Example 24.2.1 gives immediately that the spectrum of A is $\sigma(A) = \overline{\{\lambda_j : j \in \mathbb{N}\}} = \{\lambda_1, \lambda_2, \ldots, 0\}$ according to Part b). □

Corollary 25.1.1 (Hilbert–Schmidt theorem) *The orthonormal system of eigenfunctions e_j of a compact self-adjoint operator A in a complex Hilbert space is complete if, and only if, A has a trivial null space: $N(A) = \{0\}$.*

Proof. Because of Part d) of Theorem 25.1.1 the system of eigenfunctions is complete if, and only if, the closure of the range of A is the whole Hilbert space: $\overline{\mathrm{ran}\, A} = \mathcal{H}$. Taking the orthogonal decomposition $\mathcal{H} = N(A) \oplus N(A)^\perp$ and $N(A) = N(A^*) = (\mathrm{ran}\, A)^\perp$ into account we conclude. □

25.2 The Fredholm alternative

Given a compact self-adjoint operator A on a complex Hilbert space $\mathcal{H}$ and an element $g \in \mathcal{H}$, consider the equation

$$f - \mu A f = (I - \mu A)f = g. \tag{25.1}$$

Depending on the parameter $\mu \in \mathbb{C}$ one wants to find a solution $f \in \mathcal{H}$. Our starting point is the important

Lemma 25.2.1 (Lemma of Riesz) *If A is a compact operator on the Hilbert space $\mathcal{H}$ and $\mu \neq 0$ a complex number, then the range of $I - \mu A$ is closed in $\mathcal{H}$.*

Proof. Since a scalar multiple of a compact operator is again compact we can and will assume $\mu = 1$. As an abbreviation we introduce the operator $B = I - A$ and have to show that its range is closed. Given an element $f \neq 0$ in the closure of the range of B, there is a sequence $(g_n)_{n \in \mathbb{N}}$ in $\mathcal{H}$ such that $f = \lim_{n \to \infty} B g_n$. According to the decomposition $\mathcal{H} = N(B) \oplus N(B)^\perp$ we can and will assume that $g_n \in N(B)^\perp$ and $g_n \neq 0$ for all $n \in \mathbb{N}$.

Suppose that the sequence $(g_n)_{n \in \mathbb{N}}$ is bounded in $\mathcal{H}$. Then there is a subsequence which converges weakly to some element $g \in \mathcal{H}$. We denote this subsequence in the same way as the original one. Compactness of A implies that the sequence $(Ag_n)_{n \in \mathbb{N}}$ converges strongly to some $h \in \mathcal{H}$. Weak convergence of the sequence $(g_n)_{n \in \mathbb{N}}$ ensures that $h = Ag$ ($\langle u, Ag_n \rangle = \langle A^*u, g_n \rangle \to \langle A^*u, g \rangle = \langle u, Ag \rangle$ for all $u \in \mathcal{H}$ implies $h = Ag$). Since $Bg_n \to f$ as $n \to \infty$ it follows that $g_n = Bg_n + Ag_n \to f + Ag$. Thus $(g_n)_{n \in \mathbb{N}}$ converges strongly to g and the identity $g = f + Ag$ holds. This proves $f \in \mathrm{ran}\,(I - A)$.

Now consider the case that the sequence $(g_n)_{n \in \mathbb{N}}$ is not bounded. By taking a subsequence which we denote in the same way, we can assume $\lim_{n \to \infty} \|g_n\| = \infty$. Form the auxiliary sequence of elements $u_n = \frac{1}{\|g_n\|} g_n$. This sequence is certainly bounded and thus contains a weakly convergent subsequence which we denote again in the same way. Denote the weak limit of this sequence by u. Since A is compact we conclude that $Au_n \to Au$ as $n \to \infty$. Now recall $Bg_n \to f$ and $\|g_n\| \to \infty$ as $n \to \infty$. We deduce $Bu_n = \frac{1}{\|g_n\|} Bg_n \to 0$ as $n \to \infty$ and therefore the sequence $u_n = Bu_n + Au_n$ converges strongly. Since the weak limit of the sequence is u it converges strongly to u. On the other side $Bu_n + Au_n$ converges to Au as we have shown. We deduce $u = Au$, i.e., $u \in N(B)$. By construction, $g_n \in N(B)^\perp$, hence $u_n \in N(B)^\perp$, and this implies $u \in N(B)^\perp$ since $N(B)^\perp$ is closed. This shows $u \in N(B) \cap N(B)^\perp = \{0\}$, and we conclude that $u_n \to 0$ as $n \to \infty$. This contradicts $\|u_n\| = 1$ for all $n \in \mathbb{N}$. Thus the case of an unbounded sequence $(g_n)_{n \in \mathbb{N}}$ does not occur. This completes the proof. □

Now we can formulate and prove

Theorem 25.2.1 (Fredholm alternative) *Suppose A is a self-adjoint compact operator on a complex Hilbert space $\mathcal{H}$, g a given element in $\mathcal{H}$ and μ a complex number. Then* **either** $\mu^{-1} \notin \sigma(A)$ *and the equation*

$$(I - \mu A)f = g$$

has the unique solution

$$f = (I - \mu A)^{-1}g$$

or $\mu^{-1} \in \sigma(A)$ *and the equation $(I - \mu A)f = g$ has a solution if, and only if, $g \in \operatorname{ran}(I - \mu A)$. In this case, given a special solution f_0, the general solution is of the form $f = f_0 + u$, with $u \in N(I - \mu A)$, and thus the set of all solutions is a finite dimensional affine subspace of $\mathcal{H}$.*

Proof. Lemma 25.2.1 gives

$$\operatorname{ran}(I - \mu A) = N(I - (\mu A)^*)^{\perp} = N(I - \overline{\mu}A)^{\perp}.$$

If $\mu^{-1} \notin \sigma(A)$, then $\overline{\mu}^{-1} \notin \sigma(A)$ ($\sigma(A) \subset \mathbb{R}$) and thus $\operatorname{ran}(I - \mu A) = N(I - \overline{\mu}A)^{\perp} = \{0\}^{\perp} = \mathcal{H}$ and the unique solution is $f = (I - \mu A)^{-1}g$.

Now consider the case $\mu^{-1} \in \sigma(A)$. Then $N(I - \mu A) \neq \{0\}$. $\frac{1}{\mu}$ is an eigenvalue of finite multiplicity (Theorem 25.1.1, Part c)). In this case $\operatorname{ran}(I - \mu A)$ is a proper subspace of $\mathcal{H}$ and the equation $(I - \mu A)f = g$ has a solution if, and only if, $g \in \operatorname{ran}(I - \mu A)$. Since the equation is linear it is clear that any solution f differs from a special solution f_0 by an element u in the null space of the operator $(I - \mu A)$, and we conclude. □

Remark 25.2.1 1. *The Fredholm alternative states that the eigenvalue problem for a compact self-adjoint operator in an infinite dimensional Hilbert space and that of self-adjoint operators in a finite dimensional Hilbert space have the same type of solutions. According to this theorem one has the following alternative: Either the equation $Af = \lambda f$ has a solution, i.e., $\lambda \in \sigma_p(A)$, or $(\lambda I - A)^{-1}$ exists, i.e., $\lambda \in \rho(A)$, in other words, $\sigma(A)\setminus\{0\} = \sigma_p(A) = \sigma_d(A)$. Note that for self-adjoint operators which are not compact this alternative does not hold. An example is discussed in the Exercises.*

2. *In applications one encounters the first case rather frequently. Given $r > 0$ consider those $\mu \in \mathbb{C}$ with $|\mu| < r$. Then there are only a finite number of complex numbers μ for which one cannot have existence and uniqueness of the solution.*

3. *Every complex $N \times N$ matrix has at least one eigenvalue (fundamental theorem of algebra). The corresponding statement does not hold in the infinite dimensional case. There are compact operators which are not self-adjoint and which have no eigenvalues. The Exercises offer an example.*

25.3 Exercises

1. Prove: For noncompact self-adjoint operators the Fredholm alternative does not hold: In $L^2(\mathbb{R})$ the equation $Af = f$ has no solution and $(I - A)^{-1}$ does not exist for the operator $(Af)(x) = xf(x)$, for all $f \in D(A)$ where $D(A) = \{f \in L^2(\mathbb{R}) : xf \in L^2(\mathbb{R})\}$.

2. On the Hilbert space $\mathcal{H} = \ell^2(\mathbb{C})$ consider the operator A defined by

$$A(x_1, x_2, x_3, \ldots) = (0, x_1, \frac{x_2}{2}, \ldots, \frac{x_n}{n}, \ldots,).$$

   Show that A is compact and not self-adjoint and has no eigenvalues.

3. This problem is about the historical origin of the Fredholm alternative. It was developed in the study of integral equations. We consider the *Fredholm integral equation of second kind*:

$$f(x) - \mu \int k(x, y) f(y) dy = g(x).$$

   Show: For $k \in L^2(\mathbb{R}^n \times \mathbb{R}^n)$ with $\overline{k(x, y)} = k(y, x)$ the operator A defined by $(Af)(x) = \int k(x, y) f(y) dy$ is compact and self-adjoint and the Fredholm alternative applies.

   As a concrete case of the above integral equation consider the case $n = 1$ and $k = G$ where G is the Green's function of *Sturm–Liouville problem*: On the interval $[a, b]$ find the solution of the following second order linear differential equations with the given boundary conditions:

$$y''(x) - q(x) y'(x) + \mu y(x) = f(x),$$
$$h_1 y(a) + k_1 y'(a) = 0, \quad h_2 y(b) + k_2 y'(b) = 0,$$

   with $h_j, k_j \in \mathbb{R}$, and where the h_j and k_j are not simultaneously equal to zero.

   Every solution y of the Sturm–Liouville problem is a solution of the Fredholm integral equation

$$y(x) - \mu \int_a^b G(x, z) y(z) dz = g(x)$$

   where $g(x) = -\int_a^b G(x, z) f(z) dz$ and conversely.

   Hints: See Section 20 of [Vla71] for further details.

26
The Spectral Theorem

Recall: Every symmetric $N \times N$ matrix A (i.e., every symmetric operator A in the Hilbert space $\mathbb{C}^N$) can be transformed to diagonal form, that is there are real numbers $\lambda_1, \ldots, \lambda_N$ and an orthonormal system $\{e_1, \ldots, e_N\}$ in $\mathbb{C}^N$ such that $Ae_k = \lambda_k e_k$, $k = 1, \ldots, N$. If P_k denotes the projector onto the subspace $\mathbb{C} e_k$ spanned by the eigenvector e_k, we can represent the operator A in the form

$$A = \sum_{k=1}^{N} \lambda_k P_k.$$

In this case the spectrum of the operator A is $\sigma(A) = \{\lambda_1, \ldots, \lambda_N\}$ where we use the convention that eigenvalues of multiplicity larger than one are repeated according to their multiplicity. Thus we can rewrite the above representation of the operator A as

$$A = \sum_{\lambda \in \sigma(A)} \lambda P_\lambda, \qquad (26.1)$$

where P_λ is the projector onto the subspace spanned by the eigenvector corresponding to the eigenvalue $\lambda \in \sigma(A)$. The representation (26.1) is the simplest example of the *spectral representation of a self-adjoint operator*.

We had encountered this spectral representation also for self-adjoint operators in an infinite dimensional Hilbert space, namely for the operator A defined in equation (24.9) for real s_j, $j \in \mathbb{N}$. There we determined the spectrum as $\sigma(A) = \overline{\{s_j : J \in \mathbb{N}\}}$. In this case too the representation (24.9) of the operator A can be written in the form (26.1).

Clearly the characteristic feature of these two examples is that their spectrum consists of a finite or a countable number of eigenvalues. However we have learned that there are examples of self-adjoint operators which have not only eigenvalues but also a continuous spectrum (see the second example in Section 24.2). Accordingly the general form of a spectral representation of self-adjoint operators must also include the possibility of a continuous spectrum and therefore one would expect that the general form of a spectral representation is something like

$$A = \int_{\sigma(A)} \lambda \, dP_\lambda. \qquad (26.2)$$

It is the goal of this chapter to give a precise meaning to this formula and to prove it for arbitrary self-adjoint operators in a separable Hilbert space. That such a spectral representation is possible and how this representation has to be understood was shown in 1928 by J. von Neumann. Later several different proofs of this 'spectral theorem' were given. We present a version of the proof which is not necessarily the shortest one but which only uses intrinsic Hilbert space arguments. Moreover this approach has the additional advantage of giving another important result automatically, namely this proof allows us to determine the 'maximal self-adjoint part' of any closed symmetric operator. Furthermore it gives a concrete definition of the projectors P_λ as projectors onto subspaces which are defined explicitly in terms of the given operator. This proof is due to Lengyel and Stone for the case of bounded self-adjoint operators (1936). It was extended to the general case by Leinfelder in 1979 ([Lei79]).

The starting point of this proof is the so-called 'geometric characterization of self-adjointness'. It is developed in the first section. The second section will answer the following questions: What does dP_λ mean and what type of integration is used in formula (26.2)? Finally, using some approximation procedure and the results of the preceding sections, the proof of the spectral theorem and some other conclusions are given in the third section.

26.1 Geometric characterization of self-adjointness

26.1.1 Preliminaries

Lemma 26.1.1 *Suppose A is a closed symmetric operator with domain D, in a complex Hilbert space $\mathcal{H}$, and $(P_n)_{n \in \mathbb{N}}$ a sequence of orthogonal projectors with the following properties:*
a) $P_n \leq P_{n+1}$,   b) ran $P_n \subseteq D$,   c) $AP_n = P_n A P_n$ for all $n \in \mathbb{N}$,
and d) $\lim_{n \to \infty} P_n x = x$ for all $x \in \mathcal{H}$. Then

$$D = \{x \in \mathcal{H} : (AP_n x)_{n \in \mathbb{N}} \text{ converges in } \mathcal{H}\} \qquad (26.3)$$
$$= \{x \in \mathcal{H} : (\|AP_n x\|)_{n \in \mathbb{N}} \text{ converges in } \mathbb{R}\} \qquad (26.4)$$

and for all $x \in D$,

$$Ax = \lim_{n \to \infty} AP_n x \quad \text{in } \mathcal{H}. \qquad (26.5)$$

Proof. Condition a) and Lemma 22.1.1 imply $P_m P_n = P_n$ for all $m \geq n$. Therefore, by an elementary calculation using condition c) and the symmetry of A, one finds

$$\langle AP_m x, AP_n x\rangle = \langle AP_n x, AP_n x\rangle = \langle AP_n x, AP_m x\rangle \qquad \forall m \geq n, \ \forall x \in \mathcal{H},$$

and similarly

$$\langle Ax, AP_n\rangle = \langle AP_n x, AP_n\rangle = \langle AP_n x, Ax\rangle \qquad \forall m \geq n, \ \forall x \in D.$$

Evaluating the norm in terms of the inner product gives

$$\|AP_m x - AP_n x\|^2 = \|AP_m x\|^2 - \|AP_n x\|^2 \qquad \forall m \geq n, \ \forall x \in \mathcal{H} \qquad (26.6)$$

and

$$\|Ax - AP_n x\|^2 = \|Ax\|^2 - \|AP_n x\|^2 \qquad \forall m \geq n, \ \forall x \in D. \qquad (26.7)$$

Equation (26.6) shows that the sequence $(AP_n x)_{n \in \mathbb{N}}$ converges in $\mathcal{H}$ if, and only if, the sequence $(\|AP_n x\|)_{n \in \mathbb{N}}$ converges in $\mathbb{R}$. Thus the two domains (26.3) and (26.4) are the same.

Now assume $x \in \mathcal{H}$ and $(AP_n x)_{n \in \mathbb{N}}$ converges. Condition d) and the fact that A is closed imply $x \in D$ and $Ax = \lim_{n \to \infty} AP_n x$. Hence x belongs to the set (26.3).

Conversely assume $x \in D$. The identities (26.6) and (26.7) imply that $(\|AP_n x\|)_{n \in \mathbb{N}}$ is a monotone increasing sequence which is bounded by $\|Ax\|$. We conclude that this sequence converges and the above characterization of the domain D of A is established.

The identity (26.5) results from the characterization (26.3) of the domain and the fact that A is closed. □

Lemma 26.1.2 *Suppose $\mathcal{H}$ is a finite dimensional Hilbert space and F is a linear subspace of $\mathcal{H}$. For a given symmetric operator A on $\mathcal{H}$ define the function $f(x) = \langle x, Ax\rangle$ and denote $\mu = \inf\{f(x) : x \in F, \|x\| = 1\}$. Then μ is an eigenvalue of A and the corresponding eigenvector e_0 satisfies $f(e_0) = \mu$.*

Proof. f is a continuously differentiable function on $\mathcal{H}$ with derivative $f'(x) = 2Ax$, since A is symmetric (see Chapter 30). The set $\{x \in F : \|x\| = 1\}$ is compact and hence f attains its minimum on this set. This means that there is $e_0 \in F$, $\|e_0\| = 1$ such that $f(e_0) = \mu > -\infty$.

This is a minimization problem with constraint, namely to minimize the values of the function f on the subspace F under the constraint $g(x) = \langle x, x\rangle = 1$. The theorem about the existence of a Lagrange multiplicator (see Theorem 31.3.1) implies: There is an $r \in \mathbb{R}$ such that $f'(e_0) = rg'(e_0)$. The derivative of g is $g'(x) = 2x$, hence $f'(e_0) = re_0$ and therefore $f(e_0) = \langle e_0, Ae_0\rangle = r\langle e_0, e_0\rangle = r$. The Lagrange multiplier r is equal to the minimum and we conclude. □

26.1.2 Subspaces of controlled growth

Given a closed symmetric operator A in an infinite dimensional Hilbert space $\mathcal{H}$, we introduce and characterize a certain family of subspaces on which the operator A grows in a way determined by the characteristic parameter of the subspace. To begin, we introduce the subspace of those elements on which any power of the operator can be applied

$$D^\infty = D^\infty(A) = \cap_{n \in \mathbb{N}} D(A^n). \qquad (26.8)$$

This means $A^n x \in \mathcal{H}$ for all $x \in D^\infty$. For every $r > 0$ define a function $q_r : D^\infty \to [0, \infty]$ by

$$q_r(x) = \sup_{n \in \mathbb{N}} r^{-n} \|A^n x\| \qquad \forall x \in D^\infty. \qquad (26.9)$$

This function has the following properties:

$$q_r(\alpha x) = |\alpha| q_r(x), \qquad q_r(x+y) \leq q_r(x) + q_r(y) \qquad \forall\, x, y \in D^\infty, \; \forall\, \alpha \in \mathbb{C}.$$

Therefore, for every $r > 0$, the set

$$G(A, r) = \{x \in D^\infty : q_r(x) < \infty\}$$

is a linear subspace of $\mathcal{H}$. For $r = 0$ we use $G(A, 0) = N(A)$. Next the subsets of controlled growth are introduced. For $r \geq 0$ denote

$$F(A, r) = \{x \in D^\infty : \|A^n x\| \leq r^n \|x\|, \; \forall\, n \in \mathbb{N}\}. \tag{26.10}$$

The most important properties of these sets are described in

Lemma 26.1.3 *For a closed symmetric operator A in the complex Hilbert space $\mathcal{H}$ the subsets $F(A, r)$ and $G(A, r)$ are actually closed subspaces of $\mathcal{H}$ which satisfy*

a) $F(A, r) = G(A, r)$ for all $r \geq 0$;

b) $AF(A, r) \subseteq F(A, r)$;

c) *If B is a bounded operator on $\mathcal{H}$ which commutes with A in the sense that $BA \subseteq AB$, then, for all $r \geq 0$,*

$$BF(A, r) \subseteq F(A, r), \qquad B^* F(A, r)^\perp \subseteq F(A, r)^\perp. \tag{26.11}$$

Proof. From the definition of these sets the following is evident: $G(A, r)$ is a linear subspace which contains $F(A, r)$. The set $F(A, r)$ is invariant under scalar multiplication but it is not evident that the sum of two of its elements again belongs to it. Therefore in a first step we show the equality of these two sets.

Suppose that there is a $z \in G(A, r)$ which does not belong to $F(A, r)$. We can assume $\|z\| = 1$. Then there is some $m \in \mathbb{N}$ such that $\|A^m z\| > r^m \|z\|$. Introduce the auxiliary operator $S = r^{-m} A^m$. It is again symmetric and satisfies $\|Sz\| > 1$. For every $j \in \mathbb{N}$ we estimate $\|S^j z\|^2 = \langle z, S^{2j} z \rangle \leq \|S^{2j} z\|$, hence $\|S^{2^j} z\| \geq \|Sz\|^{2^j} \to \infty$ as $j \to \infty$, but this contradicts $z \in G(A, r)$, i.e., $\|S^{2^j} z\| \leq q_r(z) < \infty$. Hence both sets are equal and Part a) holds.

For $x \in G(A, r)$ the obvious estimate $q_r(Ax) \leq r q_r(x)$ holds and it implies that $G(A, r) = F(A, r)$ is invariant under the operator A, thus Part b) holds.

Next we prove that this subspace is closed. Given $y_0 \in \overline{F(A, r)}$ there is a sequence $(x_n)_{n \in \mathbb{N}} \subset F(A, r)$ such that $y_0 = \lim_{n \to \infty} x_n$. Since $F(A, r)$ is a linear subspace, $x_n - x_m \in F(A, r)$ for all $n, m \in \mathbb{N}$ and therefore $\|A^j x_n - A^j x_m\| = \|A^j (x_n - x_m)\| \leq r^j \|x_n - x_m\|$ for every $j \in \mathbb{N}$. This shows that $(A^j x_n)_{n \in \mathbb{N}}$ is a Cauchy sequence in $\mathcal{H}$ for every j. Therefore these sequences have a limit in the Hilbert space: $y_j = \lim_{n \to \infty} A^j x_n$, $j = 0, 1, 2, \ldots$.

Now observe $(x_n)_{n \in \mathbb{N}} \subset F(A, r) \subset D(A)$ and the operator A is closed. Therefore the identities $y_j = \lim_{n \to \infty} A^j x_n$ for $j = 0$ and $j = 1$ imply: $y_0 \in D(A)$ and $y_1 = Ay_0$. Because of Part b) we know $F(A, r)$ to be invariant under A, hence $A^j x_n \in F(A, r)$ for all $n, j \in \mathbb{N}$. Hence a proof of induction with respect to j applies and proves

$$y_j \in D(A), \qquad y_j = A^j y_0, \qquad j = 0, 1, 2, \ldots.$$

We deduce $y_0 \in D^\infty$ and

$$\|A^j y_0\| = \lim_{n \to \infty} \|A^j x_n\| \leq \limsup_{n \to \infty} r^j \|x_n\| = r^j \|y_0\| \qquad \forall\, j \in \mathbb{N}.$$

It follows that $y_0 \in F(A, r)$ and this subspace is closed.

If B is a bounded operator on $\mathcal{H}$ which commutes with A in the sense of $BA \subseteq AB$ we know $A^n Bx = BA^n x$ for all $x \in G(A, r)$ and all $n \in \mathbb{N}$ and therefore $q_r(Bx) \leq \|B\| q_r(x)$ for every $r > 0$, hence $BG(A, r) \subseteq G(A, r)$ for $r > 0$. For $r = 0$ the subspace $G(A, 0)$ is by definition the null space of A which is invariant under B because of the assumed commutativity with A, therefore $BG(A, r) \subseteq G(A, r)$ for all $r \geq 0$.

Finally suppose $x \in G(A, r)^\perp$. For all $y \in G(A, r)$ we find $\langle B^* x, y \rangle = \langle x, By \rangle = 0$, since $BG(A, r) \subseteq G(A, r)$, and therefore $B^* x \in G(A, r)^\perp$. This proves Part c). □

The restriction of the operator A to the closed subspace $F(A, r)$ is bounded by r, $\|Ax\| \leq r\|x\|$ for all $x \in F(A, r)$. Hence the family of closed subspaces $F(A, r)$, $r \geq 0$ controls the growth of the operator A. It does so actually rather precisely since there are also lower bounds characterized by this family as we are going to show. These lower bounds are deduced in two steps: First they are shown for finite dimensional subspaces. Then an approximation lemma controls the general case.

Lemma 26.1.4 *For a symmetric operator A in a finite dimensional complex Hilbert space $\mathcal{H}$ one has for all $r \geq 0$ and all $x \in F(A, r)^\perp$, $x \neq 0$:*

$$\|Ax\| > r\|x\| \quad \text{and} \quad \langle x, Ax \rangle > r\|x\|^2 \quad \text{if} \quad A \geq 0.$$

Proof. The proof for the general case will be reduced to that of a positive operator. So we start with the case $A \geq 0$. Denote $S_1 = \{x \in \mathcal{H} : \|x\| = 1\}$ and consider the function $f(x) = \langle x, Ax \rangle$. Since $S_1 \cap F(A, r)^\perp$ is compact f attains its minimum $\mu = \inf \{f(x) : x \in S_1 \cap F(A, r)^\perp\}$ on this set, i.e., there is an $e_0 \in S_1 \cap F(A, r)^\perp$ such that $f(e_0) = \mu$. By Lemma 26.1.2 the minimum μ is an eigenvalue of A and e_0 is the corresponding eigenvector: $Ae_0 = \mu e_0$. This proves $e_0 \in F(A, \mu)$. If we had $\mu \leq r$, then $F(A, \mu) \subset F(A, r)$ and thus $e_0 \in F(A, r) \cap F(A, r)^\perp = \{0\}$, a contradiction since $\|e_0\| = 1$. Hence the minimum must be larger than r: $\mu > r$.

The lower bound is now obvious: For $x \in F(A, r)^\perp$, $x \neq 0$, write $\langle x, Ax \rangle = \|x\|^2 \langle y, Ay \rangle$ with $y \in S_1 \cap F(A, r)^\perp$, thus $\langle x, Ax \rangle \geq \|x\|^2 \mu > r\|x\|^2$ which is indeed the lower bound of A for $A \geq 0$.

Since A is symmetric it leaves the subspaces $F(A, r)$ and $F(A, r)^\perp$ invariant. It follows that the restriction $B = A_{|F(A,r)^\perp}$ is a symmetric operator $F(A, r)^\perp \to F(A, r)^\perp$ which satisfies $\|Ax\|^2 = \langle x, B^2 x \rangle$ for all $x \in F(A, r)^\perp$. As above we conclude that

$$\mu^2 = \inf \{\|Ax\|^2 : x \in S(r)\} = \inf \{\langle x, B^2 x \rangle : x \in S(r)\}$$

is an eigenvalue of B^2 (we use the abbreviation $S(r) = S_1 \cap F(A, r)^\perp$). Elementary rules for determinants say

$$0 = \det(B^2 - \mu^2 I) = \det(B - \mu I) \det(B + \mu I)$$

and therefore either $+\mu$ or $-\mu$ is an eigenvalue of B. As above we prove $|\mu| > r$ and for $x \in F(A, r)^\perp$, $x \neq 0$, write $\|Ax\| = \|x\| \|Ay\|$ with $y \in S(r)$ and thus $\|Ax\| \geq \|x\| |\mu| > r\|x\|$. □

Lemma 26.1.5 *Let A be a closed symmetric operator in a complex Hilbert space $\mathcal{H}$. Introduce the closed subspace of controlled growth $F(A, r)$ as above and choose any $0 \leq r < s$. Then, for every given $x \in F(A, s) \cap F(A, r)^\perp$ there are a sequence $(H_n)_{n \in \mathbb{N}}$ of finite dimensional subspaces of $\mathcal{H}$, a sequence of symmetric operators $A_n : H_n \to H_n$, and a sequence of vectors $x_n \in H_n$, $n \in \mathbb{N}$, such that*

$$x_n \in F(A_n, s) \cap F(A_n, r)^\perp \qquad \forall n \in \mathbb{N},$$

$$\lim_{n \to \infty} \|x - x_n\| = 0 = \lim_{n \to \infty} \|Ax - A_n x_n\|.$$

Proof. According to Lemma 26.1.3 the subspaces $F(A, s)$ and $F(A, r)^\perp$ are invariant under the symmetric operator A. Therefore, given $x \in F(A, s) \cap F(A, r)^\perp \subset D^\infty$, we know that

$$H_n = H_n(x) = [x, Ax, \ldots, A^n x] \subseteq F(A, s) \cap F(A, r)^\perp \subset D^\infty.$$

Clearly the dimension of H_n is smaller than or equal to $n + 1$. From Lemma 26.1.3 we also deduce

$$H_n \subseteq H_{n+1} \subseteq H_\infty = \overline{\bigcup_{n \in \mathbb{N}} H_n} \subseteq F(A, s) \cap F(A, r)^\perp.$$

Introduce the orthogonal projectors P_n onto H_n and P onto H_∞ and observe $\lim_{n \to \infty} P_n y = Py$ for every $y \in \mathcal{H}$. Next we define the reductions of the operator A to these subspaces: $A_n = (P_n A P_n)_{|H_n}$. It follows that A_n is a symmetric operator on H_n and if $A \geq 0$ is positive so is A_n.

We prepare the proof of the approximation by an important convergence property of the reduced operators A_n:

$$\lim_{n \to \infty} (P_n A P_n)^j y = (PAP)^j y \qquad \forall j \in \mathbb{N}, \quad \forall y \in \mathcal{H}. \tag{26.12}$$

Equation (26.12) is shown by induction with respect to j. Since $H_\infty \subseteq F(A, s) \subseteq D^\infty$ we know $Py \in D^\infty$ and thus

$$\|PAPy - P_n A P_n y\| \leq \|(P - P_n)APy\| + \|P_n A(Py - P_n y)\|$$
$$\leq \|(P - P_n)APy\| + \|A(Py - P_n y)\|$$
$$\leq \|(P - P_n)APy\| + s\|(Py - P_n y)\|.$$

Since $\|(P - P_n)z\| \to 0$ as $n \to \infty$, equation (26.12) holds for $j = 1$. Now suppose that equation (26.12) holds for all $j \leq k$ for some $k \geq 1$. Then we estimate as follows:

$$\|(PAP)^{k+1} y - (P_n A P_n)^{k+1} y\|$$
$$= \|(P_n A P_n)[(PAP)^k - (P_n A P_n)^k]y + (PAP - P_n A P_n)(PAP)^k y\|$$
$$\leq \|(P_n A P_n)[(PAP)^k - (P_n A P_n)^k]y\| + \|(PAP - P_n A P_n)(PAP)^k y\|$$
$$\leq s\|[(PAP)^k - (P_n A P_n)^k]y\| + \|(PAP - P_n A P_n)(PAP)^k y\|.$$

As $n \to \infty$ the upper bound in this estimate converges to zero, because of our induction hypothesis. Therefore equation (26.12) follows for all j.

After these preparations the main construction of the approximations can be done. Since H_∞ is invariant under the operator A, equation (26.12) implies for all $y \in H_\infty$,

$$\lim_{n \to \infty} (P_n A P_n)^j y = (PAP)^j y = A^j y \qquad \forall j \in \mathbb{N}. \tag{26.13}$$

The given $x \in F(A, s) \cap F(A, r)^\perp$ satisfies $x \in H_n$ for all $n \in \mathbb{N}$. Thus we can project it onto the subspaces $F(A_n, r) \equiv F(A, r) \cap H_n$ and their orthogonal complement:

$$x = x_n \oplus y_n, \qquad x_n \in F(A_n, r)^\perp, \quad y_n \in F(A_n, r), \quad \forall n \in \mathbb{N}.$$

Since $\|x\|^2 = \|x_n\|^2 + \|y_n\|^2$ the sequence $(y_n)_{n \in \mathbb{N}}$ contains a weakly convergent subsequence $(y_{n(i)})_{i \in \mathbb{N}}$ with a limit denoted by y. Since all elements of the subsequence belong to the space H_∞ which is strongly closed and thus weakly closed, this weak limit y belongs to $H_\infty \subset F(A, s) \cap F(A, r)^\perp$. We are going to show $y = 0$ by showing that this weak limit y also belongs to $F(A, r)$.

For any $k \in \mathbb{N}$ equation (26.13) implies

$$\|A^k y\|^2 = \langle y, A^{2k} y \rangle = \lim_{n \to \infty} \langle y, (P_n A P_n)^{2k} y \rangle = \lim_{i \to \infty} \langle y, (P_{n(i)} A P_{n(i)})^{2k} y_{n(i)} \rangle,$$

since $(P_{n(i)} A P_{n(i)})^{2k} y$ converges strongly to $A^{2k} y$ and $y_{n(i)}$ weakly to y. We can estimate now as follows, using $y_{n(i)} \in F(A_{n(i)}, r)$:

$$\|A^k y\|^2 \leq \limsup_{i \to \infty} \|y\| \, \|(P_{n(i)} A P_{n(i)})^{2k} y_{n(i)}\| \leq \limsup_{i \to \infty} \|y\| \, r^{2k} \|y_{n(i)}\| \leq r^{2k} \|y\| \, \|x\|,$$

hence $y \in F(A, r)$, and we conclude $y = 0$.

26.1 Geometric characterization of self-adjointness

Finally we can establish the statements of the lemma for the sequence $(x_{n(i)})_{i \in \mathbb{N}}$ corresponding to the weakly convergent subsequence $(y_{n(i)})_{i \in \mathbb{N}}$. For simplicity of notation we denote these sequences $(x_n)_{n \in \mathbb{N}}$, respectively $(y_n)_{n \in \mathbb{N}}$. The elements x_n have been defined as the projections onto $F(A_n, r)^\perp \subset H_n \subset F(A, s)$. Hence the first part of the statement follows, since $H_n \cap F(A, s) = F(A_n, s)$. Note $\|x - x_n\|^2 = \langle x - x_n, y_n \rangle = \langle x, y_n \rangle$ and recall that the sequence $(y_n)_{n \in \mathbb{N}}$ converges weakly to zero, thus $\|x - x_n\|^2 \to 0$ as $n \to \infty$. According to the construction of the spaces H_n, the elements x, Ax are contained in them, thus the identity $Ax = P_n A P_n x$ holds automatically. This gives the estimate

$$\|Ax - A_n x_n\| = \|A_n x - A_n x_n\| = \|(P_n A P_n)(x - x_n)\| \le s \|x - x_n\|,$$

and the approximation for the operator A follows. □

The combination of the two last lemmas allows us to control the growth of the operator A on the family of subspaces $F(A, r), r \ge 0$.

Theorem 26.1.1 *Let A be a closed symmetric operator on the complex Hilbert space $\mathcal{H}$ and introduce the family of subspaces $F(A, r), r \ge 0$ according to equation (26.10). Choose any two numbers $0 \le r < s$. Then for every $x \in F(A, s) \cap F(A, r)^\perp$ the following estimates hold:*

$$r\|x\| \le \|Ax\| \le s\|x\| \quad \text{and} \quad r\|x\|^2 \le \langle x, Ax \rangle \le s\|x\|^2 \quad \text{if} \quad A \ge 0. \quad (26.14)$$

Proof. If $x \in F(A, s) \cap F(A, r)^\perp$, approximate it according to Lemma 26.1.5 by elements $x_n \in F(A_n, s) \cap F(A_n, r)^\perp$ and the operator A by symmetric operators A_n in the finite dimensional Hilbert space H_n. Now apply Lemma 26.1.4 to get, for all $n \in \mathbb{N}$,

$$r\|x_n\| \le \|A_n x_n\| \qquad \text{and} \qquad r\|x_n\|^2 \le \langle x_n, A_n x_n \rangle \quad \text{if} \quad A \ge 0.$$

To conclude, take the limit $n \to \infty$ in these estimates which is possible by Lemma 26.1.5. □

The family of subspaces $F(A, r), r \ge 0$, thus controls the growth of the operator A with considerable accuracy (choose $r < s$ close to each other). This family can also be used to decide whether the operator A is self-adjoint.

Theorem 26.1.2 (Geometric characterization of self-adjointness) *A closed symmetric operator A in a complex Hilbert space $\mathcal{H}$ is self-adjoint if, and only if,*

$$\bigcup_{n \in \mathbb{N}} F(A, n)$$

is dense in $\mathcal{H}$. Here the subspaces of controlled growth $F(A, n)$ are defined in equation (26.10).

Proof. According to Lemma 26.1.3 the closed subspaces $F(A, n)$ satisfy $F(A, n) \subseteq F(A, n+1)$ for all $n \in \mathbb{N}$, hence their union is a linear subspace too. Denote by P_n the orthogonal projector onto $F(A, n)$. It follows that $(P_n)_{n \in \mathbb{N}}$ is a monotone increasing family of projectors on $\mathcal{H}$. Thus, if $\bigcup_{n \in \mathbb{N}} F(A, n)$ is assumed to be dense in $\mathcal{H}$ this sequence of projectors converges strongly to the identity operator I. In order to show that the closed symmetric operator A is self-adjoint it suffices to show that the domain $D(A^*)$ of the adjoint A^* is contained in the domain $D(A)$ of the operator A.

Consider any $x \in D(A^*)$. Since P_n projects onto $F(A, n) \subset D(A) \subset D(A^*)$, we can write $A^* x = A^*(x - P_n x) + A^* P_n x = A^*(I - P_n)x + AP_n x$. Since the subspace $F(A, n)$ is invariant under A and since $I - P_n$ projects onto $F(A, n)^\perp$, one has $\langle A^*(I - P_n)x, AP_n x \rangle = \langle (I - P_n)x, A^2 P_n x \rangle = 0$. This implies

$$\|A^* x\|^2 = \|A^*(I - P_n)x\|^2 + \|AP_n x\|^2.$$

Therefore the sequence $(AP_n x)_{n \in \mathbb{N}}$ is norm bounded, and thus there is a weakly convergent subsequence $(AP_{n(i)} x)_{i \in \mathbb{N}}$. Since $(P_{n(i)} x)_{i \in \mathbb{N}}$ is weakly convergent too and since an operator is closed if, and only if, it is weakly closed, we conclude that the weak limit x of the sequence $(P_{n(i)} x)_{i \in \mathbb{N}}$ belongs to the domain $D(A)$ of A and the sequence $(AP_{n(i)} x)_{i \in \mathbb{N}}$ converges weakly to Ax. This proves $D(A^*) \subseteq D(A)$ and thus self-adjointness of A.

Conversely assume that the operator A is self-adjoint. We assume in addition that $A \geq I$. In this case the proof is technically much simpler. At the end we comment on the necessary changes for the general case which uses the same basic ideas. As we know the space $\cup_{n \in \mathbb{N}} F(A, n)$ is dense in $\mathcal{H}$ if, and only if,
$$(\cup_{n \in \mathbb{N}} F(A, n))^\perp = \cap_{n \in \mathbb{N}} F(A, n)^\perp = \{0\}.$$
The assumption $A \geq I$ implies that A^{-1} is a bounded self-adjoint operator $\mathcal{H} \to D(A)$ which commutes with A. Form the spaces $F(A^{-1}, r)$, $r \geq 0$. Lemma 26.1.3 implies that A^{-1} maps the closed subspace $H_r = F(A^{-1}, r^{-1})^\perp$ into itself. Hence $B_r = (A^{-1})_{|H_r}$ is a well-defined bounded linear operator on H_r. Theorem 26.1.1 applies to the symmetric operator B_r. Therefore, for all $x \in F(A^{-1}, r^{-1})^\perp \cap F(A^{-1}, s)$, $s = \|B_r\|$, the lower bound $\|B_r x\| = \|A^{-1} x\| \geq \frac{1}{r} \|x\|$ is available. We conclude that $B_r : H_r \to H_r$ is bijective. Hence for every $x_0 \in H_r$ there is exactly one $x_1 \in H_r$ such that $x_0 = B_r x_1 = A^{-1} x_1$. This implies $x_0 \in D(A)$ and $x_1 = Ax_0 \in H_r$. Iteration of this argument produces a sequence $x_n = A^n x_0 \in H_r \cap D(A) = F(A^{-1}, r^{-1})^\perp \cap D(A)$, $n \in \mathbb{N}$. This implies $x_0 \in D^\infty$ and $\|x_n\| = \|A^{-1} x_{n+1}\| \geq r^{-1} \|x_{n+1}\| = r^{-1} \|A x_n\|$, hence $\|A x_n\| \leq r^n \|x_0\|$ for all $n \in \mathbb{N}$, or $x_0 \in F(A, r)$ and thus
$$F(A^{-1}, r^{-1})^\perp \subset F(A, r) \qquad \forall r > 0.$$
This holds in particular for $r = n \in \mathbb{N}$, hence
$$\bigcap_{n \in \mathbb{N}} F(A, n)^\perp \subseteq \bigcap_{n \in \mathbb{N}} F(A^{-1}, \frac{1}{n}) = N(A^{-1}) = \{0\}.$$
This concludes the proof for the case $A \geq I$.

Now we comment on the proof for the general case. For a self-adjoint operator A the resolvent $R_A(z) = (A - zI)^{-1} : \mathcal{H} \to D(A)$ is well defined for all $z \in \mathbb{C} \setminus \mathbb{R}$. Clearly, the resolvent commutes with A and is injective. In the argument given above replace the operator A^{-1} by the operator $B = R_A(z)^* R_A(z) = R_A(\bar{z}) R_A(z)$. This allows us to show, for all $r > 0$,
$$F(B, r)^\perp \subseteq F(A, |z| + \frac{1}{r} \|R_A(z)\|).$$
Now, for $n > |z|$ denote $r_n = \frac{1}{n - |z|} \|R_A(z)\|$, then $F(B, r_n)^\perp \subset F(A, n)$ and therefore
$$\bigcap_{n > |z|} F(A, n)^\perp \subset \bigcap_{n > |z|} F(B, r_n) = N(B) = \{0\},$$
and we conclude as in the case $A \geq I$. □

26.2 Spectral families and their integrals

In Proposition 22.1.1 we learned that there is a one-to-one correspondence between closed subspaces of a Hilbert space and orthogonal projections. In the previous section the family of subspaces of controlled growth were introduced for a closed symmetric operator A. Thus we have a corresponding family of orthogonal projections on the Hilbert space which will finally lead to the spectral representation of self-adjoint operators. Before this can be done the basic theory of such families of projectors and their integrals have to be studied.

26.2.1 Spectral families

The correspondence between a family of closed subspaces of a complex Hilbert space and the family of projectors onto these subspaces is investigated in this section in some detail. Our starting point is

Definition 26.2.1 *Let $\mathcal{H}$ be a complex Hilbert space and E a function on $\mathbb{R}$ with values in the space $\mathfrak{P}(\mathcal{H})$ of all orthogonal projection operators on $\mathcal{H}$. E is called a **spectral family** on $\mathcal{H}$ or **resolution of the identity** if, and only if, the following conditions are satisfied.*

a) *E is monotone: $E_t E_s = E_{t \wedge s}$ for all $t, s \in \mathbb{R}$ where $t \wedge s = \min\{t, s\}$;*

b) *E is right continuous with respect to the strong topology, i.e.,*
   $\lim_{s \to t, s > t} \|E_s x - E_t x\| = 0$ *for all $x \in \mathcal{H}$ and all $t \in \mathbb{R}$;*

c) *E is normalized, i.e., $\lim_{t \to -\infty} E_t x = 0$ and $\lim_{t \to +\infty} E_t x = x$ for every $x \in \mathcal{H}$.*

The **support of a spectral family** E is $\operatorname{supp} E = \overline{\{t \in \mathbb{R} : E_t \neq 0, \ E_t \neq I\}}$.

Given a spectral family E on $\mathcal{H}$ we get a family of closed subspaces H_t of $\mathcal{H}$ by defining
$$H_t = \operatorname{ran} E_t, \qquad \forall t \in \mathbb{R}.$$

In the following proposition the defining properties a) - c) of a spectral family are translated into properties of the family of associated closed subspaces.

Proposition 26.2.1 *Let $\{E_t\}_{t \in \mathbb{R}}$ be a spectral family on $\mathcal{H}$. Then the family of closed subspaces $H_t = \operatorname{ran} E_t$ has the following properties:*

a) *monotonicity: $H_s \subseteq H_t$ for all $s \leq t$;*

b) *right continuouity: $H_s = \cap_{t > s} H_t$;*

c) *normalization: $\cap_{t \in \mathbb{R}} H_t = \{0\}$ and $\overline{\cup_{t \in \mathbb{R}} H_t} = \mathcal{H}$.*

Conversely, given a family of closed subspaces H_t, $t \in \mathbb{R}$, of $\mathcal{H}$ with the properties a) -c) then the family of orthogonal projectors E_t onto H_t, $t \in \mathbb{R}$, is a spectral family on $\mathcal{H}$.

Proof. The monotonicity condition a) for the spectral family is easily translated into that of the family of ranges H_t by Lemma 22.1.1. This implies $H_s \subseteq H_t$ for all $s < t$ and therefore $H_s \subseteq \cap_{s < t} H_t$. For any $x \in \cap_{s < t} H_t$ we know $E_t x = x$ for all $s < t$, hence $x = \lim_{t \to s, s < t} E_t x = E_s x$, i.e., $x \in \operatorname{ran} E_s = H_s$ since a spectral family is right continuous. This proves b) for the family H_t, $t \in \mathbb{R}$.

The normalization for the spectral family $\lim_{t \to \infty} E_t x = x$ for all $x \in \mathcal{H}$ implies immediately that the closure of the union of all the subspaces H_t gives the whole Hilbert space. Next consider $x \in \cap_{t \in \mathbb{R}} H_t$, then $x = E_t x$ for all $t \in \mathbb{R}$ and thus $x = \lim_{t \to -\infty} E_t x = 0$ because of the normalization for the spectral family. This proves the normalization for the family of subspaces H_t.

If a family of closed subspaces H_t, $t \in \mathbb{R}$, with the properties a) – c) is given, define a family of orthogonal projectors by defining E_t as the orthogonal projector onto the subspace H_t for all $t \in \mathbb{R}$. Suppose $s \leq t$, then $H_s \subseteq H_t$ and Lemma 22.1.1 implies $E_s = E_s E_t = E_t E_s \leq E_t$ and thus monotonicity of the family of projectors. According to Theorem 22.1.2 a monotone increasing family of projectors has a strong limit which is again an orthogonal projector. Hence, for every $x \in \mathcal{H}$ we know $\lim_{t \to s, t > s} E_t x = Px$ for some orthogonal projector P on $\mathcal{H}$. The condition b) for the family of subspaces H_t implies

$$\operatorname{ran} P = \cap_{t>s} \operatorname{ran} E_t = \cap_{t>s} \operatorname{ran} H_t = H_s = \operatorname{ran} E_s,$$

thus $P = E_s$ by Part d) of Proposition 22.1.1. Therefore the function $t \mapsto E_t \in \mathfrak{P}(\mathcal{H})$ is right continuous.

Since $t \mapsto E_t$ is monotone the following strong limits exist (Theorem 22.1.2): $\lim_{t \to -\infty} E_t = Q_-$ and $\lim_{t \to +\infty} E_t = Q_+$ with $\operatorname{ran} Q_- = \cap_{t > -\infty} \operatorname{ran} E_t = \cap_{t > -\infty} H_t = \{0\}$ and $\operatorname{ran} Q_+ = \overline{\cup_{t \in \mathbb{R}} \operatorname{ran} E_t} = \overline{\cup_{t \in \mathbb{R}} H_t} = \mathcal{H}$ and again by Proposition 22.1.1 we conclude $Q_- = 0$ and $Q_+ = I$ which are the normalization conditions of a spectral family. $\square$

26.2.2 Integration with respect to a spectral family

Given a spectral family E_t on a complex Hilbert space $\mathcal{H}$ and a continuous function $f : [a, b] \to \mathbb{R}$, we explain the definition and the properties of the integral of f with respect to the spectral family:

$$\int_a^b f(t) dE_t. \tag{26.15}$$

The definition of this integral is done in close analogy to the Stieltjes integral. Accordingly we strongly recommend studying the construction of the Stieltjes integral first.

There is naturally a close connection of the Stieltjes integral with the integral (26.15). Given any $x \in \mathcal{H}$ define $\rho_x(t) = \langle x, E_t x \rangle$ for all $t \in \mathbb{R}$. Then ρ_x is a monotone increasing function of finite total variation and thus a continuous function f has a well-defined Stieltjes integral $\int_a^b f(t) d\rho_x(t)$ with respect to ρ_x and one finds according to the definition of the integral (26.15)

$$\int_a^b f(t) d\rho_x(t) = \langle x, \int_a^b f(t) dE_t x \rangle.$$

For a given spectral family E_t on the complex Hilbert space $\mathcal{H}$ and any $s < t$ introduce

$$E(s, t] = E_t - E_s. \tag{26.16}$$

In the Exercises we show that $E(s, t]$ is an orthogonal projector on $\mathcal{H}$ with range

$$\operatorname{ran} E(s, t] = H_t \cap H_s^\perp = H_t \ominus H_s.$$

Since a spectral family is not necessarily left continuous, the operator

$$P(t) = \lim_{s \to t, s < t} E(s, t] = E_t - E_{t-0} \tag{26.17}$$

is in general a projector $\neq 0$. Indeed, $P(t) \neq 0$ if, and only if, E_t is discontinuous at t (for the strong topology). If $(s_1, t_1]$ and $(s_2, t_2]$ are two disjoint intervals, then

$$E(s_1, t_1] E(s_2, t_2] = 0. \tag{26.18}$$

A *partition Z of the interval* $[a, b]$ is a decomposition of $[a, b]$ into a finite number of disjoint subintervals together with a choice of one point in each subinterval:

$$a = t_0 < t_1 < \cdots < t_n = b, \qquad t'_j \in (t_{j-1}, t_j], \quad j = 1, 2, \ldots, n. \quad (26.19)$$

This is denoted as $Z = Z(t_j, t'_j, n)$. The number

$$|Z| = \max\{|t_j - t_{j-1}| : j = 1, \ldots, n\}$$

is called the *width of the partition Z*. It is the length of the largest subinterval. Given two partitions $Z(t_j, t'_j, n)$ and $Z(s_i, s'_i, m)$ we can form their union

$$Z(t_j, t'_j, n) \vee Z(s_i, s'_i, m) = Z(\tau_k, \tau'_k, p)$$

where $\tau_1, \ldots, \tau_p$ is an enumeration of the points $\{t_1, \ldots, t_n, s_1, \ldots, s_m\}$ in their natural order with the corresponding selection of $\tau'_k \in \{t'_1, \ldots, s'_m\}$. Obviously the width of this union is smaller than or equal to the widths of the original partitions. Thus this union is also called the joint refinement of the two partitions.

For a partition $Z = Z(t_j, t'_j, n)$ of the interval $[a, b]$ and a continuous function $f : [a, b] \to \mathbb{R}$, form the sum

$$\Sigma(f, Z) = \sum_{j=1}^{n} f(t'_j) E(t_{j-1}, t_j]. \quad (26.20)$$

Relation (26.18) implies $E(t_{j-1}, t_j] E(t_{i-1}, t_i] = \delta_{ij} E(t_{j-1}, t_j]$ and thus for all $x \in \mathcal{H}$,

$$\sum_{j=1}^{n} \|E(t_{j-1}, t_j]x\|^2 = \|\sum_{j=1}^{n} E(t_{j-1}, t_j]x\|^2 = \|E(a, b]x\|^2 \le \|x\|^2. \quad (26.21)$$

Apply the identity (26.20) to any $x \in \mathcal{H}$. Then the orthogonality relation (26.18) implies

$$\|\Sigma(f, Z)x\|^2 = \sum_{j=1}^{n} |f(t'_j)|^2 \|E(t_{j-1}, t_j]x\|^2 \quad (26.22)$$

which according to the relation (26.21) leads to the estimate

$$\|\Sigma(f, Z)x\| \le \sup\{|f(t)| : t \in [a, b]\} \|E(a, b]x\|. \quad (26.23)$$

This proves that $\Sigma(f, Z)$ is a bounded linear operator on $\mathcal{H}$. Now we study the limit of these bounded operators when the partition Z gets finer and finer, i.e., the limit $|Z| \to 0$. Suppose $Z = Z(t_j, t'_j, n)$ and $Z' = Z(s_i, s'_i, m)$ are two given partitions and $Z(\tau_k, \tau'_k, p)$ is their joint refinement, then, for any $x \in \mathcal{H}$,

$$\Sigma(f, Z)x - \Sigma(f, Z')x = \sum_{j=1}^{n} f(t'_j) E(t_{j-1}, t_j]x - \sum_{i=1}^{m} f(s'_i) E(s_{i-1}, s_i]x$$

$$= \sum_{k=1}^{p} \epsilon_k E(\tau_{k-1}, \tau_k]x$$

where $\epsilon_k = \pm[f(\tau'_{k+1}) - f(\tau'_k)]$, and because of the orthogonality of the projectors $E(\tau_{k-1}, \tau_k]$,

$$\|\Sigma(f, Z)x - \Sigma(f, Z')x\|^2 = \sum_{k=1}^{p} |\epsilon_k|^2 \|E(\tau_{k-1}, \tau_k]x\|^2.$$

Given any $\epsilon > 0$, there is a $\delta > 0$ such that $|f(t) - f(s)| < \epsilon$ whenever $|s-t| \leq \delta$ since f is uniformly continuous. If the widths of the partitions Z, Z' are both smaller than or equal to δ, the width of their joint refinement is also smaller than or equal to δ and thus we can estimate

$$|\epsilon_k| = |f(\tau'_{k+1}) - f(\tau'_k)| \leq |f(\tau'_{k+1}) - f(\tau_k)| + |f(\tau_k) - f(\tau'_k)| < 2\epsilon,$$

since $|\tau'_{k+1} - \tau_k| \leq \delta$ and $|\tau'_k - \tau_k| \leq \delta$. As in estimate (26.21) we obtain

$$\|\Sigma(f, Z)x - \Sigma(f, Z')x\| \leq 2\epsilon \|E(a, b]x\| \tag{26.24}$$

and conclude that the bounded operators $\Sigma(f, Z)$ have a strong limit as $|Z| \to 0$ and that this limit does not depend on the particular choice of the net of partitions Z which is used in its construction. We summarize our results in

Theorem 26.2.2 *Let $E_t, t \in \mathbb{R}$, be a spectral family on the complex Hilbert space $\mathcal{H}$ and $[a, b]$ some finite interval. Then for every continuous function $f : [a, b] \to \mathbb{R}$ the integral of f with respect to the spectral family E_t,*

$$\int_a^b f(t) dE_t, \tag{26.25}$$

is well defined by

$$\int_a^b f(t) dE_t x = \lim_{|Z| \to 0} \Sigma(f, Z)x. \tag{26.26}$$

It is a bounded linear operator on $\mathcal{H}$ with the following properties:

a) $\|\int_a^b f(t) dE_t x\| \leq \sup\{|f(t)| : t \in [a, b]\} \|E(a, b]x\| \quad \forall x \in \mathcal{H}$;

b) $f \mapsto \int_a^b f(t) dE_t$ *is linear on* $\mathcal{C}([a, b]; \mathbb{R})$;

c) *for every* $a < c < b$: $\int_a^b f(t) dE_t = \int_a^c f(t) dE_t + \int_c^b f(t) dE_t$;

d) $(\int_a^b f(t) dE_t)^* = \int_a^b \overline{f(t)} dE_t$;

e) $\|\int_a^b f(t) dE_t x\|^2 = \int_a^b |f(t)|^2 d\rho_x(t)$
   for all $x \in \mathcal{H}$ *where* $\rho_x(t) = \langle x, E_t x \rangle = \|E_t x\|^2$.

Proof. We have shown above that the limit (26.26) exists for every $x \in \mathcal{H}$. Taking this limit in estimate (26.23) gives Property a). Since $\Sigma(f, Z)$ is a bounded linear operator on $\mathcal{H}$ we deduce that $\int_a^b f(t) dE_t$ is a bounded linear operator. Properties b) – d) follow from the corresponding properties of the approximations $\Sigma(f, Z)$ which are easy to establish. The details are left as an exercise.

The starting point for the proof of Part e) is equation (26.22) and the observation

$$\|E(t_{j-1},t_j]x\|^2 = \rho_x(t_j) - \rho_x(t_{j-1})$$

which allows one to rewrite this equation as

$$\|\Sigma(f,Z)x\|^2 = \sum_{j=1}^n |f(t'_j)|^2 [\rho_x(t_j) - \rho_x(t_{j-1})].$$

Now in the limit $|Z| \to 0$ the identity of Part e) follows since the right-hand side is just the approximation of the Stieltjes integral $\int_a^b |f(t)|^2 d\rho_x(t)$ for the same partition Z. □

Lemma 26.2.1 *Suppose E_t, $t \in \mathbb{R}$, is a spectral family in the complex Hilbert space $\mathcal{H}$ and $f : [a,b] \to \mathbb{R}$ a continuous function. Then for any $s < t$ the integral $\int_a^b f(t) dE_t$ commutes with the projectors $E(s,t]$ and one has*

$$E(s,t] \int_a^b f(u) dE_u = \int_a^b f(u) dE_u E(s,t] = \int_{(a,b]\cap(s,t]} f(u) dE_u. \quad (26.27)$$

Proof. Since $E(s,t]$ is a continuous linear operator equation (26.26) implies

$$E(s,t] \int_a^b f(u) dE_u = \lim_{|Z|\to 0} E(s,t] \Sigma(f,Z)$$

where Z denotes a partition of the interval $[a,b]$. The definition of these approximating sums gives $E(s,t]\Sigma(f,Z) = E(s,t]\sum_{j=1}^n f(t'_j) E(t_{j-1},t_j]$. Taking the defining properties of a spectral family into account we calculate

$$E(s,t]E(t_{j-1},t_j] = E(t_{j-1},t_j]E(s,t] = E((t_{j-1},t_j] \cap (s,t]).$$

We deduce $\lim_{|Z|\to 0} E(s,t]\Sigma(f,Z) = \lim_{|Z|\to 0} \Sigma(f,Z)E(s,t]$ and the first identity in equation (26.27) is established.

For the second identity some care has to be taken with regard to the interval to which the partitions refer. Therefore we write this explicitly in the approximating sums $\Sigma(f,Z) \equiv \Sigma(f,Z,[a,b])$ when partitions of the interval $[a,b]$ are used. In this way we write

$$\Sigma(f,Z)E(s,t] = \Sigma(f,Z,[a,b])E(s,t] = \sum_{j=1}^n f(t'_j) E(t_{j-1},t_j]E(s,t]$$

$$= \sum_{j=1}^n f(t'_j) E((t_{j-1},t_j] \cap (s,t]) = \Sigma(f,Z',[a,b] \cap (s,t])$$

where Z' is the partition induced by the given partition Z on the subinterval $[a,b] \cap (s,t]$. Clearly, $|Z| \to 0$ implies $|Z'| \to 0$ and thus

$$\lim_{|Z|\to 0} \Sigma(f,Z)E(s,t] = \lim_{|Z'|\to 0} \Sigma(f,Z',[a,b] \cap (s,t]) = \int_{(a,b]\cap(s,t]} f(u) dE_u$$

and we conclude. □

For the spectral representation of self-adjoint operators and for other problems one needs not only integrals over finite intervals but also integrals over the real line $\mathbb{R}$ which are naturally defined as the limit of integrals over finite intervals $[a,b]$ as $a \to -\infty$ and $b \to +\infty$:

$$\int_{-\infty}^\infty f(t) dE_t x = \lim_{\substack{b\to +\infty \\ a\to -\infty}} \int_a^b f(t) dE_t x \equiv \lim_{a,b} \int_a^b f(t) dE_t x \quad (26.28)$$

for all $x \in \mathcal{H}$ for which this limit exists. The existence of this vector valued integral is characterized by the existence of a numerical Stieltjes integral:

Lemma 26.2.2 *Suppose E_t, $t \in \mathbb{R}$, is a spectral family in the complex Hilbert space $\mathcal{H}$ and $f : \mathbb{R} \to \mathbb{R}$ a continuous function. For $x \in \mathcal{H}$ the integral*

$$\int_{-\infty}^{\infty} f(t) dE_t x$$

exists if, and only if, the numerical integral

$$\int_{-\infty}^{\infty} |f(t)|^2 d\|E_t x\|^2$$

exists.

Proof. The integral $\int_a^b f(t) dE_t x$ has a limit for $b \to +\infty$ if, and only if, for every $\epsilon > 0$ there is b_0 such that for all $b' > b \geq b_0$,

$$\left\| \int_b^{b'} f(t) dE_t x \right\|^2 \leq \epsilon^2.$$

Part e) of Theorem 26.2.2 implies

$$\left\| \int_b^{b'} f(t) dE_t x \right\|^2 = \int_b^{b'} |f(t)|^2 d\rho_x(t)$$

where $d\rho_x(t) = d\|E_t x\|^2$. Thus the vector valued integral has a limit for $b \to \infty$ if, and only if, the numerical, i.e., real valued integral does.

In the same way the limit $a \to -\infty$ is handled. $\square$

Finally the integral of a continuous real valued function on the real line with respect to a spectral family is defined and its main properties are investigated.

Theorem 26.2.3 *Let E_t, $t \in \mathbb{R}$, be a spectral family on the complex Hilbert space $\mathcal{H}$ and $f : \mathbb{R} \to \mathbb{R}$ a continuous function. Define*

$$D = \left\{ x \in \mathcal{H} : \int_{-\infty}^{+\infty} |f(t)|^2 d\|E_t x\|^2 < \infty \right\} \tag{26.29}$$

$$= \left\{ x \in \mathcal{H} : \int_{-\infty}^{+\infty} f(t) dE_t x \text{ exists} \right\} \tag{26.30}$$

and on this domain D define an operator A by

$$Ax = \int_{-\infty}^{+\infty} f(t) dE_t x \qquad \forall x \in D. \tag{26.31}$$

Then this operator A is self-adjoint and satisfies

$$E(s,t]A \subseteq AE(s,t] \qquad \forall s < t. \tag{26.32}$$

Proof. According to Lemma 26.2.2 the two characterizations of the set D are equivalent. The second characterization and the basic rules of calculation for limits show that the set D is a linear subspace of $\mathcal{H}$. In order to prove that D is dense in the Hilbert space we construct a subset $D_0 \subset D$ for which it is easy to show that it is dense.

Denote $P_n = E_n - E_{-n}$ for $n \in \mathbb{N}$ and recall the normalization of a spectral family: $P_n x = E_n x - E_{-n} x \to x - 0$ as $n \to \infty$, for every $x \in \mathcal{H}$. This implies that $D_0 = \cup_{n \in \mathbb{N}} P_n \mathcal{H}$ is dense in $\mathcal{H}$. Now take any $x = P_n x \in D_0$ for some fixed $n \in \mathbb{N}$. In order to prove $x \in D$ we rely on the second characterization of the space D and then have to show that $\lim_{a,b} \int_a^b f(u) d E_u x$ exists in $\mathcal{H}$. This is achieved by Lemma 26.2.1 and Theorem 26.2.2:

$$\lim_{a,b} \int_a^b f(u) d E_u x = \lim_{a,b} \int_{(a,b]} f(u) d E_u E(-n,n] x$$

$$= \lim_{a,b} \int_{(a,b] \cap (-n,n]} f(u) d E_u x = \int_{(-n,n]} f(u) d E_u x.$$

Since the last integral exists, $x = P_n x$ belongs to the space D. We conclude that A is a densely defined linear operator.

Similarly, for $x \in D$, Lemma 26.2.1 implies

$$P_n A x = P_n \lim_{a,b} \int_a^b f(u) d E_u x = \lim_{a,b} P_n \int_a^b f(u) d E_u x = \lim_{a,b} \int_a^b f(u) d E_u P_n x = A P_n x,$$

i.e., $P_n A \subset A P_n$ and thus $A P_n = P_n A P_n$ for all $n \in \mathbb{N}$. In the same way we can prove relation (26.32).

For all $x, y \in D$ one has, using self-adjointness of $\int_a^b f(u) d E_u$ according to Part d) of Theorem 26.2.2,

$$\langle x, A y \rangle = \langle x, \lim_{a,b} \int_a^b f(u) d E_u y \rangle = \lim_{a,b} \langle x, \int_a^b f(u) d E_u y \rangle$$

$$= \lim_{a,b} \langle \int_a^b f(u) d E_u x, y \rangle = \langle \lim_{a,b} \int_a^b f(u) d E_u x, y \rangle = \langle A x, y \rangle,$$

hence $A \subset A^*$ and A is symmetric.

In order to prove that A is actually self-adjoint take any element $y \in D(A^*)$. Then $y^* = A^* y \in \mathcal{H}$ and $A^* y = \lim_{n \to \infty} P_n A^* y$. For all $x \in \mathcal{H}$ we find $\langle P_n A^* y, x \rangle = \langle A^* y, P_n x \rangle = \langle y, A P_n x \rangle = \langle y, P_n A P_n x \rangle = \langle P_n A P_n y, x \rangle$ where we used $P_n x \in D$, the symmetry of A and the relation $A P_n = P_n A P_n$ established earlier. It follows that

$$P_n A^* y = P_n A P_n y = A P_n y \qquad \forall n \in \mathbb{N}.$$

According to the definition of the operator A and our earlier calculations, $A P_n y$ is expressed as

$$A P_n y = \int_{-n}^n f(u) d E_u y \qquad \forall n \in \mathbb{N}.$$

The limit $n \to \infty$ of this integral exists because of the relation $A P_n y = P_n A^* y$. The second characterization of the domain D thus states $y \in D$ and therefore $A P_n y \to A y$ as $n \to \infty$. We conclude that $A^* y = A y$ and A is self-adjoint. □

26.3 The spectral theorem

Theorem 26.3.1 (Spectral theorem) *Every self-adjoint operator A on the complex Hilbert space $\mathcal{H}$ has a unique spectral representation, i.e., there is a unique spectral family $E_t = E_t^A$, $t \in \mathbb{R}$, on $\mathcal{H}$ such that*

$$D(A) = \left\{ x \in \mathcal{H} : \int_\mathbb{R} t^2 d \|E_t x\|^2 < \infty \right\}, \quad A x = \int_\mathbb{R} t \, d E_t x \quad \forall x \in D(A).$$
(26.33)

26. The Spectral Theorem

Proof. At first we give the proof for the special case $A \geq 0$ in detail. At the end the general case is addressed by using an additional limiting procedure.

For the self-adjoint operator $A \geq 0$ introduce the subspaces of controlled growth $F(A, t)$, $t \geq$, as in equation (26.10) and then define for $t \in \mathbb{R}$,

$$H_t = \begin{cases} F(A, t) & t \geq 0, \\ \{0\} & t < 0. \end{cases} \tag{26.34}$$

According to Lemma 26.1.3 this is a family of closed linear subspaces of $\mathcal{H}$ where each subspace is invariant under the operator A. We claim that this family of subspaces satisfies conditions a) – c) of Proposition 26.2.1. Condition a) of monotonicity is evident from the definition of the spaces H_t. Condition b) of right continuity $H_s = \cap_{s<t} H_t$ is obtained in the following way: By monotonicity we know $F(A, s) \subseteq \cap_{s<t} F(A, t)$ for $s \geq 0$. Conversely suppose that $x \in \cap_{s<t} F(A, t) \subset D^\infty(A)$ is given; then $\|A^n x\| \leq t^n \|x\|$ for all $t > s$ and all $n \in \mathbb{N}$ and thus $\|A^n x\| \leq s^n \|x\|$ for all $n \in \mathbb{N}$, i.e., $x \in F(A, s)$. For $s < 0$ this is trivial.

Finally we prove the normalization condition c). $\cap_{t \in \mathbb{R}} H_t = \{0\}$ trivially holds because of the definition (26.34). The second part of the normalization condition

$$\overline{\cup_{t \in \mathbb{R}} H_t} = \mathcal{H}$$

follows from the geometric characterization of self-adjointness, Theorem 26.1.2.

Now we can use Proposition 26.2.1 to define a spectral family E_t, $t \in \mathbb{R}$, such that

$$\operatorname{ran} E_t = H_t \qquad \forall t \in \mathbb{R}. \tag{26.35}$$

In particular the choice $t = n \in \mathbb{N}$ gives a sequence of projectors E_n with strong limit I and with range $H_n = F(A, n) \subseteq D^\infty(A)$ which is invariant under the operator A. It follows that $E_n A \subseteq AE_n$, hence $AE_n = E_n AE_n$ and therefore the domain of A is characterized by Lemma 26.1.1, i.e., $x \in D(A) \Leftrightarrow (AE_n x)_{n \in \mathbb{N}}$ converges in $\mathcal{H} \Leftrightarrow (\|AE_n x\|)_{n \in \mathbb{N}}$ converges in $\mathbb{R}$ and then

$$Ax = \lim_{n \to \infty} AE_n x \qquad \forall x \in D(A).$$

Denote the restriction of the operator $E_n A E_n$ to the invariant subspace H_n by A_n. A_n is a self-adjoint positive operator for which we will show the spectral representation with respect to the spectral family $E_t^{(n)} = E_t E_n$ on H_n. Given a partition $Z = Z(t_j, t_j')$ of the interval $[0, n]$ and $x \in H_n$ introduce the points $x_j = E(t_{j-1}, t_j]x \in F(A_n, t_j) \cap F(A_n, t_{j-1})^\perp$, $j = 1, \ldots, m$. Since different subintervals of the partitions are disjoint, x is the orthogonal sum of the points x_j. Note also that the operator A_n leaves the subspaces $F(n, j) = F(A_n, t_j) \cap F(A_n, t_{j-1})^\perp$ invariant and that different of these subspaces are orthogonal to each other. This implies

$$\langle x, A_n x \rangle = \sum_{j=1}^m \langle x_j, A_n x_j \rangle \quad \text{and} \quad \|A_n x\|^2 = \sum_{j=1}^m \|A_n x_j\|^2.$$

Theorem 26.1.1 allows us to estimate $\langle x_j, A_n x_j \rangle$ and $\|A_n x_j\|^2$ as follows:

$$t_{j-1} \|x_j\|^2 \leq \langle x_j, A_n x_j \rangle \leq t_j \|x_j\|^2, \qquad t_{j-1} \|x_j\| \leq \|A_n x_j\| \leq t_j \|x_j\|.$$

These estimates hold for $j = 1, \ldots, m$ and therefore

$$|\langle x, A_n x \rangle - \sum_{j=1}^m t_j \langle x_j, x_j \rangle| = |\sum_{j=1}^m [\langle x_j, A_n x_j \rangle - t_j \langle x_j, x_j \rangle]|$$

$$\leq \sum_{j=1}^m |\langle x_j, A_n x_j \rangle - t_j \langle x_j, x_j \rangle| \leq \sum_{j=1}^m (t_j - t_{j-1}) \|x_j\|^2 \leq |Z| \sum_{j=1}^m \|x_j\|^2 \leq |Z| \|x\|^2$$

26.3 The spectral theorem

and

$$|\,\|A_n x\|^2 - \sum_{j=1}^m t_j^2\|x_j\|^2\,| = |\sum_{j=1}^m [\|A_n x_j\|^2 - t_j^2\|x_j\|^2]|$$

$$\leq \sum_{j=1}^m |\,\|A_n x_j\|^2 - t_j^2\|x_j\|\,| \leq \sum_{j=1}^m (t_j^2 - t_{j-1}^2)\|x_j\|^2 \leq 2n|Z| \sum_{j=1}^m \|x_j\|^2 \leq 2n|Z|\,\|x\|^2.$$

Since $x_j = E(t_{j-1}, t_j]x$ we have $\|x_j\|^2 = \rho_x(t_j) - \rho_x(t_{j-1})$ and thus $\sum_{j=1}^m t_j^2 \|x_j\|^2 = \Sigma(t^2, Z, \rho_x)$ is the approximating sum for the Stieltjes integral $\int_0^n t^2 d\rho_x(t)$. The above estimate implies

$$\|A_n x\|^2 = \lim_{|Z|\to 0} \Sigma(t^2, Z, \rho_x) = \int_0^n t^2 d\rho_x(t) = \int_0^n t^2 d\|E_t x\|^2$$

and similarly

$$\langle x, A_n x\rangle = \lim_{|Z|\to 0} \Sigma(t, Z, \rho_x) = \int_0^n t\, d\rho_x(t) = \int_0^n t\, d\langle x, E_t x\rangle.$$

The polarization identity (see Proposition 14.1.2) implies $\langle y, A_n x\rangle = \int_0^n t\, d\langle y, E_t x\rangle$ for all $y \in \mathcal{H}$ and therefore

$$A_n x = \int_0^n t\, dE_t x \qquad \forall x \in H_n.$$

Recall $A_n x = E_n A E_n x = A E_n x$ for all $x \in H_n$ and thus the above calculations show that

$$\|A E_n x\|^2 = \int_0^n t^2 d\|E_t x\|^2 \qquad A E_n x = \int_0^n t\, dE_t x \qquad (26.36)$$

for all $n \in \mathbb{N}$ and all $x \in \mathcal{H}$. For the sequence of projectors E_n the hypotheses of Lemma 26.1.1 have been verified. Hence, $x \in D(A)$ if, and only if, $A E_n x$ has a limit and if $x \in D(A)$ then the limit for $n \to \infty$ is Ax. Therefore the vector valued integral $\int_0^n t\, dE_t x$ has a limit for $n \to \infty$, and we conclude by equation (26.36) that equation (26.33) holds for the spectral family defined by the family of subspaces of controlled growth.

Finally we show that there is only one spectral family which represents the self-adjoint operator A according to equation (26.33) by showing: If E'_t, $t \in \mathbb{R}$, is a spectral family on $\mathcal{H}$ which represents the operator A according to this equation, then

$$\operatorname{ran} E'_t = F(A, t) \qquad \forall t \geq 0.$$

Suppose $x \in \operatorname{ran} E'_t$ for some $t \geq 0$. Then $x = E'_t x$ and thus $E'_s x = E'_{s \wedge t} x$ for all $s \geq 0$. Now calculate for any $n \in \mathbb{N}$,

$$\|A^n x\|^2 = \int_0^\infty s^{2n} d\|E'_s x\|^2 = \int_0^\infty s^{2n} d\|E'_{s \wedge t} x\|^2 = \int_0^t s^{2n} d\|E'_s x\|^2$$

$$\leq t^{2n} \int_0^\infty d\|E'_s x\|^2 = t^{2n}\|x\|^2.$$

It follows that $x \in F(A, t)$ and thus $\operatorname{ran} E'_t \subseteq F(A, t)$. Since $F(A, 0) = N(A)$ it suffices to consider the case $t > 0$. Thus suppose $t > 0$ and $x \in F(A, t) \cap \operatorname{ran}(I - E'_t)$; then $x = (I - E'_t)x = \lim_{N \to \infty} E'(t, N]x$. As earlier we find

$$A^n E(t, N]x = \int_0^\infty s^n dE'_s E'(t, N]x = \int_t^N s^n dE'_s x$$

and therefore

$$\|A^n x\|^2 = \int_t^\infty s^{2n} d\|E'_s x\|^2 \geq t^{2n} \int_t^\infty d\|E'_s x\|^2 = t^{2n}\|x\|^2$$

where in the last step $x = (I - E'_t)x$ was taken into account. $x \in F(A, t)$ implies $\|A^n x\|^2 \leq t^{2n}\|x\|^2$ for all $n \in \mathbb{N}$. We conclude $\|A^n x\|^2 = t^{2n}\|x\|^2$ for all $n \in \mathbb{N}$. In terms of the spectral family this reads $\int_t^\infty (s^{2n} - t^{2n}) d\|E'_s x\| = 0$, hence $x = (I - E'_t)x = 0$ and $\operatorname{ran} E'_t = F(A, t)$ follows.

This concludes the proof for the case $A \geq 0$.

Comments on the proof for the general case: a) If A is a lower bounded self-adjoint operator, i.e., for some $c \in \mathbb{R}$ one has $A \geq -cI$, then $A_c = A + cI$ is a positive self-adjoint operator for which the above proof applies and produces the spectral representation $A_c = \int_{\mathbb{R}} t \, dE_t$. In the Exercises we deduce the spectral representation for the operator A itself.

b) The proof of the spectral representation of a self-adjoint operator A which is not lower bounded needs an additional limit process which we indicate briefly.

As in the case of lower bounded self-adjoint operators the subspaces of controlled growth $F(A, t)$ are well defined and have the properties as stated in Lemma 26.1.3. In particular for $t = n \in \mathbb{N}$ we have closed subspaces of $\mathcal{H}$ which are contained in the domain of A and which are invariant under the operator A. Hence the orthogonal projectors P_n onto these subspaces satisfy ran $P_n \subset D(A)$ and $AP_n = P_n A P_n$. Furthermore by the geometric characterization of self-adjointness (Theorem 26.1.2) the union of the ranges of these projectors is dense in $\mathcal{H}$ and therefore $\lim_{n \to \infty} P_n x = x$ for all $x \in \mathcal{H}$. Under the inner product of the Hilbert space $\mathcal{H}$ the closed subspaces $F(A, n)$ are Hilbert spaces too and $A_n = A_{|F(A,n)}$ is a self-adjoint operator which is bounded from below: $A_n + nI \geq 0$. Hence for the operator A_n our earlier results apply.

In the Hilbert space $F(A, n)$ define a spectral family $E_n(t)$, $t \in \mathbb{R}$, by

$$\text{ran } E_n(t) = \begin{cases} \{0\} & t < -n, \\ F(A_n + nI, t + n) & -n \leq t. \end{cases}$$

Then the spectral representation for the operator A_n in the space $F(A, n)$ reads

$$A_n = \int_{\mathbb{R}} t \, dE_n(t). \tag{26.37}$$

This holds for each $n \in \mathbb{N}$. A suitable limit of the spectral families $E_n(\cdot)$, $n \in \mathbb{N}$ will produce the spectral representation for the operator A. To this end one observes

$$E_k(t) P_n = E_n(t) P_n = P_n E_n(t) = E_n(t), \qquad \text{ran } E_k([-n, n]) = F(A_k, n) = F(A, n)$$

for all $t \in \mathbb{R}$, and all $k \geq n$ and then proves that the sequence of spectral families E_n has a strong limit $E(\cdot)$ which is a spectral family in the Hilbert space $\mathcal{H}$:

$$E(t)x = \lim_{n \to \infty} E_n(t)x \qquad \forall x \in \mathcal{H},$$

uniformly in $t \in \mathbb{R}$. And this spectral family satisfies $E([-n, n]) = P_n$.

The spectral representation (26.37) implies for all $x, y \in \mathcal{H}$ and all $n \in \mathbb{N}$ and all $k \geq n$,

$$\langle x, AP_n y \rangle = \langle P_n x, A_n P_n y \rangle = \int_{\mathbb{R}} t \, d\langle P_n x, E_n(t) P_n y \rangle$$

$$= \int_{\mathbb{R}} t \, d\langle P_n x, E_k(t) P_n y \rangle = \int_{[-n,n]} t \, d\langle x, E(t) y \rangle$$

and similarly, for all $x \in \mathcal{H}$ and all $n \in \mathbb{N}$

$$\|AP_n x\|^2 = \int_{[-n,n]} t^2 \, d\langle x, E(t) x \rangle.$$

Finally, another application of Lemma 26.1.1 proves the spectral representation (26.33) for the general case.

c) The proof that the spectral family is uniquely determined by the self-adjoint operator A uses also in the general case the same basic idea as in the case of positive self-adjoint operators. In this case one proves (see Exercises): If a spectral family E' represent the self-adjoint operator A, then

$$\text{ran } E'(s, t] = F(A - \frac{s+t}{2}I, \frac{t-s}{2})$$

for all $s < t$. Since projectors are determined by their range, uniqueness of the spectral family follows. □

26.4 Some applications

For a closed symmetric operator A in the complex Hilbert space $\mathcal{H}$ we can form the subspaces $F(A, r), r \geq 0$, of controlled growth. These subspaces are all contained in the domain $D(A)$ and are invariant under the operator A. The closure of the union $M = \cup_{n \in \mathbb{N}} F(A, n)$ of these subspaces is a subspace $\mathcal{H}_0$ of the Hilbert space and according to the geometric characterization of self-adjointness, the operator A is self-adjoint if, and only if, $\mathcal{H}_0 = \mathcal{H}$.

Now suppose that A is not self-adjoint. Then $\mathcal{H}_0$ is a proper subspace of $\mathcal{H}$. Since the space M is invariant under A and dense in $\mathcal{H}_0$ one would naturally expect that the restriction A_0 of the operator A to the subspace $\mathcal{H}_0$ is a self-adjoint operator in the Hilbert space $\mathcal{H}_0$. This is indeed the case and the self-adjoint operator A_0 defined in this way is called the *maximal self-adjoint part* of the closed symmetric operator A. With the help of the geometric characterization of self-adjointness the proof is straightforward but some terminology has to be introduced.

Definition 26.4.1 *Let A be a linear operator in the complex Hilbert space $\mathcal{H}$ and $\mathcal{H}_0$ a closed linear subspace. $\mathcal{H}_0$ is an **invariant subspace of the operator A** if, and only if, the operator A maps $D(A) \cap \mathcal{H}_0$ into $\mathcal{H}_0$.*

*A closed linear subspace $\mathcal{H}_0$ is a **reducing subspace of the operator A** if, and only if, $\mathcal{H}_0$ is an invariant subspace of A and the orthogonal projector P_0 onto the subspace $\mathcal{H}_0$ maps the domain $D(A)$ into itself, $P_0 D(A) \subseteq D(A)$.*

Thus a closed linear subspace $\mathcal{H}_0$ reduces the linear operator A if, and only if, a) $P_0 x \in D(A)$ and b) $A P_0 x \in \mathcal{H}_0$ for all $x \in D(A)$. Both conditions can be expressed through the condition that $A P_0$ is an extension of the operator $P_0 A$, i.e.,

$$P_0 A \subseteq A P_0. \tag{26.38}$$

Clearly, if $\mathcal{H}_0$ is a reducing subspace of the operator A, the restriction A_0 to the subspace $\mathcal{H}_0$ is a well-defined linear operator in the Hilbert space $\mathcal{H}_0$.

Definition 26.4.2 *Let A be a linear operator in the Hilbert space $\mathcal{H}_0$. The restriction A_0 of A to a reducing subspace $\mathcal{H}_0$ is called the **maximal self-adjoint part of A** if, and only if, A_0 is self-adjoint in the Hilbert space $\mathcal{H}_0$, and if $\mathcal{H}_1$ is any other reducing subspace of A on which the restriction A_1 of A is self-adjoint, then $\mathcal{H}_1 \subseteq \mathcal{H}_0$ and $A_1 \subseteq A_0$.*

Theorem 26.4.1 (Maximal self-adjoint part) *Every closed symmetric operator A in a complex Hilbert space $\mathcal{H}$ has a maximal self-adjoint part A_0. A_0 is defined as the restriction of A to the closure $\mathcal{H}_0$ of the union $M = \cup_{n \in \mathbb{N}} F(A, n)$ of the subspaces of controlled growth.*

Proof. Denote by P_n the orthogonal projector onto the closed invariant subspace $F(A, n), n \in \mathbb{N}$. The sequence of projectors is monotone increasing and thus has a strong limit Q_0, $Q_0 x = \lim_{n \to \infty} P_n x$ for all $x \in \mathcal{H}$. The range of Q_0 is the closure of the union of the ranges of the projectors P_n, i.e., ran $Q_0 = \mathcal{H}_0$.

In order to show that $\mathcal{H}_0$ is a reducing subspace of A, recall that $P_n A x = A P_n x$ for all $x \in D(A)$, a property which has been used before on several occasions. For $n \to \infty$ the left-hand side of this

352   26. The Spectral Theorem

identity converges to $Q_0 Ax$, hence the right-hand side $AP_n x$ converges too for $n \to \infty$. We know $\lim_{n\to\infty} P_n x = Q_0 x$ and A is closed, hence $Q_0 x \in D(A)$ and $AQ_0 x = Q_0 Ax$ for all $x \in D(A)$. Thus $\mathcal{H}_0$ is a reducing subspace.

Consider the restriction $A_0 = A_{|D(A) \cap \mathcal{H}_0}$ of A to the reducing subspace $\mathcal{H}_0$. It follows easily that A_0 is closed and symmetric and that the subspaces of controlled growth coincide: $F(A_0, n) = F(A, n)$ for all $n \in \mathbb{N}$ (see Exercises). Hence the geometric characterization of self-adjointness proves A_0 to be self-adjoint.

Now let $\mathcal{H}_1$ be another reducing subspace of A on which the restriction $A_1 = A_{|D(A) \cap \mathcal{H}_1}$ is self-adjoint. Theorem 26.1.2 implies

$$\mathcal{H}_1 = \overline{\cup_{n \in \mathbb{N}} F(A_1, n)}.$$

Since A_1 is a restriction of A we know $F(A_1, n) \subseteq F(A, n)$ for all $n \in \mathbb{N}$ and thus $\mathcal{H}_1 \subseteq \mathcal{H}_0$ and therefore $A_1 \subseteq A_0$. We conclude that A_0 is the maximal self-adjoint part of A. □

Another powerful application of the geometric characterization of self-adjointness are convenient sufficient conditions for a symmetric operator to be essentially self-adjoint. The idea is to use lower bounds for the subspaces of controlled growth. And here considerable flexibility is available. This is very important since in practice it is nearly impossible to determine the subspaces of controlled growth explicitly.

Theorem 26.4.2 *Let $A \subset A^*$ be a symmetric operator in the Hilbert space $\mathcal{H}$. If for every $n \in \mathbb{N}$ there is a subset $D(A, n) \subset F(A, n)$ of the subspaces of controlled growth such that their union $D_0(A) = \cup_{n \in \mathbb{N}} D(A, n)$ is a total subset of the Hilbert space $\mathcal{H}$, then A is essentially self-adjoint.*

Proof. The closure $\overline{A}$ of A is a closed symmetric operator. It is self-adjoint if, and only if, $H_0(\overline{A}) = \cup_{n \in \mathbb{N}} F(\overline{A}, n)$ is a dense subspace of $\mathcal{H}$. Obviously one has $F(A, n) \subset F(\overline{A}, n)$ and thus $D_0(A) \subseteq H_0(\overline{A})$. By assumption the set $D_0(A)$ is total in $\mathcal{H}$, hence $H_0(\overline{A})$ is dense and thus $\overline{A}$ is self-adjoint. □

We conclude this section by pointing out an interesting connection of the geometric characterization of self-adjointness with a classical result of Nelson which is discussed in detail in the book [RS75].

Let A be a symmetric operator in the complex Hilbert space $\mathcal{H}$. $x \in D^\infty(A)$ is called an *analytic vector* of A if, and only if, there is a constant $C_x < \infty$ such that

$$\|A^n x\| \leq C_x^n n! \quad \forall n \in \mathbb{N}.$$

Denote by $D^\omega(A)$ the set of all analytic vectors of A. Then Nelson's theorem states that A is essentially self-adjoint if, and only if, the space of all analytic vectors $D^\omega(A)$ is dense in $\mathcal{H}$. Furthermore, a closed symmetric operator A is self-adjoint if, and only if, $D^\omega(A)$ is dense in $\mathcal{H}$. In the Exercises we show

$$H_0(A) = \cup_{n \in \mathbb{N}} F(A, n) \subseteq D^\omega(A).$$

Thus Nelson's results are easily understood in terms of the geometric characterization of self-adjointness.

26.5 Exercises

1. For a spectral family $E_t, t \in \mathbb{R}$, on a Hilbert space $\mathcal{H}$ prove
$$E(I_1)E(I_2) = E(I_1 \cap I_2)$$
for any intervals $I_j = (s_j, t_j]$. Here we use $E(\emptyset) = 0$.

2. Prove parts b) – d) of Theorem 26.2.2.

3. Suppose a self-adjoint operator A in a complex Hilbert space $\mathcal{H}$ has the spectral representation $A = \int_{\mathbb{R}} t\, dE_t$ with spectral family $E_t, t \in \mathbb{R}$. Let $c \in \mathbb{R}$ be a constant. Then find the spectral representation for the operator $A + cI$.

4. Let A be a self-adjoint operator in the complex Hilbert space $\mathcal{H}$ which is represented by the spectral family $E'_t, t \in \mathbb{R}$. Prove:
$$\operatorname{ran} E'(s, t] = F(A - \frac{s+t}{2}I, \frac{t-s}{2})$$
for all $s < t$ and conclude uniqueness of the spectral family.
Hints: Recall that $\operatorname{ran} E'(s, t] = \operatorname{ran} E'_t \cap (\operatorname{ran} E'_s)^\perp$ and prove first
$$\|(A - \frac{s+t}{2}I)^n\|^2 = \int_s^t (\tau - \frac{s+t}{2})^{2n} d\|E'_\tau x\|^2$$
for all $x \in \operatorname{ran} E'(s, t]$. Then one can proceed as in the case of a positive operator.

5. Let A be a closed symmetric operator in the Hilbert space $\mathcal{H}$ and let $\mathcal{H}_0$ be the closure of the union of the subspaces of controlled growth $F(A, n), n \in \mathbb{N}$. $\mathcal{H}_0$ is known to be a reducing subspace of A. Prove: a) The restriction A_0 of A to $\mathcal{H}_0$ is a closed and symmetric operator in $\mathcal{H}_0$; b) $F(A, n) = F(A_0, n)$ for all $n \in \mathbb{N}$.

6. Denote by M_g the operator of multiplication with the real valued piecewise continuous function g in the Hilbert space $\mathcal{H} = L^2(\mathbb{R})$. Assume that for every $n \in \mathbb{N}$ there are nonnegative numbers r_n, R_n such that $[-r_n, R_n] \subseteq \{x \in \mathbb{R} : |g(x)| \leq n\}$. Prove: M_g is self-adjoint on $D(M_g) = \{f \in L^2(\mathbb{R}) : g \cdot f \in L^2(\mathbb{R})\}$ if $r_n \to \infty$ and $R_n \to \infty$ as $n \to \infty$.
Hints: In Theorem 26.4.2 try the sets
$$D(M_g, n) = \left\{ f \in L^2(\mathbb{R}) : \operatorname{supp} f \subseteq [-r_n, R_n] \right\}.$$

7. Consider the free Hamilton operator H_0 in momentum representation in the Hilbert space $\mathcal{H} = L^2(\mathbb{R}^3)$, i.e., $(H_0 \psi)(\underline{p}) = \frac{p^2}{2m} \psi(\underline{p})$ for all $\underline{p} \in \mathbb{R}^3$ and all $\psi \in D(H_0)$. Since we had shown earlier that H_0 is self-adjoint we know that it has a spectral representation. Determine this spectral representation explicitly.

27
Some Applications of the Spectral Representation

For a self-adjoint operator A in a complex Hilbert space $\mathcal{H}$ the spectral representation

$$A = \int_{\mathbb{R}} t\, dE_t$$

has many interesting consequences. Some of these we discuss in this chapter.

In Theorem 26.2.3 we learned to integrate functions with respect to a spectral family $E_t, t \in \mathbb{R}$. This applies in particular to the spectral family of a self-adjoint operator and thus allows us to define quite general functions $f(A)$ of a self-adjoint operator A. Some basic facts of this functional calculus are presented in the first section.

The next section introduces a detailed characterization of the different parts of the spectrum of a self-adjoint operator in terms of its spectral family. The different parts of the spectrum are distinguished by the properties of the measure $d\rho_x(t) = d\langle x, E_t x\rangle$ in relation to the Lebesgue measure and this leads to the different spectral subspaces of the operator.

Finally we discuss the physical interpretation of the different parts of the spectrum for a self-adjoint Hamilton operator.

27.1 Functional calculus

We restrict ourselves to the functional calculus for continuous functions though it can be extended to a much wider class, the Borel functions, through an additional limit process.

Theorem 27.1.1 *Let A be a self-adjoint operator in the complex Hilbert space $\mathcal{H}$ and E_t, $t \in \mathbb{R}$, its spectral family. Denote by $\mathcal{C}_b(\mathbb{R})$ the space of bounded continuous functions $g : \mathbb{R} \to \mathbb{C}$. Then for every $g \in \mathcal{C}_b(\mathbb{R})$,*

$$g(A) = \int_\mathbb{R} g(u) dE_u \tag{27.1}$$

is a well-defined bounded linear operator on $\mathcal{H}$ and $g \mapsto g(A)$ is a continuous algebraic *-homomorphism $\mathcal{C}_b(\mathbb{R}) \to \mathcal{B}(\mathcal{H})$, i.e.,

a) $(a_1 g_1 + a_2 g_2)(A) = a_1 g_1(A) + a_2 g_2(A)$ *for all* $g_j \in \mathcal{C}_b(\mathbb{R})$ *and all* $a_j \in \mathbb{C}$;

b) $1(A) = I$;

c) $g(A)f(A) = (g \cdot f)(A)$ *for all* $f, g \in \mathcal{C}_b(\mathbb{R})$;

d) $g(A)^* = \overline{g}(A)$ *for all* $g \in \mathcal{C}_b(\mathbb{R})$;

e) $\|g(A)\| \leq \|g\|_\infty$ *for all* $g \in \mathcal{C}_b(\mathbb{R})$.

In addition the following holds:

1) $id(A) = A$;

2) $g \in \mathcal{C}_b(\mathbb{R})$ *and* $g \geq 0$ *implies* $g(A) \geq 0$;

3) *If* $g \in \mathcal{C}_b(\mathbb{R})$ *is such that* $\frac{1}{g} \in \mathcal{C}_b(\mathbb{R})$, *then* $g(A)^{-1} = \frac{1}{g}(A)$;

4) $\sigma(g(A)) = g(\sigma(A)) = \{g(\lambda) : \lambda \in \sigma(A)\}$ **(spectral mapping theorem)**.

Proof. Theorem 26.2.2 and Lemma 26.2.2 easily imply that for every $g \in \mathcal{C}_b(\mathbb{R})$ the operator $g(A)$ is a well-defined bounded linear operator on $\mathcal{H}$, since for all $x \in \mathcal{H}$ one has $\int_\mathbb{R} |g(u)|^2 d\|E_u x\|^2 \leq \|g\|_\infty^2 \int_\mathbb{R} d\|E_u x\|^2 = \|g\|_\infty^2 \|x\|^2$. Part e) follows immediately. Parts a), b), and d) also follow easily from a combination of Theorem 26.2.2 and Lemma 26.2.2. The proof of Part c) is left as an exercise where some hints are given.

The first of the additional statements is just the spectral theorem. For the second we observe that for all $x \in \mathcal{H}$ one has $\langle x, g(A)x \rangle = \int_\mathbb{R} g(u) d\|E_u x\|^2 \geq 0$ and hence $g(A) \geq 0$. The third follows from the combination of b) and c). The proof of the spectral mapping theorem is left as an exercise for the reader. □

Corollary 27.1.1 *Let A be a self-adjoint operator in the Hilbert space $\mathcal{H}$ and E_t, $t \in \mathbb{R}$ its spectral family. Define*

$$V(t) = e^{itA} = \int_\mathbb{R} e^{itu} dE_u \quad \forall t \in \mathbb{R}. \tag{27.2}$$

Then $V(t)$ is a strongly continuous one-parameter group of unitary operators on $\mathcal{H}$ with generator A.

Proof. $V(t)$ is defined as $e_t(A)$ where e_t is the continuous bounded functions $e_t(u) = e^{itu}$ for all $u \in \mathbb{R}$. These exponential functions e_t satisfy $\overline{e_t} = e_{-t} = \frac{1}{e_t}$. Hence parts d) and 3) imply $(e_t(A))^* e_t(A) = e_t(A)(e_t(A))^* = I$. Furthermore these functions satisfy $e_t \cdot e_s = e_{t+s}$ for all $t, s \in \mathbb{R}$ and $e_0 = 1$. Hence parts b) and c) imply $V(t)V(s) = V(t+s)$ and $V(0) = I$, thus $V(t)$ is a one-parameter group of unitary operators on $\mathcal{H}$.

For $x \in \mathcal{H}$ and $s, t \in \mathbb{R}$ we have $\|V(t+s)x - V(t)x\| = \|V(t)(V(s)x - x)\| = \|V(s)x - x\|$ and

$$\|V(s)x - x\|^2 = \int_{\mathbb{R}} |(e^{isu} - 1)|^2 d\|E_u x\|^2.$$

Since $|(e^{isu} - 1)| \leq 2$ and $|(e^{isu} - 1)| \to 0$ as $s \to 0$ for every u, a simple application of Lebesgue's dominated convergence theorem implies $\|V(s)x - x\| \to 0$ for $s \to 0$. Therefore the group $V(t)$ is strongly continuous.

According to Stone's Theorem 22.2.3 this group has a self-adjoint generator B defined on $D = \left\{ x \in \mathcal{H} : \exists \lim_{t \to 0} \frac{1}{t}(V(t)x - x) \right\}$ by $iBx = \lim_{t \to 0} \frac{1}{t}(V(t)x - x)$.

According to the spectral theorem 26.3.1 a vector $x \in \mathcal{H}$ belongs to the domain of A if, and only if, $\int_{\mathbb{R}} u^2 d\|E_u x\|^2 < \infty$. Thus by another application of Lebesgue's dominated convergence theorem we find that

$$\left\| \frac{1}{t}[V(t)x - x] \right\|^2 = \int_{\mathbb{R}} \left| \frac{e^{itu} - 1}{t} \right|^2 d\|E_u x\|^2$$

has a limit for $t \to 0$ since $\left| \frac{e^{itu}-1}{t} \right|^2 \leq u^2$. We conclude that $D(A) \subseteq D$ and $A \subseteq B$. Since A is self-adjoint this implies $A = B$. $\square$

The following corollary completes the proof of Stone's theorem.

Corollary 27.1.2 *Let $U(t)$ be a strongly continuous one-parameter group of unitary operators on the complex Hilbert space $\mathcal{H}$ and A its self-adjoint generator. Then*

$$U(t) = e_t(A) \equiv e^{itA} \quad \forall t \in \mathbb{R}.$$

Proof. We know already that the strongly continuous one-parameter group of unitary operators $V(t) = e_t(A)$ has the generator A and that both $U(t)$ and $V(t)$ leave the domain D of the generator A invariant. For $x \in D$ introduce $x(t) = U(t)x - V(t)x \in D$ for all $t \in \mathbb{R}$. Thus this function has the derivative $\frac{d}{dt} x(t) = iAU(t)x - iAV(t)x = iAx(t)$ and therefore

$$\frac{d}{dt} \|x(t)\|^2 = \langle \frac{d}{dt} x(t), x(t) \rangle + \langle x(t), \frac{d}{dt} x(t) \rangle = \langle iAx(t), x(t) \rangle + \langle x(t), iAx(t) \rangle = 0$$

for all $t \in \mathbb{R}$. Since $x(0) = 0$ we conclude that $x(t) = 0$ for all $t \in \mathbb{R}$ and therefore the groups $U(t)$ and $V(t)$ agree on D. Since D is dense this proves that $U(t)$ and $V(t)$ agree on $\mathcal{H}$. $\square$

27.2 Decomposition of the spectrum – Spectral subspaces

According to Weyl's criterium (Theorem 27.2.6) a real number λ belongs to the spectrum of a self-adjoint operator A if, and only if, there is a sequence of unit vectors x_n such that $\|(A - \lambda I)x_n\| \to 0$ as $n \to \infty$. The spectral theorem allows us to translate this criterium into a characterization of the points of the spectrum of A into properties of its spectral family E. This will be our starting point for this section. Then a number of consequences are investigated. When we relate the

358   27. Some Applications of the Spectral Representation

spectral measure $d\rho_x$ associated to the spectral family of A and a vector $x \in \mathcal{H}$ to the Lebesgue measure $d\lambda$ on the real numbers we will obtain a finer decomposition of the spectrum $\sigma(A)$.

Theorem 27.2.1 *Let A be a self-adjoint operator in a complex Hilbert space $\mathcal{H}$ and E_t, $t \in \mathbb{R}$, its spectral family. Then the following holds:*

a) $\mu \in \sigma(A) \Leftrightarrow E_{\mu+\epsilon} - E_{\mu-\epsilon} \neq 0 \quad \forall \epsilon > 0;$

b) $\mu \in \mathbb{R}$ *is an eigenvalue of* $A \Leftrightarrow E(\{\mu\}) = E_\mu - E_{\mu-0} \neq 0.$

Proof. Suppose that there is an $\epsilon > 0$ such that $P = E_{\mu+\epsilon} - E_{\mu-\epsilon} = 0$. Then for any $x \in D(A)$ with $\|x\| = 1$ we find by the spectral theorem that

$$\|(A - \mu I)x\|^2 = \int_{|t-\mu|\geq\epsilon} |t-\mu|^2 d\|E_t x\|^2 \geq \epsilon^2 \int_{|t-\mu|\geq\epsilon} d\|E_t x\|^2 = \epsilon^2 \|x\|^2 = \epsilon^2 > 0$$

since we can write $x = Px + (I - P)x = (I - P)x$. Thus no sequence of unit vectors in $D(A)$ can satisfy Weyl's criterium, hence $\mu \notin \sigma(A)$.

Conversely, if $P_n = E_{\mu+\frac{1}{n}} - E_{\mu-\frac{1}{n}} \neq 0$ for all $n \in \mathbb{N}$, then there is a sequence $x_n = P_n x_n$ in $D(A)$ with $\|x_n\| = 1$. For this sequence we have by the spectral theorem

$$\|(A - \mu I)x_n\|^2 = \int_{|t-\mu|\leq\frac{1}{n}} |t-\mu|^2 d\|E_t x_n\|^2 \leq \frac{1}{n^2} \|x_n\|^2 = \frac{1}{n^2}$$

and thus this sequence satisfies Weyl's criterium and therefore μ belongs to the spectrum of A. This proves Part a).

Next suppose that $\mu \in \mathbb{R}$ is an eigenvalue of A. Let $x \in D(A)$ be a normalized eigenvector. Again by the spectral representation the identity

$$0 = \|(A - \mu I)x\|^2 = \int_\mathbb{R} |t-\mu|^2 d\|E_t x\|^2$$

holds. In particular, for all $N \in \mathbb{N}$ and all $\epsilon > 0$,

$$0 = \int_{\mu+\epsilon}^N |t-\mu|^2 d\|E_t x\|^2 \geq \epsilon^2 \int_{\mu+\epsilon}^N d\|E_t x\|^2 = \epsilon^2 \|E(\mu+\epsilon, N]x\|^2.$$

We conclude that $0 = E_N x - E_{\mu+\epsilon} x$ and similarly $0 = E_{-N} x - E_{\mu-\epsilon} x$ for all $N \in \mathbb{N}$ and all $\epsilon > 0$. Now apply the normalization condition of a spectral family to conclude $x = E_{\mu+\epsilon} x$ and $0 = E_{\mu-\epsilon} x$ for all $\epsilon > 0$. This implies that, using right continuity of a spectral family, $x = (E_\mu - E_{\mu-0})x$ and the projector $E_\mu - E_{\mu-0}$ is not zero.

When we know that the projector $P = E_\mu - E_{\mu-0}$ is not zero, then there is a $y \in \mathcal{H}$ such that $y = Py$ and $\|y\| = 1$. It follows that $y \in D(A)$ and $E_t y = y$ for $t > \mu$ and $E_t y = 0$ for $t < \mu$, hence $\|(A - \mu I)y\|^2 = \int_\mathbb{R} |t-\mu|^2 d\|E_t y\|^2 = 0$, i.e., $(A - \mu I)y = 0$ and μ is an eigenvalue of A. □

The set $D_e = \{x \in D(A) : x \neq 0, \ Ax = \lambda x \text{ for some } \lambda \in \mathbb{R}\}$ of all eigenvectors of the self-adjoint operator A generates the closed subspace $[D_e] = \mathcal{H}_p = \mathcal{H}_p(A)$ called the *discontinuous subspace* of A. Its orthogonal complement $\mathcal{H}_p^\perp$ is the *continuous subspace* $\mathcal{H}_c(A)$ of A, and thus one has the decomposition

$$\mathcal{H} = \mathcal{H}_p(A) \oplus \mathcal{H}_c(A)$$

of the Hilbert space.

With every spectral family E_t, $t \in \mathbb{R}$, one associates a family of spectral measures $(d\rho_x)_{x \in \mathcal{H}}$ on the real line $\mathbb{R}$ which are defined by $\int_a^b d\rho_x(t) = \langle x, E(a,b]x \rangle$. In terms of these spectral measures the continuous and discontinuous subspaces are characterized by

27.2 Decomposition of the spectrum – Spectral subspaces

Proposition 27.2.2 *Let A be a self-adjoint operator in the complex Hilbert space $\mathcal{H}$ with spectral family E_t, $t \in \mathbb{R}$. For $x \in \mathcal{H}$ denote by $d\rho_x$ the spectral measure defined by the spectral family of A. Then*

a) $x \in \mathcal{H}_p(A)$ if, and only if, there is a countable set $a \subset \mathbb{R}$ such that $E(a)x = x$ or equivalently $\rho_x(a^c) = 0$;

b) $x \in \mathcal{H}_c(A)$ if, and only if, $t \mapsto \|E_t x\|^2$ is continuous on $\mathbb{R}$ or equivalently $\rho_x(\{t\}) = 0$ for every $t \in \mathbb{R}$.

Proof. If $a \subset \mathbb{R}$ is a Borel set, then $E(a)x = x$ if, and only if, $E(a^c)x = 0$, if, and only if, $\rho_x(a^c) = \|E(a^c)x\|^2 = 0$. Therefore the two characterizations of $\mathcal{H}_p(A)$ are equivalent. Since $\mathcal{H}_p(A)$ is defined as the closure of the set of all eigenvectors of A, every point $x \in \mathcal{H}_p(A)$ is of the form $x = \lim_{n \to \infty} \sum_{j=1}^n c_j e_j$ with coefficients $c_j \in \mathbb{C}$ and eigenvectors e_j of A corresponding to eigenvalues λ_j. The list of all different eigenvalues is a countable set $a = \{\lambda_{j(i)} : i \in \mathbb{N}\}$ and the corresponding projectors $E(\{\lambda_j\})$ are orthogonal and satisfy $E(\{\lambda_j\})e_j = e_j$ according to Theorem 27.2.1. For every $k \in \mathbb{N}$ we thus find $E(a)e_k = e_k$ and therefore $E(a)x = \lim_{n \to \infty} E(a)\sum_{k=1}^n c_k e_k = x$.

Conversely, if $x \in \mathcal{H}$ satisfies $E(a)x = x$ for some countable set $a = \{\lambda_j : j \in \mathbb{N}\}$, then $x = \lim_{n \to \infty} \sum_{j=1}^n E(\{\lambda_j\})x$ and $E(\{\lambda_j\})$ is not zero if, and only if, λ_j is an eigenvalue (Theorem 27.2.1). This proves Part a).

For every $x \in \mathcal{H}_p(A)^{\perp}$ and every $\lambda \in \mathbb{R}$ we find $\rho_x(\{\lambda\}) = \langle x, E(\{\lambda\})x\rangle = 0$, since by the first part $E(\{\lambda\})x \in \mathcal{H}_p(A)$.

If for $x \in \mathcal{H}$ we know $\rho_x(\{\lambda\}) = 0$ for every $\lambda \in \mathbb{R}$, then $\|E(a)x\|^2 = \rho_x(a) = 0$ for every countable set $a \subset \mathbb{R}$. For every $y \in \mathcal{H}_p(A)$ there is a countable set $a \subset \mathbb{R}$ such that $E(a)y = y$, hence $\langle x, y\rangle = \langle x, E(a)y\rangle = \langle E(a)x, y\rangle = 0$ and thus $x \in \mathcal{H}_p(A)^{\perp}$. The definition of the spectral measure $d\rho_x$ implies easily that the two characterizations of $\mathcal{H}_c(A)$ are equivalent. □

A further decomposition of the continuous subspace of a self-adjoint operator A is necessary for an even finer analysis.

Definition 27.2.1 *For a self-adjoint operator A in a complex Hilbert space $\mathcal{H}$ with spectral family E_t, $t \in \mathbb{R}$, the following spectral subspaces are distinguished:*

*a) **singularly continuous subspace** $\mathcal{H}_{sc}(A)$ of A: $x \in \mathcal{H}_{sc}(A)$ if, and only if, there exists a Borel set $a \subset \mathbb{R}$ of Lebesgue measure zero ($|a| = 0$) such that $E(a)x = x$;*

*b) **absolutely continuous subspace** $\mathcal{H}_{ac}(A)$ of A: $\mathcal{H}_{ac}(A) = \mathcal{H}_c(A) \ominus \mathcal{H}_{sc}(A)$;*

*c) **singular subspace** $\mathcal{H}_s(A) = \mathcal{H}_p(A) \oplus \mathcal{H}_{sc}(A)$.*

In the Exercises we show that $\mathcal{H}_{sc}(A)$ is indeed a closed linear subspace of $\mathcal{H}$. Evidently these definitions imply the following decomposition of the Hilbert space into spectral subspaces of the self-adjoint operator A.

$$\mathcal{H} = \mathcal{H}_p(A) \oplus \mathcal{H}_c(A) = \mathcal{H}_p(A) \oplus \mathcal{H}_{sc}(A) \oplus \mathcal{H}_{ac}(A) = \mathcal{H}_s(A) \oplus \mathcal{H}_{ac}(A). \quad (27.3)$$

Again these spectral subspaces have a characterization in terms of the associated spectral measures.

Proposition 27.2.3 *For a self-adjoint operator A in the complex Hilbert space $\mathcal{H}$ the singular and the absolutely continuous subspace are characterized by*

$$\mathcal{H}_s(A) = \{x \in \mathcal{H} : \exists \text{ Borel set } a \subset \mathbb{R} \text{ such that } |a| = 0 \text{ and } \rho_x(a^c) = 0\}$$
$$= \{x \in \mathcal{H} : \rho_x \text{ is singular with respect to the Lebesgue measure }\},$$
$$\mathcal{H}_{ac}(A) = \{x \in \mathcal{H} : \text{ for every Borel set } a \subset \mathbb{R} \text{ with } |a| = 0 \text{ one has } \rho_x(a) = 0\}$$
$$= \{x \in \mathcal{H} : \rho_x \text{ is absolutely continuous w. resp. to the L-measure }\}.$$

Proof. Every $x \in \mathcal{H}_s(A)$ is the sum of a unique $y \in \mathcal{H}_p(A)$ and a unique $z \in \mathcal{H}_{sc}(A)$. According to Proposition 27.2.2 there is a countable set $a \subset \mathbb{R}$ such that $E(a)y = y$ and by defintion of the singularly continuous subspace there is a Borel set $b \subset \mathbb{R}$ with $|b| = 0$ and $E(b)z = z$. $m = a \cup b$ is again a Borel set with Lebesgue measure zero and we have $E(m)x = E(m)E(a)y + E(m)E(b)z = E(a)y + E(b)z = x$. Then clearly $\rho_x(m^c) = 0$.

Conversely, if $x \in \mathcal{H}$ satisfies $\rho_x(m^c) = 0$ for some Borel set m of measure zero, then $E(m)x = x$. Recall that $t \mapsto \|E_t\|^2$ is a monotone increasing function of bounded total variation. Thus it has a jump at, at most, countably many points t_j. Introduce the set $a = \{t_j : j \in \mathbb{N}\}$. The last proposition implies that $E(a)x \in \mathcal{H}_p(A)$. In the Exercises we show

$$E(\{t\})E(a^c)x = 0 \quad \forall t \in \mathbb{R}.$$

We deduce $E(b)E(a^c)x = 0$ for every countable set $b \subset \mathbb{R}$. If $y \in \mathcal{H}_p(A)$ is given, there is a countable set $b \subset \mathbb{R}$ such that $E(b)y = y$; we calculate $\langle y, E(a^c)x \rangle = \langle E(b)y, E(a^c)x \rangle = \langle y, E(b)E(a^c)x \rangle = 0$ and see $E(a^c)x \in \mathcal{H}_p(A)^\perp = \mathcal{H}_c(A)$. Furthermore the identity $E(m)x = x$ implies $E(m)E(a^c)x = E(m)x - E(m)E(a)x = E(a^c)x$. Therefore the vector $E(a^c)x$ belongs to the singularly continuous subspace. The identity $x = E(a)x + E(a^c)x \in \mathcal{H}_p(A) \oplus \mathcal{H}_{sc}(A)$ finally proves the first part.

To prove the second part take any $x \in \mathcal{H}$ and suppose that for every Borel set $a \subset \mathbb{R}$ with $|a| = 0$ we know $\rho_x(a) = 0$ and therefore $E(a)x = 0$. For every $y \in \mathcal{H}_p(A)$ there is a countable set $a \subset \mathbb{R}$ such that $E(a)y = y$ and for every $z \in \mathcal{H}_{sc}(A)$ there is a Borel set $b \subset \mathbb{R}$ such that $|b| = 0$ and $E(b)z = z$. This implies $\langle x, y + z \rangle = \langle x, E(a)y \rangle + \langle x, E(b)z \rangle = \langle E(a)x, y \rangle + \langle E(b)x, z \rangle = 0 + 0 = 0$, hence $x \in \mathcal{H}_s(A)^\perp = \mathcal{H}_{ac}(A)$.

For $x \in \mathcal{H}$ and any Borel set $b \subset \mathbb{R}$ with $|b| = 0$, one knows $E(b)x \in \mathcal{H}_s(A)$ according to the first part. If now $x \in \mathcal{H}_{ac}(A)$ is given and $b \subset \mathbb{R}$ any Borel set with $|b| = 0$, we find $\rho_x(b) = \|E(b)x\|^2 = \langle x, E(b)x \rangle = 0$ which proves the characterization of $\mathcal{H}_{ac}(A)$. $\square$

There is another way to introduce these spectral subspaces of a self-adjoint operator A in a Hilbert space $\mathcal{H}$. As we know, for every $x \in \mathcal{H}$ the spectral measure $d\rho_x$ is a Borel measure on the real line $\mathbb{R}$. Lebesgue's decomposition theorem (see for instance [Rud80]) for such measures states that $d\rho_x$ has a unique decomposition into pairwise singular measures

$$d\rho_x = d\rho_{x,pp} + d\rho_{x,sc} + d\rho_{x,ac} \tag{27.4}$$

with the following specification of the three measures: $d\rho_{x,pp}$ is a pure point measure, i.e., there are at most countably many points t_j such that $\rho_{x,pp}(\{t_j\}) \neq 0$. $d\rho_{x,sc}$ is a continuous measure, i.e., $\rho_{x,ac}(\{t\}) = 0$ for all $t \in \mathbb{R}$, which is singular with respect to the Lebesgue measure, i.e., there is a Borel set $a \subset \mathbb{R}$ such that $\rho_{x,sc}(a) = 0$ while $|a^c| = 0$. Finally, $d\rho_{x,ac}$ is a Borel measure which is absolutely continuous with respect to the Lebesgue measure, i.e., for every Borel set $b \subset \mathbb{R}$ with $|b| = 0$, one has $\rho_{x,ac}(b) = 0$.

27.2 Decomposition of the spectrum – Spectral subspaces

As a consequence we have the following decomposition of the corresponding L^2-space:

$$L^2(\mathbb{R}, d\rho_x) = L^2(\mathbb{R}, d\rho_{x,pp}) \oplus L^2(\mathbb{R}, d\rho_{x,sc}) \oplus L^2(\mathbb{R}, d\rho_{x,ac}). \quad (27.5)$$

In the terminology of Lebesgue's decomposition theorem we can reformulate the definition of the various spectral subspaces:

$\mathcal{H}_p(A) = \{x \in \mathcal{H} : d\rho_x \text{ is a pure point measure on } \mathbb{R}\}$;
$\mathcal{H}_{sc}(A) = \{x \in \mathcal{H} : d\rho_x \text{ is continuous and singular w. resp. to the L-measure}\}$;
$\mathcal{H}_{ac}(A) = \{x \in \mathcal{H} : d\rho_x \text{ is absolutely continuous w. resp. to the L-measure}\}$.

Therefore, because of the spectral theorem and our previous characterization of the spectral subspaces, the decompositions (27.3) and (27.5) correspond to each other and thus in the sense of Lebesgue measure theory this decomposition is natural.

We proceed by showing that the given self-adjoint operator A has a restriction $A_i = A_{|D_i}$, $D_i = D(A) \cap \mathcal{H}_i$, to its spectral subspace $\mathcal{H}_i = \mathcal{H}_i(A)$ where i stands for p, c, sc, ac, s. This is done by proving that these spectral subspaces are reducing for the operator A.

Theorem 27.2.4 *Let A be a self-adjoint operator in the complex Hilbert space $\mathcal{H}$. Then the restriction A_i of A to the spectral subspace $\mathcal{H}_i$ is a self-adjoint operator in the Hilbert space $\mathcal{H}_i$, $i = p, c, sc, ac, s$.*

Proof. Denote by P_i the orthogonal projector from $\mathcal{H}$ onto the spectral subspace $\mathcal{H}_i$. Recall that $\mathcal{H}_i$ is a reducing subspace for the operator A if

$$P_i D(A) \subset D(A) \quad \text{and} \quad A P_i x = P_i A x \quad \forall x \in D(A).$$

We verify this condition explicitly for the case $i = p$, i.e., for the restriction to the discontinuous subspace.

According to Proposition 27.2.2 a point $x \in \mathcal{H}$ belongs to the discontinuous subspace $\mathcal{H}_p(A)$ if, and only if, there is a countable set $a \subset \mathbb{R}$ such that $E(a)x = x$. The projector $E(a)$ commutes with all the projectors E_t, $t \in \mathbb{R}$, of the spectral family E of A. Thus $x \in \mathcal{H}_p(A)$ implies $E_t x \in \mathcal{H}_p(A)$ for all $t \in \mathbb{R}$ and therefore $E_t P_p = P_p E_t$ for all $t \in \mathbb{R}$.

The spectral theorem says: $x \in D(A)$ if, and only if, $\int_\mathbb{R} t^2 d\|E_t x\|^2 < \infty$. For $x \in D(A)$ we thus find

$$\int_\mathbb{R} t^2 d\|E_t P_p x\|^2 = \int_\mathbb{R} t^2 d\|P_p E_t x\|^2 \le \int_\mathbb{R} t^2 d\|E_t x\|^2 < \infty.$$

This proves $P_p x \in D(A)$. Now we apply again the spectral theorem to calculate for $x \in D(A)$

$$A P_p x = \int_\mathbb{R} t\, dE_t P_p x = \int_\mathbb{R} t\, dP_p E_t x = \int_\mathbb{R} t\, P_p dE_t x = P_p \int_\mathbb{R} t\, dE_t x = P_p A x.$$

It follows that $\mathcal{H}_p(A)$ is a reducing subspace for the self-adjoint operator A. We conclude that the restriction of A to this reducing subspace is self-adjoint.

In the Exercises the reader is asked to fill in some details and to prove the remaining cases. □

The last result enables the definition of those parts of the spectrum of a self-adjoint operator A which correspond to the various spectral subspaces.

$$\left.\begin{array}{rcccl} \sigma_c(A) &=& \sigma(A_c) &=& \text{continuous} \\ \sigma_{sc}(A) &=& \sigma(A_{sc}) &=& \text{singularly continuous} \\ \sigma_{ac}(A) &=& \sigma(A_{ac}) &=& \text{absolutely continuous} \\ \sigma_s(A) &=& \sigma(A_s) &=& \text{singular} \end{array}\right\} \text{spectrum of } A.$$

362   27. Some Applications of the Spectral Representation

The *point spectrum* $\sigma_p(A)$ however is defined as the set of all eigenvalues of A. This means that in general we only have

$$\overline{\sigma_p(A)} = \sigma(A_p).$$

Corresponding to the definition of the various spectral subspaces (Definition 27.2.1) the spectrum of a self-adjoint operator A can be decomposed as follows:

$$\sigma(A) = \overline{\sigma_p(A) \cup \sigma_{sc}(A) \cup \sigma_{ac}(A)} = \sigma_s(A) \cup \sigma_{ac}(A) = \overline{\sigma_p(A)} \cup \sigma_c(A). \quad (27.6)$$

There is a third way to decompose the spectrum of a self-adjoint operator into two parts. Denote by $\sigma_d(A)$ the set of those isolated points of $\sigma(A)$ which are eigenvalues of finite multiplicity. This set is the *discrete spectrum* $\sigma_d(A)$. The remaining set $\sigma_e(A) = \sigma(A) \setminus \sigma_d(A)$ is called the *essential spectrum* of A,

$$\sigma(A) = \sigma_d(A) \cup \sigma_e(A). \quad (27.7)$$

As we are going to show, the essential spectrum has remarkable stability properties with regard to certain changes of the operator. But first the essential spectrum has to be characterized more explicitly.

Theorem 27.2.5 *For a self-adjoint operator A in a complex Hilbert space $\mathcal{H}$ with spectral family E, the following statements are equivalent.*

a) $\lambda \in \sigma_e(A)$;

b) *there is a sequence* $(x_n)_{n \in \mathbb{N}} \subset D(A)$ *such that*

b_1) $(x_n)_{n \in \mathbb{N}}$ *converges weakly to 0;*

b_2) $\liminf_{n \to \infty} \|x_n\| > 0$;

b_3) $\lim_{n \to \infty} (A - \lambda I) x_n = 0$;

c) $\dim(\operatorname{ran}(E_{\lambda+r} - E_{\lambda-r} 0)) = \infty$ *for every* $r > 0$.

Proof. Suppose $\lambda \in \sigma_e(A)$. If λ is an eigenvalue of infinite multiplicity, then there is an infinite orthonormal system of eigenvectors x_n. Such a system is known to converge weakly to 0 and thus b) holds in this case. Next suppose that λ is an accumulation point of the spectrum of A. Then there is a sequence $(\lambda_n)_{n \in \mathbb{N}} \subset \sigma(A)$ with the following properties:

$$\lim_{n \to \infty} \lambda_n = \lambda, \quad \lambda_n \neq \lambda, \quad \lambda_n \neq \lambda_m \quad \forall n, n \in \mathbb{N}, \ n \neq m.$$

Hence there is a sequence of numbers $r_n > 0$ which converges to zero such that the intervals $(\lambda_n - r_n, \lambda_n + r_n)$ are pair-wise disjoint. Points of the spectrum have been characterized in Theorem 27.2.1. Thus we know for $\lambda_n \in \sigma(A)$ that $E_{\lambda_n + r_n} - E_{\lambda_n - r_n} \neq 0$. Therefore we can find a normalized vector x_n in the range of the projector $E_{\lambda_n + r_n} - E_{\lambda_n - r_n}$ for all $n \in \mathbb{N}$. Since the intervals $(\lambda_n - r_n, \lambda_n + r_n)$ are pair-wise disjoint, the projectors $E_{\lambda_n + r_n} - E_{\lambda_n - r_n}$ are pair-wise orthogonal and we deduce $\langle x_n, x_m \rangle = \delta_{nm}$. The identity

$$\|(A - \lambda I) x_n\|^2 = \int_\mathbb{R} (t - \lambda)^2 d\|E_t x_n\|^2 = \int_{\lambda_n - r_n}^{\lambda_n + r_n} (t - \lambda)^2 d\|E_t x_n\|^2$$

implies $\lim_{n \to \infty} (A - \lambda I) x_n = 0$ since $\lim_{n \to \infty} \lambda_n = \lambda$ and $\lim_{n \to \infty} r_n = 0$. Again, since infinite orthonormal systems converge weakly to 0, statement b) holds in this case too. Thus a) implies b).

27.2 Decomposition of the spectrum – Spectral subspaces

Now assume b). An indirect proof will show that then c) holds. Suppose that there is some $r > 0$ such that the projector $E_{\lambda+r} - E_{\lambda-r}$ has a finite dimensional range. Then this projector is compact. Since compact operators map weakly convergent sequences onto strongly convergent ones, we know for any sequence $(x_n)_{n\in\mathbb{N}}$ satisfying b) that $\lim_{n\to\infty}(E_{\lambda+r} - E_{\lambda-r})x_n = 0$. Now observe the lower bound

$$\|(A - \lambda I)x_n\|^2 = \int_{\mathbb{R}} (t - \lambda)^2 d\|E_t x_n\|^2 \geq r^2 \left(\int_{\mathbb{R}} d\|E_t x_n\|^2 - \int_{\lambda-r}^{\lambda+r} d\|E_t x_n\|^2 \right)$$
$$= r^2 (\|x_n\|^2 - \|(E_{\lambda+r} - E_{\lambda-r})x_n\|^2)$$

which gives

$$\|x_n\|^2 \leq \|(E_{\lambda+r} - E_{\lambda-r})x_n\|^2 + \frac{1}{r^2}\|(A - \lambda I)x_n\|^2,$$

and thus a contradiction between b_2), b_3) and the implication of b_1) given above.

Finally suppose c). We have to distinguish two cases:

α) $\dim(\mathrm{ran}\,(E_\lambda - E_{\lambda-0})) = \infty$,   β) $\dim(\mathrm{ran}\,(E_\lambda - E_{\lambda-0})) < \infty$.

In the first case we know by Theorem 27.2.1 that λ is an eigenvalue of infinite multiplicity and therefore $\lambda \in \sigma_e(A)$.

Now consider the second case. By assumption we know that

$$E_{\lambda+r} - E_{\lambda-r} = (E_{\lambda+r} - E_\lambda) + (E_\lambda - E_{\lambda-0}) + (E_{\lambda-0} - E_{\lambda-r})$$

is a projector of infinite dimensional range for every $r > 0$. The three projectors of this decomposition are orthogonal to each other since the corresponding intervals are disjoint. Therefore the sum of the projectors $(E_{\lambda+r} - E_\lambda) + (E_{\lambda-0} - E_{\lambda-r})$ has an infinite dimensional range and thus (Theorem 27.2.1) in particular $[(\lambda-r, \lambda) \cup (\lambda, \lambda+r)] \cap \sigma(A) \neq \emptyset$ for every $r > 0$. This means that λ is an accumulation point of the spectrum of A, i.e., $\lambda \in \sigma_e(A)$. We conclude that c) implies a). □

Remark 27.2.1 *From the proof of this theorem it is evident that condition b) could be reformulated as*

> *There is an infinite orthonormal system $\{x_n : n \in \mathbb{N}\}$ with the property $\lim_{n\to\infty}(A - \lambda I)x_n = 0$.*

This characterization b) of the points of the essential spectrum is the key to the proof of the following theorem on the 'invariance' of the essential spectrum under 'perturbations' of the operator A.

Theorem 27.2.6 (Theorem of Weyl) *Suppose that A and B are two self-adjoint operators in the complex Hilbert space $\mathcal{H}$. If there is a $z \in \rho(A) \cap \rho(B)$ such that*

$$T = (A - zI)^{-1} - (B - zI)^{-1}$$

is a compact operator, then the essential spectra of A and B agree: $\sigma_e(A) = \sigma_e(B)$.

Proof. We show first $\sigma_e(A) \subset \sigma_e(B)$. Take any $\lambda \in \sigma_e(A)$. Then there is a sequence $(x_n)_{n\in\mathbb{N}}$ which satisfies condition b) of Theorem 27.2.5 for the operator A. For all $n \in \mathbb{N}$ define $y_n = (A - zI)x_n = (A - \lambda I)x_n + (\lambda - z)x_n$. It follows that this sequence converges weakly to 0 and the estimate $\|y_n\| \geq |\lambda - z|\|x_n\| - \|(A - \lambda I)x_n\|$, valid for sufficiently large $n \in \mathbb{N}$, implies $\liminf_{n\to\infty}\|y_n\| > 0$. Next we take the identity

$$[(B - zI)^{-1} - (\lambda - z)^{-1}]y_n = -Ty_n - (\lambda - z)^{-1}(A - \lambda I)x_n$$

into account. Since T is compact and the sequence $(y_n)_{n \in \mathbb{N}}$ converges weakly to 0, we deduce from condition b_3) that
$$\lim_{n \to \infty}[(B - zI)^{-1} - (\lambda - z)^{-1}]y_n = 0.$$
Now introduce the sequence $z_n = (B - zI)^{-1}y_n$, $n \in \mathbb{N}$. Clearly $z_n \in D(B)$ for all $n \in \mathbb{N}$ and this sequence converges weakly to 0. From the limit relation given above we see $\liminf_{n \to \infty} \|z_n\| > 0$. This limit relation also implies
$$\lim_{n \to \infty}(B - \lambda I)z_n = 0$$
since $(B - \lambda I)z_n = (B - zI)z_n + (z - \lambda)z_n = y_n + (z - \lambda)(B - zI)^{-1}y_n$ and since $y_n = (A - zI)x_n$ converges to 0 by condition b_3).

Therefore the sequence $(z_n)_{n \in \mathbb{N}}$ satisfies condition b) for the operator B and our previous theorem implies that λ is a point of the essential spectrum of the operator B.

Since with T also the operator $-T$ is compact, we can exchange in the above proof the role of the operators A and B. Then we get $\sigma_e(B) \subset \sigma_e(A)$ and thus equality of the essential spectra. □

27.3 Interpretation of the spectrum of a self-adjoint Hamiltonian

For a self-adjoint operator A in a complex Hilbert space one can form the one-parameter group of unitary operators $U(t) = e^{-itA}$, and one can identify several spectral subspaces $\mathcal{H}_i(A)$ for this operator. It follows that this unitary group leaves the spectral subspaces invariant but it behaves quite differently on different spectral subspaces. This behaviour we study in this section, but for the more concrete case of a self-adjoint Hamiltonian in the Hilbert space $\mathcal{H} = L^2(\mathbb{R}^3)$ where a concrete physical and intuitive interpretation is available. These investigations lead naturally to the quantum mechanical scattering theory for which there are quite a number of detailed expositions, for instance the books [RS79, BW83]. Certainly we cannot give a systematic presentation of scattering here, we just mention a few basic and important facts in a special context, thus indicating some of the major difficulties.

In quantum mechanics the dynamics of a free particle of mass $m > 0$ is governed by the free Hamilton operator $H_0 = \frac{1}{2m}P^2$. Its spectrum has been determined to be $\sigma(H_0) = \sigma_c(H_0) = [0, \infty)$. In case of an interaction the dynamic certainly is changed. If $V(Q)$ is the interaction operator the dynamic is determined by the Hamilton operator
$$H = H_0 + V(Q).$$
We have discussed several possibilities to ensure that this Hamilton operator is self-adjoint (see Theorem 23.2.1). Here we work under the following assumptions:

$V(Q)$ is defined and symmetric on the domain D of the free Hamilton operator.
$H = H_0 + V(Q)$ is self-adjoint and lower bounded on D.

These two self-adjoint operators generate two one-parameter groups of unitary operators in $L^2(\mathbb{R}^3)$:
$$U_t^0 = e^{-\frac{i}{\hbar}tH_0}, \qquad U_t = U_t(V) = e^{-\frac{i}{\hbar}tH}, \qquad \forall t \in \mathbb{R}.$$

27.3 Interpretation of the spectrum of a self-adjoint Hamiltonian

Recall: If $\phi_0 \in D(H)$, then $\phi(t) = U_t\phi_0$ is the solution of the Schrödinger equation

$$i\hbar \frac{d}{dt}\phi(t) = H\phi(t)$$

for the initial condition $\phi(t = 0) = \phi_0$. Quantum scattering theory studies the long term behaviour of solutions of the Schrödinger equation. If λ is an eigenvalue of H with eigenvector ϕ_0, then by functional calculus $U_t\phi_0 = e^{-\frac{i}{\hbar}t\lambda}\phi_0$ and the localization properties of this eigenvector do not change under the dynamics.

For potentials $V \neq 0$ which decay to 0 for $|x| \to \infty$ one expects that the particle can 'escape to infinity' for certain initial states ϕ_0 and that its time evolution $U_t(V)\phi_0$ approaches that of the free dynamics $U_t(V=0)\psi_0$ for a certain initial state ψ_0, since 'near infinity' the effect of the potential should be negligible. This expectation can be confirmed, in a suitable framework.

According to classical mechanics we expect to find two classes of states for the dynamics described by the Hamilton operator H:

a) In some states the particle remains localized in a bounded region of $\mathbb{R}^3$, for all times $t \in \mathbb{R}$ (as the eigenstate mentioned above). States describing such behaviour are called *bound states*.

b) In certain states ϕ the particle can 'escapes to infinity' under the time evolution U_t. Such states are called *scattering states*.

Certainly, we have to give a rigorous meaning to these two heuristic concepts of a bound and of a scattering state. This is done in terms of Born's probability interpretation of quantum mechanics. Given $\phi \in L^2(\mathbb{R}^3)$ with $\|\phi\| = 1$ define

$$m(U_t\phi, \Delta) = \int_\Delta |(U_t\phi)(x)|^2 dx = \|\chi_\Delta U_t\phi\|_2^2. \tag{27.8}$$

$m(U_t\phi, \Delta)$ is the probability of finding the particle at time t in the region $\Delta \subset \mathbb{R}^3$. χ_Δ is the characteristic function of the set Δ.

Definition 27.3.1 $\phi \in L^2(\mathbb{R}^3)$ is called a **bound state** for the Hamilton operator H if, and only if, for every $\epsilon > 0$ there is a compact set $K \subset \mathbb{R}^3$ such that $m(U_t\phi, K) \geq 1 - \epsilon$ for all $t \in \mathbb{R}$.

$\psi \in L^2(\mathbb{R}^3)$ is called a **scattering state** for the Hamilton operator H if, and only if, for every compact set $K \subset \mathbb{R}^3$ one has $m(U_t\psi, K) \to 0$ as $|t| \to \infty$.

Bound states and scattering states have an alternative characterization which in most applications is more convenient to use.

Lemma 27.3.1 *a)* $\phi \in L^2(\mathbb{R}^3)$ *is a bound state for the Hamiltonian H if, and only if,*

$$\lim_{R \to \infty} \sup_{t \in \mathbb{R}} \|F_{>R} U_t\phi\|_2 = 0 \tag{27.9}$$

where $F_{>R}$ is the characteristic function of the set $\{x \in \mathbb{R}^3 : \|x\| > R\}$.

b) $\phi \in L^2(\mathbb{R}^3)$ is a scattering state for the Hamiltonian H if, and only if, for every $R \in (0, \infty)$,
$$\lim_{|t| \to \infty} \|F_{\leq R} U_t \phi\|_2 = 0 \qquad (27.10)$$

where $F_{\leq R}$ is the characteristic function of the set $\{x \in \mathbb{R}^3 : \|x\| \leq R\}$.

Proof. The proof is a straightforward exercise. □

Denote the set of all bound states for a given Hamiltonian H in $L^2(\mathbb{R}^3)$ by $M_b(H)$ and by $M_s(H)$ the set of all scattering states for this Hamilton operator. The following lemma describes some basic facts about these sets.

Lemma 27.3.2 *The sets of all bound states, respectively of all scattering states of a Hamilton operator H are closed subspaces in $L^2(\mathbb{R}^3)$ which are orthogonal to each other: $M_b(H) \perp M_s(H)$. Both subspaces are invariant under the group U_t.*

Proof. The characterization (27.9) of bound states and the basic rules of calculation for limits immediately imply that $M_b(H)$ is a linear subspace. The same applies to the $M_s(H)$. Also invariance under the group U_t is evident from the defining identities for these subspaces.

Suppose that $\phi \in L^2(\mathbb{R}^3)$ is an element of the closure of $M_b(H)$. Then there is a sequence $(\phi_n)_{n \in \mathbb{N}} \subset M_b(H)$ with limit ϕ. For $R > 0$ and $t \in \mathbb{R}$ we estimate with arbitrary $n \in \mathbb{N}$,

$$\|F_{>R} U_t \phi\|_2 \leq \|F_{>R} U_t (\phi - \phi_n)\|_2 + \|F_{>R} U_t \phi_n\|_2 \leq \|\phi - \phi_n\|_2 + \|F_{>R} U_t \phi_n\|_2.$$

For a given $\epsilon > 0$ there is an $n \in \mathbb{N}$ such that $\|\phi - \phi_n\|_2 < \epsilon/2$, and since $\phi_n \in M_b(H)$ there is an $R_n \in (0, \infty)$ such that $\|F_{>R} U_t \phi_n\|_2 < \epsilon/2$ for all $R > R_n$ and all $t \in \mathbb{R}$. Therefore $\|F_{>R} U_t \phi\|_2 < \epsilon$ for all $t \in \mathbb{R}$ and all $R > R_n$ and thus condition (27.9) holds. This proves that the linear space of all bound states is closed. The proof that the space of all scattering states is closed is similar (See Exercises).

Since U_t is unitary we find for $\phi \in M_b(H)$ and $\psi \in M_s(H)$,

$$\langle \phi, \psi \rangle_2 = \langle U_t \phi, U_t \psi \rangle_2 = \langle F_{>R} U_t \phi, U_t \psi \rangle_2 + \langle U_t \phi, F_{\leq R} U_t \psi \rangle_2$$

and thus for all $t \in \mathbb{R}$ and all $0 < R < \infty$,

$$|\langle \phi, \psi \rangle_2| \leq \|F_{>R} U_t \phi\|_2 \|\psi\|_2 + \|\phi\|_2 \|F_{\leq R} U_t \psi\|_2.$$

In the first term take the limit $R \to \infty$ and in the second term the limit $|t| \to \infty$ and observe equation (27.9), respectively equation (27.10) to conclude $\langle \phi, \psi \rangle_2 = 0$. This proves orthogonality of the spaces $M_b(H)$ and $M_s(H)$. □

There is a fundamental connection between the spaces of bound states, respectively scattering states, on one side and the spectral subspaces of the Hamiltonian on the other side. A first step in establishing this connection is taken in the following proposition.

Proposition 27.3.1 *For a self-adjoint Hamilton operator H in $L^2(\mathbb{R}^3)$ every normalized vector of the discontinuous subspace is a bound state and every scattering state belongs to the continuous subspace, i.e.,*

$$\mathcal{H}_p(H) \subseteq M_b(H), \qquad M_s(H) \subseteq \mathcal{H}_c(H). \qquad (27.11)$$

27.3 Interpretation of the spectrum of a self-adjoint Hamiltonian

Proof. For an eigenvector ϕ of the Hamiltonian H with eigenvalue E, the time dependence is $U_t\phi = e^{-\frac{i}{\hbar}Et}\phi$ and thus $\|F_{>R}U_t\phi\|_2^2 = \int_{|x|>R} |\phi(x)|^2 dx \to 0$ as $R \to \infty$, for every $t \in \mathbb{R}$ and condition (27.9) follows, i.e., $\phi \in M_b(H)$. Since $M_b(H)$ is closed this proves $\mathcal{H}_p(H) \subseteq M_b(H)$. By taking the orthogonal complements we find $M_b(H)^\perp \subseteq \mathcal{H}_p(H)^\perp = \mathcal{H}_c(H)$. Finally recall $M_s(H) \subseteq M_b(H)^\perp$. And the proof is complete. □

Heuristic considerations seem to indicate that the state of a quantum mechanical particle should be either a bound state or a scattering state, i.e., that the total Hilbert spaces $\mathcal{H} = L^2(\mathbb{R}^3)$ has the decomposition

$$\mathcal{H} = M_b(H) \oplus M_s(H).$$

Unfortunately this is not true in general. Nevertheless a successful strategy is known which allows us to establish this decomposition under certain assumptions on the Hamilton operator.

Suppose that we can show

$$\text{A.} \quad \mathcal{H}_{ac}(H) \subseteq M_s(H), \qquad \text{B.} \quad \mathcal{H}_{sc}(H) = \emptyset. \tag{27.12}$$

Then, because of $\mathcal{H} = \mathcal{H}_p(H) \oplus \mathcal{H}_c(H)$, $\mathcal{H}_c = \mathcal{H}_{ac}(H) \oplus \mathcal{H}_{sc}(H)$, and the general relations shown above, one has indeed

$$\mathcal{H}_p(H) = M_b(H), \qquad \mathcal{H}_{ac}(H) = M_s(H), \qquad \mathcal{H} = M_b(H) \oplus M_s(H). \tag{27.13}$$

While the verification of Part A) of (27.12) is relatively straightforward, the implementation of Part B) is quite involved. Thus for this part we just mention some basic results and have to refer to the specialized literature on (mathematical scattering) for the proofs.

The starting point for the proof of $\mathcal{H}_{ac}(H) \subseteq M_s(H)$ is the following lemma.

Lemma 27.3.3 *For all $\psi \in \mathcal{H}_{ac}(H)$ the time evolution $U_t\psi$ converges weakly to 0 for $|t| \to \infty$.*

Proof. The strategy of the proof is to show with the help of the spectral theorem and the characterization of elements ψ in $\mathcal{H}_{ac}(H)$ in terms of properties of the spectral measure $d\rho_\psi$ that for every $\phi \in \mathcal{H}$ the function $t \mapsto \langle \phi, U_t\psi \rangle$ is the Fourier transform of a function $F_{\phi,\psi} \in L^1(\mathbb{R})$ and then to apply the Riemann–Lebesgue Lemma (which states that the Fourier transform of a function in $L^1(\mathbb{R})$ is a continuous function which vanishes at infinity, see Lemma 10.1.1).

For arbitrary $\phi \in \mathcal{H}$ spectral calculus allows us to write

$$\langle \phi, U_t\psi \rangle = \int_{\mathbb{R}} e^{-\frac{i}{\hbar}ts} d\langle \phi, E_s\psi \rangle$$

for all $t \in \mathbb{R}$. Let $\Delta \subset \mathbb{R}$ be a Borel set. Then $\int_\Delta d\langle \phi, E_s\psi \rangle = \langle \phi, E(\Delta)\psi \rangle$. Denote by P_{ac} the orthogonal projector onto the subspace $\mathcal{H}_{ac}(H)$. It is known to commute with $E(\Delta)$ and therefore we have $\langle \phi, E(\Delta)\psi \rangle = \langle \phi, E(\Delta)P_{ac}\psi \rangle = \langle P_{ac}\phi, E(\Delta)\psi \rangle$. Thus the estimate $|\langle \phi, E(\Delta)\psi \rangle| \leq \|E(\Delta)P_{ac}\phi\| \|E(\Delta)\psi\|$ follows. According to Proposition 27.2.3, $\psi \in \mathcal{H}_{ac}(H)$ is characterized by the fact that the spectral measure $d\rho_\psi(s) = d\|E_s\psi\|^2$ is absolutely continuous with respect to the Lebesgue measure on $\mathbb{R}$, i.e., there is a nonnegative function f_ψ such that $d\rho_\psi(s) = f_\psi(s)ds$. Since $\int_\mathbb{R} d\rho_\psi(s) = \|\psi\|^2$ we find $0 \leq f_\psi \in L^1(\mathbb{R})$.

The estimate $|\langle\phi, E(\Delta)\psi\rangle| \leq \|E(\Delta)P_{ac}\phi\| \|E(\Delta)\psi\|$ implies that the measure $d\langle\phi, E_s\psi\rangle$ too is absolutely continuous with respect to the Lebesgue measure; hence there is a function $F_{\phi,\psi}$ on $\mathbb{R}$ such that $d\langle\phi, E_s\psi\rangle = F_{\phi,\psi}(s)ds$. The above estimate also implies $|F_{\phi,\psi}(s)| \leq \sqrt{f_{P_{ac}\phi}(s)f_\psi(s)}$, thus $F_{\phi,\psi} \in L^1(\mathbb{R})$. We conclude that

$$\langle\phi, U_t\psi\rangle = \int_{\mathbb{R}} e^{-\frac{i}{\hbar}ts} F_{\phi,\psi}(s)ds$$

is the Fourier transform of an absolutely integrable function and therefore is a continuous function which vanishes for $|t| \to \infty$. $\square$

Lemma 27.3.4 *Let E be the spectral family of the self-adjoint operator H and introduce the projector $P_n = E_n - E_{-n}$. If all the operators*

$$F_{>R}P_n, \qquad n \in \mathbb{N}, \quad 0 < R < \infty$$

are compact in $\mathcal{H} = L^2(\mathbb{R}^3)$, then $\mathcal{H}_{ac}(H) \subseteq M_s(H)$.

Proof. Suppose $\psi \in \mathcal{H}_{ac}(H)$ is given. Then by the previous lemma $U_t\psi$ converges weakly to 0. Since $F_{>R}P_n$ is assumed to be a compact operator, it maps this sequence onto a strongly convergent sequence, therefore

$$\lim_{|t|\to\infty} \|F_{>R}P_n U_t\psi\| = 0.$$

Given $\epsilon > 0$ there is an $n \in \mathbb{N}$ such that $\|(I - P_n)\psi\| < \epsilon/2$. This number n we use in the following estimate, for any $0 < R < \infty$:

$$\|F_{>R}U_t\psi\| \leq \|F_{>R}(I - P_n)U_t\psi\| + \|F_{>R}P_n U_t\psi\|$$
$$\leq \|(I - P_n)\psi\| + \|F_{>R}P_n U_t\psi\| \leq \epsilon/2 + \|F_{>R}P_n U_t\psi\|.$$

Now we see that ψ satisfies the characterization (27.10) of scattering states and we conclude. $\square$

Certainly, it is practically impossible to verify the hypothesis of the last lemma directly. But this lemma can be used to arrive at the same conclusion under more concrete hypotheses. The following theorem gives a simple example for this.

Theorem 27.3.2 *Suppose that for the self-adjoint Hamiltonian H in $\mathcal{H} = L^2(\mathbb{R}^3)$ there are $q \in \mathbb{N}$ and $z \in \rho(H)$ such that the operator*

$$F_{>R}(H - z)^{-q}$$

is compact for every $0 < R < \infty$. Then $\mathcal{H}_{ac}(H) \subseteq M_s(H)$ holds.

Proof. Write

$$F_{>R}P_n = F_{>R}(H - zI)^{-q}(H - zI)^q P_n$$

and observe that $(H - zI)^q P_n$ is a bounded operator (this can be seen by functional calculus). The product of a compact operator with a bounded operator is compact (Theorem 22.3.4). Thus we can apply Lemma 27.3.4 and conclude. $\square$

There are by now quite a number of results available which give sufficient conditions on the Hamiltonian H which ensure that the singular continuous subspace $\mathcal{H}_{sc}(H)$ is empty. But the proof of these results is usually quite involved and is beyond the scope of this introduction. A successful strategy is to use restrictions on H which imply estimates for the range of its spectral projections, for instance

$$\operatorname{ran} E(a, b) \subseteq \mathcal{H}_{ac}(H).$$

A detailed exposition of this and related theories is given in the books [RS78, RS79, AS77, Pea88]. We mention without proof one of the earliest results in this direction.

Theorem 27.3.3 *For the Hamilton operator $H = H_0 + V$ in the Hilbert space $\mathcal{H} = L^2(\mathbb{R}^3)$ assume*

a) $\|V\|_R^2 = \int\int \frac{|V(x)V(y)|}{|x-y|^2} dx dy < (4\pi)^2$ *or*

b) $\|e^{a|\cdot|}V\|_R < \infty$ *for some $a > 0$.*

Then the singular subspace of H is empty: $\mathcal{H}_s(H) = \emptyset$, hence in particular $\mathcal{H}_{sc}(H) = \emptyset$, and there are no eigenvalues.

A more recent and fairly comprehensive discussion of the existence of bound states and on the number of bound states of Schrödinger operators is given in [BS96, GM97].

27.4 Exercises

1. Prove Part c) of Theorem 27.1.1.

    Hints: Given $f, g \in \mathcal{C}_b(\mathbb{R})$ and $x, y \in \mathcal{H}$ show first that $\langle x, g(A)f(A)y \rangle = \langle \overline{g}(A)x, f(A)y \rangle = \lim_{n \to \infty} \langle \overline{g_n}(A)x, f_n(A)y \rangle$ with continuous functions f_n, g_n with support in $[-n, n]$. Then prove

    $$\langle \overline{g_n}(A)x, f_n(A)y \rangle = \lim_{|Z| \to 0} \langle \Sigma(\overline{g_n}, Z)x, \Sigma(f_n, Z)y \rangle$$

    where the approximations $\Sigma(f_n, Z)$ are defined in equation (26.20). Z is a partition of the interval $[-n, n]$. Then use orthogonality of different projectors $E(t_{j-1}, t_j]$ to show

    $$\langle \Sigma(\overline{g_n}, Z)x, \Sigma(f_n, Z)y \rangle = \langle x, \Sigma(g_n \cdot f_n, Z)y \rangle.$$

2. Prove the spectral mapping theorem, Part 4) of Theorem 27.1.1.

    Hints: For $z \notin \sigma(g(A))$ the resolvent has the representation $R_{g(A)}(z) = \int_{\mathbb{R}} \frac{1}{g(t)-z} dE_t$.

3. Denote by a the countable set of all points t_j at which the spectral family E has a jump. Show: $E(\{t\})E(a^c)x = 0 \quad \forall t \in \mathbb{R}$.

4. Let A be a self-adjoint operator in the complex Hilbert space $\mathcal{H}$. Show that $\mathcal{H}_{sc}(H)$ is a closed linear subspace of $\mathcal{H}$.

5. Let A be a self-adjoint operator in the complex Hilbert space $\mathcal{H}$ and $\mathcal{H}_0$ a reducing subspace. Prove that the restriction A_0 of A to this subspace is a self-adjoint operator in $\mathcal{H}_0$.

6. Complete the proof of Theorem 27.2.4.

7. Prove Lemma 27.3.1.

8. For a self-adjoint Hamiltonian H in the Hilbert space $L^2(\mathbb{R}^3)$ prove that the set of all scattering states is a closed linear subspace.

9. Let E be the spectral family of a self-adjoint operator H in the complex Hilbert space $\mathcal{H}$. Prove *Stone's formula*:

$$\frac{\pi}{2} \langle x, [E[a,b] + E(a,b)]x \rangle = \lim_{r \to 0, r > 0} \int_a^b \operatorname{Im} \langle x, (H - (t+ir)I)^{-1} x \rangle dt$$

for all $x \in \mathcal{H}$ and all $-\infty < a < b < \infty$.

Hints: Prove first that the functions g_r, $r > 0$, defined by

$$g_r(t) = \frac{1}{2i\pi} \int_a^b \left(\frac{1}{s - t - ir} - \frac{1}{s - t + ir} \right) ds$$

have the following properties: This family is uniformly bounded and

$$\lim_{r \to 0, r > 0} g_r(t) = \begin{cases} 0 & \text{if } t \notin [a,b], \\ 1/2 & \text{if } t \in \{a,b\}, \\ 1 & \text{if } t \in (a,b). \end{cases}$$

Part III

Variational Methods

28
Introduction

The first two parts of this book were devoted to generalized functions and Hilbert spaces whose operators are primarily of importance for quantum mechanics and quantum field theory. These two physical theories were born and developed in the 20th century. In sharp contrast to this are the variational methods which have a much longer history. In 1744, L. Euler published a first textbook on what soon after was called the *calculus of variations*, with the title 'A method for finding curves enjoying certain maximum or minimum properties'. In terms of the calculus which had recently been invented by Leibniz and Newton, optimal curves were determined by Euler. Depending on the case which is under investigation optimal means "maximal" or "minimal". Though not under the same name the calculus of variations is actually older and closely related to the invention and development of differential calculus, since already in 1684 Leibniz' first publication on differential calculus appeared under the title *Nova methodus pro maximis et minimis itemque tangentibus*. This can be considered as the beginning of a mathematical theory which intends to solve problems of "optimization" through methods of analysis and functional analysis. Later in the 20th century methods of topology were also used for this. Here 'optimal' can mean a lot of very different things, for instance: shortest distance between two points in space, optimal shapes or forms (of buildings, of plane wings, of natural objects), largest area enclosed by a fence of given length, minimal losses (of a company in difficult circumstances), maximal profits (as a general objective of a company). And in this wider sense of 'finding optimal solutions' as part of human nature or as part of human belief that in nature an optimal solution exists and is realized there, the calculus of variations goes back more than 2000 years to ancient Greece. In short, the calculus of variations has a long and fascinating history. However 'variational methods' are not a mathematical theory of the past, related to classical physics, but an active area of modern mathematical research as the numerous publications in this field show, with many practical or potential applications in science, engineering and economics. Clearly this means for us that in this short third part we will be able to present only the basic aspects of one direction of the modern developments in the calculus of variations, namely those with close links to the previous parts, mainly to Hilbert space methods.

28.1 Roads to Calculus of Variations

According to legend Queen Dido, fleeing from Tyre, a Phoenician city ruled by King Pygmalion, her tyrannical brother, and arriving at the site that was called later Carthage, sought to purchase land from the natives. They asserted that they were willing to sell only as much ground as she could surround with a bull's hide. Dido accepted the deal and cut a bull's hide into very narrow strips which she pieced together to form the longest possible strip. She reasoned that the maximal area should be obtained by shaping the strip into the circumference of a circle. A complete mathematical proof of Dido's claim as the best possible choice, was not achieved until the nineteenth century. Today one still speaks of the general problem of Dido as an *isoperimetric problem* but where this adjective has the much wider interpretation as referring to any problem in which an extremum is to be determined subject to one or more constraints, for instance the problem of finding the form which will give the greatest volume within a fixed surface area.

Heron of Alexandria postulated a minimum principle for optics and deduced the law of reflection of light for a straight mirror. In 1662 Fermat generalized Heron's principle by postulating a principle of least time for the propagation of light. Later several other principles of optimality (minima or maxima) were formulated about fundamental physical quantities such as energy, action, entropy, separation in the space-time of special relativity. In other fields of science one knows such principles too. In probability and statistics we have 'least square' and 'maximum likelihood' laws. Minimax principles are fundamental in game theory, statistical decision theory and mathematical economics.

In short, the calculus of variations can be described as the generalization of the solution of problems of minima and maxima by elementary calculus, a generalization to the case of infinitely many variables, i.e., to infinite dimensional spaces. In 1744 Euler explained and extended the maxi - minimal notions of Newton, the Bernoulli's and Maupertuis. His 1753 "Dissertatio de principio minimae actionis" associates him with Lagrange as one of the inventors of the calculus of variations, in its analytic form. In 1696 Jean Bernoulli posed the problem of determining the path of fastest descent of a point mass, i.e., the brachystochrone problem. This problem was typical for the problems considered at this time since it requires one to find an unknown function $y = f(x)$ which minimizes or maximizes an integral of the form

$$S(f) = \int_a^b L(x, f(x), f'(x))dx.$$

Such an integral is a function on a function space or a functional, a name introduced by J. Hadamard and widely used nowadays.

Another famous problem whose solution has been a paradigm in the calculus of variations now for about a century is the so-called *Dirichlet problem*. In this problem one is asked to find a differentiable function f whose derivatives are square integrable over a domain $\Omega \subset \mathbb{R}^3$ and which has prescribed values on the boundary $\partial \Omega$, i.e., $f_{|\partial \Omega} = g$ where g is some given function on $\partial \Omega$, so that the 'Dirichlet integral'

$$I(f) = \int_\Omega |Df(x)|^2 dx$$

is minimal. (Such a problem arises for instance in electrostatics for the electric potential f.) The existence of a solution of the Dirichlet problem was first taken for granted, since the integrand is nonnegative. It was only Weierstrass around 1870 who pointed out that there are variational problems without a solution, i.e. in modern language for which there is no minimum though the functional has a finite infimum.

Under natural technical assumptions the existence of a mimimizing function f of the Dirichlet integral was proven by D. Hilbert in 1899. The decisive discoveries which allowed Hilbert to prove this result were the notion of the 'weak topology' on spaces which today are called Hilbert spaces and pre-compactness with respect to this weak topology of bounded sets (compare the introduction to Part B).

For readers who are interested in a more extensive exposition of the fascinating history of the calculus of variations we recommend the books [Gol80, BB92] for a start. An impressive account of the great diversity of variational methods is given in an informal way in the recent book [HT96].

28.2 Classical approach versus direct methods

Historically the calculus of variations started with one dimensional problems. In these cases one tries to find an extremal point (minimum or maximum) of functionals of the form

$$f(u) = \int_a^b F(t, u(t), u'(t)) dt \qquad (28.1)$$

over all functions $u \in M = \{v \in C^2([a,b], \mathbb{R}^m) : u(a) = u_0, u(b) = u_1\}$ where $u_0, u_1 \in \mathbb{R}^m$ are given points and $m \in \mathbb{N}$. The integrand $F : [a,b] \times \mathbb{R}^m \times \mathbb{R}^m \to \mathbb{R}$ is typically assumed to be of class C^2 in all variables. A familiar example is the action functional of Lagrangian mechanics. In this case the integrand F is just the Lagrange function L which for a particle of mass m moving in the force field of a potential V is $L(t, u, u') = \frac{m}{2} u'^2 - V(u)$.

There is a counterpart in dimensions $d > 1$. Let $\Omega \subseteq \mathbb{R}^d$ be an open nonempty set and $F : \Omega \times \mathbb{R}^m \times M_{md} \to \mathbb{R}$ a function of class C^2 where M_{md} is the space of all $m \times d$ matrices; for $u : \Omega \to \mathbb{R}^m$ denote by $Du(x)$ the $m \times d$ matrix of first derivatives of u. Then under suitable integrability assumptions a functional $f(u)$ is well defined by the integral

$$f(u) = \int_\Omega F(x, u(x), Du(x))dx. \qquad (28.2)$$

Such functionals are usually studied under some restrictions on u on the boundary $\partial\Omega$ of Ω, for instance the so-called *Dirichlet boundary condition* $u_{|\partial\Omega} = u_0$ where u_0 is some given function $\partial\Omega \to \mathbb{R}^m$.

In elementary calculus we find extremal points of a function $f(x)$, $x \in \mathbb{R}$, of class C^2 by determining first the points x_i at which the derivative of f vanishes, and then deciding whether a point x_i gives a local minimum or a local maximum or a stationary point of f according to value $f^{(2)}(x_i)$ of the second derivative.

The classical approach for functionals of the form (28.1) follows in principle the same strategy, though the concepts of differentiation are more involved since differentiation with respect to variables in an infinite dimensional function space is required. The necessary definitions and basic results about this differential calculus in Banach spaces is developed in the next chapter. Thus in a first step we have to find the zeros of the first derivative f', i.e., solutions of the *Euler–Lagrange equation*

$$f'(u) = 0. \qquad (28.3)$$

For functionals of the form (28.1) this equation is equivalent to a second order ordinary differential equation for the unknown function u (see for instance [JLJ98] or [BB92]). If the second derivative $f^{(2)}(u)$ is positive (in a sense which has to be defined), then the functional f has a local minimum at the function u. If this applies to the functional $-f$, then f has a local maximum at u.

If only the problem of existence of an extremal point is considered there is another strategy available. In order to understand it, it is important to recall Weierstrass' theorem and its proof: A lower (upper) semi-continuous function f has a finite minimum (a finite maximum) on a closed and bounded interval $[a, b]$. Here it is essential that closed and bounded sets in $\mathbb{R}$ are compact, i.e., infinite sequences in $[a, b]$ have a convergent subsequence.

This strategy too has a very successful counterpart for functionals of the form (28.1) or (28.2). It is called the *direct method of the calculus variations*. We give a brief description of its basic steps.

1. Suppose M is a subset of the domain of the functional f and we want to find a minimum of f on M.

2. Through assumptions on f and/or M, assure that f has a finite infimum on M, i.e.,

$$\inf_{u \in M} f(u) = I(f, M) = I > -\infty. \qquad (28.4)$$

   Then there is a *minimizing sequence* $(u_n)_{n \in \mathbb{N}} \subset M$, i.e., a sequence in M such that

$$\lim_{n \to \infty} f(u_n) = I. \qquad (28.5)$$

28.2 Classical approach versus direct methods

3. Suppose that we can find **one** minimizing sequence $(u_n)_{n \in \mathbb{N}} \subset M$ such that

$$u = \lim_{n \to \infty} u_n \in M, \tag{28.6}$$

$$f(u) \leq \liminf_{n \to \infty} f(u_n); \tag{28.7}$$

then the minimization problem is solved since then we have

$$I \leq f(u) \leq \liminf_{n \to \infty} f(u_n) = I$$

where the first inequality holds because of $u \in M$ and where the second identity holds because $(u_n)_{n \in \mathbb{N}}$ is a minimizing sequence. Obviously, for equation (28.6) a topology has to be specified on M.

4. Certainly, it is practically impossible to find one minimizing sequence with the two properties given above. Thus in explicit implementations of this strategy one works under conditions where the two properties hold for all convergent sequences, with respect to a suitable topology. If one looks at the proof of Weierstrass' theorem one expects to get a convergent minimizing sequence by taking a suitable subsequence of a given minimizing sequence. Recall: The coarser the topology is, the easier it is for a sequence to have a convergent subsequence and to have a limit point, i.e., to have equation (28.6). On the other hand, the stronger the topology is the easier it is to satisfy inequality (28.7) which is a condition of lower semi-continuity.

5. The paradigmatic solution of this problem in infinite dimensional spaces is due to Hilbert who suggested using the weak topology, the main reason being that in a Hilbert space bounded sets are relatively sequentially compact for the weak topology while for the norm topology there are not too many compact sets of interest. Thus suppose that M is a weakly closed subset of a reflexive Banach space and that minimizing sequences are bounded (with respect to the norm). Then there is a weakly convergent subsequence whose weak limit belongs to M. Thus in order to conclude one verifies that inequality (28.7) holds for all weakly convergent sequences, i.e., that f is lower semi-continuous for the weak topology.

In the following chapter the concepts and results which have been used above will be explained and some concrete existence results for extremal points will be formulated where the above strategy is implemented.

Suppose that with the direct methods of the calculus of variations we managed to show the existence of a local minimum of the functional f and that this functional is differentiable (in the sense of the classical methods). Then, if the local minimum occurs at an interior point u_0 of the domain of f, the Euler–Lagrange equation $f'(u_0) = 0$ holds and thus we have found a solution of this equation. If the functional f has the form (28.1) (or 28.2), then the equation $f'(u_0) = 0$ is a nonlinear ordinary (partial) differential equation and thus the direct methods become a powerful tool for solving nonlinear ordinary and partial differential equations.

Some modern implementations of this strategy with many new results on nonlinear (partial) differential equations is described in good detail in the following books [Dac82, Dac89, BB92, JLJ98, Str00], in a variety of directions.

Note that a functional f can have other critical points than local extrema. These other critical points are typically not obtained by the direct methods as described above. However there are other, often topological methods of global analysis by which the existence of these other critical points can be established. We mention the minimax methods, index theory and mountain pass lemmas. These methods are developed and applied in [Zei85, BB92, Str00]. But we cannot present them here.

28.3 The objectives of the following chapters

The overall strategy of Part III has been explained in the Introduction. The next chapter on direct methods is the abstract core of this part of the book. There we present some general existence results for extrema of functionals which one can call *generalized Weierstrass theorems*. Since the realization of all the hypotheses in these results is not obvious some concrete ways of implementing them are discussed in some detail.

The following chapter introduces differential calculus on Banach spaces and proves those results which are needed for the 'classical approach' of the variational methods.

On the basis of the differential calculus on Banach spaces, the third chapter formulates in great generality the Lagrange multiplier method and proves in a fairly general setting the existence of such a multiplier. When applied to linear or nonlinear partial differential operators the existence of a Lagrange multiplier is equivalent to the existence of an eigenvalue. Thus this chapter is of particular importance for spectral theory of linear and nonlinear partial differential operators. In the fourth chapter we continue this topic and determine explicitly the spectrum of some linear second order partial differential operators. In particular the spectral theorem for compact self-adjoint operators is proven.

The final chapter presents the mathematical basis of the Hohenberg–Kohn density functional theory, which is the starting point of various concrete methods used mainly in chemistry. It is based on the theory of Schrödinger operators for N-particle systems which was introduced and discussed in Part II for $N = 1$.

29
Direct Methods in the Calculus of Variations

29.1 General existence results

From the Introduction we know that semi-continuity plays a fundamental role in direct methods in the calculus of variations. Accordingly we recall the definition and the basic characterization of lower semi-continuity. Upper semi-continuity of a function f is just lower semi-continuity of $-f$.

Definition 29.1.1 *Let M be a Hausdorff space. A function $f : M \to \mathbb{R} \cup \{+\infty\}$ is called **lower semi-continuous at a point** $x_0 \in M$ if, and only if, x_0 is an interior point of the set $\{x \in M : f(x) > f(x_0) - \epsilon\}$ for every $\epsilon > 0$. f is called **lower semi-continuous on** M if, and only if, f is lower semi-continuous at every point $x_0 \in M$.*

Lemma 29.1.1 *Let M be a Hausdorff space and $f : M \to \mathbb{R} \cup \{+\infty\}$ a function on M.*

a) *If f is lower semi-continuous at $x_0 \in M$, then for every sequence $(x_n)_{n \in \mathbb{N}} \subset M$ converging to x_0, one has*
$$f(x_0) \leq \liminf_{n \to \infty} f(x_n). \tag{29.1}$$

b) *If M satisfies the first axiom of countability, i.e., if every point of M has a countable neighborhood basis, then the converse of a) holds.*

Proof. For the simple proof we refer to the Exercises. □

In the Introduction we also learned that *compactness* plays a fundamental role too, more precisely, the direct methods use *sequential compactness* in a decisive way.

Definition 29.1.2 *Let M be a Hausdorff space. A subset $K \subset M$ is called **sequentially compact** if, and only if, every infinite sequence in K has a subsequence which converges in K.*

The following fundamental results proves the existence of a minimum. Replacing f by $-f$ it can easily be translated into a result on the existence of a maximum.

Theorem 29.1.1 (Existence of a minimizer) *Let $f : M \to \mathbb{R} \cup \{+\infty\}$ be a lower semi-continuous function on the Hausdorff space M. Suppose that there is a real number r such that*

a) $[f \leq r] = \{x \in M : f(x) \leq r\} \neq \emptyset$ and

b) $[f \leq r]$ is sequentially compact.

Then there is a minimizing point x_0 for f on M:

$$f(x_0) = \inf_{x \in M} f(x). \tag{29.2}$$

Proof. We begin by showing indirectly that f is lower bounded. If f is not bounded from below there is a sequence $(x_n)_{n \in \mathbb{N}}$ such that $f(x_n) < -n$ for all $n \in \mathbb{N}$. For sufficiently large n the elements of the sequence belong to the set $[f \leq r]$, hence there is a subsequence $y_j = x_{n(j)}$ which converges to a point $y \in M$. Since f is lower semi-continuous we know $f(y) \leq \liminf_{j \to \infty} f(y_j)$, a contradiction since $f(y_j) < -n(j) \to -\infty$. We conclude that f is bounded from below and thus has a finite infimum:

$$-\infty < I = I(f, M) = \inf_{x \in M} f(x) \leq r.$$

Therefore there is a minimizing sequence $(x_n)_{n \in \mathbb{N}}$ whose elements belong to $[f \leq r]$ for all sufficiently large n. Since $[f \leq r]$ is sequentially compact there is again a subsequence $y_j = x_{n(j)}$ which converges to a unique point $x_0 \in [f \leq r]$. Since f is lower semi-continuous we conclude

$$I \leq f(x_0) \leq \liminf_{j \to \infty} f(y_j) = \lim_{j \to \infty} f(y_j) = I.$$

□

Sometimes one can prove

Theorem 29.1.2 (Uniqueness of minimizer) *Suppose M is a convex set in a vector space E and $f : M \to \mathbb{R}$ is a strictly convex function on M. Then f has at most one minimizing point in M.*

Proof. Suppose there are two different minimizing points x_0 and y_0 in M. Since M is convex all points $x(t) = tx_0 + (1-t)y_0$, $0 < t < 1$, belong to M and therefore $f(x_0) = f(y_0) \leq f(x(t))$. Since f is strictly convex we know $f(x(t)) < tf(x_0) + (1-t)f(y_0) = f(x_0)$ and therefore the contradiction $f(x_0) < f(x_0)$. Thus there is at most one minimizing point. □

29.2 Minimization in Banach spaces

In interesting minimization problems we typically have at our disposal much more information about the set M and the function f than we have assumed in Theorem 29.1.1. If for instance one is interested in minimizing the functional (28.2) one would prefer to work in a suitable Banach space of functions, usually a Sobolev space. These function spaces and their properties are an essential input for applying them in the direct methods. A concise introduction to the most important of these function spaces can be found in [LL01].

Concerning the choice of a topology on Banach spaces which is suitable for the direct methods (compare our discussion in the Introduction) we begin by recalling the wellknown result of Riesz (see Theorem 18.1.1): The closed unit ball of a normed space is compact (for the norm topology) if, and only if, this space is finite dimensional. Thus, in infinite dimensional Banach spaces compact sets have an empty interior and therefore are not of much interest for must purposes of analysis, in particular not for the direct methods. Which other topology can be used? Recall that Weierstrass' result on the existence of extrema of continuous functions on closed and bounded sets uses in an essential way that in finite dimensional Euclidean spaces a set is compact if, and only if, it is closed and bounded. A topology with such a characterization of closed and bounded sets is known for infinite dimensional Banach spaces too, the weak topology. Suppose E is a Banach space and E' is its topological dual space. Then the weak topology $\sigma = \sigma(E, E')$ on E is defined by the system $\{q_u(\cdot) : u \in E'\}$ of semi-norms q_u, $q(x) = |u(x)|$ for all $x \in E$. In most applications one can actually use reflexive Banach space and there the following important result is available.

Lemma 29.2.1 *In a reflexive Banach space E every bounded set (for the norm) is relatively compact for the weak topology $\sigma(E, E')$.*

A fairly detailed discussion about compact and weakly compact sets in Banach spaces, as they are relevant for the direct methods, is given in the Appendix of [BB92]. Prominent examples of reflexive Banach spaces are Hilbert spaces (see Chapter 18), the Lebesgue spaces L^p for $1 < p < \infty$, and the corresponding Sobolev spaces $W^{m,p}$, $m = 1, 2, \ldots$, $1 < p < \infty$.

Accordingly we decide to use mainly reflexive Banach spaces for the direct methods, whenever this is possible. Then, with the help of Lemma 29.2.1, we always get weakly convergent minimizing sequences whenever we can show that bounded minimizing sequences exist. Thus the problem of lower semi-continuity of the functional f for the weak topology remains. This is unfortunately not a simple problem. Suppose we consider a functional of the form (28.2) and, according to the growth restrictions on the integrand F, we decide to work in a Sobolev space $E = W^{1,p}(\Omega)$ or in a closed subspace of this space, $\Omega \subseteq \mathbb{R}^d$ open. Typically, the restrictions on F, which assure that f is well defined on E, imply that f is continuous (for the norm topology). However the question when such a functional is lower semi-continuous for the weak topology is quite involved, nevertheless a fairly comprehensive answer is known (see [Dac82]). Under certain technical assumptions on the integrand F the functional f is lower semi-continuous for the weak topology on $E = W^{1,p}(\Omega)$ if, and only if, for (almost) all $(x, u) \in \Omega \times \mathbb{R}^m$ the function $y \mapsto F(x, u, y)$ is convex (if $m = 1$), respectively quasi-convex (if $m > 1$).

Though in general continuity of a functional for the norm topology does not imply its continuity for the weak topology, there is a large and much used class of functionals where this implication holds. This is the class of convex functionals and for this reason *convex minimization* is relatively easy. We prepare the proof of this important result with a lemma.

Lemma 29.2.2 *Let E be a Banach space and M a weakly (sequentially) closed subset. A function $f : M \to \mathbb{R}$ is (sequentially) lower semi-continuous on M for the weak topology if, and only if, the sub-level sets $[f \leq r]$ are weakly (sequentially) closed for every $r \in \mathbb{R}$.*

Proof. We give the proof explicitly for the case of sequential convergence. For the general case one proceeds in the same way using nets.

Let f be weakly sequentially lower semi-continuous and for some $r \in \mathbb{R}$ let $(x_n)_{n \in \mathbb{N}}$ be a sequence in $[f \leq r]$ which converges weakly to some point $x \in M$ (since M is weakly sequentially closed). By Lemma 29.1.1 we know $f(x) \leq \liminf_{n \to \infty} f(x_n)$ and therefore $f(x) \leq r$, i.e., $x \in [f \leq r]$. Therefore $[f \leq r]$ is closed.

Conversely assume that all the sub-level sets $[f \leq r]$, $r \in \mathbb{R}$, are weakly sequentially closed. Suppose f is not weakly sequentially lower semi-continuous on M. Then there is a weakly convergent sequence $(x_n)_{n \in \mathbb{N}} \subset M$ with limit $x \in M$ such that $\liminf_{n \to \infty} f(x_n) < f(x)$. Choose a real number r such that $\liminf_{n \to \infty} f(x_n) < r < f(x)$. Then there is a subsequence $y_j = x_{n(j)} \subset [f \leq r]$. This subsequence too converges weakly to x and, since $[f \leq r]$ is weakly sequentially closed, we know $x \in [f \leq r]$, a contradiction. We conclude that f is sequentially lower semi-continuous for the weak topology. □

Lemma 29.2.3 *Let E be a Banach space, M a convex closed subset and $f : M \to \mathbb{R}$ a continuous convex function. Then f is lower semi-continuous on M for the weak topology.*

Proof. Because f is continuous (for the norm topology) the sub-level sets $[f \leq r]$, $r \in \mathbb{R}$, are all closed. Since f is convex these sub-level sets are convex subsets of E ($x, y \in [f \leq r]$, $0 \leq t \leq 1 \Rightarrow f(tx + (1-t)y) \leq tf(x) + (1-t)f(y) \leq tr + (1-t)r = r$). As in Hilbert spaces one knows that a convex subset is closed if, and only if, it is weakly closed. We deduce that all the sub-level sets are weakly closed and conclude by Lemma 29.2.2. □

As a conclusion to this section we present a summary of our discussion in the form of two explicit results on the existence of a minimizer in reflexive Banach spaces.

Theorem 29.2.1 (Generalized Weierstrass theorem I) *A weakly sequentially lower semi-continuous function f attains its infimum on a bounded and weakly sequentially closed subset M of a real reflexive Banach space E, i.e., there is $x_0 \in M$ such that*

$$f(x_0) = \inf_{x \in M} f(x).$$

Proof. All the sub-level sets $[f \le r]$, $r \in \mathbb{R}$, are bounded and therefore relatively weakly compact since we are in a reflexive Banach space (see Lemma 29.2.1). Now Lemma 29.2.2 implies that all hypotheses of Theorem 29.1.1 are satisfied. Thus we conclude by this theorem. □

In Theorem 29.2.1 one can replace the assumption that the set M is bounded by an assumption on the function f which implies that the sub-level sets of f are bounded. Then one obtains another generalized Weierstrass theorem.

Theorem 29.2.2 (Generalized Weierstrass theorem II) *Let E be a reflexive Banach space, $M \subset E$ a weakly (sequentially) closed subset, and $f : M \to \mathbb{R}$ a weakly (sequentially) lower semi-continuous function on M. If f is coercive, i.e., if $\|x\| \to \infty$ implies $f(x) \to +\infty$, then f has a finite minimum on M, i.e., there is a $x_0 \in M$ such that*

$$f(x_0) = \inf_{x \in M} f(x).$$

Proof. Since f is coercive the sub-level sets $[f \le r]$ are not empty for sufficiently large r and are bounded. We conclude as in the previous result. □

For other variants of generalized Weierstrass theorems we refer to [Zei85]. Detailed results on the minimization of functionals of the form (28.2) can be found in [Dac89, JLJ98, Str00].

29.3 Minimization of special classes of functionals

For a self-adjoint compact operator A in the complex Hilbert space $\mathcal{H}$ consider the sesquilinear function $Q : \mathcal{H} \times \mathcal{H} \to \mathbb{C}$ defined by $Q(x, y) = \langle x, Ay \rangle + r \langle x, y \rangle$ for $r = \|A\| + c$ for some $c > 0$. This function has the following properties: $Q(x, x) \ge c\|x\|^2$ for all $x \in \mathcal{H}$ and for fixed $x \in \mathcal{H}$ the function $y \mapsto Q(x, y)$ is weakly continuous (since a compact operator maps weakly convergent sequences onto norm convergent ones). Then $f(x) = Q(x, x)$ is a concrete example of a *quadratic functional* on $\mathcal{H}$ which has a unique minimum on closed balls B_r of $\mathcal{H}$. This minimization is actually a special case of the following result on the minimization of quadratic functionals on reflexive Banach spaces.

Theorem 29.3.1 (Minimization of quadratic forms) *Let E be a reflexive Banach space and Q a symmetric sesquilinear form on E having the following properties: There is a constant $c > 0$ such that $Q(x, x) \ge c\|x\|^2$ for all $x \in E$ and for fixed $x \in E$ the functional $y \mapsto Q(x, y)$ is weakly continuous on E. Then, for every $u \in E'$ and every $r > 0$, there is exactly one point $x_0 = x_0(u, r)$ which minimizes the functional*

$$f(x) = Q(x, x) - \operatorname{Re} u(x), \qquad x \in E$$

on the closed ball $B_r = \{x \in E : \|x\| \le r\}$, i.e.,

$$f(x_0) = \inf_{x \in B_r} f(x).$$

Proof. Consider $x, y \in E$ and $0 < t < 1$, then a straightforward calculation gives

$$f(tx + (1-t)y) = tf(x) + (1-t)f(y) - t(1-t)Q(x-y, x-y) < tf(x) + (1-t)f(y).$$

for all $x, y \in E$, $x \neq y$, since then $t(1-t)Q(x-y, x-y) > 0$, hence the functional f is strictly convex and thus has at most one minimizing point by Theorem 29.1.2.

Suppose a sequence $(x_n)_{n \in \mathbb{N}}$ in E converges weakly to $x_0 \in E$. Since $Q(x_n, x_n) = Q(x_0, x_0) + Q(x_0, x_n - x_0) + Q(x_n - x_0, x_0) + Q(x_n - x_0, x_n - x_0)$ and since Q is strictly positive it follows that

$$Q(x_n, x_n) \geq Q(x_0, x_0) + Q(x_n - x_0, x_0) + Q(x_0, x_n - x_0)$$

for all $n \in \mathbb{N}$. Since Q is symmetric and weakly continuous in the second argument the last two terms converge to 0 as $n \to \infty$ and this estimate implies

$$\liminf_{n \to \infty} Q(x_n, x_n) \geq Q(x_0, x_0).$$

Therefore the function $x \mapsto Q(x, x)$ is weakly lower semi-continuous, thus, for every $u \in E'$, $x \mapsto f(x) = Q(x, x) - \operatorname{Re} u(x)$ is weakly lower semi-continuous on E and we conclude by Theorem 29.2.1 (Observe that the closed balls B_r are weakly closed, as closed convex sets). □

Corollary 29.3.1 *Let A be a bounded symmetric operator in complex Hilbert space $\mathcal{H}$ which is strictly positive, i.e., there is a constant $c > 0$ such that $\langle x, Ax \rangle \geq c \langle x, x \rangle$ for all $x \in \mathcal{H}$. Then, for every $y \in \mathcal{H}$ the function $x \mapsto f(x) = \langle x, Ax \rangle - \operatorname{Re} \langle y, x \rangle$ has a unique minimizing point $x_0 = x_0(y, r)$ on every closed ball B_r, i.e., there is exactly one $x_0 \in B_r$ such that*

$$f(x_0) = \inf_{x \in B_r} f(x).$$

Proof. Using the introductory remark to this section one verifies easily that $Q(x, y) = \langle x, Ay \rangle$ satisfies the hypothesis of Theorem 29.3.1. □

29.4 Exercises

1. Prove Lemma 29.1.1.

2. Show without the use of Lemma 29.2.2 that the norm $\|\cdot\|$ on a Banach space E is weakly lower semi-continuous.

   Hints: Recall that $\|x_0\| = \sup_{u \in E', \|u\|' \leq 1} |u(x_0)|$ for $x_0 \in E$. If a sequence $(x_n)_{n \in \mathbb{N}}$ converges weakly to x_0, then for every $u \in E'$ one knows $u(x_0) = \lim_{n \to \infty} u(x_n)$.

3. Prove: The functional

$$f(u) = \int_0^1 (tu'(t))^2 dt,$$

   defined on all continuous functions on $[0, 1]$ which have a weak derivative $u' \in L^2(0, 1)$ and which satisfy $u(0) = 0$ and $u(1) = 1$, has 0 as infimum and there is no function in this class at which the infimum is attained.

4. On the space $E = C^1([-1, 1], \mathbb{R})$ define the functional

$$f(u) = \int_{-1}^{1} (tu'(t))^2 dt$$

and show that it has no minimum under the boundary conditions $u(\pm 1) = \pm 1$.

Hints: This variation of the previous problem is due to Weierstrass. Show first that on the class of functions u_ϵ, $\epsilon > 0$, defined by

$$u_\epsilon(x) = \frac{\arctan \frac{x}{\epsilon}}{\arctan \frac{1}{\epsilon}},$$

the infimum of f is zero.

30
Differential Calculus on Banach Spaces and Extrema of Functions

As is well known from calculus on finite dimensional Euclidean spaces, the behavior of a sufficiently smooth function f in a neighborhood of some point x_0 is determined by the first few derivatives $f^{(n)}(x_0), n \leq m$, of f at this point, $m \in \mathbb{N}$ depending on f and the intended accuracy. For example, if f is a twice continuously differentiable real valued function on the open interval $\Omega \subset \mathbb{R}$ and $x_0 \in \Omega$, the Taylor expansion of order 2

$$f(x) = f(x_0) + f^{(1)}(x_0)(x - x_0) + \frac{1}{2!} f^{(2)}(x_0)(x - x_0)^2 + (x - x_0)^2 R_2(x, x_0) \tag{30.1}$$

with $\lim_{x \to x_0} R_2(x, x_0) = 0$ is available, and on the basis of this representation the values of $f^{(1)}(x_0)$ and $f^{(2)}(x_0)$ determine whether x_0 is a critical point of the function f, or a local minimum, or a local maximum, or an inflection point.

In variational problems too one has to determine whether a function f has critical points, local minima or maxima or inflection points, but in these problems the underlying spaces are typically infinite dimensional Banach spaces. Accordingly an expansion of the form (30.1) in this infinite dimensional case can be expected to be an important tool too. Obviously one needs differential calculus on Banach spaces to achieve this goal.

Recall that differentiability of a real valued function f on an open interval Ω at a point $x_0 \in \Omega$ is equivalent to the existence of a proper tangent to the graph of the function through the point $(x_0, f(x_0)) \in \mathbb{R}^2$. A proper tangent means that the difference between the values of the tangent and of the function f at a point $x \in \Omega$ is of higher order in $x - x_0$ than the linear term. Since the tangent has the

equation $y(x) = f^{(1)}(x_0)(x - x_0) + f(x_0)$ this approximation means
$$f(x) - y(x) = f(x) - f^{(1)}(x_0)(x - x_0) - f(x_0) = o(x - x_0) \qquad (30.2)$$
where o is some function on $\mathbb{R}$ with the properties $o(0) = o$ and $\lim_{h \to 0, h \neq 0} \frac{o(h)}{h}$. In the case of a real valued function of several variables the tangent plane takes the role of the tangent line. As we are going to show, this way to look at differentiability has a natural counterpart for functions defined on infinite dimensional Banach spaces.

30.1 The Fréchet derivative

Let E, F be two real Banach spaces with norms $\|\cdot\|_E$, respectively $\|\cdot\|_F$. As usual $\mathcal{L}(E, F)$ denotes the space of all continuous linear operators from E into F. By Theorem 21.3.3 the space $\mathcal{L}(E, F)$ is a real Banach space too. The symbol o denotes any function $E \to F$ which is of higher than linear order in its argument, i.e., any function satisfying

$$o(0) = 0, \qquad \lim_{h \to 0, h \in E \setminus \{0\}} \frac{\|o(h)\|_F}{\|h\|_E} = 0. \qquad (30.3)$$

Definition 30.1.1 *Let $U \subset E$ be a nonempty open subset of the real Banach space E and $f : U \to F$ a function from U into the real Banach space F. f is called **Fréchet differentiable at a point** $x_0 \in U$ if, and only if, there is an $\ell \in \mathcal{L}(E, F)$ such that*

$$f(x) = f(x_0) + \ell(x - x_0) + o(x_0; x - x_0) \qquad \forall x \in U. \qquad (30.4)$$

*If f is differentiable at $x_0 \in U$ the continuous linear operator $\ell \in \mathcal{L}(E, F)$ is called the **derivative of** f **at** x_0 and is denoted by*

$$f'(x_0) \equiv D_{x_0} f \equiv Df(x_0) \equiv \ell. \qquad (30.5)$$

*If f is differentiable at every point $x_0 \in U$, f is called **differentiable on** U and the function $Df : U \to \mathcal{L}(E, F)$ which assigns to every point $x_0 \in U$ the derivative $Df(x_0)$ of f at x_0 is called the **derivative of the function** f.*

*If the derivative $Df : U \to \mathcal{L}(E, F)$ is continuous, the function f is called **continuously differentiable on** U or of class $\mathcal{C}^1$, also denoted by $f \in \mathcal{C}^1(U, F)$.*

This definition is indeed meaningful because of the following

Lemma 30.1.1 *Under the assumptions of Definition 30.1.1 there is at most one $\ell \in \mathcal{L}(E, F)$ satisfying equation (30.4).*

Proof. Suppose there are $\ell_1, \ell_2 \in \mathcal{L}(E, F)$ satisfying equation (30.4). Then, for all $h \in B_r$ where B_r denotes an open ball in E with center 0 and radius $r > 0$ such that $x_0 + B_r \subset U$, we have $f(x_0) + \ell_1(h) + o_1(x_0, h) = f(x_0 + h) = f(x_0) + \ell_2(h) + o_2(x_0, h)$ and hence the linear functional $\ell = \ell_2 - \ell_1$ satisfies $\ell(h) = o_1(x_0, h) - o_2(x_0, h)$ for all $h \in B_r$. A continuous linear operator can be of higher than linear order on an open ball only if it is the null operator (see Exercises). This proves $\ell = 0$ and thus uniqueness. □

30.1 The Fréchet derivative 389

Definition 30.1.1 is easy to apply. Suppose $f : U \to F$ is constant, i.e., for some $a \in F$ we have $f(x) = a$ for all $x \in U \subset E$. Then $f(x) = f(x_0)$ for all $x, x_0 \in U$ and with the choice of $\ell = 0 \in \mathcal{L}(E, F)$ condition (30.4) is satisfied. Thus f is continuously Fréchet differentiable on U with derivative zero.

As another simple example consider the case were E is some real Hilbert space with inner product $\langle \cdot, \cdot \rangle$ and $F = \mathbb{R}$. For a continuous linear operator $A : E \to E$ define a function $f : E \to \mathbb{R}$ by $f(x) = \langle x, Ax \rangle$ for all $x \in E$. For $x, h \in E$ we calculate $f(x+h) = f(x) + \langle A^*x + Ax, h \rangle + f(h)$. $h \mapsto \langle A^*x + Ax, h \rangle$ is certainly a continuous linear functional $E \to \mathbb{R}$ and $f(h) = o(h)$ is obviously of higher than linear order (actually second order) in h. Hence f is Fréchet differentiable on E with derivative $f'(x) \in \mathcal{L}(E, \mathbb{R})$ given by $f'(x)(h) = \langle A^*x + Ax, h \rangle$ for all $h \in E$.

In the Exercises the reader will be invited to show that the above definition of differentiability reproduces the wellknown definitions of differentiability for functions of finitely many variables.

The Fréchet derivative has all the properties which are well known for the derivative of functions of one real variable. Indeed the following results hold.

Proposition 30.1.1 *Let $U \subset E$ be an open nonempty subset of the Banach space E and F some other real Banach space.*

a) *The Fréchet derivative D is a linear mapping $\mathcal{C}^1(U, F) \to \mathcal{C}(U, F)$, i.e., for all $f, g \in \mathcal{C}^1(U, F)$ and all $a, b \in \mathbb{R}$ one has*

$$D(af + bg) = aDf + bDg.$$

b) *The chain rule holds for the Fréchet derivative D: Let $V \subset F$ be an open set containing $f(U)$ and G a third real Banach space. Then for all $f \in \mathcal{C}^1(U, F)$ and all $g \in \mathcal{C}^1(V, G)$ we have $g \circ f \in \mathcal{C}^1(U, G)$ and for all $x \in U$*

$$D(g \circ f)(x) = (Dg)(f(x)) \circ (Df)(x).$$

Proof. The proof of the first part is left as an exercise.
Since f is differentiable at $x \in U$ we know

$$f(x + h) - f(x) = f'(x)(h) + o_1(h) \qquad \forall h \in B_r, \quad x + B_r \subset U$$

and similarly, since g is differentiable at $y = f(x) \in V$,

$$g(y + k) - g(y) = g'(y)(k) + o_2(k) \qquad \forall k \in B_\rho, \quad y + B_\rho \subset V.$$

Since f is continuous one can find, for the radius $\rho > 0$ in the differentiability condition for g, a radius $r > 0$ such that $f(B_r) \subseteq B_\rho$ and such that the differentiability condition for f holds. Then, for all $h \in B_r$, the following chain of identities holds, taking the above differentiability conditions into account:

$$\begin{aligned} g \circ f(x+h) - g \circ f(x) &= g[f(x+h)] - g[f(x)] \\ &= g[f(x) + f'(x)(h) + o_1(h)] - g[f(x)] \\ &= g'(y)(f'(x)(h) + o_1(h)) + o_2(f'(x)(h) + o_1(h)) \\ &= g'(y)(f'(x)(h)) + o(h) \end{aligned}$$

where
$$o(h) = g'(y)(o_1(h)) + o_2(f'(x)(h) + o_1(h))$$
is indeed a higher order term as shown in the Exercises. Thus we conclude. □

Higher order derivatives can be defined in the same way. Suppose E, F are two real Banach spaces and $U \subset E$ is open and nonempty. Given a function $f \in \mathcal{C}^1(U, F)$ we know $f' \in \mathcal{C}(U, \mathcal{L}(E, F))$, i.e., the derivative is a continuous function on U with values in the Banach space $\mathcal{L}(E, F)$. If this function f' is differentiable at $x_0 \in U$ (on U), the function f is called *twice differentiable* at $x_0 \in U$ (on U) and is denoted by

$$D^2 f(x_0) = f^{(2)}(x_0) = D_{x_0}^2 f \equiv D(f')(x_0). \tag{30.6}$$

According to Definition 30.1.1 and equation (30.6) the second derivative of $f : U \to F$ is a continuous linear operator $E \to \mathcal{L}(E, F)$, i.e., an element of the space $\mathcal{L}(E, \mathcal{L}(E, F))$. There is a natural isomorphism of the space of continuous linear operators from E into the space of continuous linear operators from E into F and the space $\mathcal{B}(E \times E, F)$ of continuous bilinear operators from $E \times E$ into F,

$$\mathcal{L}(E, \mathcal{L}(E, F)) \cong \mathcal{B}(E \times E, F). \tag{30.7}$$

This natural isomorphism is defined and studied in the Exercises. Thus the second derivative $D^2 f(x_0)$ at a point $x_0 \in U$ is considered as a continuous bilinear map $E \times E \to F$. If the second derivative $D^2 f : U \to \mathcal{B}(E \times E, F)$ exists on U and is continuous, the function f is said to be *of class $\mathcal{C}^2$* and we write $f \in \mathcal{C}^2(U, F)$.

The derivatives of higher order are defined in the same way. The derivative of order $n \geq 3$ is the derivative of the derivative of order $n - 1$, according to Definition 30.1.1:

$$D^n f(x_0) = D(D^{n-1} f)(x_0). \tag{30.8}$$

In order to describe $D^n f(x_0)$ conveniently we extend the isomorphism (30.7) to higher orders. Denote by $E^{\times n} = E \times \cdots \times E$ (n factors) and by $\mathcal{B}(E^{\times n}, F)$ the Banach space of all continuous n-linear operators $E^{\times n} \to F$. In the Exercises one shows for $n = 3, 4, \ldots$

$$\mathcal{L}(E, \mathcal{B}(E^{\times n-1}, F)) \cong \mathcal{B}(E^{\times n}, F). \tag{30.9}$$

Under this isomorphism the third derivative at some point $x_0 \in U$ is then a continuous 3-linear map $E^{\times 3} \to F$, $D^3 f(x_0) \in \mathcal{B}(E^{\times 3}, F)$. Using the isomorphisms (30.9) the higher order derivatives are

$$D^n f(x_0) \in \mathcal{B}(E^{\times n}, F) \tag{30.10}$$

if they exist. If $D^n f : U \to \mathcal{B}(E^{\times n}, F)$ is continuous the function f is called *n-times continuously differentiable* or *of class $\mathcal{C}^n$*. Then we write $f \in \mathcal{C}^n(U, F)$.

As an illustration we calculate the second derivative of the function $f(x) = \langle x, Ax \rangle$ on a real Hilbert space E with inner product $\langle \cdot, \cdot \rangle$, A a bounded linear operator on E. The first Fréchet derivative has been calculated, $f'(x_0)(h) = \langle (A + A^*)x_0, y \rangle$ for all $y \in E$. In order to determine the second derivative we evaluate $f'(x_0 + h) - f'(x_0)$. For all $y \in E$ one finds through a simple calculation

$$(f'(x_0 + h) - f'(x_0))(y) = \langle (A + A^*)h, y \rangle.$$

Hence the second derivative of f exists and is given by the continuous bilinear form $(D^2 f)(x_0)(y_1, y_2) = \langle (A + A^*)y_1, y_2 \rangle$, $y_1, y_2 \in E$. We see in this example that the second derivative is actually a symmetric bilinear form. With some effort this can be shown for every twice differentiable function.

As we have mentioned, the first few derivatives of a differentiable function $f : U \to F$ at a point $x_0 \in U$ control the behavior of the function in a sufficiently small neighborhood of this point. The key to this connection is the Taylor expansion with remainder. In order to be able to prove this fundamental result in its strongest form we need the fundamental theorem of calculus for functions with values in a Banach space. And this in turn requires the knowledge of the Riemann integral for functions on the real line with values in a Banach space.

Suppose E is a real Banach space and $u : [a, b] \to E$ a continuous function on the bounded interval $[a, b]$. In the section on the integration of spectral families we had introduced partitions Z of the interval $[a, b]$. Roughly, a partition Z of the interval $[a, b]$ is an ordered family of points $a = t_0 < t_1 < t_2 < \cdots < t_n = b$ and of some points $t'_j \in (t_{j-1}, t_j]$, $j = 1, \ldots, n$. For each partition we introduce the approximating sums

$$\Sigma(u, Z) = \sum_{j=1}^{n} u(t'_j)(t_j - t_{j-1}).$$

By forming the joint refinement of two partitions one shows, as in Section 26.2 on the integration of spectral families, the following result: Given $\epsilon > 0$ there is $\delta > 0$ such that

$$\|\Sigma(u, Z) - \Sigma(u, Z')\|_E < \epsilon$$

for all partitions Z, Z' with $|Z'|, |Z| < \delta$, $|Z| = \max\{t_j - t_{j-1} : j = 1, \ldots, n\}$. This estimate implies that the approximating sums $\Sigma(u, Z)$ have a limit with respect to partitions Z with $|Z| \to 0$.

Theorem 30.1.2 *Suppose E is a real Banach space and $u : [a, b] \to E$ a continuous function. Then u has an integral over this finite interval, defined by the following limit in E:*

$$\int_a^b u(t)dt = \lim_{|Z| \to 0} \Sigma(u, Z). \tag{30.11}$$

This integral of functions with values in a Banach space has the standard properties, i.e., it is linear in the integrand, additive in the interval of integration, and is bounded by the maximum of the function multiplied by the length of the integration interval:

$$\left\| \int_a^b u(t)dt \right\|_E \leq (b - a) \max_{a \leq t \leq b} \|u(t)\|_E.$$

Proof. It is straightforward to verify that the approximating sums $\Sigma(u, Z)$ are linear in u and additive in the interval of integration. The basic rules of calculation for limits then prove the statements for the integral. For the estimate observe

$$\|\Sigma(u, Z)\|_E \leq \sum_{j=1}^{n} \|u(t'_j)\|_E (t_j - t_{j-1}) \leq \sup_{a \leq t \leq b} \|u(t)\|_E \sum_{j=1}^{n} (t_j - t_{j-1})$$

which implies the above estimate for the approximating sums. Thus we conclude. □

Corollary 30.1.1 (Fundamental theorem of calculus) *Let E be a real Banach space, $[a, b]$ a finite interval and $u : [a, b] \to E$ a continuous function. For some $e \in E$ define a function $v : [a, b] \to E$ by*

$$v(t) = e + \int_a^t u(s)ds \qquad \forall s \in [a, b]. \tag{30.12}$$

Then v is continuously differentiable with derivative $v'(t) = u(t)$ and one thus has for all $a \leq c < d \leq b$,

$$v(d) - v(c) = \int_c^d v'(t)dt. \tag{30.13}$$

Proof. We prove differentiability of v at some interior point $t \in (a, b)$. At the end points of the interval the usual modifications apply. Suppose $\tau > 0$ such that $t + \tau \in [a, b]$. Then, by definition of v,

$$v(t + \tau) - v(t) = \int_a^{t+\tau} u(s)ds - \int_a^t u(s)ds = \int_t^{t+\tau} u(s)ds$$

since

$$\int_a^{t+\tau} u(s)ds = \int_a^t u(s)ds + \int_t^{t+\tau} u(s)ds.$$

The basic bound for integrals gives

$$\left\| \int_t^{t+\tau} [u(s) - u(t)]ds \right\|_E \leq \tau \sup_{t \leq s \leq t+\tau} \|u(s) - u(t)\|_E$$

and thus proves that this integral is of higher order in τ. We deduce $v(t + \tau) = v(t) + \tau u(t) + o(\tau)$ and conclude that v is differentiable at t with derivative $v'(t) = u(t)$. The rest of the proof is standard. □

Theorem 30.1.3 (Taylor expansion with remainder) *Suppose E, F are real Banach spaces, $U \subset E$ an open and nonempty subset, and $f \in C^n(U, F)$. Given $x_0 \in U$ choose $r > 0$ such that $x_0 + B_r \subset U$ where B_r is the open ball in E with center 0 and radius r. Then for all $h \in B_r$ we have, using the abbreviation $(h)^k = (h, \ldots, h)$, k terms,*

$$f(x_0 + h) = \sum_{k=0}^{n} \frac{1}{k!} f^{(k)}(x_0)(h)^k + R_n(x_0; h) \tag{30.14}$$

where the remainder R_n has the form

$$R_n(x_0; h) = \frac{1}{(n-1)!} \int_0^1 (1-t)^{n-1} [f^{(n)}(x_0 + th) - f^{(n)}(x_0)](h)^n dt \tag{30.15}$$

and thus is of order $o((h)^n)$, i.e.,

$$\lim_{h\to 0, h\in E\setminus\{0\}} \frac{\|R_n(x_0; h)\|_F}{\|h\|_E^n} = 0.$$

Proof. Basically the Taylor formula is obtained by applying the fundamental theorem of calculus repeatedly (n times) and transforming the multiple integral which is generated in this process by a change of the integration order into a one-dimensional integral.

However there is a simplification of the proof based on the following observation (see [YCB82]). Let v be a function on $[0, 1]$ which is n times continuously differentiable, then

$$\frac{d}{dt}\sum_{k=0}^{n-1}\frac{(1-t)^k}{k!}v^{(k)}(t) = \frac{(1-t)^{n-1}}{(n-1)!}v^{(n)}(t) \qquad \forall t \in [0,1].$$

The proof of this identity follows simply by differentiation and grouping terms together appropriately.

Integrate this identity for the function $v(t) = f(x_0 + th)$. Since $f \in C^n(U, F)$ the application of the chain rule yields for $h \in B_r$,

$$v^{(k)}(t) = f^{(k)}(x_0 + th)(h)^k$$

and thus the result of this integration is, using Equation 30.13,

$$f(x_0 + h) = \sum_{k=0}^{n-1}\frac{1}{k!}f^{(k)}(x_0)(h)^k + R$$

with remainder

$$R = \frac{1}{(n-1)!}\int_0^1 (1-t)^{n-1}f^{(n)}(x_0 + th)(h)^n\,dt$$

which can be written as

$$R = \frac{1}{n!}f^{(n)}(x_0)(h)^n + \frac{1}{(n-1)!}\int_0^1 (1-t)^{n-1}[f^{(n)}(x_0 + th) - f^{(n)}(x_0)](h)^n\,dt.$$

The differentiability assumption for f implies that the function $h \mapsto f^{(n)}(x_0 + th)$ from B_r into $\mathcal{B}(E^{\times n}, F)$ is continuous, hence

$$\|[f^{(n)}(x_0 + th) - f^{(n)}(x_0)]\|_{\mathcal{B}(E^{\times n}, F)} \to 0$$

as $h \to 0$. Thus we conclude. $\square$

30.2 Extrema of differentiable functions

Taylor's formula (30.14) says that a function $f : U \to F$ of class C^n is approximated at each point of a neighborhood of some point $x_0 \in U$ by a polynomial of degree n, and the error is of order $o((x - x_0)^n)$. We apply now this approximation for $n = 2$ to characterize local extrema of a function of class C^2 in terms of the first and second derivative of f. We begin with the necessary definitions.

Definition 30.2.1 *Let E be a real Banach space, $M \subseteq E$ a nonempty subset, and $f : M \to \mathbb{R}$ a real valued function on M. A point $x_0 \in M$ is called a **local minimum (maximum)** of f on M if there is some $r > 0$ such that*

$$f(x_0) \le f(x), \quad (f(x_0) \ge f(x)) \qquad \forall x \in M \cap (x_0 + B_r).$$

A *local minimum (maximum)* is **strict** if

$$f(x_0) < f(x), \quad (f(x_0) > f(x)) \quad \forall x \in M \cap (x_0 + B_r), \ x \neq x_0.$$

If $f(x_0) \leq f(x)$, $(f(x_0) \geq f(x))$ holds for all $x \in M$, we call x_0 a **global minimum (maximum)**.

Definition 30.2.2 *Suppose E, F are two real Banach spaces, $U \subset E$ an open nonempty subset, and $f : U \to F$ a function of class C^1. A point $x_0 \in U$ is called a **regular** (**critical**) **point** of the function f if, and only if, the Fréchet derivative $Df(x_0)$ of f at x_0 is surjective (not surjective).*

Remark 30.2.1 *For the case $F = \mathbb{R}$ the Fréchet derivative $Df(x_0) = f'(x_0) \in \mathcal{L}(E, \mathbb{R})$ is not surjective, if and only if, $f'(x_0) = 0$; hence the notion of a critical point introduced above is nothing else than the generalization of the corresponding notion introduced in elementary calculus.*

For extremal points which are interior points of the domain M of the function f a fairly detailed description can be given. In this situation we can assume that the domain $M = U$ is an open set.

Theorem 30.2.1 (Necessary condition of Euler–Lagrange) *Suppose U is an open nonempty subset of the real Banach space E and $f \in C^1(U, \mathbb{R})$. Then every extremal point (i.e., every local or global minimum and every local or global maximum) is a critical point of f.*

Proof. Suppose that $x_0 \in U$ is a local minimum of f. Then there is an $r > 0$ such that $x_0 + B_r \subset U$ and $f(x_0) \leq f(x_0 + h)$ for all $h \in B_r$. Since $f \in C^1(U, \mathbb{R})$ Taylor's formula applies, thus

$$f(x_0) \leq f(x_0 + h) = f(x_0) + f'(x_0)(h) + R_1(x_0, h) \quad \forall h \in B_r$$

or

$$0 \leq f'(x_0)(h) + R_1(x_0, h) \quad \forall h \in B_r.$$

Choose any $h \in B_r$, $h \neq 0$. Then all $th \in B_r$, $0 < t \leq 1$ and therefore $0 \leq f'(x_0)(th) + R_1(x_0, th)$. Since $\lim_{t \to 0} t^{-1} R_1(x_0, th) = 0$ we can divide this inequality by $t > 0$ and take the limit $t \to 0$. This gives $0 \leq f'(x_0)(h)$. This argument applies to any $h \in B_r$, thus in particular to $-h$ and therefore $0 \leq f'(x_0)(-h) = -f'(x_0)(h)$. We conclude that $0 = f'(x_0)(h)$ for all $h \in B_r$. The open nonempty ball B_r absorbs the points of E, i.e., every point $x \in E$ can be written as $x = \lambda h$ with some $h \in B_r$ and some $\lambda \in \mathbb{R}$. It follows that $0 = f'(x_0)(x)$ for all $x \in E$ and therefore $f'(x_0) = 0 \in \mathcal{L}(E, \mathbb{R}) = E'$.

If $x_0 \in U$ is a local maximum of f, then this point is a local minimum of $-f$ and we conclude as above. $\square$

Theorem 30.2.2 (Necessary and sufficient conditions for local extrema) *Suppose $U \subset E$ is a nonempty open subset of the real Banach space E and $f \in C^2(U, \mathbb{R})$.*

a) If f has a local minimum at $x_0 \in U$, then the first Fréchet derivative of f vanishes at x_0, $f'(x_0) = 0$, and the second Fréchet derivative of f is nonnegative at x_0, $f^{(2)}(x_0)(h, h) \geq 0$ for all $h \in E$.

b) *If conversely $f'(x_0) = 0$ and if the second Fréchet derivative of f is strictly positive at x_0, i.e., if $\inf \{ f^{(2)}(x_0)(h, h) : h \in E, \|h\|_E = 1 \} = c > 0$, then f has a local minimum at x_0.*

Proof. Suppose $x_0 \in U$ is a local minimum of f. Then by Theorem 30.2.1 $f'(x_0) = 0$. Since $f \in C^2(U, \mathbb{R})$ Taylor's formula implies

$$f(x_0) \leq f(x_0 + h) = f(x_0) + \frac{1}{2!} f^{(2)}(x_0)(h, h) + R_2(x_0, h) \quad \forall h \in B_r \quad (30.16)$$

for some $r > 0$ such that $x_0 + B_r \subset U$. Choose any $h \in B_r$. Then for all $0 < t \leq 1$ we know $0 \leq \frac{1}{2!} f^{(2)}(x_0)(th, th) + R_2(x_0, th)$ or, after division by $t^2 > 0$,

$$0 \leq f^{(2)}(x_0)(h, h) + \frac{2}{t^2} R_2(x_0, th) \quad \forall 0 < t \leq 1.$$

Since $R_2(x_0, th)$ is a higher order term we know $t^{-2} R_2(x_0, th) \to 0$ as $t \to 0$. This gives $0 \leq f^{(2)}(x_0)(h, h)$ for all $h \in B_r$ and since open balls are absorbing, $0 \leq f^{(2)}(x_0)(h, h)$ for all $h \in E$. This proves Part a).

Conversely assume that $f'(x_0) = 0$ and that $f^{(2)}(x_0)$ is strictly positive. Choose $r > 0$ such that $x_0 + B_r \subset U$. The second order Taylor expansion gives

$$f(x_0 + h) - f(x_0) = \frac{1}{2!} f^{(2)}(x_0)(h, h) + R_2(x_0, h) \quad \forall h \in B_r,$$

and thus for all $h \in E$ with $\|h\|_E = 1$ and all $0 < s < r$,

$$f(x_0 + sh) - f(x_0) = \frac{1}{2!} f^{(2)}(x_0)(sh, sh) + R_2(x_0, sh)$$

$$= s^2 [\frac{1}{2!} f^{(2)}(x_0)(h, h) + s^{-2} R_2(x_0, sh)].$$

Since $R_2(x_0, sh)$ is a higher order term there is an $s_0 \in (0, r)$ such that $|s^{-2} R_2(x_0, sh)| < c/2$ for all $0 < s \leq s_0$, and since $\frac{1}{2!} f^{(2)}(x_0)(h, h) \geq c/2$ for all $h \in E$, $\|h\|_E = 1$, we get $[\frac{1}{2!} f^{(2)}(x_0)(h, h) + s^{-2} R_2(x_0, sh)] \geq 0$ for all $0 < s < s_0$ and all $h \in E$, $\|h\|_E = 1$. It follows that $f(x_0 + h) - f(x_0) \geq 0$ for all $h \in B_{s_0}$ and therefore the function f has a local minimum at x_0. □

As we mentioned before a function f has a local maximum at some point x_0 if, and only if, the function $-f$ has a local minimum at this point. Therefore Theorem 30.2.2 easily implies necessary and sufficient conditions for a local maximum.

30.3 Convexity and monotonicity

We begin with the discussion of an interesting connection between convexity of a functional and monotonicity of its first Fréchet derivative which has far-reaching implications for optimization problems. For differentiable real valued functions of one real variable these results are well known.

The following theorem states this connection in detail and provides the relevant definitions.

Theorem 30.3.1 (Convexity–Monotonicity) *Let U be a convex open subset of the real Banach space E and $f \in C^1(U, \mathbb{R})$. Then the following statements are equivalent:*

a) f is **convex**, i.e., for all $x, y \in U$ and all $0 \le t \le 1$ one has

$$f(tx + (1-t)y) \le tf(x) + (1-t)f(y); \tag{30.17}$$

b) The Fréchet derivative $f' : E \to E'$ is **monotone**, i.e., for all $x, y \in U$ one has

$$\langle f'(x) - f'(y), x - y \rangle \ge 0 \tag{30.18}$$

where $\langle \cdot, \cdot \rangle$ denotes the canonical bilinear form on $E' \times E$.

Proof. If f is convex inequality (30.17) implies, for $x, y \in U$ and $0 < t \le 1$,

$$f(y + t(x-y)) - f(y) \le tf(x) + (1-t)f(y) - f(y) = t(f(x) - f(y)).$$

If we divide this inequality by $t > 0$ and then take the limit $t \to 0$ the result is

$$\langle f'(y), x - y \rangle \le f(x) - f(y).$$

If we exchange the roles of x and y in this argument we obtain

$$\langle f'(x), y - x \rangle \le f(y) - f(x).$$

Now add the two inequalities to get

$$\langle f'(y), x - y \rangle + \langle f'(x), y - x \rangle \le 0,$$

thus condition (30.18) follows and therefore f' is monotone.

Suppose conversely that the Fréchet derivative $f' : E \to E'$ is monotone. For $x, y \in U$ and $0 \le t \le 1$ consider the function $p : [0, 1] \to \mathbb{R}$ defined by $p(t) = f(tx + (1-t)y) - tf(x) - (1-t)f(y)$. This function is differentiable with derivative $p'(t) = \langle f'(x(t)), x - y \rangle - f(x) + f(y), x(t) = tx + (1-t)y$, and satisfies $p(0) = 0 = p(1)$. The convexity condition is equivalent to the condition $p(t) \le 0$ for all $t \in [0, 1]$. We prove this condition indirectly. Thus we assume that there is some point in $(0, 1)$ at which p is positive. Then there is some point $t_0 \in (0, 1)$ at which p attains its positive maximum. For $t \in (0, 1)$ calculate

$$(t - t_0)(p'(t) - p'(t_0)) = (t - t_0)\langle f'(x(t)) - f'(x(t_0)), x - y \rangle$$
$$= \langle f'(x(t)) - f'(x(t_0)), x(t) - x(t_0) \rangle.$$

Since f' is monotone it follows that $(t - t_0)(p'(t) - p'(t_0)) \ge 0$. Since p attains its maximum at t_0, $p'(t_0) = 0$, and thus $(t - t_0)p'(t) \ge 0$, hence $p'(t) \ge 0$ for all $t_0 < t \le 1$, a contradiction. We conclude $p(t) \le 0$ and thus condition (30.17). □

Corollary 30.3.1 *Let U be a nonempty convex open subset of the real Banach space E and $f \in C^1(U, \mathbb{R})$. If f is convex, then every critical point of f is actually a minimizing point, i.e., a point at which f has a local minimum.*

Proof. If $x_0 \in U$ is a critical point, there is an $r > 0$ such that $x_0 + B_r \subset U$. Then for every $h \in B_r$ the points $x(t) = x_0 + th$, $0 \le t \le 1$, belong to $x_0 + B_r$. Since f is differentiable we find

$$f(x_0 + h) - f(x_0) = \int_0^1 \frac{d}{dt} f(x(t)) dt = \int_0^1 \langle f'(x(t)), h \rangle dt.$$

Since $x(t) - x_0 = th$ the last integral can be written as:

$$= \lim_{\epsilon \downarrow 0} \int_\epsilon^1 \langle f'(x(t)) - f'(x_0), x(t) - x_0 \rangle \frac{dt}{t}.$$

Theorem 30.3.1 implies that the integrand of this integral is non-negative, hence $f(x_0 + h) - f(x_0) \ge 0$ for all $h \in B_r$ and f has a local minimum at the critical point x_0. □

Corollary 30.3.2 *Let U be a nonempty convex open subset of the real Banach space E and $f \in C^1(U, \mathbb{R})$. If f is convex, then f is weakly lower semi-continuous.*

Proof. Suppose that a sequence $(x_n)_{n \in \mathbb{N}} \subset U$ converges weakly to $x_0 \in U$. Again differentiability of f implies

$$f(x_n) - f(x_0) = \int_0^1 \frac{d}{dt} f(x_0 + t(x_n - x_0)) dt = \int_0^1 \langle f'(x_0 + t(x_n - x_0)), x_n - x_0 \rangle dt$$

$$= \int_0^1 \langle f'(x_0 + t(x_n - x_0)) - f'(x_0), x_n - x_0 \rangle dt + \langle f'(x_0), x_n - x_0 \rangle.$$

As in the proof of the previous corollary, monotonicity of f' implies that the integral is not negative, hence

$$f(x_n) - f(x_0) \geq \langle f'(x_0), x_n - x_0 \rangle.$$

As $n \to \infty$ the righthand side of this estimate converges to 0 and thus $\liminf_{n \to \infty} f(x_n) - f(x_0) \geq 0$ or $\liminf_{n \to \infty} f(x_n) \geq f(x_0)$. This shows that f is weakly lower semi-continuous at x_0. Since $x_0 \in U$ was arbitrary, we conclude. □

Corollary 30.3.3 *Let U be a nonempty convex open subset of the real Banach space E and $f \in C^2(U, \mathbb{R})$. Then f is convex if, and only if, $f^{(2)}(x_0)$ is non-negative for all $x_0 \in U$, i.e., $f^{(2)}(x_0)(h, h) \geq 0$ for all $h \in E$.*

Proof. By Theorem 30.3.1 we know that f is convex if, and only if, its Fréchet derivative f' is monotone. Suppose f' is monotone and $x_0 \in U$. Then there is an $r > 0$ such that $x_0 + B_r \subset U$ and $\langle f'(x_0 + h) - f'(x_0), h \rangle \geq 0$ for all $h \in B_r$. Since $f \in C^2(U, \mathbb{R})$, Taylor's Theorem implies that

$$\langle f'(x_0 + h) - f'(x_0), h \rangle = f^{(2)}(x_0)(h, h) + R_2(x_0, h),$$

hence

$$0 \leq f^{(2)}(x_0)(h, h) + R_2(x_0, h) \qquad \forall h \in B_r.$$

Since $R_2(x_0, h) = o((h)^2)$ we deduce, as in the proof of Theorem 30.2.2, that $0 \leq f^{(2)}(x_0)(h, h)$ for all $h \in E$. Thus $f^{(2)}$ is nonnegative at $x_0 \in U$.

Conversely assume that $f^{(2)}$ is nonnegative on U. For $x, y \in U$ we know

$$\langle f'(x) - f'(y), x - y \rangle = \int_0^1 f^{(2)}(y + t(x - y))(x - y, x - y) dt.$$

By assumption the integrand is nonnegative, and it follows that f' is monotone. □

30.4 Gâteaux derivatives and variations

For functions $f : \mathbb{R}^n \to \mathbb{R}$ one has the concepts of the total differential and that of partial derivatives. The Fréchet derivative has been introduced as the generalization of the total differential to the case of infinite dimensional Banach spaces. Now we introduce the Gâteaux derivatives as the counterpart of the partial derivatives.

Definition 30.4.1 *Let E, F be two real Banach spaces, $U \subseteq E$ a nonempty open subset, and $f : U \to F$ a mapping from U into F. The **Gâteaux differential of** f **at a point** $x_0 \in U$ is a mapping $\delta f(x_0, \cdot) : E \to F$ such that, for all $h \in E$,*

$$\lim_{t \to 0, t \neq 0} \frac{1}{t} (f(x_0 + th) - f(x_0)) = \delta f(x_0, h). \tag{30.19}$$

$\delta f(x_0, h)$ is called the **Gâteaux differential of** f **at the point** x_0 **in the direction** $h \in E$. If the Gâteaux differential of f at x_0 is a continuous linear map $E \to F$, one writes

$$\delta f(x_0, h) = \delta_{x_0} f(h)$$

and calls $\delta_{x_0} f$ the **Gâteaux derivative of** f **at the point** x_0.

Basic properties of the Gâteaux differential, respectively derivative, are collected in the following

Lemma 30.4.1 *Let E, F be two real Banach spaces, $U \subseteq E$ a nonempty open subset, and $f : U \to F$ a mapping from U into F.*

a) *If the Gâteaux differential of f exists at a point $x_0 \in U$, it is a homogeneous map $E \to F$, i.e., $\delta f(x_0, \lambda h) = \lambda \delta f(x_0, h)$ for all $\lambda \in \mathbb{R}$ and all $h \in E$;*

b) *If the Gâteaux derivatives exist at a point $x \in U$, they are linear in f, i.e., for $f, g : U \to F$ and $\alpha, \beta \in \mathbb{R}$ one has $\delta_x(\alpha f + \beta g) = \alpha \delta_x f + \beta \delta_x g$;*

c) *If f is Gâteaux differentiable at a point $x \in U$, then f is continuous at x in every direction $h \in E$;*

d) *Suppose G is a third real Banach space, $V \subseteq F$ a nonempty open subset such that $f(U) \subseteq V$ and $g : V \to G$ a mapping from V into G. If f has a Gâteaux derivative at $x \in U$ and g has a Gâteaux derivative at $y = f(x)$, then $g \circ f : U \to G$ has a Gâteaux derivative at $x \in U$ and the chain rule*

$$\delta_x(g \circ f) = \delta_y g \circ \delta_x f$$

holds.

Proof. Parts a) and b) follow easily from the basic rules of calculation for limits. Part c) is obvious from the definitions. The proof of the chain rule is similar but easier than the proof of this rule for the Fréchet derivative and thus we leave it as an exercise. □

The following result establishes the important connection between Fréchet and Gâteaux derivatives, as a counterpart of the connection between total differential and partial derivatives for functions of finitely many variables.

Lemma 30.4.2 *Let E, F be two real Banach spaces, $U \subseteq E$ a nonempty open subset, and $f : U \to F$ a mapping from U into F.*

a) *If f is Fréchet differentiable at a point $x \in U$, then f is Gâteaux differentiable at x and both derivatives are equal: $\delta_x f = D_x f$.*

b) *Suppose that f is Gâteaux differentiable at all points in a neighborhood V of the point $x_0 \in U$ and that $x \mapsto \delta_x f \in \mathcal{L}(E, F)$ is continuous on V. Then f is Fréchet differentiable at x_0 and $\delta_{x_0} f = D_{x_0} f$.*

30.4 Gâteaux derivatives and variations

Proof. If f is Fréchet differentiable at $x \in U$ we know, for all $h \in E$, $f(x+th) = f(x) + (D_x f)(th) + o(th)$, hence

$$\lim_{t \to 0} \frac{1}{t}(f(x+th) - f(x)) = (D_x f)(h) + \lim_{t \to 0} \frac{o(th)}{t} = (D_x f)(h),$$

and Part a) follows.

If f is Gâteaux differentiable in the neighborhood V of $x_0 \in U$, there is an $r > 0$ such that f is Gâteaux differentiable at all points $x_0 + h$, $h \in B_r$. Given $h \in B_r$ it follows that $g(t) = f(x_0 + th)$ is differentiable at all points $t \in [0, 1]$ and $g'(t) = (\delta_{x_0+th} f)(h)$. This implies

$$g(1) - g(0) = \int_0^1 g'(t)dt = \int_0^1 (\delta_{x_0+th} f)(h)dt$$

and thus

$$f(x_0 + h) - f(x_0) - (\delta_{x_0} f)(h) = g(1) - g(0) - (\delta_{x_0} f)(h) = \int_0^1 [(\delta_{x_0+th} f)(h) - (\delta_{x_0} f)(h)]dt.$$

The integral can be estimated in norm by

$$\sup_{0 \leq t \leq 1} \|(\delta_{x_0+th} f) - (\delta_{x_0} f)\|_{\mathcal{L}(E,F)} \|h\|_E$$

and therefore

$$\|f(x_0 + h) - f(x_0) - (\delta_{x_0} f)(h)\|_F \leq \sup_{0 \leq t \leq 1} \|(\delta_{x_0+th} f) - (\delta_{x_0} f)\|_{\mathcal{L}(E,F)} \|h\|_E.$$

Continuity of $(\delta_x f)$ in $x \in x_0 + B_r$ implies $f(x_0 + h) - f(x_0) - (\delta_{x_0} f)(h) = o(h)$ and thus f is Fréchet differentiable at x_0 and $(D_{x_0} f)(h) = (\delta_{x_0} f)(h)$ for all $h \in B_r$ and therefore for all $h \in E$. □

Lemma 30.4.2 can be very useful in finding the Fréchet derivative of functions. We give a simple example. On the Banach space $E = L^p(\mathbb{R}^n)$, $1 < p < 2$, consider the functional

$$f(u) = \int_{\mathbb{R}^n} |u(x)|^p dx, \qquad \forall u \in E.$$

To prove directly that f is continuously Fréchet differentiable on E is not so simple. If however Lemma 30.4.2 is used the proof becomes a straightforward calculation. We only need to verify the hypotheses of this lemma. In the Exercises the reader is asked to show that there are constants $0 < c < C < \infty$ such that

$$c|s|^p \leq |1+s|^p - 1 - ps \leq C|s|^p \qquad \forall s \in \mathbb{R}.$$

Insert $s = t\frac{h(x)}{u(x)}$, for all points $x \in \mathbb{R}^n$ with $u(x) \neq 0$ and multiply with $|u(x)|^p$. The result is

$$c|th(x)|^p \leq |u(x) + th(x)|^p - |u(x)|^p - pth(x)|u(x)|^{p-1}\text{sgn}(u(x)) \leq C|th(x)|^p.$$

Integration of this inequality gives

$$c|t|^p f(h) \leq f(u+th) - f(u) - pt \int_{\mathbb{R}^n} h(x)v(x)dx \leq C|t|^p f(h)$$

where

$$v(x) = |u(x)|^{p-1}\text{sgn}(u(x)).$$

Note that $v \in L^q(\mathbb{R}^n)$, $\frac{1}{q} + \frac{1}{p} = 1$ and that $L^q(\mathbb{R}^n)$ is (isomorphic to) the topological dual of $E = L^p(\mathbb{R}^n)$. This estimate allows us to determine easily the Gâteaux derivative of f:

$$\delta f(u, h) = \lim_{t \to 0} \frac{1}{t}[f(u+th) - f(u)] = p \int_{\mathbb{R}^n} v(x)h(x)dx.$$

Hölder's inequality implies that the absolute value of this integral is bounded by $\|v\|_q \|h\|_p$, hence $h \mapsto \delta f(u, h)$ is a continuous linear functional on E and

$$\|\delta_u f\|_{\mathcal{L}(E,\mathbb{R})} = \|v\|_q = \|u\|_p^{p/q}.$$

Therefore $u \mapsto \delta_u f$ is a continuous map from $E \to \mathcal{L}(E, \mathbb{R})$ and Lemma 30.4.2 implies that f is Fréchet differentiable with derivative $D_u f(h) = \delta_u f(h)$.

Suppose that M is a nonempty subset of the real Banach space E which is not open, for instance M has a nonempty interior and part of the boundary of M belongs to M. Suppose furthermore that a function $f : M \to \mathbb{R}$ attains a local minimum at the boundary point x_0. Then we cannot investigate the behavior of f in terms of the first few Fréchet or Gâteaux derivatives of f at the point x_0 as we did previously since this required that a whole neighborhood of x_0 is contained in M. In such situations the *variations of the function* in suitable directions are a convenient tool to study the local behavior of f.

Assume that $h \in E$ and that there is some $r = r_h > 0$ such that $x(t) = x_0 + th \in M$ for all $0 \leq t < r$. Then we can study the function $f_h(t) = f(x(t))$ on the interval $[0, r)$. Certainly, if f has a local minimum at x_0, then $f_h(0) \leq f_h(t)$ for all $t \in [0, r)$ (if necessary we can decrease the value of r) and this gives restrictions on the first few derivatives of f_h, if they exist. These derivatives are then called the variations of f.

Definition 30.4.2 *Let $M \subset E$ be a nonempty subset of the real Banach space E and $x_0 \in M$. For $h \in E$ suppose that there is an $r > 0$ such that $x_0 + th \in M$ for all $0 \leq t < r$. Then the nth **variation** of f in the direction h is defined as*

$$\Delta^n f(x_0, h) = \frac{d^n}{dt^n} f(x_0 + th)|_{t=0} \qquad n = 1, 2, \ldots \qquad (30.20)$$

if these derivatives exist.

In favorable situations obviously the first variation is just the Gâteaux derivative:

Lemma 30.4.3 *Suppose that M is a nonempty subset of the real Banach space E, x_0 an interior point of M, and f a real valued function on M. Then the Gâteaux derivative $\delta_{x_0} f$ of f at x_0 exists if, and only if, the first variation $\Delta f(x_0, h)$ exists for all $h \in E$ and $h \mapsto \Delta f(x_0, h)$ is a continuous linear functional on E.*

In this case one has $\Delta f(x_0, h) = \delta_{x_0} f$.

Proof. A straightforward inspection of the respective definitions easily proves this lemma. □

30.5 Exercises

1. Complete the proof of Lemma 30.1.1.

2. Let E and F be two real normed spaces and $A : E \to F$ a continuous linear operator such that $Ax = o(x)$ for all $x \in E$, $\|x\| < 1$. Prove: $A = 0$.

3. For a function $f : U \to \mathbb{R}^m$, $U \subset \mathbb{R}^n$ open, assume that it is differentiable at a point $x_0 \in U$. Use Definition 30.1.1 to determine the Fréchet derivative $f'(x_0)$ of f at x_0 and relate it to the Jabobi matrix $\frac{\partial f}{\partial x}(x_0)$ of f at x_0.

4. Prove Part a) of Proposition 30.1.1.

5. Prove that $o(h) = g'(y)(o_1(h)) + o_2(f'(x)(h) + o_1(h))$ is a higher order term, under the assumptions of Proposition 30.1.1, Part b).

6. Let $I = [a, b]$ be some finite closed interval. Equip the space $E = \mathcal{C}^1(I, \mathbb{R})$ of all continuously differentiable real valued functions (one-sided derivatives at the end points of the interval) with the norm

$$\|u\|_{I,1} = \sup\left\{|u^{(j)}(t)| : t \in I, \ j = 0, 1\right\}.$$

Under this norm $E = \mathcal{C}^1(I, \mathbb{R})$ is a Banach space. For a given continuously differentiable function $F : I \times \mathbb{R} \times \mathbb{R} \to \mathbb{R}$, define a function $f : E \to \mathbb{R}$ by

$$f(u) = \int_a^b F(t, u(t), u'(t))dt$$

and show that f is Fréchet differentiable on E. Show in particular

$$f'(u)(v) = \int_a^b [F_{,u}(t, u(t), u'(t)) - \frac{d}{dt}F_{,u'}(t, u(t), u'(t))]v(t)dt$$
$$+ F_{,u'}(t, u(t), u'(t))v(t)|_a^b.$$

for all $v \in E$. $F_{,u}$ denotes the derivative of F with respect to the second argument and similarly, $F_{,u'}$ denotes the partial derivative with respect to the third argument.

Now consider $M = \{u \in E : u(a) = c, \ u(b) = d\}$ for some given values $c, d \in \mathbb{R}$ and show that the derivative of the restriction of f to M is

$$f'(u)(v) = \int_a^b [F_{,u}(t, u(t), u'(t)) - \frac{d}{dt}F_{,u'}(t, u(t), u'(t))]v(t)dt \quad (30.21)$$

for all $v \in E$, $v(a) = 0 = v(b)$. Deduce the *Euler–Lagrange equation*

$$F_{,u}(t, u(t), u'(t)) - \frac{d}{dt}F_{,u'}(t, u(t), u'(t)) = 0 \quad (30.22)$$

Hints: Use the Taylor expansion with remainder for F and the arguments for the proof of Theorem 3.2.2.

7. Suppose that E, F are two real Banach spaces. Prove the existence of the natural isomorphism $\mathcal{L}(E, \mathcal{L}(E, F)) \cong \mathcal{B}(E \times E, F)$.

   Hints: For $h \in \mathcal{L}(E, \mathcal{L}(E, F))$ define $\hat{h} \in \mathcal{B}(E \times E, F)$ by $\hat{h}(e_1, e_2) = h(e_1)(e_2)$ for all $e_1.e_2 \in E$ and for $b \in \mathcal{B}(E \times E, F)$ let us define $\check{b} \in \mathcal{L}(E, \mathcal{L}(E, F))$ by $\check{b}(e_1)(e_2) = b(e_1, e_2)$ and then show that these mappings are inverse to each other. Write the definition of the norms of the spaces $\mathcal{L}(E, \mathcal{L}(E, F))$ and $\mathcal{B}(E \times E, F)$ explicitly and show that the mappings $h \mapsto \hat{h}$ and $b \mapsto \check{b}$ both have a norm ≤ 1.

8. Prove the existence of the natural isomorphism (30.9) for $n = 2, 3, 4, \ldots$.

9. Prove the chain rule for the Gâteaux derivative.

10. Complete the proof of Part d) of Lemma 30.4.2.

11. Let V be a function $\mathbb{R}^3 \to \mathbb{R}$ of class C^1. Find the Euler–Lagrange equation (30.22) explicitly for the functional $I(u) = \int_a^b (\frac{m}{2}(u'(t))^2 - V(u(t)))dt$ on differentiable functional $u : [a, b] \to \mathbb{R}^3$, $u(a) = x$, $u(b) = y$ for given points $x, y \in \mathbb{R}^3$.

12. Consider the function $g(s) = \frac{|1+s|^p - 1 - ps}{|s|^p}$, $s \in \mathbb{R} \setminus \{0\}$, and show $g(s) \to 1$ as $|s| \to \infty$, $g(s) \to 0$ as $s \to 0$. Conclude that there are constants $0 < c < C < \infty$ such that $c|s|^p \leq g(s)|s|^p \leq C|s|^p$ for all $s \in \mathbb{R}$.

31
Constrained Minimization Problems (Method of Lagrange Multipliers)

In the calculus of variations we have often to do with the following problem: Given a real valued function f on a nonempty open subset U of a real Banach space E, find the minimum (maximum) of f on all those points x in U which satisfy a certain restriction or constraint. A very important example of such a constraint is that the points have to belong to a level surface of some function g, i.e., have to satisfy $g(x) = c$ where the constant c distinguishes the various level surfaces of the function g. In elementary situations, and typically also in Lagrangian mechanics, one introduces a so-called *Lagrange multiplier* λ as a new variable and proceeds to minimize the function $f(\cdot)+\lambda(g(\cdot)-c)$ on the set U. In simple problems (typically finite dimensional) this strategy is successful. The problem is the existence of a Lagrange multiplier.

As numerous successful applications have shown the following setting is an appropriate framework for such constrained minimization problems:

> Let E, F be two real Banach spaces, $U \subseteq E$ an open nonempty subset, $g : U \to F$ a mapping of class C^1, $f : U \to \mathbb{R}$ a function of class C^1, and y_0 some point in F. The *optimization problem for the function f under the constraint $g(x) = y_0$* is the problem of finding extremal points of the function $f_{|M} : M \to \mathbb{R}$ where $M = [g = y_0]$ is the level surface of g through the point y_0.

In this chapter we present a comprehensive solution for the infinite dimensional case, mainly based on ideas of Ljusternik [Lju34]. A first section explains in a simple setting the geometrical interpretation of the existence of a Lagrange multiplier. As an important preparation for the main results the existence of tangent spaces to level surfaces of C^1-functions is shown in substantial generality. Finally the existence of a Lagrange multiplier is proven and some simple applications are discussed.

In the following chapter, after the necessary preparations, we will use the results on the existence of a Lagrange multiplier to solve eigenvalue problems, for linear and nonlinear partial differential operators.

31.1 Geometrical interpretation of constrained minimization

In order to develop some intuition about constrained minimization problems and the rôle of the Lagrange multiplier we consider such a problem first on a space of dimension two and discuss heuristically in geometrical terms how to obtain the solution. Let $U \subset \mathbb{R}^2$ be a nonempty open subset. Our goal is to determine the minimum of a continuous function $f : U \to \mathbb{R}$ under the constraint $g(x) = c$ where the constraint function $g : U \to \mathbb{R}$ is continuous. This means: Find $x_0 \in U$ satisfying $g(x_0) = c$ and $f(x_0) \leq f(x)$ for all $x \in U$ such that $g(x) = c$. In this generality the problem does not have a solution. If however both f and g are continuously differentiable on U, then the level surfaces of both functions have well-defined tangents, and then we expect a solution to exist, because of the following heuristic considerations.

Introduce the level surface

$$[g = c] = \{x \in U : g(x) = c\}$$

and similarly the family of level surfaces $[f = d]$, $d \in \mathbb{R}$, for the function f. If a level surface $[f = d]$ does not intersect the level surface $[g = c]$, then no point on this level surface of f satisfies the constraint and is thus not relevant for our problem. If for a certain value of d the level surfaces $[f = d]$ and $[g = c]$ intersect in exactly one point (at some finite angle), then for all values d' close to d the level surfaces $[g = c]$ and $[f = d']$ also intersect at exactly one point, and thus d is not the minimum of f under the constraint $g(x) = c$. Next consider a value of d for which the level surfaces $[g = c]$ and $[f = d]$ intersect in at least two distinct points (at finite angles). Again for all values d' sufficiently close to d the level surfaces $[f = d']$ and $[g = c]$ intersect in at least two distinct points and therefore d is not the minimum of f under the given constraint. Finally consider a value d_0 for which the level surfaces $[g = c]$ and $[f = d_0]$ 'touch' in exactly one point x_0, i.e., $[g = c] \cap [f = d_0] = \{x_0\}$ and the tangents to both level surfaces at this point coincide. In this situation small changes of the value of d lead to an intersection which is either empty or consists of at least two points, hence these values $d' \neq d_0$ do not produce a minimum under the constraint $g(x) = c$. We conclude that d_0 is the minimum value of f under the given constraint and that x_0 is the minimizing point. The following figure shows in a two dimensional problem three of the cases discussed above. Given the level surface $[g = c]$ of the constraint function g, three different level surfaces of the function f are considered.

Figure 31.1: Level surface $[g = c]$ and $[f = d_i]$, $i = 0, 1, 2$; $d_1 < d_0 < d_2$; $i = 1$ two points of intersection, $i = 0$ touching level surfaces; $i = 2$ no intersection.

Recall that the level surfaces $[g = c]$ and $[f = d]$ are level surfaces of smooth functions over an open set $U \subset \mathbb{R}^2$. Assume (or prove under appropriate assumptions with the help of the implicit function theorem) that in a neighborhood of the point $x_0 = (x_1^0, x_2^0)$ these level surfaces have the explicit representation $x_2 = y(x_1)$, respectively $x_2 = \xi(x_1)$. Under these assumptions it is shown in the Exercises that the tangent to these touching level surfaces coincide if, and only if,

$$(Df)(x_0) = \lambda(Dg)(x_0) \tag{31.1}$$

for some $\lambda \in \mathbb{R} \cong \mathcal{L}(\mathbb{R}, \mathbb{R})$.

31.2 Tangent spaces of level surfaces

In our setting a constraint minimization problem is a problem of analysis on level surfaces of C^1 mappings. It requires that we can do differential calculus on these surfaces which in turn relies on the condition that these level surfaces are differential manifolds. The following approach does not assume this but works under the hypothesis that one has, at the points of interest on these level surfaces, the essential element of a differential manifold, namely a proper tangent space.

Recall that in infinite dimensional Banach spaces E a closed subspace K does not always have a topological complement, i.e., a closed subspace L such that E is the direct sum of these two subspaces (see for instance [RR73]). Thus in our fundamental result on the existence of a proper tangent space this property is assumed but later we will show when and how it holds.

Theorem 31.2.1 (Existence of a tangent space) *Let E, F be real Banach spaces, $U \subseteq E$ a nonempty open subset, and $g : U \to F$ a mapping of class C^1. Suppose that x_0 is a point of the level surface $[g = y_0]$ of the mapping g. If x_0 is a regular point of g at which the null-space $N(g'(x_0))$ of the derivative of g has a topological complement in E, then the set*

$$T_{x_0}[g = y_0] = \{x \in E : \exists u \in N(g'(x_0)), \ x = x_0 + u\} = x_0 + N(g'(x_0)) \tag{31.2}$$

is a proper tangent space of the level surface $[g = y_0]$ at the point x_0, i.e., there is a homeomorphism χ of a neighborhood U' of x_0 in $T_{x_0}[g = y_0]$ onto a neighborhood V of x_0 in $[g = y_0]$ with the following properties:

a) $\chi(x_0 + u) = x_0 + u + \varphi(u)$ for all $x_0 + u \in U'$;

b) φ is continuous and of higher than linear order in u, $\varphi(u) = o(h)$.

Proof. Since x_0 is a regular point of g, the derivative $g'(x_0)$ is a surjective continuous linear mapping from E onto F. By assumption the null-space $K = N(g'(x_0))$ of the mapping has a topological complement L in E so that the Banach space E is the direct sum of these two closed subspaces, $E = K + L$. It follows (see [RR73]) that there are continuous linear mappings p and q of E onto K and L, respectively, which have the following properties: $K = \operatorname{ran} p = N(q)$, $L = N(p) = \operatorname{ran} q$, $p^2 = p$, $q^2 = q$, $p + q = id$.

Since U is open there is $r > 0$ such that the open ball B_r in E with center 0 and radius r satisfies $x_0 + B_r + B_r \subset U$. Now define a mapping $\psi : K \cap B_r \times L \cap B_r \to F$ by

$$\psi(u, v) = g(x_0 + u + v) \qquad \forall u \in K \cap B_r, \quad \forall v \in L \cap B_r. \tag{31.3}$$

By the choice of the radius r this map is well defined. The chain rule implies that it has the following properties: $\psi(0, 0) = g(x_0) = y_0$, ψ is continuously differentiable and

$$\psi_{,u}(0, 0) = g'(x_0)_{|K} = 0 \in \mathcal{L}(K, F), \quad \psi_{,v}(0, 0) = g'(x_0)_{|L} \in \mathcal{L}(L, F).$$

On the complement L of its null-space the surjective mapping $g'(x_0) : E \to F$ is bijective, thus $\psi_{,v}(0, 0)$ is a bijective continuous linear mapping of the Banach space L onto the Banach space F. The inverse mapping theorem (see Appendix 34.5) implies that the inverse $\psi_{,v}(0, 0)^{-1} : F \to L$ is a continuous linear operator too. Thus all hypotheses of the implicit function theorem (see, for example, [Die69]) are satisfied for the problem

$$\psi(u, v) = y_0.$$

This theorem implies that there is $0 < \delta < r$ and a unique function $\varphi : K \cap B_\delta \to L$ which is continuously differentiable such that

$$y_0 = \psi(u, \varphi(u)) \quad \forall u \in K \cap B_\delta \quad \text{and} \quad \varphi(0) = 0.$$

Since in general $\varphi'(0) = -\psi_{,v}(0, 0)^{-1} \psi_{,u}(0, 0)$ we have here $\varphi'(0) = 0$ and thus $\varphi(u) = o(u)$.

Define a mapping $\chi : x_0 + K \cap B_\delta \to M$ by $\chi(x_0 + u) = x_0 + u + \varphi(u)$. Clearly χ is continuous. By construction, $y_0 = \psi(u, \varphi(u)) = g(x_0 + u + \varphi(u))$, hence χ maps into $M = [g = y_0]$. By construction, u and $\varphi(u)$ belong to complementary subspaces of E, therefore χ is injective and thus invertible on

$$V = \{x_0 + u + \varphi(u) : u \in K \cap B_\delta\} \subset M.$$

Its inverse is $\chi^{-1}(x_0 + u + \varphi(u)) = x_0 + u$. Since $\operatorname{ran} p = K$ and $N(p) = L$ the inverse can be represented as

$$\chi^{-1}(x_0 + u + \varphi(u)) = x_0 + p(u + \varphi(u))$$

and this shows that χ^{-1} is continuous too. Therefore χ is a homeomorphism from $U' = x_0 + K \cap B_\delta$ onto $V \subset M$. This concludes the proof. □

Apart from the natural assumption about the regularity of the point x_0 this theorem uses the technical assumption that the nullspace $K = N(g'(x_0))$ of $g'(x_0) \in \mathcal{L}(E, F)$ has a topological complement in E. We show now that this assumption is quite adequate for the general setting by proving that it is automatically satisfied for three large and frequent classes of special cases.

Proposition 31.2.2 *Let E, F be real Banach spaces and $A : E \to F$ a surjective continuous linear operator. The nullspace $K = N(A)$ has a topological complement in E, in the following three cases:*

a) E is a Hilbert space;

b) F is a finite dimensional Banach space;

c) $N(A)$ is finite dimensional, for instance $A : E \to F$ is a Fredholm operator (i.e., an operator with finite dimensional null-space and closed range of finite codimension).

Proof. If K is a closed subspace of the Hilbert space E, the projection theorem guarantees existence of the topological complement $L = K^\perp$ and thus proves Part a).

If F is a finite dimensional Banach space, there exist linearly independent vectors $e_1, \ldots, e_m \in E$ such that $\{f_1 = Ae_1, \ldots, f_m = Ae_m\}$ is a basis of F. The vectors $e_1, \ldots, e_m$ generate a linear subspace V of E of dimension m and it follows that A now is represented by $Ax = \sum_{j=1}^{m} a_j(x) f_j$ with continuous linear functionals $a_j : E \to \mathbb{R}$. Define $px = \sum_{j=1}^{m} a_j(x) e_j$ and $qx = x - px$. One proves easily that $p^2 = p, q^2 = q, p + q = id, V = pE$ and that both maps are continuous. Thus $V = pE$ is the topological complement of $N(A) = qE$. This proves b).

Suppose $\{e_1, \ldots, e_m\}$ is a basis of $N(A)$. There are continuous linear functionals a_j on E such that $a_i(e_j) = \delta_{ij}$ for $i, j = 1, \ldots, m$. (Use the Hahn–Banach theorem). As above define $px = \sum_{j=1}^{m} a_j(x) e_j$ and $qx = x - px$ for all $x \in E$. Now we conclude as in Part b). (See the Exercises) □

Corollary 31.2.1 *Suppose that E, F are real Banach spaces, $U \subset E$ a nonempty open set and $g : U \to F$ a map of class C^1. In each of the three cases mentioned in Proposition 31.2.2 for $A = g'(x_0)$ the tangent space of the level surface $[g = y_0]$ at every regular point $x_0 \in [g = y_0]$ of g is given by equation (31.2).*

Proof. Proposition 31.2.2 ensures the hypotheses of Theorem 31.2.1. □

31.3 Existence of Lagrange multipliers

The results on the existence of the tangent spaces of level surfaces allow us to translate the heuristic considerations on the existence of a Lagrange multiplier into precise statements. The result which we present now is primarily useful for the explicit calculation of the extremal points once their existence has been established, say as a consequence of the direct methods discussed earlier.

Theorem 31.3.1 (Existence of Lagrange multipliers) *Let E, F be real Banach spaces, $U \subset E$ open and nonempty, $g : U \to F$ and $f : U \to \mathbb{R}$ of class C^1. Suppose that f has a local extremum at the point $x_0 \in U$ subject to the constraint $g(x) = y_0 = g(x_0)$. If x_0 is a regular point of the map g and if the null-space $K = N(g'(x_0))$ of $g'(x_0)$ has a topological complement L in E, then there exists a continuous linear functional $\ell : F \to \mathbb{R}$ such that x_0 is a critical point of the function $F = f - \ell \circ g : U \to \mathbb{R}$, that is*

$$f'(x_0) = \ell \circ g'(x_0). \tag{31.4}$$

Proof. The restriction H of $g'(x_0)$ to the topological complement L of its kernel K is a continuous injective linear map from the Banach space L onto the Banach space F since x_0 is a regular point of g. The inverse mapping theorem (see Appendix) implies that H has an inverse H^{-1} which is a continuous linear operator $F \to L$.

According to Theorem 31.2.1 the level surface $[g = y_0]$ has a proper tangent space at x_0. Thus the points x of this level surface, in a neighborhood V of x_0, are given by $x = x_0 + u + \varphi(u)$, $u \in K \cap B_\delta$ where $\delta > 0$ is chosen as in the proof of Theorem 31.2.1. Suppose that f has a local minimum at x_0 (otherwise consider $-f$). Then there is an $r \in (0, \delta)$ such that $f(x_0) \leq f(x_0 + u + \varphi(u))$ for all $u \in K \cap B_r$, hence by Taylor's theorem

$$0 \leq f'(x_0)(u) + f'(x_0)(\varphi(u)) + o(u + \varphi(u)) \quad \forall u \in K \cap B_r.$$

Since we know that $\varphi(u) = o(u)$, this implies $f'(x_0)(u) = 0$ for all $u \in K \cap B_r$. But $u \in K \cap B_r$ is absorbing in K, therefore $f'(x_0)(u) = 0$ for all $u \in K$, i.e.,

$$K = N(g'(x_0)) \subseteq N(f'(x_0)). \tag{31.5}$$

By assumption, E is the direct sum of the closed subspaces $K, L, E = K + L$. Denote the canonical projections onto K and L by p respectively q. If $x_1, x_2 \in E$ satisfy $q(x_1) = q(x_2)$, then $x_1 - x_2 \in K$ and thus equation (31.5) implies $f'(x_0)(x_1) = f'(x_0)(x_2)$. Therefore a continuous linear functional $\hat{f}'(x_0) : L \to \mathbb{R}$ is well defined by $\hat{f}'(x_0)(qx) = f'(x_0)(x)$ for all $x \in E$. This functional is used to define

$$\ell = \hat{f}'(x_0) \circ H^{-1} : F \to \mathbb{R}$$

as a continuous linear functional on the Banach space F which satisfies equation (31.4), since for every $x \in E$

$$\ell \circ g'(x_0)(x) = \ell \circ g'(x_0)(qx) = \ell \circ H(qx) = \hat{f}'(x_0)(qx) = f'(x_0)(x).$$

We conclude that x_0 is a critical point of the function $F = f - \ell \circ g$, by using the chain rule. □

To illustrate some of the strengths of this theorem we consider a simple example. Suppose E is a real Hilbert space with inner product $\langle \cdot, \cdot \rangle$ and A a bounded self-adjoint operator on E. The problem is to minimize the function $f(x) = \langle x, Ax \rangle$ under the constraint $g(x) = \langle x, x \rangle = 1$. Obviously both functions are of class C^1. Their derivatives are given by $f'(x)(u) = 2\langle Ax, u \rangle$, respectively by $g'(x) = 2\langle x, u \rangle$ for all $u \in E$. It follows that all points of the level surface $[g = 1]$ are regular points of g. Corollary 31.2.1 implies that Theorem 31.3.1 can be used to infer the existence of a Lagrange multiplier $\lambda \in \mathbb{R}$ if x_0 is a minimizing point of f under the constraint $g(x) = 1$: $f'(x_0) = \lambda g'(x_0)$ or $Ax_0 = \lambda x_0$, i.e., the Lagrange multiplier λ is an eigenvalue of the operator A and x_0 is the corresponding normalized eigenvector. This simple example suggests a strategy to determine eigenvalues of operators. Later we will explain this powerful strategy in some detail, not only for linear operators.

In the case of finite dimensional Banach spaces we know that the technical assumptions of Theorem 31.3.1 are naturally satisfied. In this theorem assume that $E = \mathbb{R}^n$ and $F = \mathbb{R}^m$. Every continuously linear functional ℓ on $\mathbb{R}^m$ is characterized uniquely by some m-tuple $(\lambda_1, \ldots, \lambda_m)$ of real numbers. Explicitly Theorem 31.3.1 takes now the form

Corollary 31.3.1 *Suppose that $U \subset \mathbb{R}^n$ is open and nonempty, and consider two mappings $f : U \to \mathbb{R}$ and $g : U \to \mathbb{R}^m$ of class C^1. Furthermore assume that the function f attains a local extremum at a regular point $x_0 \in U$ of the mapping g (i.e., the Jacobi matrix $g'(x_0)$ has maximal rank m) under the constraint $g(x) = y_0 \in \mathbb{R}^m$. Then there exist real numbers $\lambda_1, \ldots, \lambda_m$ such that*

$$\frac{\partial f}{\partial x_i}(x_0) = \sum_{j=1}^m \lambda_j \frac{\partial g_j}{\partial x_i}(x_0), \quad i = 1, \ldots, n. \tag{31.6}$$

Note that equation (31.6) of Corollary 31.3.1 and the equation $g(x_0) = y_0 \in \mathbb{R}^m$ give us exactly $n+m$ equations to determine the $n+m$ unknowns $(\lambda, x_0) \in \mathbb{R}^m \times U$.

Theorem 31.3.1 can also be used to derive necessary and sufficient conditions for extremal points under constraints. For more details we have to refer to chapter 4 of the book [BB92].

31.3.1 Comments on Dido's problem

According to the brief discussion in the introduction to Part C Dido's original problem is a paradigmatic example of constrained minimization. Though intuitively the solution is clear (a circle where the radius is determined by the given length) a rigorous proof is not very simple even with the help of the abstract results which we have developed in this section. Naturally Dido's problem and its solution have been discussed much in the history of the calculus of variations (see [Gol80]). Weierstrass solved this problem in his lectures in 1872 and 1879. There is also an elegant geometrical solution based on symmetry considerations due to Steiner.

In the Exercises we invite the reader to find the solution by two different methods. The first method suggests parametrizing the curve we are looking for by its arc length and using Parseval's relation in the Hilbert space $\mathcal{H} = L^2([0, 2\pi])$. This means that we assume that this curve is given in parametric form by a parametrization $(x(t), y(t)) \in \mathbb{R}^2$, $0 \le t \le 2\pi$ where x, y are differentiable functions satisfying $\dot{x}(t)^2 + \dot{y}(t)^2 = 1$ for all $t \in [0, 2\pi]$. With this normalization and parametrization the total length of the curve is $L = \int_0^{2\pi} \sqrt{\dot{x}(t)^2 + \dot{y}(t)^2} dt = 2\pi$ and the area enclosed by this curve is

$$A = \int_0^{2\pi} x(t)\dot{y}(t)dt.$$

Proposition 31.3.2 *For all parametrizations of the form described above one has $A \le \pi$. Furthermore, $A = \pi$ if, and only if, the curve is a circle of radius 1.*

Proof. See the Exercises. □

The second approach uses the Lagrange multiplier method as explained above. Suppose that the curve is to have the total length $2L_0$. Choose a parameter a such that $2a < L_0$. In a suitable coordinate system the curve we are looking for is given as $y = u(x)$, $-a \le x \le a$, and $u(x) \ge 0$, $u(\pm a) = 0$ with a function u of class C^1. Its length is $\int_{-a}^{a} \sqrt{1 + u'(x)^2} dx = L(u)$ and the area enclosed by the x-axis and this curve is $A(u) = \int_{-a}^{a} u(x) dx$. The problem then is to determine u such that $A(u)$ is maximal under the constraint $L(u) = L_0$.

Proposition 31.3.3 *For the constrained minimization problem for $A(u)$ under the constraint $L(u) = L_0$ there is a Lagrange multiplier λ satisfying $\frac{s}{\sqrt{1+s^2}} = \frac{a}{\lambda}$ for some $s \in \mathbb{R}$ and a solution $u(x) = \lambda[\sqrt{1 - (\frac{x}{\lambda})^2} - \sqrt{1 - (\frac{a}{\lambda})^2}]$, $-a \le x \le a$.*

One has $L_0 = 2\lambda\theta(a)$ with $\theta(a) = \arcsin\frac{a}{\lambda} \in [0, \frac{\pi}{2}]$. For this curve the area is

$$A(u) = \lambda^2 \theta(a) - a\sqrt{\lambda^2 - a^2}.$$

Proof. See the Exercises. □

Since $L_0 = 2\lambda\theta(a)$ the Lagrange multiplier λ is a function of a and hence one can consider $A(u)$ as a function of a. Now it is not difficult to determine a so that the enclosed area $A(u)$ is maximal. For $a = \lambda = \frac{L_0}{\pi}$ this area is maximal and is given by $A(u) = a^2\pi/2$. This is the area enclosed by a half-circle of radius $a = \frac{L_0}{\pi}$.

Remark 31.3.1 *There is an interesting variation of Dido's problem which has found important applications in modern probability theory (see [LT91]) and which we mention briefly. Let $A \subset \mathbb{R}^n$ be a bounded domain with a sufficiently smooth boundary and for $t > 0$ consider the set*

$$A_t = \{x \in \mathbb{R}^n \setminus A : \|x - y\| \leq t, \ \forall y \in A\}.$$

Now minimize the volume $|A_t|$ of the set A_t under the constraint that the volume $|A|$ of A is fixed. The answer is known: This minimum is attained when A is a ball in $\mathbb{R}^n$. This is of particular interest in the case of very high dimensions $n \to \infty$ since then it is known that practically the volume of $A_t \cup A$ is equal to the volume of A_t. For the proof of this result we refer to the book [BZ88] and the article [Oss78].

31.4 Exercises

1. Let $U \subset \mathbb{R}^2$ be open and nonempty. Suppose $f, g \in \mathcal{C}^1(U, \mathbb{R})$ have level surfaces $[g = c]$ and $[f = d]$ which touch in a point $x_0 \in U$ in which the functions f, g have nonvanishing derivatives with respect to the second argument. Prove Equation 31.1.

2. Prove in detail: A finite dimensional subspace V of a Banach space E has a topological complement.

3. Prove Corollary 31.3.1.

4. Prove Proposition 31.3.2.

    Hints: Use the Fourier expansion for x, y:

    $$x(t) = \frac{a_0}{2} + \sum_{k=1}^{\infty}(a_k \cos kt + b_k \sin kt)$$

    $$y(t) = \frac{\alpha_0}{2} + \sum_{k=1}^{\infty}(\alpha_k \cos kt + \beta_k \sin kt).$$

Calculate $\dot{x}(t), \dot{y}(t)$ and calculate $\int_0^{2\pi}[\dot{x}(t)^2 + \dot{y}(t)^2]dt$ as

$$\langle \dot{x}(t), \dot{x}(t)\rangle_2 + \langle \dot{y}(t), \dot{y}(t)\rangle_2$$

using $\langle \cos kt, \sin jt \rangle_2 = 0$ and $\langle \cos kt, \cos kt \rangle_2 = \langle \sin kt, \sin kt \rangle_2 = \pi$. Similarly one can calculate $A = \langle x, \dot{y} \rangle_2 = \pi \sum_{k=1}^{\infty} k(a_k\beta_k - b_k\alpha_k)$. This gives

$$2\pi - 2A = \pi \sum_{k=1}^{\infty}(k^2 - k)[a_k^2 + b_k^2 + \alpha_k^2 + \beta_k^2] + \pi \sum_{k=1}^{\infty} k[(a_k - \beta_k)^2 + (\alpha_k + b_k)^2].$$

Now it is straightforward to conclude.

5. Prove Proposition 31.3.3.

   Hints: 1. Calculate the Fréchet derivative of the constraint functional $L(u)$ and show that all points of a level surface $[L = L_0]$ are regular points of the mapping L, for $2a < L_0$. 2. Prove that $|u(x)| \leq L(u)$ for all $x \in [-a, a]$ and hence $A(u) \leq 2aL(u) = 2aL_0$. 3. Prove that $A(u)$ is (upper semi-)continuous for the weak topology on $E = H_0^1(-a, a)$. 4. Conclude that a maximizing element $u \in E$ and a Lagrange multiplier λ exist. 5. Solve the differential equation $A'(u) = \lambda L'(u)$ under the boundary condition $u(-a) = u(a) = 0$. 6. Calculate $L(u)$ for this solution and equate the result to L_0. 7. Calculate the area $A(u)$ for this solution.

32
Boundary and Eigenvalue Problems

One of the first areas in which variational concepts and methods have been applied were linear boundary and eigenvalue problems. They can typically be solved in concrete Hilbert spaces of functions. These and related problems will be the topic of this chapter. Before we turn to these concrete problems we discuss several abstract minimization problems in Hilbert spaces, some of them have already been mentioned in Part II on Hilbert spaces.

In order to prepare for the solution of linear boundary and eigenvalue problems the connection of linear partial differential operators and quadratic forms is established in Section 2. Since on a general level minimization of quadratic forms has been well discussed, it is fairly easy then to solve these concrete boundary and eigenvalue problems.

32.1 Minimization in Hilbert spaces

According to our outline of the general strategy of the direct methods in the calculus of variations, Hilbert spaces are well suited for problems of minimization since they are spaces in which bounded sets are relatively compact for the weak topology. Their additional geometric structure is often very helpful too. This advantage will be evident from the following

Theorem 32.1.1 (Projection theorem for convex sets) *Suppose that K is a convex closed subset of a Hilbert space $\mathcal{H}$. Then, for any $x \in \mathcal{H}$, there is a unique*

$u \in K$ which satisfies

$$\|x - u\| = \inf_{v \in K} \|x - v\| = d(x, K). \tag{32.1}$$

The element $u \in K$ is called the **projection of** x **onto** K: $u = \operatorname{proj}_K x$. It is characterized by the inequality

$$\langle x - u, v - u \rangle \leq 0 \qquad \forall v \in K.$$

The mapping $\operatorname{proj}_K : \mathcal{H} \to K$ defined in this way is continuous, and one has

$$\|\operatorname{proj}_K x - \operatorname{proj}_K y\| \leq \|x - y\| \qquad \forall x, y \in \mathcal{H}.$$

Proof. For fixed $x \in \mathcal{H}$ consider the function $\psi_x : K \to \mathbb{R}$ defined by $\psi_x(z) = \|x - z\|$ for all $z \in K$. It is certainly continuous and as a small calculation shows (see Exercises), strictly convex. Lemma 29.2.2 implies that ψ_x is weakly lower semi-continuous on the closed convex and thus weakly closed convex set. Therefore, in the case that K is bounded, Theorem 29.2.1 applies. If K is not bounded, then certainly ψ_x is coercive, i.e., $\psi_x(z) \to \infty$ as $\|z\| \to \infty$, and thus Theorem 29.2.2 applies. In both cases we conclude that there is a point $u_x \in K$ which minimizes ψ_x on K,

$$\psi_x(u_x) = \|x - u_x\| = \inf_{z \in K} \|x - z\|.$$

The minimizing point is unique since ψ_x is strictly convex (Theorem 29.1.2). These arguments apply to any $x \in \mathcal{H}$. Because $x \mapsto u_x$ is one-to-one we get a well-defined map $p_K : \mathcal{H} \to K$ by

$$p_K(x) = \begin{cases} u_x, & x \in \mathcal{H} \setminus K, \\ x, & x \in K. \end{cases}$$

In order to prove the characteristic inequality for $u = u_x \in K$ take any $z \in K$. Since K is convex $tz + (1-t)u \in K$ and we know $\|x - u\| \leq \|x - tz - (1-t)u\|$ or

$$0 \leq \|x - u + t(u - z)\|^2 - \|x - u\|^2 = 2t\langle x - u, u - z\rangle + t^2\|u - z\|^2$$

for all $0 < t < 1$. A standard argument implies $\langle x - u, u - z \rangle \geq 0$.

Conversely assume that $u \in K$ is a point for which this inequality holds for all $z \in K$. Then on the basis of this inequality we estimate as follows:

$$\begin{aligned}\|x - u\|^2 &= \langle x - u, z - u\rangle = \langle x - u, x - z\rangle + \langle x - u, x - u - (x - z)\rangle \\ &= \langle x - u, x - z\rangle + \langle x - u, z - u\rangle \leq \langle x - u, x - z\rangle \\ &= \langle x - z, x - z\rangle + \langle z - u, x - z\rangle \\ &\leq \|x - z\|^2 + \langle z - u, x - u\rangle - \langle z - u, z - u\rangle \leq \|x - z\|^2,\end{aligned}$$

hence $\|x - u\| \leq \|x - z\|$ for all $z \in K$ and thus $\|x - u\| = \inf_{z \in K} \|x - z\|$. We conclude that the minimizing element $u_x = u$ is indeed characterized by the above variational inequality.

Finally we prove Lipschitz continuity of the mapping p_K. Given $x, y \in \mathcal{H}$ denote $u = p_K(x)$ and $v = p_K(y)$. Now apply the variational inequality first for u and $z = v \in K$ and then for v and $z = u \in K$. This gives the two inequalities

$$\langle x - u, v - u \rangle \leq 0, \qquad \langle y - v, u - v \rangle \leq 0$$

which we add to get

$$\langle x - y + v - u, v - u \rangle \leq 0$$

or

$$\|v - u\|^2 \leq \langle y - x, v - u \rangle \leq \|y - x\| \|v - u\|,$$

and the estimate $\|v - u\| \leq \|y - x\|$ follows, which is just the continuity estimate and we conclude. □

In Part II the spectral theory for compact operators (Riesz–Schauder theory, Theorem 25.1.1) has been developed by using mainly Hilbert space intrinsic arguments. We discuss this proof now as an application of the direct methods.

Let $\mathcal{H}$ be a separable Hilbert space and $A \neq 0$ a self-adjoint compact operator on $\mathcal{H}$. Denote by $B_1 = \{x \in \mathcal{H} : \|x\| \leq 1\}$ the closed unit ball of $\mathcal{H}$ and by S_1 the unit sphere of this space. Recall that the norm of the operator can be expressed as

$$\|A\|_{\mathfrak{B}(\mathcal{H})} = \sup_{u \in B_1} \|Au\| = \sup_{u \in S_1} |\langle u, Au \rangle|.$$

For simplicity of notation we write $\|A\|_{\mathfrak{B}(\mathcal{H})}$ simply as $\|A\|$.

Thus the calculation of the norm can be regarded as a problem of finding a maximum of the function $u \mapsto \|Au\|$ on the closed unit ball B_1 or of the function $u \mapsto |\langle u, Au \rangle|$ on the unit sphere S_1. Since A is compact, it maps weakly convergent sequences into norm convergent sequences and thus both functions are weakly continuous. Since the closed unit ball B_1 is weakly compact we can apply Theorem 29.2.1 and thus get a point $e_1 \in B_1$ such that

$$\|A\| = \|Ae_1\| = \sup_{u \in B_1} \|Au\|.$$

Since $A \neq 0$ we know $e_1 \neq 0$ and thus $\|e_1\| = 1$ as a simple scaling argument shows.

Consider the function $f(u) = \langle u, Au \rangle$ on $\mathcal{H}$. It is Fréchet differentiable with derivative $f'(u)(x) = 2\langle Au, x \rangle$ for all $x \in \mathcal{H}$. As we have shown above this function has a maximum on S_1 given by

$$\sup_{u \in S_1} f(u) = \pm \|A\| = \pm f(e_1).$$

The unit sphere S_1 is the level surface $[g = 1]$ of the constraint function $g(x) = \langle x, x \rangle$ on $\mathcal{H}$ which is Fréchet differentiable too, and its derivative is $g'(u)(x) = 2\langle u, x \rangle$ for all $x \in \mathcal{H}$. Therefore all points $u \in S_1$ are regular points of g. The results on the existence of a Lagrange multiplier (Theorem 31.3.1 and Corollary 31.2.1) apply, i.e., there is an $\lambda_1 \in \mathbb{R}$ such that $f'(e_1) = \lambda_1 g'(e_1)$ or $Ae_1 = \lambda_1 e_1$. It follows that $|\lambda_1| = \|A\|$.

Then the proof is completed as it has been shown in Part B, Section 25.1. This ends our remarks on the proof of the spectral theorem for compact operators as an application of variational methods.

Theorem 25.1.1 establishes the existence and some of the properties of eigenvalues of a compact self-adjoint operator. The above comments on the proof hint at a method for calculating these eigenvalues. And indeed this method has been worked out in full detail and leads to the *classical minimax principle of Courant–Weyl–Fischer–Poincaré–Rayleigh–Ritz*.

Theorem 32.1.2 (Minimax Principle) *Let $\mathcal{H}$ be a real separable Hilbert space and $A \geq 0$ a self-adjoint operator on $\mathcal{H}$ with spectrum $\sigma(A) = \{\lambda_m : m \in \mathbb{N}\}$ ordered according to size, $\lambda_m \leq \lambda_{m+1}$. For $m = 1, 2, \ldots$ denote by $\mathcal{E}_m$ the family* of all m-dimensional subspaces E_m of $\mathcal{H}$. Then the eigenvalue λ_m can be calculated as

$$\lambda_m = \min_{E_m \in \mathcal{E}_m} \max_{v \in E_m} \frac{\langle v, Av \rangle}{\langle v, v \rangle}. \tag{32.2}$$

Proof. The proof is obtained by determining the lower bound for the values of the Rayleigh quotient $R(v) = \frac{\langle v, Av \rangle}{\langle v, v \rangle}$. In order to do this we expand every $v \in \mathcal{H}$ in terms of eigenvectors e_j of A. This gives $v = \sum_{i=1}^{\infty} a_i e_i$ and $\langle v, v \rangle = \sum_{i=1}^{\infty} a_i^2$. In this form the Rayleigh quotient reads

$$R(v) = \frac{\sum_{i=1}^{\infty} \lambda_i a_i^2}{\sum_{i=1}^{\infty} a_i^2}.$$

Denote by V_m the linear subspace generated by the first m eigenvectors of A. It follows that

$$\max_{v \in V_m} R(v) = \max_{(a_1,\ldots,a_m) \in \mathbb{R}^m} \frac{\sum_{i=1}^{m} \lambda_i a_i^2}{\sum_{i=1}^{m} a_i^2} = \lambda_m = R(e_m),$$

and thus we are left with showing $\max_{v \in E_m} R(v) \geq \lambda_m$ for every other subspace $E_m \in \mathcal{E}_m$. Let $E_m \neq V_m$ be such a subspace; then $E_m \cap V_m^\perp \neq \{0\}$ and therefore

$$\max_{v \in E_m} R(v) \geq \max_{v \in E_m \cap V_m^\perp} R(v).$$

Every $v \in E_m \cap V_m^\perp$ is of the form $v = \sum_{i \geq m+1} a_i e_i$ and for such vectors we have

$$R(v) = \frac{\sum_{i \geq m+1} \lambda_i a_i^2}{\sum_{i \geq m+1} a_i^2} \geq \lambda_{m+1} \geq \lambda_m.$$

This then completes the proof. $\square$

Theorem 32.1.2 implies for the smallest eigenvalue of the operator A the simple formula

$$\lambda_1 = \min_{v \in E,\, v \neq 0} \frac{\langle v, Av \rangle}{\langle v, v \rangle}. \tag{32.3}$$

32.2 The Dirichlet–Laplace operator and other elliptic differential operators

The goal of this section is to illustrate the application of the general strategy and the results developed thus far. This is done by solving several relatively simple linear boundary and eigenvalue problems. The typical example is the Laplace operator with Dirichlet boundary conditions on a bounded domain Ω. Naturally, for these concrete problems we have to use concrete function spaces, and we need to know a number of basic facts about them. In this brief introduction we have to refer the reader to the literature for the proof of these facts. We recommend the books [LL01, JLJ98, BB92].

For a bounded domain $\Omega \subset \mathbb{R}^n$ with smooth boundary $\partial\Omega$ consider the real Hilbert space $L^2(\Omega)$ with inner product $\langle \cdot, \cdot \rangle_2$. Then define a space $H^1(\Omega)$ as

$$H^1(\Omega) = \left\{ u \in L^2(\Omega) : \partial_j u \in L^2(\Omega),\ j = 1, \ldots, n \right\}. \tag{32.4}$$

32.2 The Dirichlet–Laplace operator and other elliptic differential operators

Here naturally the partial derivatives $\partial_j u$ are understood in the weak (distributional) sense. One shows that $H^1(\Omega)$ is a Hilbert space with the inner product

$$\langle u, v \rangle = \langle u, v \rangle_2 + \langle Du, Dv \rangle_2 \qquad \forall u, v \in H^1(\Omega) \tag{32.5}$$

where $Du = (\partial_1 u, \ldots, \partial_n u)$ and where in the second term the natural inner product of $L^2(\Omega)^{\times n}$ is used. This space is the *Sobolev* space $W^{1,2}(\Omega)$. Next define a subspace of this space:

$$H_0^1(\Omega) = \text{closure of } \mathcal{D}(\Omega) \text{ in } H^1(\Omega). \tag{32.6}$$

Intuitively, $H_0^1(\Omega)$ is the subspace of those $u \in H^1(\Omega)$ whose restriction to the boundary $\partial \Omega$ vanishes, $u_{|\partial\Omega} = 0$.

The Sobolev space $H^1(\Omega)$ is by definition contained in the Hilbert space $L^2(\Omega)$, however for us of much greater importance are the following compact embeddings for $2 \leq n$,

$$H^1(\Omega) \hookrightarrow L^q(\Omega), \quad 1 \leq q < 2^* = \frac{2n}{n-2}, \quad 2 < n \tag{32.7}$$

and

$$H^1(\Omega) \hookrightarrow L^q(\Omega), \quad 1 \leq q < \infty, \quad 2 = n. \tag{32.8}$$

This means that every weakly convergent sequence in $H^1(\Omega)$ converges strongly in $L^q(\Omega)$. In addition we are going to use the important *Sobolev inequality*

$$\|u\|_q \leq S \|Du\|_p = S \left(\sum_{j=1}^n \|\partial_j u\|_p^p \right)^{1/p} \qquad \forall u \in H^1(\Omega), \tag{32.9}$$

where S is the Sobolev constant depending on q, n and where q is in the range indicated in (32.7) respectively (32.8).

Now we are in the position to show that the famous Dirichlet problem has a solution.

Theorem 32.2.1 (Dirichlet problem) *Let $\Omega \subset \mathbb{R}^n$ be a bounded open set with smooth boundary and $v_0 \in H^1(\Omega)$ some given element. Then the Dirichlet integral*

$$f(v) = \int_\Omega |Dv(x)|^2 dx = \int_\Omega \sum_{j=1}^n |\partial_j v(x)|^2 dx \tag{32.10}$$

is minimized on $M = v_0 + H_0^1(\Omega)$ by an element $v \in M$ satisfying

$$\Delta v = 0 \text{ in } \Omega \quad \text{and} \quad v_{|\partial\Omega} = v_{0|\partial\Omega}. \tag{32.11}$$

Proof. Observe that $f(u) = Q(u, u)$ with the quadratic functional $Q(u, v) = \langle Du, Dv \rangle_2$. This quadratic form satisfies, because of inequality (32.9), the estimate

$$c\|u\|^2 \leq Q(u, u) \leq \|u\|^2$$

for some $c > 0$. It follows that Q is a strictly positive continuous quadratic form on $H^1(\Omega)$ and thus f is a strictly convex continuous function on this space (see the proof of Theorem 29.3.1). We conclude, by Lemma 29.2.2 or Theorem 29.3.1, that f is weakly lower semi-continuous on $H^1(\Omega)$.

As a Hilbert space, $H_0^1(\Omega)$ is weakly complete and thus the set $M = v_0 + H_0^1(\Omega)$ is weakly closed. Therefore Theorem 29.2.2 applies and we conclude that there is a minimizing element v for the functional f on M.

Since the minimizing element $v \in M$ satisfies $f(v) = f(v_0 + u) \le f(v_0 + w)$ for all $w \in H_0^1(\Omega)$ we deduce as earlier that $f'(v)(w) = 0$ for all $w \in H_0^1(\Omega)$ and thus

$$0 = f'(v)(w) = \int_\Omega Dv(x) \cdot Dw(x)dx \qquad \forall w \in \mathcal{D}(\Omega).$$

Recalling the definition of differentiation in the sense of distributions, this means $-\Delta v = 0$ in the sense of $\mathcal{D}'(\Omega)$. Now the Lemma of Weyl (see [JLJ98, BB92]) implies that $-\Delta v = 0$ also holds in the classical sense, i.e., as an identity for functions of class $\mathcal{C}^2$.

Because for $u \in H_0^1(\Omega)$ one has $u_{|\partial\Omega} = 0$ the minimizer v satisfies the boundary condition too. Thus we conclude. □

As a simple application of the theory of constrained minimization we solve the eigenvalue problem for the Laplace operator on an open bounded domain Ω with Dirichlet boundary conditions, i.e., the problem is to find a number λ and a function $u \ne 0$ satisfying

$$-\Delta u = \lambda u \quad \text{in} \quad \Omega, \qquad u_{|\partial\Omega} = 0. \tag{32.12}$$

The strategy is simple. On the Hilbert space $H_0^1(\Omega)$ we minimize the functional $f(u) = \frac{1}{2}\langle Du, Du\rangle_2$ under the constraint $g(u) = \frac{1}{2}$ for the constraint functional $g(u) = \frac{1}{2}\langle u, u\rangle_2$. The derivative of g is easily calculated; it is $g'(u)(v) = \langle u, v\rangle_2$ for all $v \in H_0^1(\Omega)$ and thus the level surface $[g = \frac{1}{2}]$ consists only of regular points of the mapping g.

Since we know that f is weakly lower semi-continuous and coercive on $H_0^1(\Omega)$ we can prove the existence of a minimizer for the functional f on $[g = \frac{1}{2}]$ by verifying that $[g = \frac{1}{2}]$ is weakly closed and then to apply Theorem 29.2.2.

Suppose a sequence $(u_j)_{j\in\mathbb{N}}$ converges to u weakly in $H_0^1(\Omega)$. According to the Sobolev embedding (32.7) the space $H_0^1(\Omega)$ is compactly embedded into the space $L^2(\Omega)$ and thus this sequence converges strongly in $L^2(\Omega)$ to u. It follows that $g(u_j) \to g(u)$ as $j \to \infty$, i.e., g is weakly continuous on $H_0^1(\Omega)$ and its level surfaces are weakly closed.

Theorem 29.2.2 implies the existence of a minimizer of f under the constraint $g(u) = 1/2$. Using Corollary 31.2.1 and Theorem 31.3.1 we deduce that there is a Lagrange multiplier $\lambda \in \mathbb{R}$ for this constrained minimization problem, i.e., a real number λ satisfying $f'(u) = \lambda g'(u)$. In detail this identity reads

$$\int_\Omega Du(x) \cdot Dv(x)dx = \lambda \int_\Omega u(x)v(x)dx \qquad \forall v \in H_0^1(\Omega),$$

and in particular for all $v \in \mathcal{D}(\Omega)$, thus $-\Delta u = \lambda u$ in $\mathcal{D}'(\Omega)$; and by elliptic regularity theory (see for instance Section 9.3 of [BB92]) we conclude that this identity holds in the classical sense. Since the solution u belongs to the space $H_0^1(\Omega)$ it satisfies the boundary condition $u_{|\partial\Omega} = 0$. This proves

32.2 The Dirichlet–Laplace operator and other elliptic differential operators

Theorem 32.2.2 (Dirichlet Laplacian) *Let $\Omega \subset \mathbb{R}^n$ be a bounded open set with smooth boundary $\partial \Omega$. Then the eigenvalue problem for the Laplace operator with Dirichlet boundary conditions (32.12) has a solution.*

The above argument which proved the existence of the lowest eigenvalue λ_1 of the Dirichlet–Laplace operator can be repeated on the orthogonal complement of the eigenfunction u_1 of the first eigenvalue and thus gives an eigenvalue $\lambda_2 \geq \lambda_1$ (some additional arguments show $\lambda_2 > \lambda_1$). In this way one proves actually the existence of an infinite sequence of eigenvalues for the Dirichlet–Laplace operator. By involving some refined methods of the theory of Hilbert space operators it can be shown that these eigenvalues are of the order $\lambda_k \approx$ constant $(\frac{k}{|\Omega|})^{\frac{2}{n}}$ (see for instance [LL01]).

Next we consider more generally the following class of second order linear partial differential operators A defined on sufficiently smooth functions u by

$$Au = A_0 u - \sum_{j=1}^{n} \partial_j \left(\sum_{i=1}^{n} a_{ji} \partial_i u \right). \tag{32.13}$$

The matrix a of coefficient functions $a_{ji} = a_{ij} \in L^\infty(\Omega)$ satisfies for almost all $x \in \Omega$ and all $\xi \in \mathbb{R}^n$,

$$m \sum_{j=1}^{n} \xi_j^2 \leq \sum_{i,j=1}^{n} \xi_j a_{ji}(x) \xi_i \leq M \sum_{j=1}^{n} \xi_j^2 \tag{32.14}$$

for some constants $0 < m < M$. A_0 is a bounded symmetric operator in $L^2(\Omega)$ which is bounded from below, $\langle u, A_0 u \rangle_2 \geq -r \|u\|_2^2$ for some positive number r satisfying $0 \leq r < \frac{m}{c^2}$. Here m is the constant in condition (32.14) and c is the smallest constant for which $\|u\|_2 \leq c \|Du\|_2$ holds for all $u \in H_0^1(\Omega)$.

As we are going to show, under these assumptions, the arguments used for the study of the Dirichlet problem and the eigenvalue problem for the Dirichlet–Laplace operator still apply. The associated quadratic form

$$Q(u, v) = \langle u, A_0 v \rangle_2 + \sum_{i,j=1}^{n} \langle \partial_j v, a_{ji} \partial_i u \rangle_2 \quad \forall u, v \in H_0^1(\Omega)$$

is strictly positive since the ellipticity condition (32.14) and the lower bound for A_0 imply

$$Q(u, u) = \langle u, A_0 u \rangle_2 + \int_\Omega \sum_{i,j=1}^{n} \partial_j v(x) a_{ji}(x) \partial_i u(x) dx$$

$$\geq -r \|u\|_2^2 + \int_\Omega m \sum_{j=1}^{n} (\partial_j u(x))^2 dx = -r \|u\|_2^2 + m \|Du\|_2^2$$

$$\geq (-rc^2 + m) \|Du\|_2^2 = c_0 \|Du\|_2^2, \quad c_0 = -rc^2 + m > 0.$$

As earlier we deduce that the functional $f(u) = Q(u, u)$ is coercive and weakly lower semi-continuous on $H^1(\Omega)$. Hence Theorem 29.2.2 allows us to minimize f on $M = v_0 + H_0^1(\Omega)$ and thus to solve the boundary value problem for a given $v_0 \in H^1(\Omega)$ or on the level surface $[g = \frac{1}{2}]$ for the constraint function $g(u) = \frac{1}{2}\langle u, u \rangle_2$ on $H_0^1(\Omega)$. The conclusion is that the *linear elliptic partial differential operator* (32.13) with Dirichlet boundary conditions has an increasing sequence of eigenvalues, as it is the case for the Laplace operator.

32.3 Nonlinear convex problems

In order to be able to minimize functionals of the general form (28.2) we first have to find a suitable domain of definition and then to have enough information about it. We begin with the description of several important aspects from the theory of Lebesgue spaces. A good reference for this are paragraphs 18–20 of [Vai64].

Let $\Omega \subset \mathbb{R}^n$ be a nonempty open set and $h : \Omega \times \mathbb{R} \to \mathbb{R}$ a function such that $h(\cdot, y)$ is measurable on Ω for every $y \in \mathbb{R}$ and $y \mapsto h(x, y)$ is continuous for almost every $x \in \Omega$. Such functions are often called *Carathéodory functions*. If now $u : \Omega \to \mathbb{R}$ is (Lebesgue) measurable, define $\hat{h}(u) : \Omega \to \mathbb{R}$ by $\hat{h}(u)(x) = h(x, u(x))$ for almost every $x \in \Omega$. Then $\hat{h}(u)$ is measurable too. For our purposes it is enough to consider $\hat{h}$ on Lebesgue integrable functions $u \in L^p(\Omega)$ and we need that the image $\hat{h}(u)$ is Lebesgue integrable too, for instance $\hat{h}(u) \in L^q(\Omega)$ for some exponents $1 \le p, q$. Therefore the following lemma will be useful.

Lemma 32.3.1 *Suppose that $\Omega \subset \mathbb{R}^n$ is a bounded open set and $h : \Omega \times \mathbb{R} \to \mathbb{R}$ a Carathéodory function. Then $\hat{h}$ maps $L^p(\Omega)$ into $L^q(\Omega)$ if, and only if, there are $0 \le a \in L^q(\Omega)$ and $b \ge 0$ such that for almost all $x \in \Omega$ and all $y \in \mathbb{R}$,*

$$|h(x, y)| \le a(x) + b|y|^{p/q}. \tag{32.15}$$

If this condition holds the map $\hat{h} : L^p(\Omega) \to L^q(\Omega)$ is continuous.

This result extends naturally to Carathéodory functions $h : \Omega \times \mathbb{R}^{n+1} \to \mathbb{R}$. For $u_j \in L^{p_j}(\Omega)$, $j = 0, 1, \ldots, n$ define $\hat{h}(u_0, \ldots, u_n)(x) = h(x, u_0(x), \ldots, u_n(x))$ for almost every $x \in \Omega$. Then $\hat{h} : L^{p_0}(\Omega) \times \cdots \times L^{p_n}(\Omega) \to L^q(\Omega)$ if, and only if, there are $0 \le a \in L^q(\Omega)$ and $b \ge 0$ such that

$$|h(x, y_0, \ldots, y_n)| \le a(x) + b \sum_{j=0}^n |y_j|^{p_j/q}. \tag{32.16}$$

And $\hat{h}$ is continuous if this condition holds.

As a last preparation define, for every $u \in W^{1,p}(\Omega)$, the functions $y(u) = (y_0(u), y_1(u), \ldots, y_n(u))$ where $y_0(u) = u$ and $y_j(u) = \partial_j u$ for $j = 1, \ldots, n$. By definition of the Sobolev space $W^{1,p}(\Omega)$ we know that

$$y : W^{1,p}(\Omega) \to L^p(\Omega) \times \cdots \times L^p(\Omega) = L^p(\Omega)^{\times(n+1)}$$

is a continuous linear map.

Now suppose that the integrand in formula (28.2) is a Carathéodory function and satisfies the bound

$$|F(x, y)| \leq a(x) + b \sum_{j=0}^{n} |y_j|^p, \tag{32.17}$$

for all $y \in \mathbb{R}^{n+1}$ and almost all $x \in \Omega$, for some $0 \leq a \in L^1(\Omega)$ and some constant $b \geq 0$. Then, as a composition of continuous mappings, $\hat{F} \circ y$ is a well-defined continuous mapping $W^{1,p}(\Omega) \to L^1(\Omega)$. We conclude that under the growth restriction (32.17) the Sobolev space $W^{1,p}(\Omega)$ is a suitable domain for the functional

$$f(u) = \int_\Omega F(x, u(x), Du(x)) dx. \tag{32.18}$$

For $1 < p < \infty$ the Sobolev spaces $W^{1,p}(\Omega)$ are known to be separable reflexive Banach spaces, and thus well suited for the direct methods ([LL01]).

Proposition 32.3.1 *Let $\Omega \subset \mathbb{R}^n$ be a bounded open set and $F : \Omega \times \mathbb{R}^{n+1} \to \mathbb{R}$ a Carathéodory function.*

a) *If F satisfies the growth restriction (32.17), then a functional $f : W^{1,p}(\Omega) \to \mathbb{R}$ is well defined by (32.18). It is polynomially bounded according to*

$$|f(u)| \leq \|a\|_1 + b\|u\|_p^p + b\|Du\|_p^p \quad \forall u \in W^{1,p}(\Omega). \tag{32.19}$$

b) *If F satisfies a lower bound of the form*

$$F(x, y) \geq -\alpha(x) - \beta|y_0|^r + c|\underline{y}|^p \tag{32.20}$$

for all $y = (y_0, \underline{y}) \in \mathbb{R}^{n+1}$ and almost all $x \in \Omega$, for some $0 \leq \alpha \in L^1(\Omega)$, $\beta \geq 0$, $c > 0$ and $0 \leq r < p$, then the functional f is coercive.

c) *If $y \mapsto F(x, y)$ is convex for almost all $x \in \Omega$, then f is lower semi-continuous for the weak topology on $W^{1,p}(\Omega)$.*

Proof. To complete the proof of Part a) we note that the assumed bound for F implies that $|F \circ y(u)(x)| \leq a(x) + b \sum_{j=0}^n |y_j(u)(x)|^p$ and thus by integration the polynomial bound follows.

Integration of the lower bound $F(x, u(x), Du(x)) \geq -\alpha(x) - \beta|u(x)|^r + c|Du(x)|^p$ for almost all $x \in \Omega$ gives $f(u) \geq -\|\alpha\|_1 - \beta\|u\|_r^r + c\|Du\|_p^p$. By inequality (32.9), $\|u\|_r^r \leq S^r \|Du\|_p^r$, hence $f(u) \to \infty$ as $\|Du\|_p \to \infty$ since $r < p$ and $c > 0$.

For any $u, v \in W^{1,p}(\Omega)$ and $0 \leq t \leq 1$ we have $\hat{F}(y(tu + (1-t)v)) = \hat{F}(ty(u) + (1-t)y(v)) \leq t\hat{F}(y(u)) + (1-t)\hat{F}(y(v))$ since F is assumed to be convex with respect to y. Hence integration over Ω gives $f(tu + (1-t)v) \leq tf(u) + (1-t)f(v)$. This shows that f is a convex functional. According to Part a), f is continuous on $W^{1,p}(\Omega)$, therefore Lemma 29.2.2 implies that f is weakly lower semi-continuous on $W^{1,p}(\Omega)$. □

Let us remark that the results presented in Part c) of Proposition 32.3.1 are not optimal (see for instance [Dac82, JLJ98, Str00]). But certainly the result given above has the advantage of a very simple proof. The above result uses stronger assumptions insofar as convexity with respect to u and Du is used whereas in fact convexity with respect to Du is sufficient.

Suppose we are given a functional f of the form (32.18) for which parts a) and c) of Proposition 32.3.1 apply. Then, by Theorem 29.2.1 we can minimize f on any bounded weakly closed subset $M \subset W^{1,p}(\Omega)$. If in addition f is coercive, i.e., if Part b) of Proposition 32.3.1 applies too, then we can minimize f on any weakly closed subset $M \subset W^{1,p}(\Omega)$.

In order to relate these minimizing points to solutions of nonlinear partial differential operators we need differentiability of the functional f. For this we will not consider the most general case but make assumptions which are typical and allow a simple proof.

Let us assume that the integrand F of the functional f is of class C^1 and that all derivatives $F_j = \frac{\partial F}{\partial y_j}$ are again Carathéodory functions. Assume furthermore that there are functions $0 \leq a_j \in L^{p'}(\Omega)$ and constants $b_j > 0$ such that for all $y \in \mathbb{R}^{n+1}$ and almost all $x \in \Omega$,

$$|F_j(x,y)| \leq a_j(x) + b_j \sum_{j=0}^{n} |y_j|^{p-1}, \qquad j = 0, 1, \ldots, n \qquad (32.21)$$

where p' denotes the Hölder conjugate exponent, $\frac{1}{p} + \frac{1}{p'} = 1$. Since $(p-1)p' = p$ we get for all $u \in W^{1,p}(\Omega)$ the simple identity $\|y_j(u)\|_{p'}^{p'} = \|y_j(u)\|_p^p$ and it follows that $\hat{F}_j(y(u)) \in L^{p'}(\Omega)$ for all $u \in W^{1,p}(\Omega)$ and $j = 0, 1, \ldots, n$. This implies the estimates, for all $u, v \in W^{1,p}(\Omega)$,

$$\|\hat{F}_j(y(u))y_j(v)\|_1 \leq \|\hat{F}_j(y(u))\|_{p'}\|y_j(v)\|_p, \qquad j = 0, 1, \ldots, n$$

and thus

$$v \mapsto \int_\Omega \sum_{j=0}^n F_j(x, y(u)(x)) y_j(v)(x) dx \qquad (32.22)$$

is a continuous linear functional on $W^{1,p}(\Omega)$, for every $u \in W^{1,p}(\Omega)$. Now it is straightforward (see Exercises) to calculate the derivative of the functional f, by using Taylor's Theorem. The result is the functional

$$f'(u)(v) = \int_\Omega \sum_{j=0}^n F_j(x, y(u)(x)) y_j(v)(x) dx \quad \forall u, v \in W^{1,p}(\Omega). \qquad (32.23)$$

As further preparation for the solution of nonlinear eigenvalue problems we specify the relevant properties of the class of constraint functionals

$$g(u) = \int_\Omega G(x, u(x)) dx, \qquad u \in W^{1,p}(\Omega) \qquad (32.24)$$

which we are going to use. Here G is a Carathéodory function which has a derivative $G_0 = \frac{\partial G}{\partial u}$ which itself is a Carathéodory function. Since we are working on the space $W^{1,p}(\Omega)$ we assume the following growth restrictions. There are functions $0 \leq \alpha \in L^1(\Omega)$ and $0 \leq \alpha_0 \in L^{p'}(\Omega)$ and constants $0 \leq \beta, \beta_0$ such that for all $u \in \mathbb{R}$ and almost all $x \in \Omega$,

$$|G(x,u)| \leq \alpha(x) + \beta |u|^q, \qquad |G_0(x,u)| \leq \alpha_0(x) + \beta_0 |u|^{q-1} \qquad (32.25)$$

with an exponent q satisfying $2 \leq q < p^*$. Because of Sobolev's inequality (32.9) the functional g is well defined and continuous on $W^{1,p}(\Omega)$ and its absolute values are bounded by $|g(u)| \leq \|\alpha\|_1 + \beta \|u\|_q^q$.

Since $2 \leq q < p^*$ there is an exponent $1 \leq r < p^*$ such that $(q-1)r' < p^*$ (in the Exercises the reader is asked to show that any choice of r with $\frac{p^*}{p^*+1-q} < r < p^*$ satisfies this requirement). Then Hölder's inequality implies $\| |u|^{q-1} v \|_1 \leq \| |u|^{q-1} \|_{r'} \|v\|_r$. Therefore the bound for G_0 shows that for every $u \in W^{1,p}(\Omega)$ the functional $v \mapsto \int_\Omega G_0(x, u(x)) v(x) dx$ is well defined and continuous on $W^{1,p}(\Omega)$. Now it is straightforward to show that the functional g is Fréchet differentiable on $W^{1,p}(\Omega)$ with derivative

$$g'(u)(v) = \int_\Omega G_0(x, u(x)) v(x) dx \qquad \forall u, v \in W^{1,p}(\Omega). \qquad (32.26)$$

Finally we assume that g has a level surface $[g = c]$ with the property that $g'(u) \neq 0$ for all $u \in [g = c]$.

A simple example of a function G for which all the assumptions formulated above are easily verified is $G(x, u) = au^2$ for some constant $a > 0$. Then all level surfaces $[g = c]$, $c > 0$, only contain regular points of g.

The nonlinear eigenvalue problems which can be solved by the strategy indicated above are those of divergence type, i.e., those which are of the form (32.27) below.

Theorem 32.3.2 (Nonlinear eigenvalue problem) *Let $\Omega \subset \mathbb{R}^n$ be a bounded open set with smooth boundary $\partial \Omega$ and $F : \Omega \times \mathbb{R}^{n+1} \to \mathbb{R}$ a Carathéodory function which satisfies all the hypotheses of Proposition 32.3.1 and in addition the growth restrictions (32.21) for its derivatives F_j. Furthermore let $G : \Omega \times \mathbb{R} \to \mathbb{R}$ be a Carathéodory function with derivative G_0 which satisfies the growth conditions (32.25). Finally assume that the constraint functional g defined by G has a level surface $[g = c]$ which consists of regular points of g. Then the nonlinear eigenvalue problem*

$$F_0(x, u(x), Du(x)) - \sum_{j=1}^n \partial_j F_j(x, u(x), Du(x)) = \lambda G_0(x, u(x)) \qquad (32.27)$$

with Dirichlet boundary conditions has a nontrivial solution $u \in W_0^{1,p}(\Omega)$.

Proof. Because of the Dirichlet boundary conditions we consider the functionals f and g on the closed subspace

$$E = W_0^{1,p}(\Omega) = \text{ closure of } \mathcal{D}(\Omega) \text{ in } W^{1,p}(\Omega). \qquad (32.28)$$

Proposition 32.3.1 implies that f is a coercive continuous and weakly lower semi-continuous functional on E. The derivative of f is given by the restriction of the identity (32.23) to E.

Similarly, the functional g is defined and continuous on E and its derivative is given by the restriction of the identity (32.26) to E. Furthermore the bound (32.25) implies that g is defined and thus continuous on $L^q(\Omega)$.

Now consider a level surface $[g = c]$ consisting of regular points of g. Suppose $(u_n)_{n \in \mathbb{N}}$ is a weakly convergent sequence in E, with limit u. Because of the compact embedding of E into $L^q(\Omega)$ this sequence converges strongly in $L^q(\Omega)$. Since g is continuous on $L^q(\Omega)$ we conclude that $(g(u_n))_{n \in \mathbb{N}}$ converges to $g(u)$, thus g is weakly continuous on E. Therefore all level surface of g are weakly closed.

Theorem 29.2.2 implies that the functional f has a minimizing element $u \in [g = c]$ on the level surface $[g = c]$. By assumption, u is a regular point of g, hence Theorem 31.3.1 on the existence of a Lagrange multiplier applies and assures the existence of a number $\lambda \in \mathbb{R}$ such that

$$f'(u) = \lambda g'(u). \tag{32.29}$$

In detail this equations reads: $f'(u)(v) = \lambda g'(u)(v)$ for all $v \in E$ and thus for all v in the dense subspace $\mathcal{D}(\Omega)$ of $E = W_0^{1,p}(\Omega)$.

For $v \in \mathcal{D}(\Omega)$ we calculate

$$f'(u)(v) = \int_\Omega F_0(x, u(x), Du(x))v(x)dx + \int_\Omega \sum_{j=1}^n F_j(x, u(x), Du(x))\partial_j v(x)dx$$

$$= \int_\Omega F_0(x, u(x), Du(x))v(x)dx + \int_\Omega \sum_{j=1}^n \partial_j[F_j(x, u(x), Du(x))v(x)]dx$$

$$- \int_\Omega \sum_{j=1}^n (\partial_j F_j(x, u(x), Du(x)))v(x)dx$$

$$= \int_\Omega [F_0(x, u(x), Du(x)) - \sum_{j=1}^n (\partial_j F_j(x, u(x), Du(x)))]v(x)dx$$

since the second integral vanishes because of the Gauss divergence theorem and $v \in \mathcal{D}(\Omega)$. Hence equation (32.29) implies

$$\int_\Omega [F_0(x, u(x), Du(x)) - \sum_{j=1}^n (\partial_j F_j(x, u(x), Du(x))) - \lambda G_0(x, u(x))]v(x)dx = 0$$

for all $v \in \mathcal{D}(\Omega)$. We conclude that u solves the eigenvalue equation (32.27). □

Remark 32.3.1 *1. A very important assumption in the problems we solved in this section was that the domain $\Omega \subset \mathbb{R}^n$ on which we studied differential operators is bounded so that compact Sobolev embeddings can be used. Certainly, this strategy breaks down if Ω is not bounded. Nevertheless there are many important problems on unbounded domains Ω and one has to modify the strategy presented above. In the last twenty years considerable progress has been made in solving these global problems. The interested reader is referred to the books [BB92, LL01] and in particular to the book [Str00] for a comprehensive presentation of the new strategies used for the global problems.*

2. As is well known, a differentiable function can have other critical points than minima or maxima for which we have developed a method to prove their existence and in favorable situations to calculate them. For these other critical points of functionals (saddle points or mountain passes) a number of other, mainly topological methods have been shown to be quite effective in proving their existence, such as index theories, mountain pass lemmas, perturbation theory). Modern books which treat these topics are [Str00, JLJ98] where one also finds many references to original articles.

3. *The well-known mountain pass lemma of Ambrosetti and Rabinowitz is a beautiful example of results in variational calculus where elementary intuitive considerations have lead to a powerful analytical tool for finding critical points of functionals f on infinite dimensional Banach spaces E.*

   To explain this lemma in intuitive terms consider the case of a function f on $E = \mathbb{R}^2$ which has only positive values. We can image that f gives the height of the surface of the earth over a certain reference plane. Imagine further a town T_0 which is surrounded by a mountain chain. Then, in order to get to another town T_1 beyond this mountain chain, we have to cross the mountain chain at some point S. Certainly we want to climb as little as possible, i.e., at a point S with minimal height $f(S)$. Such a point is a mountain pass of minimal height which is a saddle point of the function f. All other mountain passes M have a height $f(M) \geq f(S)$. Furthermore we know $f(T_0) < f(S)$ and $f(T_1) < f(S)$. In order to get from town T_0 to town T_1 we go along a continuous path γ which has to wind through the mountain chain, $\gamma(0) = T_0$ and $\gamma(1) = T_1$. As described above we know $\sup_{0 \leq t \leq 1} f(\gamma(t)) \geq f(S)$ and for one path γ_0 we know $\sup_{0 \leq t \leq 1} f(\gamma_0(t)) = f(S)$. Thus, if we denote by Γ the set of all continuous paths γ from T_0 to T_1 we get

$$f(S) = \inf_{\gamma \in \Gamma} \sup_{0 \leq t \leq 1} f(\gamma(t)),$$

   i.e., the saddle point S of f is determined by a 'minimax' principle.

4. *If $u \in E$ is a critical point of a differentiable functional f of the form (32.18) on a Banach space E, then this means that u satisfies $f'(u)(v) = 0$ for all $v \in E$. This means that u is a weak solution of the (nonlinear) differential equation $f'(u) = 0$. But in most cases we are actually interested in a strong solution of this equation, i.e., a solution which satisfies the equation $f'(u) = 0$ at least point-wise almost everywhere. For a classical solution this equation should be satisfied in the sense of functions of class $\mathcal{C}^2$. For the linear problems which we have discussed in some detail we have used the special form of the differential operator to argue that for these problems a weak solution is automatically a classical solution. The underlying theory is the theory of elliptic regularity. The basic results of this theory are presented in the books [BB92, JLJ98].*

32.4 Exercises

1. Let $\mathcal{H}$ be a real Hilbert space and $K \subset \mathcal{H}$ a closed convex subset. For every $x \in \mathcal{H}$ show: The functional $\psi_x : K \to \mathbb{R}$ defined by $\psi_x(z) = \|x - z\|$ is strictly convex and coercive (if K is not bounded).

2. Calculate the derivative of the functional f, equation (32.18) by using the assumptions (32.17) and (32.21).

3. For the Sobolev space $H_0^1([a, b])$ prove:

    (a) $|u(x) - u(y)| \le \|u'\|_2 \sqrt{|x - y|}$ for all $u \in H_0^1([a, b])$ and all $x, y \in [a, b]$;

    (b) $\|u\|_2 \le \frac{b-a}{\sqrt{2}} \|u'\|_2$ for all $u \in H_0^1([a, b])$.

4. Suppose functions $p, q \in \mathcal{C}([a, b])$ are given which satisfy the lower bounds $p(x) \ge c > 0$ and $q(x) \ge -r$ with $r \ge 0$ such that $c_0 = c - r(b - a) > 0$. Prove:

    (a) Given any $g \in L^2([a, b])$, the functional
    $$f(u) = \frac{1}{2} \int_a^b p(x) u'(x)^2 dx + \frac{1}{2} \int_a^b q(x) u(x)^2 dx - \int_a^b g(x) u(x) dx$$
    $$= \frac{1}{2} \langle u', pu' \rangle_2 + \frac{1}{2} \langle u, qu \rangle_2 - \langle g, u \rangle_2$$
    has a unique minimum u_0 on the Sobolev space $H_0^1([a, b])$;

    (b) this unique minimum u_0 solves the *Sturm–Liouville problem* for the interval $[a, b]$ and the coefficient functions p, q, g, i.e., the problem of solving the equation
    $$-\frac{d}{dx}[p(x) \frac{du}{dx}(x)] + q(x) u(x) = g(x) u(x) \quad \forall x \in (a, b) \quad (32.30)$$
    for the boundary conditions $u(a) = 0 = u(b)$.

    Hints: Observe the previous problem and show that f is a strictly convex coercive functional on the Sobolev space $H_0^1([a, b])$. Conclude by our general results. Deduce that under the assumptions $g \in \mathcal{C}([a, b])$ and $p \in \mathcal{C}^1([a, b])$ the weak solution u_0 is actually a classical solution of the Sturm–Liouville problem (32.30).

5. Given an exponent $1 < p < n$ and an exponent q satisfying $2 \le q < p^*$ find an exponent r, $1 \le r < p^*$, such that $(q - 1)p' < p^*$ where p' is the Hölder conjugate exponent of the exponent p. Show that the Sobolev space $W^{1,p}([a, b])$ is contained in the space of continuous functions $\mathcal{C}([a, b])$ on the closed interval $[a, b]$ and that the identical embedding is completely continuous, i.e., continuous and compact.

Hints: For $u \in W^{1,p}([a, b])$ and $x, y \in (a, b)$ show first that $|u(x) - u(y)| \le |x - y| \, \|u'\|_p$.

6. For a bounded open set $\Omega \subset \mathbb{R}^n$ with smooth boundary $\partial\Omega$ and an exponent p, $2 < p < 2^* = \frac{2n}{n-2}$, find a solution of the following nonlinear boundary value problem:

$$-\Delta u + \lambda u = u|u|^{p-2}, \qquad u > 0 \quad \text{in } \Omega,$$

and $u = 0$ on $\partial\Omega$. Assume $\lambda > -\lambda_1$, λ_1 the smallest eigenvalue of the Dirichlet–Laplace operator on Ω.

Hints: Consider the functional $f(u) = \frac{1}{2}\langle Du, Du \rangle_2 + \frac{\lambda}{2}\langle u, u \rangle_2$ on the Sobolev space $E = H_0^1(\Omega)$ and minimize it under the constraint $g(u) = 1$ with the constraint functional $g(u) = \frac{1}{p}\int_\Omega |u(x)|^p dx$. Apply the theorem on the existence of a Lagrange multiplier and show that the Lagrange multiplier is positive, using the lower bound for the parameter λ. Finally use a rescaling argument.

7. For a bounded open set $\Omega \subset \mathbb{R}^n$ with smooth boundary $\partial\Omega$ and an exponent p, $2 < p < 2^* = \frac{2n}{n-2}$ solve the nonlinear eigenvalue problem

$$-\Delta u + Au = \lambda \beta(x) u |u|^{p-2} \qquad \text{in } \Omega$$

under Dirichlet boundary conditions. A is a bounded symmetric operator in $L^2(\Omega)$ with lower bound $A \ge -\lambda_1$, λ_1 the smallest eigenvalue of the Dirichlet–Laplace operator on Ω. β is a nonnegative essentially bounded function on Ω, $\beta \ne 0$.

Hints: Minimize the functional $f(u) = \frac{1}{2}\langle Du, Du \rangle_2 + \frac{1}{2}\langle u, Au \rangle_2$ on the Sobolev space $E = H_0^1(\Omega)$ on $[g = 1]$ for the constraint functional $g(u) = \frac{1}{p}\int_\Omega \beta(x)|u(x)|^p dx$. Apply the theorem on the existence of a Lagrange multiplier.

8. For a bounded open set $\Omega \subset \mathbb{R}^n$ with smooth boundary $\partial\Omega$ and an exponent p, $2 \le p < \infty$, show that there exists a weak solution $u \in W_0^{1,p}(\Omega)$ of the boundary value problem

$$-\nabla \cdot (|\nabla u|^{p-2} \nabla u) = g \qquad \text{in } \Omega,$$
$$u = 0 \qquad \text{on } \partial\Omega$$

in the sense that u satisfies the equation

$$\int_\Omega [\nabla u |\nabla u|^{p-2} \nabla v - gv] dx = 0 \quad \forall v \in \mathcal{D}(\Omega) \tag{32.31}$$

where g is any given element in $W_0^{1,p}(\Omega)'$.

Hints: Consider the functional

$$f(u) = \frac{1}{p}\int_\Omega |\nabla u|^p dx - \int_\Omega gu\, dx$$

and show that it is well defined and of class C^1 on the Banach space $E = W_0^{1,p}(\Omega)$. Show furthermore that the left-hand side of equation (32.31) is just the directional derivative of f in the direction v. Now verify the hypotheses of one of the generalized Weierstrass theorems, i.e., show that f is weakly lower semi-continuous on E and coercive. Deduce that a minimizer u of f on E exists and that it satisfies equation (32.31).

33
Density Functional Theory of Atoms and Molecules

The Schrödinger equation is a (linear) partial differential equation that can be solved exactly only in very few special cases such as the Coulomb potential or the harmonic oscillator potential. For more general potentials or for problems with more than two particles the quantum mechanical problem is no easier to solve than the corresponding classical one. In these situations variational methods are one of the most powerful tools for deriving approximate eigenvalues E and eigenfunctions ψ. These approximations are done in terms of a theory of density functionals as proposed by Thomas, Fermi, Hohenberg and Kohn. This chapter explains briefly the basic facts of this theory.

33.1 Introduction

Suppose that the spectrum $\sigma(H)$ of a Hamilton operator H is purely discrete and can be ordered according to the size of the eigenvalues, i.e., $E_1 < E_2 < E_3 < \cdots$. The corresponding eigenfunctions ψ_i form an orthonormal basis of the Hilbert space $\mathcal{H}$. Consider a trial function

$$\overline{\psi} = \sum_{i=1}^{\infty} c_i \psi_i, \quad \sum_{i=1}^{\infty} |c_i|^2 = 1.$$

The expectation value of H in the mixed state $\overline{\psi}$ is

$$\overline{E} = \langle \overline{\psi}, H\overline{\psi} \rangle = \sum_{i=1}^{\infty} |c_i|^2 E_i.$$

It can be rewritten as

$$\overline{E} = E_1 + |c_2|^2(E_2 - E_1) + |c_3|^2(E_3 - E_1) + \cdots \geq E_1.$$

Hence $\overline{E}$ is an upper bound for the eigenvalue E_1 which corresponds to the ground state of the system. One basic idea of the variational calculations concerning spectral properties of atoms and molecules is to choose trial functions depending on some parameters and then to adjust the parameters so that the corresponding expectation value $\overline{E}$ is minimized.

Application of this method to the helium atom by Hylleras played an important role in 1928–1929 when it provided the first test of the Schrödinger equation for a system that is more complicated than the hydrogen atom. In the limit of infinite nuclear mass the Hamilton operator for the helium atom is

$$H = -\frac{\hbar^2}{2m}(\Delta_1 + \Delta_2) - \frac{e^2}{r_1} - \frac{e^2}{r_2} + \frac{e^2}{r_{12}}$$

where $r_i = |x_i|$ and where $r_{12} = |x_1 - x_2|$ is the electron – electron separation. The term $\frac{e^2}{r_{12}}$ describes the Coulomb repulsion between two electrons. Hylleras introduced trial functions of the form $\overline{\psi} = \sum_{i,j,k} a_{ijk} r_1^i r_2^j r_{12}^k e^{-\alpha r_1 - \beta r_2}$ depending on the parameters a_{ijk}, α and β.

The history of the density functional theory dates back to the pioneering work of Thomas [Tho27] and Fermi [Fer27]. In the sixties Hohenberg and Kohn [HK64] and Kohn and Sham [KS65] made substantial progress to give the density functional theory a foundation based on the quantum mechanics of atoms and molecules. Since then an enormous number of results has been obtained, and this method of studying solutions of many electron problems for atoms and molecules has become competitive in accuracy with up to date quantum chemical methods.

The following section gives a survey of the most prominent of these density functional theories. These density functional theories are of considerable mathematical interest since they present challenging minimization problems of a type which has not been attended to before. In these problems one has to minimize certain functionals over spaces of functions defined on unbounded domains (typically on $\mathbb{R}^3$) and where nonreflexive Banach spaces are involved.

The last section reports on the progress in relating these density functional theories to the quantum mechanical theory of many electron systems for atoms and molecules. Here the results on self-adjoint Schrödinger operators obtained in Part B will be the mathematical basis. The results on the foundation of density functional theories are mainly due to Hohenberg–Kohn [HK64] and Kohn–Sham [KS65]. The original paper by Hohenberg–Kohn has generated a vast literature, see for instance [Dav76, PY89, DG90, Nag98, Esc96].

33.2 Semi-classical theories of density functionals

The main goal of these semi-empirical models is to describe correctly the ground state energy by minimizing various types of density functionals.

In all these density functional theories we are looking for the energy and the charge density of the ground state by solving directly a minimization problem of the form

$$\min\left\{F(\rho) + \int \rho(x)v(x)dx : \rho \in D_F\right\}.$$

Here F is a functional of the charge density and depends only on the number N of electrons but not on the potential v generated by the nuclei. The minimum has to be calculated over a set D_F of densities which is either equal to or a subset of $D_N = \{\rho \in L^1(\mathbb{R}^3) : 0 \leq \rho, \|\rho\|_1 = N\}$ depending on the specific theory considered. Let us mention some of the prominent models.

- The model of *Thomas and Fermi* uses the functional

$$F_{TF}(\rho) = c_{TF} \int_{\mathbb{R}^3} \rho(x)^{5/3} dx + D(\rho, \rho)$$

on the domain $D_{TF} = D_N \cap L^{5/3}(\mathbb{R}^3)$. In the simplest models of this theory the potential v is given by $v(x) = -\frac{Z}{|x|}$ where $Z > 0$ is a fixed parameter representing the charge of the atomic nucleus and

$$D(\rho, \rho) = \frac{1}{2} \int_{\mathbb{R}^3} \int_{\mathbb{R}^3} \frac{\rho(x)\rho(y)}{|x-y|} dx dy$$

is nothing else than the Coulomb energy for the charge density ρ. The constant c_{TF} has the value $3/5$.

- The model of *Thomas–Fermi–von Weizsäcker* is associated with the functional

$$F_{TFW}(\rho) = c_W \int_{\mathbb{R}^3} (\nabla \rho(x)^{1/2})^2 dx + F_{TF}(\rho)$$

on the domain $D_{TFW} = D_{TF} \cap H^1(\mathbb{R}^3)$.

- The model of *Thomas–Fermi–Dirac–von Weizsäcker* leads to the functional

$$F_{TFDW}(\rho) = F_{TFW}(\rho) - c_D \int_{\mathbb{R}^3} \rho(x)^{4/3} dx$$

on the same domain D_{TFW}. Note that for $1 \leq p_1 < q < p_2$ one has $L^q(\mathbb{R}^3) \subset L^{p_1}(\mathbb{R}^3) \cap L^{p_2}(\mathbb{R}^3)$ and $\|u\|_q \leq \|u\|_{p_1}^t \|u\|_{p_2}^{1-t}$ with $t = \frac{\frac{1}{q}-\frac{1}{p_2}}{\frac{1}{p_1}-\frac{1}{p_2}}$ which we apply for $p_1 = 1$, $p_2 = 5/3$ and $q = 4/3$. It follows that $\|\rho\|_{4/3}$ is finite on D_{TFW}. Therefore the domain of F_{TFDW} is D_{TFW}.

All these models describe partially some observed natural phenomena but are nevertheless rather rudimentary and are no longer in use in the practice of quantum chemistry. From a theoretical point of view these models are quite interesting since we are confronted with the same type of (mathematical) difficulties as in more realistic approaches.

Though the Thomas–Fermi theory is quite old, a mathematically rigorous solution of the minimization problem has been found only in 1977 by Lieb and Simon ([LS77]. The basic aspects of this solution are discussed in [BB92, LL01].

33.3 Hohenberg–Kohn theory

The Hohenberg–Kohn theory is a successful attempt to link these semi-classical density functional theories to the quantum mechanics of atoms and molecules. Nevertheless from a mathematical point of view there remain several challenging problems as we will see later.

The N-particle Hamilton operators which are considered are assumed to be of the form

$$H_N = H_N(v) = -\sum_{j=1}^N \Delta_j + \sum_{j<k} u(x_j - x_k) + \sum_{j=1}^N v(x_j) \equiv H_0 + V \quad (33.1)$$

where $v(x)$ is a real-valued function on $\mathbb{R}^3$ and $V = \sum_{j=1}^N v(x_j)$. In typical situations u denotes the Coulomb interaction, but many other interactions can be used in this approach too. We restrict ourselves to the Coulomb case $u(x_j - x_k) = \frac{e^2}{|x_j - x_k|}$. In this case the operator H_0 is well defined and self-adjoint on the domain $D(T)$ of operator $T = -\sum_{j=1}^N \Delta_j$ of the kinetic energy (compare Theorem 23.2.1 and the exercises for this theorem). For the one-particle potential v we assume in the following always $v \in L^2(\mathbb{R}^3) + L^\infty(\mathbb{R}^3)$ so that for these potentials too Kato's perturbation theory applies and assures that H_N is self-adjoint on $D(T)$. Note that $L^2(\mathbb{R}^3) + L^\infty(\mathbb{R}^3)$ is a Banach space when equipped with the norm

$$\|v\| = \inf\left\{\|v_1\|_2 + \|v_2\|_\infty : v_1 \in L^2(\mathbb{R}^3), v_2 \in L^\infty(\mathbb{R}^3), v = v_1 + v_2\right\}.$$

However this Banach space is not reflexive. It is actually the topological dual of the Banach space $X = L^1(\mathbb{R}^3) \cap L^2(\mathbb{R}^3)$ for the norm $\|u\| = \|u\|_1 + \|u\|_3$, i.e.,

$$X' = L^2(\mathbb{R}^3) + L^\infty(\mathbb{R}^3).$$

In 1964 Hohenberg and Kohn proposed a method to solve the problem of finding the ground state energy of H_N through a varational principle. To explain this method we need some preparation. The single-particle reduced density matrix γ of an N-particle wave function ψ is given by the kernel

$$\gamma(z, z') = \int \overline{\psi}(z, z_2, \ldots, z_N) \psi(z', z_2, \ldots, z_N) dz_2 \cdots dz_N \quad (33.2)$$

33.3 Hohenberg–Kohn theory

where $z_i = (x_i, \sigma_i)$ denotes the space variable x_i and the spin variable σ_i. This formula defines a mapping $\psi \to \gamma$. This density matrix allows us to express the single particle density as

$$\rho(x) = N \sum_\sigma \gamma((x,\sigma),(x,\sigma)) \qquad (33.3)$$

which defines a mapping $\gamma \to \rho$ and thus a mapping $v \to \rho_v = R(v)$ from potentials v to one-particle densities ρ when ψ is a ground state of $H_N(v)$. This mapping R plays a fundamental rôle in the Hohenberg–Kohn theory. Denote by G_N the set of all those potentials v for which the Hamiltonian $H_N(v)$ has a (unique) ground state $\psi \in D(T)$. Then we consider R as a mapping

$$R : G_N \cap X' \to \left\{\rho \in L^1(\mathbb{R}^3) : 0 \leq \rho\right\}, \qquad (33.4)$$

and one wants to know when this mapping has an inverse. In order to be able to make progress in this problem one has to have a characterization of the range of the mapping R, i.e., one has to know: Under which conditions on ρ there is a potential $v \in G_N \cap X'$ such that the Hamilton operator $H_N(v)$ has a ground state ψ which defines $\rho = \rho_\psi$ through equations (33.2) and (33.3).

Up to now this problem has found only a partial solution which nevertheless allows us to proceed. There are two conditions which are obviously necessary, namely $0 \leq \rho(x)$ for all $x \in \mathbb{R}^3$ and $\|\rho\|_1 = N$, i.e., $\rho \in L^1(\mathbb{R}^3)$. The following lemma gives additional necessary conditions.

Lemma 33.3.1 *Suppose $\rho = \rho_\psi$ is obtained by equations (33.2) and (33.3) from a state ψ which belongs to the domain of the kinetic energy T. Then*

a) $\rho^{1/2} \in H^1(\mathbb{R}^3)$ *and* $\|\nabla \rho^{1/2}\|_2^2 \leq T(\psi)$;

b) $\rho \in L^3(\mathbb{R}^3) \cap L^1(\mathbb{R}^3)$ *and* $\|\rho_\psi\|_3 \leq $ *constant* $T(\psi)$.

Proof. The kinetic energy is defined by

$$T(\psi) = \sum_{i=1}^N \int |\nabla_i \psi(x_1, \ldots, x_i, \ldots, x_N)|^2 dx_1 \cdots dx_N$$

$$= N \int |\nabla_1 \psi(x_1, \ldots, x_N)|^2 dx_1 \cdots dx_N.$$

For the density we calculate

$$\nabla \rho(x) = N(\int [(\overline{\nabla \psi} \psi)(x, x_2, \ldots, x_N) + (\overline{\psi} \nabla \psi)(x, x_2, \ldots, x_N)]dx_2 \cdots dx_N,$$

and Schwarz' inequality implies

$$|\nabla \rho(x)|^2 \leq 4N \int |(\nabla_1 \psi)(x, x_2, \ldots, x_N)|^2 dx_2 \cdots dx_N \rho(x).$$

We deduce

$$\|\nabla \rho^{1/2}\|_2^2 = \frac{1}{4} \int (\nabla \rho(x))^2 \frac{dx}{\rho(x)} \leq T(\psi).$$

This implies Part a).

Sobolev's inequality in $\mathbb{R}^3$ states (see (32.9)) $\|u\|_6^2 \leq S \|\nabla u\|_2^2$ which we apply for $u = \rho^{1/2}$ to get $\|\rho\|_3 = \|\rho^{1/2}\|_6^2 \leq S \|\nabla \rho^{1/2}\|_2^2 \leq S T(\psi) < \infty$. Thus Part b) follows. □

Corollary 33.3.1

$$\operatorname{ran} R \subseteq \left\{ \rho \in L^1(\mathbb{R}^3) \cap L^3(\mathbb{R}^3) : 0 \leq \rho,\ \rho^{1/2} \in H^1(\mathbb{R}^3) \right\} \equiv \mathbb{D}$$

and for $\rho \in \mathbb{D}$ there is a state ψ in the domain $D(T)$ such that $\rho = \rho_\psi$.

Proof. The first part of the corollary is just a summary of the previous lemma. Given $\rho \in \mathbb{D}$ define ψ as a normalized symmetric N-fold tensor product of $\rho^{1/2}$. Since $\int (\nabla \rho(x))^2 \frac{dx}{\rho(x)} < \infty$ it follows that $\psi \in D(T)$. □

Note that this corollary only gives some estimate of the set of those densities ρ for which there is $v \in G_N \cap X'$ such that ρ is the density of a ground state ψ of $H_N(v)$. The problem is that the set G_N is not known explicitly and thus the range of the map R is not known precisely.

The map $\psi \mapsto \rho$ is clearly not bijective and different ψ can give the same ρ. However one can prove continuity though the proof is not too easy (see the appendix of [Lie83]). Part of the difficulty comes from the fact that this map is not linear. Observe that the space $H^1(\mathbb{R}^{3N})$ is the form domain of the kinetic energy T.

Theorem 33.3.1 *$\psi \mapsto \rho^{1/2}$ is a continuous map $H^1(\mathbb{R}^{3N}) \to H^1(\mathbb{R}^3)$.*

Recall that we only consider one-particle potentials $v \in X'$ so that the domain of the N-particle Hamiltonian $H_N(v)$ is the domain

$$W_N = \left\{ \psi \in L^2(\mathbb{R}^{3N}) : T(\psi) < \infty \right\} = D(T)$$

of the kinetic energy T. This allows us to determine the ground state energy of $H_N(v)$ as the solution of a minimization problem:

$$E(v) = \inf_{\psi \in W_N \setminus \{0\}} \frac{\langle \psi, H_N(v) \psi \rangle}{\langle \psi, \psi \rangle}. \tag{33.5}$$

There may or may not be a minimizing element ψ for the minimization problem (33.5) for the ground state energy. And if there exists one we do not always have uniqueness. Accordingly, any minimizing element ψ of (33.5) is called a ground state of $H_N(v)$. It satisfies $H_N(v)\psi = E(v)\psi$ at least in the sense of distributions. $E(v)$ has some important properties.

Theorem 33.3.2 *The ground state energy $E(v)$ defined by (33.5) has the following properties.*

a) *$E(v)$ is concave in $v \in X'$, i.e., for all $v_1, v_2 \in X'$ and all $0 \leq t \leq 1$ one has $E(tv_1 + (1-t)v_2) \geq tE(v_1) + (1-t)E(v_2)$;*

b) *$E(v)$ is monotone increasing, i.e., if $v_1, v_2 \in X'$ and $v_1(x) \leq v_2(x)$ for all $x \in \mathbb{R}^3$, then $E(v_1) \leq E(v_2)$;*

c) *$E(v)$ is continuous with respect to the norm of X' and it is locally Lipschitz.*

Proof. See the Exercises. □

The key result of the Hohenberg–Kohn theory is the observation that under certain conditions different potentials $v_1, v_2 \in G_N \cap X'$ lead to different densities ρ_1, ρ_2, thus proving injectivity of the map R.

Theorem 33.3.3 (Uniqueness theorem) *Suppose $v_1, v_2 \in G_N \cap X'$ are potentials for which the Hamilton operators $H_N(v_1)$ and $H_N(v_2)$ respectively have different ground states ψ_1, ψ_2. Then the densities $\rho_{\psi_1}, \rho_{\psi_2}$ defined by these states are different, $\rho_{\psi_1}(x) \neq \rho_{\psi_2}(x)$ for all points x in a set of positive Lebesgue measure.*

Proof. We give the proof for the case where the ground state energies for both operators $H_N(v_1)$ and $H_N(v_2)$ are not degenerate. For the general case we refer to the literature [DG90].
According to our definitions we know $E(v_i) = \langle \psi_i, H_N(v_i)\psi_i \rangle$, $\psi_i \in W_N$, $\|\psi_i\| = 1$ and $E(v_i) \leq \langle \psi, H_N(v_i)\psi \rangle$ for all $\psi \in W_N$, $\|\psi\| = 1$ and $E(v_i) < \langle \psi, H_N(v_i)\psi \rangle$ for all $\psi \in W_N$, $\|\psi\| = 1$, $\psi \neq \psi_i$, $i = 1, 2$. Equations (33.1)–(33.3) imply $\langle \psi, H_N(v_i)\psi \rangle = \langle \psi, H_0\psi \rangle + N \int v_i(x) \rho_\psi(x) dx$, hence

$$E(v_1) = \langle \psi_1, H_0\psi_1 \rangle + N \int v_2(x) \rho_{\psi_1}(x) dx + N \int (v_1(x) - v_2(x)) \rho_{\psi_1}(x) dx$$
$$> E(v_2) + N \int (v_1(x) - v_2(x)) \rho_{\psi_1}(x) dx$$

and similarly $E(v_2) > E(v_1) + N \int (v_2(x) - v_1(x)) \rho_{\psi_2}(x) dx$. By adding these two inequalities we get

$$0 > N \int (v_1(x) - v_2(x))(\rho_{\psi_1}(x) - \rho_{\psi_1}(x)) dx.$$

All the above integrals are well defined because of Part b) of Lemma 33.3.1 and the interpolation estimate $\|\rho\|_2 \leq \|\rho\|_1^{1/4} \|\rho\|_3^{3/4}$. □

Note that the assumption that $H_N(v_1)$ and $H_N(v_2)$ have different ground states excludes the case that the potentials differ by a constant. This assumption was originally used by Hohenberg–Kohn.

Certainly one would like to have stronger results based on conditions on the potentials v_1, v_2 which imply that the Hamilton operators $H_N(v_1)$ and $H_N(v_2)$ have different ground states ψ_1 and ψ_2. But such conditions are not available here.

The basic Hohenberg–Kohn uniqueness theorem is an existence theorem. It claims that there exists a bijective map $R : v \to \rho$ between an unknown set of potentials v and a corresponding set of densities which is unknown as well. Nevertheless this result implies that the ground state energy E can in principle be obtained by using $v = R^{-1}(\rho)$, i.e., the potential v as a functional of the ground state density ρ. However there is a serious problem since nobody knows this map explicitly.

33.3.1 Hohenberg–Kohn variational principle

Hohenberg and Kohn assume that every one-particle density ρ is defined in terms of a ground state ψ for some potential v, i.e., $H_N(v)\psi = E(v)\psi$. Accordingly they introduce the set

$$A_N = \left\{ \rho \in L^1 \cap L^3(\mathbb{R}^3) : 0 \leq \rho, \sqrt{\rho} \in H^1(\mathbb{R}^3), \exists \text{ ground state } \psi : \psi \mapsto \rho \right\}$$

and on A_N they considered the functional

$$F_{HK}(\rho) = E(v) - \int v(x)\rho(x)dx. \tag{33.6}$$

This definition of F_{HK} requires Theorem 33.3.3 according to which there is a one-particle potential v associated with ρ, $v = R^{-1}(\rho)$. Using this functional the Hohenberg–Kohn variational principle reads

Theorem 33.3.4 (Hohenberg–Kohn variational principle) *For any $v \in G_N \cap X'$ the ground state energy is*

$$E(v) = \min_{\rho \in A_N} [F_{HK}(\rho) + \int v(x)\rho(x)dx]. \tag{33.7}$$

It must be emphasized that this variational principle holds only for $v \in G_N \cap X'$ and $\rho \in A_N$. But we have three major problems: The sets G_N and A_N and the form of the functional F_{HK} are unknown. On one hand the Hohenberg–Kohn theory is an enormous conceptual simplification since it gives some hints that the semi-classical density functional theories are reasonable approximations. On the other hand the existence Theorem 33.3.3 does not provide any practical method for calculating physical properties of the ground state from the one electron density ρ. In experiments we measure ρ but we do not know what Hamilton operator $H_N(v)$ it belongs to.

The contents of the uniqueness theorem can be illustrated by an example. Consider the N_2 and CO molecules. They have exactly the same numbers of electrons and nuclei, but whereas the former has a symmetric electron density this is not the case for the latter. We are therefore able to distinguish between the molecules. Imagine now that we add an external electrostatic potential along the bond for the N_2 molecule. The electron density becomes polarized and it is no more obvious to distinguish between N_2 and CO. But according to the Hohenberg–Kohn uniqueness theorem it is possible to distinguish between the two molecules in a unique way.

The Hohenberg–Kohn variational principle provides the justification for the variational principle of Thomas Fermi in the sense that $E_{TF}(\rho)$ is an approximation to the functional $E(\rho)$ associated with the total energy. Let us consider the functional $E_v(\rho) = F_{HK}(\rho) + \int v(x)\rho(x)dx$. The Hohenberg–Kohn variational principle requires that the ground state density is a stationary point of the functional $E_v(\rho) - \mu[\int \rho(x)dx - N]$ which gives the Euler–Lagrange equation (assuming differentiability)

$$\mu = DE_v(\rho) = v + DF_{HK}(\rho) \tag{33.8}$$

where μ denotes the chemical potential of the system.

If we were able to know the exact functional $F_{HK}(\rho)$ we would obtain by this method an exact solution for the ground state electron density. It must be noted that $F_{HK}(\rho)$ is defined independently of the external potential v; this property means that $F_{HK}(\rho)$ is a universal functional of ρ. As soon as we have an explicit form (approximate or exact) for $F_{HK}(\rho)$ we can apply this method to any system and the Euler–Lagrange equation (33.8) will be the basic working equation of the Hohenberg–Kohn density functional theory. A serious difficulty here is that the functional $F_{HK}(\rho)$ is defined only for those densities which are in the range of the map R, a condition which, as already explained, is still unknown.

33.3.2 The Kohn–Sham equations

The Hohenberg–Kohn uniqueness theorem states that all the physical properties of a system of N interacting electrons are uniquely determined by its one-electron ground state density ρ. This property holds independently of the precise form of the electron – electron interaction. In particular when the strength of this interaction vanishes the functional $F_{HK}(\rho)$ defines the ground state kinetic energy of a system of noninteracting electrons as a functional of its ground state density $T_0(\rho)$. This fact was used by Kohn and Sham [KS65] in 1965 to map the problem of interacting electrons for which the form of the functional $F_{HK}(\rho)$ is unknown onto an equivalent problem for noninteracting particles. To this end $F_{HK}(\rho)$ is written in the form

$$F_{HK}(\rho) = T_0(\rho) + \frac{1}{2} \int \frac{\rho(x)\rho(y)}{|x-y|} dx dy + E_{xc}(\rho). \tag{33.9}$$

The second term is nothing else than the classical electrostatic self-interaction, and the term $E_{xc}(\rho)$ is called the exchange-correlation energy.

Variations with respect to ρ under the constraint $\|\rho\|_1 = N$ leads formally to the same equation which holds for a system of N noninteracting electrons under the influence of an effective potential V_{scf}, also called the self-consistent field potential whose form is explicitly given by

$$v_{scf}(x) = v(x) + (\rho * \frac{1}{|x|})(x) + v_{xc}(x), \tag{33.10}$$

where the term $v_{xc}(x) = D_\rho E_{xc}(\rho)$ is called the exchange – correlation potential, as the functional derivative of the exchange – correlation energy.

There have been a number of attempts to remedy the shortcomings of the Hohenberg–Kohn theory. One of the earliest and best known is due to E. Lieb [Lie83]. The literature we have mentioned before offers a variety of others. Though some progress is achieved major problems are still unresolved. Therefore we can not discuss them here in our short introduction.

A promising direction seems to be the following. By Theorem 33.3.2 we know that $-E(v)$ is a convex continuous functional on X'. Hence (see [ET83]) it can be represented as the polar functional of its polar functional $(-E)^*$:

$$-E(v) = \sup_{u \in X''} [\langle v, u \rangle - (-E)^*(u)] \qquad \forall v \in X'] \tag{33.11}$$

where the polar functional $(-E)^*$ is defined on X'' by

$$(-E)^*(u) = \sup_{v \in X'} [\langle v, u \rangle - (-E)(v)] \qquad \forall u \in X''. \tag{33.12}$$

Now $X = L^2(\mathbb{R}^3) \cap L^1(\mathbb{R}^3)$ is contained in the bi-dual X'' but this bi-dual is much larger ($L^1(\mathbb{R}^3)$ is not a reflexive Banach space) and $L^3(\mathbb{R}^3) \cap L^1(\mathbb{R}^3) \subset L^2(\mathbb{R}^3) \cap L^1(\mathbb{R}^3)$. But one would like to have a representation of this form in terms of densities $\rho \in A_N \subset L^3(\mathbb{R}^3) \cap L^1(\mathbb{R}^3)$, not in terms of $u \in X''$.

Remark 33.3.1 *In Theorem 33.3.4 the densities are integrable functions on all of $\mathbb{R}^3$ which complicates the minimization problem in this theorem considerably, as we had mentioned before in connection with global boundary- and eigenvalue problems. However having the physical interpretation of the functions ρ in mind as one-particle densities of atoms or molecules, it is safe to assume that all the relevant densities have a compact support contained in some finite ball in $\mathbb{R}^3$. Thus in practice one considers this minimization problem over a bounded domain B with the benefit that compact Sobolev embeddings are available. As an additional advantage we can then work in the reflexive Banach space $L^3(B)$ since $L^1(B) \subset L^3(B)$ instead of $L^1(\mathbb{R}^3) \cap L^3(\mathbb{R}^3)$.*

33.4 Exercises

1. Prove Theorem 33.3.2.

    Hints: For $v_1, v_2 \in X'$ and $0 \le t \le 1$ show first that $H_N(tv_1 + (1-t)v_2) = tH_N(v_1) + (1-t)H_N(v_2)$. Part a) now follows easily. For Part b) consider $v_1, v_2 \in X'$ such that $v_1(x) \le v_2(x)$ for almost all $x \in \mathbb{R}^3$ and show as a first step: $\langle \psi, H_N(v_1)\psi \rangle \le \langle \psi, H_N(v_2)\psi \rangle$ for all $\psi \in W_N$, $\|\psi\| = 1$.

    For Part c) proceed similarly and show $|\langle \psi, (H_N(v_1) - H_N(v_2))\psi \rangle| \le N\|v_1 - v_2\|_\infty$ for all $\psi \in W_N$, $\|\psi\| = 1$. This implies $\pm(E(v_1) - E(v_2)) \le N\|v_1 - v_2\|_\infty$.

2. Show that the Coulomb energy functional D is weakly lower semi-continuous on the Banach space $L^{6/5}(\mathbb{R}^3)$.

3. Prove: The Thomas – Fermi energy functional E_{TF} is well defined on the cone $D_{TF} = \{\rho \in L^{5/3} \cap L^1(\mathbb{R}^3) : \rho \ge 0\}$.

Part IV
Appendix

Appendix A
Completion of Metric Spaces

A *metric* on a set X is a function $d : X \times X \to \mathbb{R}$ with these properties:

(D_1) $d(x_1, x_2) \geq 0$,
(D_2) $d(x_1, x_2) = d(x_2, x_1)$,
(D_3) $d(x_1, x_2) \leq d(x_1, x) + d(x, x_2)$,
(D_4) $d(x_1, x_2) = 0 \Leftrightarrow x_1 = x_2$.

for all $x, x_1, x_2 \in X$. A set X on which a metric d is given is called a *metric space* (X, d). Sets of the form

$$B(x, r) = \{y \in X : d(y, x) < r\}$$

are called *open balls* in (X, d) with center x and radius $r > 0$. These balls are used to define the *topology* $\mathcal{T}_d$ on X.

A sequence $(x_n)_{n \in \mathbb{N}}$ in (X, d) is called a *Cauchy sequence* if, and only if, the distance $d(x_n, x_m)$ of the elements x_n and x_m of this sequence goes to zero as $n, m \to \infty$. A metric space (X, d) is called *complete* if, and only if, every Cauchy sequence has a *limit* x in (X, d), i.e., if, and only if, for every Cauchy sequence $(x_n)_{n \in \mathbb{N}}$ there is a point $x \in X$ such that $\lim_{n \to \infty} d(x, x_n) = 0$.

In the text we encountered many examples of metric spaces and in many applications it was very important that these metric spaces were complete, respectively could be extended to complete metric spaces. We are going to describe in some detail the much used construction which enables one to 'complete' every incomplete space by 'adding the missing points'. The model for this construction is the construction of the space of real numbers as the space of equivalence classes of

Cauchy sequences of rational numbers. A complete metric space (Y, D) is called a *completion* of the metric space (X, d) if, and only if, (Y, D) contains a subspace (Y_0, D_0) which is dense in (Y, D) and which is isometric to (X, d). The following results ensure the *existence of a completion*.

Theorem A.0.1 *Every metric space (X, d) has a completion (Y, D). Every two completions of (X, d) are isomorphic under an isometry which leaves the points of X invariant.*

Proof. Denote by $\mathcal{S} = \mathcal{S}(X, d)$ the set of all Cauchy sequences $\underline{x} = (x_n)_{n \in \mathbb{N}}$ in the metric space (X, d). Given $\underline{x}, \underline{y} \in \mathcal{S}$ one has the estimate

$$|d(x_n, y_n) - d(x_m, y_m)| \leq d(x_n, x_m) + d(y_n, y_m)$$

which shows that $(d(x_n, y_n))_{n \in \mathbb{N}}$ is a Cauchy sequence in the field $\mathbb{R}$ and thus converges. This allows one to define a function $d_1 : \mathcal{S} \times \mathcal{S} \to \mathbb{R}$ by

$$d_1(\underline{x}, \underline{y}) = \lim_{n \to \infty} d(x_n, y_n).$$

Obviously the function d_1 has the properties (D_1) and (D_2) of a metric. To verify the triangle inequality (D_3) observe that for any $\underline{x}, \underline{y}, \underline{z} \in \mathcal{S}$ and all $n \in \mathbb{N}$ we have

$$d(x_n, y_n) \leq d(x_n, z_n) + d(z_n, y_n).$$

The standard calculation rules for limits imply that this inequality also holds in the limit $n \to \infty$ and thus proves (D_3) for the function d_1. The separation property (D_4) however does not hold for the function d_1. Therefore we introduce in $\mathcal{S}$ an equivalence relation which expresses this separation property.

Two Cauchy sequences $\underline{x}, \underline{y} \in \mathcal{S}$ are called equivalent if, and only if, $d_1(\underline{x}, \underline{y}) = 0$. We express this equivalence relation by $\underline{x} \sim \underline{y}$. The properties established thus far for the function d_1 imply that this is indeed an equivalence relation on $\mathcal{S}$. The equivalence class determined by the element $\underline{x} \in \mathcal{S}$ is denoted by $[\underline{x}]$, i.e., $[\underline{x}] = \{\underline{y} \in \mathcal{S} : \underline{y} \sim \underline{x}\}$. The space of all these equivalence classes is called Y, $Y = \{[\underline{x}] : \underline{x} \in \mathcal{S}\}$. Next define a function $D : Y \times Y \to \mathbb{R}$ by

$$d([\underline{x}], [\underline{y}]) = d_1(\underline{x}, \underline{y})$$

where $\underline{x}, \underline{y}$ are any representatives of their respective classes. One shows that D is well defined, i.e., independent of the chosen representative: Suppose $\underline{x}' \sim \underline{x}$, then the triangle inequality for the function d_1 gives

$$d_1(\underline{x}, \underline{y}) \leq d_1(\underline{x}, \underline{x}') + d_1(\underline{x}', \underline{y}) = d_1(\underline{x}', \underline{y}) \leq d_1(\underline{x}', \underline{x}) + d_1(\underline{x}, \underline{y}) = d_1(\underline{x}, \underline{y}),$$

which shows that $d_1(\underline{x}, \underline{y}) = d_1(\underline{x}', \underline{y})$ whenever $\underline{x} \sim \underline{x}'$. By definition, the function D satisfies the separation property (D_4):

$$d([\underline{x}], [\underline{y}]) = 0 \Leftrightarrow [\underline{x}] = [\underline{y}].$$

We conclude that D is a metric on the set Y, hence (Y, D) is a metric space.

Next we embed the given metric space into (Y, D). For every $x \in X$ consider the constant sequence $\underline{x}^0 = (x, x, x, \ldots)$. Clearly $\underline{x}^0 \in \mathcal{S}$ and thus a map $\tau : X \to Y$ is well defined by

$$\tau(x) = [\underline{x}^0] \quad \forall x \in X.$$

By the definition of D, respectively d_1, we have

$$D(\tau(x), \tau(y)) = d_1(\underline{x}^0, \underline{y}^0) = d(x, y)$$

for all $x, y \in X$, hence the map τ is isometric.

Given $[\underline{x}] \in Y$ choose a representative $\underline{x} = (x_1, x_2, x_3, \ldots)$ of this class. Then the sequence $(\tau(x_n))_{n \in \mathbb{N}}$ converges to $[\underline{x}]$:

$$\lim_{n \to \infty} D(\tau(x_n), [\underline{x}]) = \lim_{n \to \infty} d_1(\underline{x_n}^0, \underline{x}) = \lim_{n \to \infty} \lim_{m \to \infty} d(x_n, x_m) = 0.$$

We conclude that the image $Y_0 = \tau(X)$ of X under the isometry τ is dense in (Y, D).

Finally we prove completeness of the metric space (Y, D). Suppose $([\underline{y_n}])_{n \in \mathbb{N}}$ is a Cauchy sequence in (Y, D). Since Y_0 is dense in (Y, D) there is a sequence $(\tau(x_n))_{n \in \mathbb{N}} \subset \overline{Y_0}$ such that $D(\tau(x_n), [\underline{y_n}]) \leq \frac{1}{n}$ for each n. It is easy to see that the sequences $(\tau(x_n))_{n \in \mathbb{N}}$ and $([\underline{y_n}])_{n \in \mathbb{N}}$ either both converge or both diverge. Now observe that $\underline{x} = (x_1, x_2, x_3, \ldots)$ is a Cauchy sequence in the given metric space (X, d):

$$d(x_n, x_m) = D(\tau(x_n), \tau(x_m))$$
$$\leq D(\tau(x_n), [\underline{y_n}]) + D([\underline{y_n}], [\underline{y_m}]) + D([\underline{y_m}], \tau(x_m))$$
$$\leq \frac{1}{n} + D([\underline{y_n}], [\underline{y_m}]) + \frac{1}{m}.$$

Since $([\underline{y_n}])_{n \in \mathbb{N}}$ is a Cauchy sequence in (Y, D) the statement follows immediately and therefore $[\underline{x}] \in Y$. The identity

$$\lim_{n \to \infty} D(\tau(x_n), [\underline{x}]) = \lim_{n \to \infty} \lim_{m \to \infty} d(x_n, x_m) = 0$$

proves that the sequence $(\tau(x_n))_{n \in \mathbb{N}}$ converges to $[\underline{x}]$ in the metric space in (Y, D). The construction of the points x_n implies that the given Cauchy sequence too converges to $[\underline{x}]$ in (Y, D). Hence this space is complete.

Since we do not use the second part of the theorem we leave its proof as an exercise. □

Corollary A.0.1 *Every normed space $(X_0, \|\cdot\|_0)$ has a completion which is a Banach space $(X, \|\cdot\|)$. Every inner product space $(\mathcal{H}_0, \langle\cdot,\cdot\rangle_0)$ has a completion which is a Hilbert space $(\mathcal{H}, \langle\cdot,\cdot\rangle)$.*

Proof. We only comment on the proof. It is a good exercise to fill in the details.

According to Theorem A.0.1 we only know that the normed space, respectively the inner product space, have a completion as a metric space. But since the original space X_0, respectively $\mathcal{H}_0$, carries a vector space structure, the space of Cauchy sequences of elements of these spaces too can be given a natural vector space structure. The same applies to the space of equivalence classes of such Cauchy sequences. Finally one has to show that the given norm, respectively the given inner product, has a natural extension to this space of equivalence classes of Cauchy sequences which is again a norm, respectively an inner product. Then the proof of completeness of these spaces is as above. □

Appendix B
Metrizable Locally Convex Topological Vector Spaces

A Hausdorff locally convex topological vector space $(X, \mathcal{P})$ is called *metrizable* if, and only if, there is a metric d on X which generates the given topology $\mathcal{T}_\mathcal{P}$, i.e., if $\mathcal{T}_d$ denotes the topology generated by the metric d, one has $\mathcal{T}_\mathcal{P} = \mathcal{T}_d$. Recall that two different metrics might generate the same topologies. In such a case the two metrics are called *equivalent*. Important and big classes of Hausdorff locally convex topological vector spaces are indeed metrizable.

Theorem B.0.2 *Every Hausdorff locally convex topological vector space $(X, \mathcal{P})$ with countable system $\mathcal{P} = \{p_j : j \in \mathbb{N}\}$ of continuous semi-norms p_j is metrizable. A translation invariant metric which generates the given topology is*

$$d(x, y) = \sum_{j=1}^{\infty} \frac{1}{2^j} \frac{p_j(x-y)}{1 + p_j(x-y)} \quad \forall x, y \in X. \tag{B.1}$$

Proof. All the semi-norms p_j are continuous for the topology $\mathcal{T}_\mathcal{P}$ and the series (B.1) converges uniformly on $X \times X$. Therefore this series defines a continuous function d on $(X, \mathcal{T}_\mathcal{P}) \times (X, \mathcal{T}_\mathcal{P})$. This function d obviously satisfies the defining conditions (D_1) and (D_2) of a metric. The separation property (D_4) holds since the space $(X, \mathcal{P})$ is Hausdorff.

In order to show the triangle inequality (D_3) observe first that for any $x, y, z \in X$ one has

$$\frac{p_j(x-y)}{1 + p_j(x-y)} \leq \frac{p_j(x-z)}{1 + p_j(x-z)} + \frac{p_j(z-y)}{1 + p_j(z-y)}$$

since all terms are nonnegative and $p_j(x-y) \leq p_j(x-z) + p_j(z-y)$. Summation now implies the triangle inequality for the function d which thus is a metric on X. Obviously this metric d is translation invariant:

$$d(x+z, y+z) = d(x, y) \quad \forall x, y, z \in X.$$

Since the metric d is continuous for the topology $\mathcal{T_P}$, the open balls $B_d(x, r)$ for the metric d are open in $(X, \mathcal{T_P})$. Since these open balls generate the topology $\mathcal{T}_d$, we conclude that the topology $\mathcal{T_P}$ is finer than the metric topology $\mathcal{T}_d$, $\mathcal{T}_d \subseteq \mathcal{T_P}$.

In order to show the converse $\mathcal{T_P} \subseteq \mathcal{T}_d$ we prove that every element V of a neighborhood basis of zero for the topology $\mathcal{T_P}$ contains an open ball $B_d(0, r)$ with respect to the metric d. Suppose

$$V = \cap_{i=1}^{k} B_{p_i}(0, r_i)$$

with $r_i > 0$ for $i = 1, \ldots, k$ is given. Choose some number r_0,

$$0 < r_0 < \min\left\{2^{-k}, \frac{2^{-i} r_i}{1 + r_i}, i = 1, \ldots, k\right\}.$$

Then for fixed r, $0 < r \leq r_0$, and every $x \in B_d(0, r)$ we know by equation (B.1) that

$$2^{-i} \frac{p_i(x)}{1 + p_i(x)} \leq d(x, 0) < r, \qquad i = 1, \ldots, k.$$

These inequalities together with the choice of $r \leq r_0$ imply immediately $p_i(x) < r_i$, $i = 1, \ldots, k$, hence $x \in V$ and thus $B_d(0, r) \subseteq V$ which proves $\mathcal{T_P} \subseteq \mathcal{T}_d$. □

Two examples of Hausdorff locally convex topological vector spaces which are metrizable and which were used in the text are the spaces $\mathcal{D}_K(\Omega)$ and $\mathcal{S}(\mathbb{R}^n)$.

Recall that for an open and nonempty set $\Omega \subset \mathbb{R}^n$ and a compact subset $K \subset \Omega$, the space $\mathcal{D}_K(\Omega)$ consists of all $\mathcal{C}^\infty$-functions on Ω which have their support in K. The topology of this space is generated by the countable system of norms $q_{K,m}$: $m = 0, 1, 2, \ldots$; the space $\mathcal{D}_K(\Omega)$ is metrizable according to Theorem B.0.2. In Proposition 2.1.8 we had indicated a proof of its completeness. Hence $\mathcal{D}_K(\Omega)$ is a complete metrizable Hausdorff locally convex topological vector space.

The space $\mathcal{S}(\mathbb{R}^n)$ is defined as the space of all those $\mathcal{C}^\infty$-functions on $\mathbb{R}^n$ which, with all their derivatives, decay faster than constant $\times (1 + x^2)^{-k/2}$ for any $k = 0, 1, 2, \ldots$. The countable system $\{p_{m,k} : k, m = 0, 1, 2, \ldots\}$ of norms defines the topology of $\mathcal{S}(\mathbb{R}^n)$. It is a good exercise to prove that this space too is complete. Therefore the space $\mathcal{S}(\mathbb{R}^n)$ is a complete metrizable Hausdorff topological vector space.

Appendix C
The Theorem of Baire

On an open nonempty set $\Omega \subseteq \mathbb{R}^n$ consider a sequence of continuous functions $f_n : \Omega \to \mathbb{R}$ and suppose that this sequence has a 'pointwise' limit f, i.e., for every $x \in \Omega$ the limit $\lim_{n \to \infty} f_n(x) = f(x)$ exists. Around 1897, Baire investigated the question whether the limit function f is continuous on Ω. He found that this is not the case in general and he found that the set of points in Ω at which the limit function f is not continuous is a 'rather small subset of Ω'. Naturally a precise meaning had to be given to the expression of a 'rather small subset of Ω'. In this context Baire suggested the concept of *subset of first category in* Ω, i.e., subsets of Ω which can be represented as a *countable union of nowhere dense sets*. And a subset $A \subset \Omega$ is called *nowhere dense in* Ω if, and only if, the closure $\overline{A}$ in Ω has no interior points. Later the subsets of first category in Ω were given the more intuitive name of a *meager subset*. All subsets which are not of the first category are called *subsets of the second category* or *nonmeager subsets*.

Note the following simple implication of the definition of a nowhere dense subset. If $B \subset \Omega$ is nowhere dense, then $A = \Omega \setminus \overline{B}$ is an open and dense subset of Ω, $\Omega = \overline{A}$. Thus Baire reduced the above statement about the set of points of continuity of the limit function f to the following statement.

Theorem C.0.3 (Theorem of Baire, Version 1) *If A_j, $j \in \mathbb{N}$, is a countable family of open and dense subsets of an open nonempty subset $\Omega \subset \mathbb{R}^n$, then the intersection*
$$A = \cap_{j=1}^{\infty} A_j \tag{C.1}$$
is also dense in Ω.

Proof. Given an open ball $B_0 = B(x_0, r_0) = \{x \in \mathbb{R}^n : \|x - x_0\| < r_0\}$ in Ω we have to show that $A \cap B_0$ is not empty.

Since A_1 is an open and dense subset of Ω we know that $A_1 \cap B_0$ is an open nonempty subset of B_0. Hence there is an open ball
$$B_1 = B(x_1, r_1) = \{x \in \mathbb{R}^n : \|x - x_1\| < r_1\}$$
with $\overline{B_1} \subset A_1 \cap B_0$. We can and will assume that $0 < r_1 \leq r_0/2$. By the same reasoning $A_2 \cap B_1$ is an open nonempty subset of B_1. Hence there is an open ball $B_2 = B(x_2, r_2) \subset \Omega$ with the property $\overline{B_2} \subset A_2 \cap B_1$ and $0 < r_2 \leq r_1/2$.

These arguments can be iterated and thus produce a sequence of open balls $B_k = B(x_k, r_k)$ satisfying
$$\overline{B_{k+1}} \subset A_k \cap B_k, \qquad r_{k+1} \leq \frac{r_k}{2}, \ k = 1, 2, \ldots.$$

Per construction $r_k \leq 2^{-k} r_0$ and $x_{k+m} \in B_k$ for all $m \geq 0$, hence $\|x_{k+m} - x_k\| \leq 2^{-k} r_0$ for all $k, m = 0, 1, 2, \ldots$. We conclude that the sequence of centers x_k of these balls B_k is a Cauchy sequence in $\mathbb{R}^n$ and thus converges to a unique point
$$y = \lim_{k \to \infty} x_k \in \cap_{k=1}^{\infty} \overline{B_k} = M.$$

According to the construction of these balls we get that $M \subset B_0$ and $M \subset A_k$ for all $k \in \mathbb{N}$ and thus we conclude that $M \subset A \cap B_0$. □

Some years later Banach and Steinhaus realized that Baire's proof did not use the special structure of the Euclidean space $\mathbb{R}^n$. The proof relies on two properties of $\mathbb{R}^n$: $\mathbb{R}^n$ is a metric space and $\mathbb{R}^n$ is complete (with respect to the metric). Thus Banach and Steinhaus formulated the following result which nowadays usually is called the theorem of Baire.

Theorem C.0.4 (Theorem of Baire, Version 2) *Suppose that (X, d) is a complete metric space and $\Omega \subset X$ an open nonempty subset of X. Then the intersection $A = \cap_{i \in \mathbb{N}} A_i$ of a countable family of open and dense subsets A_i of Ω is again dense in Ω.*

Proof. The proof of Version 1 applies when we replace the Euclidean balls $B(x, r)$ by the open balls $B_d(x, r) = \{y \in X : d(y, x) < r\}$ of the metric space (X, d). □

In most applications of Baire's theorem however the following 'complementary' version is used.

Theorem C.0.5 (Theorem of Baire, Version 3) *Suppose (X, d) is a complete metric space and B_i, $i \in \mathbb{N}$, is a countable family of closed subsets of X such that*
$$X = \cup_{i=1}^{\infty} B_i. \tag{C.2}$$

Then at least one of the sets B_i has a nonempty interior.

Proof. If all the closed sets B_i had an empty interior, then $A_i = X \backslash B_i, i \in \mathbb{N}$, would be a countable family of open and dense subsets of X. The second version of Baire's theorem implies that $A = \cap_{i \in \mathbb{N}}$ is dense in X, thus $A^c = X \backslash A \neq X$, a contradiction since $A^c = \cup_{i \in \mathbb{N}} B_i$. Therefore, at least one of the sets B_i must have a nonempty interior. □

Definition C.0.1 *A topological space X in which the third version of Baire's theorem holds, is called a* **Baire space**.

Thus a Baire space X can be exhausted by a countable family of closed subsets B_i only when at least one of the subsets B_i has a nonempty interior. Then the third version of Baire's theorem can be restated as saying that all complete metric spaces are Baire spaces. It follows immediately that all complete metrizable Hausdorff locally convex topological vector spaces are Baire spaces. In particular, every Banach space is a Baire space. The spaces of functions $\mathcal{D}_K(\Omega)$ and $\mathcal{S}(\mathbb{R}^n)$ which play a fundamental rôle in the theory of distributions are Baire spaces.

C.1 The uniform boundedness principle

The results of Baire and Banach–Steinhaus have found many very important applications in functional analysis. The most prominent one is the uniform boundedness principle which we are going to discuss in this section. It has been used in the text for many important conclusions.

Definition C.1.1 *Suppose $(X, \mathcal{P})$ and $(Y, \mathcal{Q})$ are two Hausdorff locally convex topological vector spaces over the field $\mathbb{K}$. Denote the set of linear functions $T : X \to Y$ with $L(X, Y)$. A subset $\Lambda \subset L(X, Y)$ is called*

a) **pointwise bounded** *if, and only if, for every $x \in X$ the set $\{Tx : T \in \Lambda\}$ is bounded in $(Y, \mathcal{Q})$, i.e., for every semi-norm $q \in \mathcal{Q}$,*

$$\sup\{q(Tx) : T \in \Lambda\} = C_{x,q} < \infty;$$

b) **equi-continuous** *if, and only if, for every semi-norm $q \in \mathcal{Q}$ there is a semi-norm $p \in \mathcal{P}$ and a constant $C \geq 0$ such that*

$$q(Tx) \leq Cp(x) \quad \forall x \in X, \ \forall T \in \Lambda.$$

Obviously, the elements of an equi-continuous family of linear mappings are continuous and such a family is pointwise bounded. For an important class of spaces $(X, \mathcal{P})$ the converse holds too, i.e., a pointwise bounded family of continuous linear mappings $\Lambda \subset L(X, Y)$ is equi-continuous.

Theorem C.1.1 (Theorem of Banach–Steinhaus) *Assume that two Hausdorff locally convex topological vector spaces over the field $\mathbb{K}$, $(X, \mathcal{P})$ and $(Y, \mathcal{Q})$, are given and assume that $(X, \mathcal{P})$ is a Baire space. Then every*

bounded family Λ of continuous linear mappings $T : X \to Y$ is equi-continuous.

Proof. For an arbitrary semi-norm $q \in \mathcal{Q}$ introduce the sets

$$U_{T,q} = \{x \in X : q(Tx) \leq 1\}, \quad T \in \Lambda.$$

Since T is a continuous linear map, the set $U_{T,q}$ is a closed absolutely convex subset of $(X, \mathcal{P})$. Hence

$$U_q = \bigcap_{T \in \Lambda} U_{T,q}$$

is closed and absolutely convex too.

Now given a point $x \in X$ the family Λ is bounded in this point, i.e., for every $q \in \mathcal{Q}$ there is a $C_{x,q} < \infty$ such that $q(Tx) \leq C_{x,q}$ for all $T \in \Lambda$. Choose $n \in \mathbb{N}, n \geq C_{x,q}$; then for all $T \in \Lambda$ we find $q(T(x/n)) = \frac{1}{n}q(Tx) \leq \frac{1}{n}C_{x,q} \leq 1$, hence $\frac{1}{n}x \in U_q$ or $x \in nU_q$. Clearly, with U_q also the set nU_q is closed (and absolutely convex); thus X is represented as the countable union of the closed sets nU_q:

$$X = \bigcup_{n \in \mathbb{N}} nU_q.$$

Since $(X, \mathcal{P})$ is assumed to be a Baire space, at least one of the sets nU_q must have a nonempty interior, hence U_q has a nonempty interior i.e., there are a point $x_0 \in U_q$, semi-norms $p_1, \ldots, p_N \in \mathcal{P}$ and positive numbers $r_1, \ldots, r_N$ such that $V = \cap_{j=1}^n B_{p_j}(x_0, r_j) \subset U_q$. Now choose $p = \max\{p_1, \ldots, p_n\}$ and $r = \min\{r_1, \ldots, r_N\}$. We have $p \in \mathcal{P}, r > 0$, and $B_p(x_0, r) \subset V \subset U_q$. The definition of U_q implies $q(Tx) \leq 1$ for all $x \in B_p(x_0, r)$ and all $T \in \Lambda$. Hence for every $\xi \in B_p(0, r)$ and every $T \in \Lambda$: $q(T\xi) \leq q(T(x_0 + \xi)) + q(Tx_0) \leq 1 + q(Tx_0) = C$. Lemma 2.1.2 now implies

$$q(Tx) \leq \frac{C}{r}p(x) \qquad \forall x \in X, \quad \forall T \in \Lambda$$

which proves that the family Λ is equi-continuous. □

The Banach–Steinhaus theorem has many applications in functional analysis. We mention some of them which are used in our text. They are just special cases for the choice of the domain space $(X, \mathcal{P})$ which has to be a Baire space.

Every Banach space X is a Baire space. Therefore Theorem C.1.1 applies. Given a family $\{T_\alpha : \alpha \in A\}$ of continuous linear maps $T_\alpha : X \to \mathbb{K}$ which is

or weakly bounded, then, for every $x \in X$, there is a constant C_x such that

$$\sup\{|T_\alpha(x)| : \alpha \in A\} \leq C_x < \infty.$$

According to the Banach–Steinhaus theorem such a family is equi-continuous, i.e., there is a constant $C < \infty$ such that

$$|T_\alpha(x)| \leq C\|x\| \qquad \forall x \in X, \quad \forall \alpha \in A$$

and therefore

$$\sup_{\alpha \in A} \|T_\alpha\| \leq C.$$

Hence the family $\{T_\alpha : \alpha \in A\}$ is not only

bounded, it is uniformly or norm bounded: This is the *uniform boundedness principle in Banach spaces*, see also Theorem C.1.1.

Earlier in this appendix we had argued that the spaces $\mathcal{D}_K(\Omega), \Omega \subseteq \mathbb{R}^n, K \subset \Omega$ compact, are Baire spaces. Thus the Banach–Steinhaus theorem applies to them. Suppose $\{T_\alpha : \alpha \in A\} \subset \mathcal{D}'_K(\Omega)$ is a

bounded family of continuous linear forms on $\mathcal{D}_K(\Omega)$, i.e., for every $f \in \mathcal{D}_K(\Omega)$ there is a $C_f < \infty$ such that $|T_\alpha(f)| \leq C_f$ for all $\alpha \in A$. Theorem C.1.1 implies that this family is equi-continuous, i.e., there is an $m \in \mathbb{N}$ and a constant C such that

$$|T_\alpha(f)| \leq Cq_{K,m}(f) \qquad \forall f \in \mathcal{D}_K(\Omega), \quad \forall \alpha \in A.$$

Now we come back to the problem of continuity of the

limit of continuous functions which were the starting point of Baire's investigations. We consider continuous linear functions on Hausdorff locally convex topological vector spaces. For the case of continuous nonlinear functions on finite dimensional spaces we refer to the Exercises (this case is more involved).

Theorem C.1.2 *Suppose $(T_j)_{j \in \mathbb{N}}$ is a sequence of continuous linear functionals on a Hausdorff topological vector space $(X, \mathcal{P})$ with the property that for every $x \in X$ the numerical sequence $(T_j(x))_{j \in \mathbb{N}}$ is a Cauchy sequence. Then:*

a) A linear functional T is well defined on X by

$$T(x) = \lim_{j \to \infty} T_j(x) \quad \forall x \in X.$$

b) If $(X, \mathcal{P})$ is a Baire space, then the functional T defined in a) is continuous.

Proof. Since the field $\mathbb{K}$ is complete, the Cauchy sequence $(T_j(x))_{j \in \mathbb{N}}$ converges in $\mathbb{K}$. We call its limit $T(x)$. Thus a function $T: X \to \mathbb{K}$ is well defined. Basic rules of calculations for limits now prove linearity of this function T.

Cauchy sequences in the field $\mathbb{K}$ are bounded, hence, for every $x \in X$ there is a finite constant C_x such that $\sup \{|T_j(x)| : j \in \mathbb{N}\} \leq C_x$. The theorem of Banach–Steinhaus implies that this sequence is equi-continuous, i.e., there is some $p \in \mathcal{P}$ and there is a finite constant C such that $|T_j(x)| \leq Cp(x)$ for all $x \in X$ and all $j \in \mathbb{N}$. Taking the limit $j \to \infty$ in this estimate we get $|T(x)| \leq Cp(x)$ for all $x \in X$ and thus T is continuous. □

Part b) of this theorem is often formulated in the following way.

Corollary C.1.1 *The topological dual space X' of a Hausdorff locally convex Baire space $(X, \mathcal{P})$ is weakly sequentially complete.*

And as a special case of this result we have:

Corollary C.1.2 *The spaces of distributions $\mathcal{D}'(\Omega)$, $\Omega \subset \mathbb{R}^n$ open and nonempty, and $\mathcal{S}'(\mathbb{R}^n)$ are weakly sequentially complete.*

Proof. The main point of the proof is to establish that the spaces $\mathcal{D}_K(\Omega)$, $K \subset \Omega$ compact, and $\mathcal{S}(\mathbb{R}^n)$ are complete metrizable and thus Baire spaces. But this has already been done. □

Finally we use Baire's theorem to show in a relatively simple way that the test function spaces $\mathcal{D}(\Omega)$, $\Omega \subset \mathbb{R}^n$ open and nonempty, are not metrizable.

To this end we recall that the spaces $\mathcal{D}_K(\Omega)$, $K \subset \Omega$ compact, are closed in $\mathcal{D}(\Omega)$. Furthermore there is a sequence of compact sets $K_j \subset K_{j+1}$ for all $j \in \mathbb{N}$ such that $\Omega = \cup_{j \in \mathbb{N}} K_j$. It follows that

$$\mathcal{D}(\Omega) = \cup_{j \in \mathbb{N}} \mathcal{D}_{K_j}(\Omega).$$

If $\mathcal{D}(\Omega)$ were metrizable, then according to the third version of Baire's theorem one of the spaces $\mathcal{D}_{K_j}(\Omega)$ must have a nonempty interior which obviously is not the case (to show this is a recommended exercise).

Proposition C.1.3 *The test function spaces $\mathcal{D}(\Omega)$, $\Omega \subset \mathbb{R}^n$ open and nonempty, are complete non-metrizable Hausdorff locally convex topological vector spaces.*

Proof. In the book [KG82] one finds a proof of this result which does not use Baire's theorem (see Theorem 28, page 71). □

C.2 The open mapping theorem

This section introduces other frequently used consequences of Baire's results. These consequences are the open mapping theorem and its immediate corollary, the inverse mapping theorem.

Definition C.2.1 *A mapping $T : E \to F$ between two topological spaces is called* **open** *if, and only if, $T(V)$ is open in F for every open set $V \subset E$.*

Our main interest here are linear open mappings between Banach spaces. Thus the following characterization of these mappings is very useful.

Lemma C.2.1 *A linear map $T : E \to F$ between two normed spaces E, F is open if, and only if,*

$$\exists r > 0 : \quad B_r^F \subseteq T(B_1^E) \tag{C.3}$$

where B_1^E is the open ball in E with radius 1 and center 0 and B_r^F the open ball in F with radius $r > 0$ and center 0.

Proof. If T is an open mapping, then $T(B_1^E)$ is an open set in F which contains $0 \in F$ since T is linear. Hence there is an $r > 0$ such that relation (C.3) holds.
   Conversely assume that relation (C.3) holds and that $V \subset E$ is open. Choose any $y = Tx \in T(V)$. Since V is open there is a $\rho > 0$ such that $x + B_\rho^E \subset V$. It follows that $y + T(B_\rho^E) = T(x + B_\rho^E) \subset T(V)$. Relation (C.3) implies that $B_{\rho r}^F = \rho B_r^F \subset \rho T(B_1^E) = T(B_\rho^E)$ and thus $y + B_{\rho r}^F \subset T(V)$. Therefore y is an interior point $T(V)$ and we conclude. $\square$

Theorem C.2.1 (Open mapping theorem) *Let E, F be two Banach spaces and $T : E \to F$ a surjective continuous linear mapping. Then T is open.*

Proof. For a proof one has to show relation (C.3). This will be done in two steps.
   For simplicity of notation the open balls in E of radius $r > 0$ and center 0 are denoted by B_r. Since obviously $B_{1/2} - B_{1/2} \subset B_1$ and since T is linear we have $T(B_{1/2}) - T(B_{1/2}) \subset T(B_1)$. In any topological vector space for any two sets A, B the relation $\overline{A} - \overline{B} \subset \overline{A - B}$ for their closures is known. This implies $\overline{T(B_{1/2})} - \overline{T(B_{1/2})} \subset \overline{T(B_1)}$.
   Surjectivity and linearity of T give

$$F = \cup_{k=1}^\infty kT(B_{1/2}).$$

As a Banach space, F is a Baire space and therefore at least one of the sets $k\overline{T(B_{1/2})}$, $k \in \mathbb{N}$, must have a nonempty interior. Since $y \mapsto ky$ is a surjective homeomorphism of F the set $\overline{T(B_{1/2})}$ has a nonempty interior, i.e., there is some open nonempty set V in F which is contained in $\overline{T(B_{1/2})}$, and hence $V - V \subset \overline{T(B_{1/2})} - \overline{T(B_{1/2})} \subset \overline{T(B_1)}$. $V - V$ is an open set in F which contains $0 \in F$. Therefore there is some $r > 0$ such that $B_r^F \subset V - V$ and we conclude

$$B_r^F \subset \overline{T(B_1)}. \tag{C.4}$$

In the second step we use relation (C.4) to deduce Relation C.3. Pick any $y \in V_r \equiv B_r^F$, then $\|y\|_F < r$ and we can choose some $R \in (\|y\|_F, r)$. Now rescale y to $y' = \frac{r}{R}y$. Clearly $\|y'\|_F < r$ and therefore $y' \in V_r \subset \overline{T(B_1)}$. Since $0 < \frac{R}{r} < 1$ there is $0 < a < 1$ such that $\frac{R}{r} + a < 1$, i.e., $\frac{R}{r}\frac{1}{1-a} < 1$. Since y' belongs to the closure of the set $T(B_1)$ there is a $y_0 \in T(B_1)$ such that $\|y' - y_0\|_F < ar$. It follows that $z_0 = \frac{1}{a}(y' - y_0) \in V_r$ and by the same reason there is a $y_1 \in T(B_1)$ such that $\|z_0 - y_1\|_F < ar$, and again $z_1 = \frac{1}{a}(z_0 - y_1) \in V_r$ and there is a $y_2 \in T(B_1)$ such that $\|z_1 - y_2\|_F < ar$. By induction this process defines a sequence of points $y_0, y_1, y_2, \ldots$ in $T(B_1)$ which satisfies

$$\|y' - \sum_{i=0}^{n} a^i y_i\|_F < a^{n+1} r, \quad n = 1, 2, \ldots. \tag{C.5}$$

Estimate (C.5) implies $y' = \sum_{i=0}^{\infty} a^i y_i$. By construction $y_i = T(x_i)$ for some $x_i \in B_1$. Since $\|a^i x_i\| < a^i$ for all i and since E is complete, the series $\sum_{i=0}^{\infty} a^i x_i$ converges in E. Call the limit x'. A standard estimate gives

$$\|x'\|_E \le \sum_{i=0}^{\infty} a^i = \frac{1}{1-a}.$$

Continuity of T implies $T(x') = \sum_{i=0}^{\infty} a^i T(x_i) = y'$ and if we introduce $x = \frac{R}{r} x'$ we get $T(x) = \frac{R}{r} y' = y$. By choice of the parameter a the limit x actually belongs to B_1. This follows from $\|x\|_E = \frac{R}{r} \|x'\|_E < \frac{R}{r} \frac{1}{1-a} < 1$. We conclude that $y \in V_r$ is the image under T of a point in B_1. Since y was any point in V_r this completes the proof. □

Corollary C.2.1 (Inverse mapping theorem) *A continuous linear map T from a Banach space E onto a Banach space F which is injective has a continuous inverse $T^{-1} : F \to E$ and there are positive numbers r and R such that*

$$r\|x\|_E \le \|Tx\|_F \le R\|x\|_E \quad \forall x \in E.$$

Proof. Such a map is open and thus T satisfies relation (C.3) which implies immediately that the inverse T^{-1} is bounded on the unit ball B_1^F by $\frac{1}{r}$, hence T^{-1} is continuous and its norm is $\le \frac{1}{r}$. The two inequalities just express continuity of T (upper bound) and of T^{-1} (lower bound). □

If E, F are two Banach spaces over the same field, then $E \times F$ is a Banach space too when the vector space $E \times F$ is equipped with the norm

$$\|(x, y)\| = \|x\|_E + \|y\|_F.$$

The proof is a straightforward exercise. If $T : E \to F$ is a linear mapping, then its *graph*

$$G(T) = \{(x, y) \in E \times F : y = Tx\}$$

is a linear subspace of $E \times F$. If the graph $G(T)$ of a linear mapping T is closed in $E \times F$ the mapping T is called *closed*. Recall that closed linear mappings or operators have been studied in some detail in the context of Hilbert spaces (Section 19.2).

Theorem C.2.2 (Closed graph theorem) *If $T : E \to F$ is a linear mapping from the Banach space E into the Banach space F whose graph $G(T)$ is closed in $E \times F$, then T is continuous.*

Proof. As a closed subspace of the Banach space $E \times F$ the graph $G(T)$ is a Banach space too, under the restriction of the norm $\|\cdot\|$ to it. Define the standard projection mappings $p : G(T) \to E$ and $q : G(T) \to F$ by $p(x, y) = x$, respectively $q(x, y) = y$. Since $G(T)$ is the graph of a linear mapping, both p and q are linear and p is injective. By definition, p is surjective too. Continuity of p and q follow easily from the definition of the norm on $G(T)$: $\|p(x, y)\|_E = \|x\|_E \le \|x\|_E + \|y\|_F$ and similarly for q. Hence p is a bijective continuous linear map of the Banach space $G(T)$ onto the Banach space E and as such has a continuous inverse, by the inverse mapping theorem. Thus T is represented as the composition $q \circ p^{-1}$ of two continuous linear mappings, $T(x) = q \circ p^{-1}(x)$, for all $x \in E$, and therefore T is continuous. □

Appendix D
Bilinear Functionals

A functional of two variables from two vector spaces is called bilinear if, and only if, the functional is linear in one variable while the other variable is kept fixed. For such functionals there are two basic types of continuity. The functional is continuous with respect to one variable while the other is kept fixed, and the functional is continuous with respect to simultaneous change of both variables. Here we investigate the important question for which Hausdorff locally convex topological vector spaces both concepts of continuity agree.

Definition D.0.2 *Let $(X, \mathcal{P})$ and $(Y, \mathcal{Q})$ be two Hausdorff locally convex vector spaces over the field $\mathbb{K}$ and $B : X \times Y \to \mathbb{K}$ a bilinear functional. B is called*

a) **separately continuous** *if, and only if, for every $x \in X$ there are a constant C_x and a semi-norm $q_x \in \mathcal{Q}$ such that $|B(x, y)| \leq C_x q_x(y)$ for all $y \in Y$, and for every $y \in Y$ there are a constant C_y and a semi-norm $p_y \in \mathcal{P}$ such that $|B(x, y)| \leq C_y p_y(x)$ for all $x \in X$.*

b) **continuous** *if, and only if, there are a constant C and semi-norms $p \in \mathcal{P}$ and $q \in \mathcal{Q}$ such that $|B(x, y)| \leq C p(x) q(y)$ for all $x \in X$ and all $y \in Y$.*

Obviously, every continuous bilinear functional is separately continuous. The converse statement does not hold in general. However for a special but very important class of Hausdorff topological vector spaces one can show that separately continuous bilinear functionals are continuous.

Theorem D.0.3 *Suppose that $(X, \mathcal{P})$ and $(Y, \mathcal{Q})$ are two Hausdorff locally convex metrizable topological vector spaces and assume that $(X, \mathcal{P})$ is complete. Then every separately continuous bilinear functional $B : X \times Y \to \mathbb{K}$ is continuous.*

Proof. For metrizable Hausdorff locally convex topological vector spaces continuity and sequential continuity are equivalent. Thus we can prove continuity of B by showing that $B(x_j, y_j) \to B(x, y)$ whenever $x_j \to x$ and $y_j \to y$ as $j \to \infty$.

Suppose such sequences $(x_j)_{j \in \mathbb{N}}$ and $(y_j)_{j \in \mathbb{N}}$ are given. Define a sequence of linear functionals $T_j : X \to \mathbb{K}$ by $T_j(x) = B(x, y_j)$ for all $j \in \mathbb{N}$. Since B is separately continuous all the functionals T_j are continuous linear functionals on $(X, \mathcal{P})$. Since the sequence $(y_j)_{j \in \mathbb{N}}$ converges in $(Y, \mathcal{Q})$ we know that $C_{q,x} = \sup_{j \in \mathbb{N}} q_x(y_j)$ is finite for every fixed $x \in X$. Hence separate continuity implies

$$\sup_{j \in \mathbb{N}} |T_j(x)| \leq C_x C_{q,x}$$

where the constant C_x refers to the constant in the definition of separate continuity. This shows that $(T_j)_{j \in \mathbb{N}}$ is a point-wise bounded sequence of continuous linear functionals on the complete metrizable Hausdorff locally convex topological vector space $(X, \mathcal{P})$. The Theorem of Banach–Steinhaus implies that this sequence is equi-continuous. Hence there are a constant C and a semi-norm $p \in \mathcal{P}$ such that $|T_j(x)| \leq Cp(x)$ for all $x \in X$. This gives

$$|B(x_j, y_j) - B(x, y)| \leq |T_j(x_j - x)| + |B(x, y_j - y)| \leq Cp(x_j - x) + |B(x, y_j - y)| \to 0$$

as $j \to \infty$. Therefore B is sequential continuous and thus continuous. □

An application which is of interest in connection with the definition of the tensor product of distributions (see Section 6.2) is the following. Suppose $\Omega_j \subset \mathbb{R}^{n_j}$ are open and nonempty subsets and $K_j \subset \Omega_j$ are compact, $j = 1, 2$. Then the spaces $\mathcal{D}_{K_j}(\Omega_j)$ are complete metrizable Hausdorff locally convex topological vector spaces. Thus every separately continuous bi-linear functional $\mathcal{D}_{K_1}(\Omega_1) \times \mathcal{D}_{K_2}(\Omega_2) \to \mathbb{K}$ is continuous.

References

[Amr81]   W. O. Amrein. *Non-relativistic quantum dynamics*. Reidel, Dordrecht, 1981.

[AS77]   W. O. Amrein and K. B. Sinha. *Scattering theory in quantum mechanics: physical principles and mathematical methods*, volume 16 of *Lecture notes and supplements in physics*. Benjamin, Reading, Mass., 1977.

[BB92]   Ph. Blanchard and E. Brüning. *Variational Methods in Mathematical Physics. A unified approach.* Texts and Monographs in Physics. Springer-Verlag, Berlin, Heidelberg, New York, London, Paris, Tokyo, Hong Kong, Barcelona, Budapest, 1992.

[BLOT90] N. N. Bogolubov, A. A. Logunov, A. I. Oksak, and I. T. Todorov. *General Principles of Quantum Field Theory*, volume 10 of *Mathematical Physics and Applied Mathematics*. Kluwer Academic Publishers, Dordrecht Boston London, 1990.

[BN89]   E. Brüning and S. Nagamachi. Hyperfunction quantum field theory: Basic structural results. *J. Math. Physics*, 30:2340–2359, 1989.

[Bre65]   H. Bremermann. *Complex Variables and Fourier transforms*. Addison-Wesley, Reading, 1965.

[BS96]   Ph. Blanchard and J. Stubbe. Bound states for Schrödinger Hamiltonians: Phase space methods and applications. *Review of Mathematical Physics*, 8:503–547, 1996.

458 References

[BW83]   H. Baumgaertel and M. Wollenberg. *Mathematical Scattering Theory*. Birkhäuser, Basel, 1983.

[BZ88]   Y. D. Burago and V. A. Zalgaller. *Geometric inequalities*, volume 285 of *Die Grundlehren der mathematischen Wissenschaften in Einzeldarstellungen*. Springer-Verlag, Berlin, 1988.

[Cha89]   K. Chandrasekharan. *Classical Fourier transforms*. Universitext. Springer-Verlag, New York, 1989.

[Dac82]   B. Dacorogna. *Weak continuity and weak lower semicontinuity of non-linear functionals*, volume 922, *Lecture Notes in Mathematics*. Springer-Verlag, Berlin, 1982.

[Dac89]   B. Dacorogna. *Direct Methods in the Calculus of Variations*, volume 78 of *Applied mathematical sciences*. Springer-Verlag, Berlin, 1989.

[Dav76]   E. R. Davidson. *Reduced Density Matrices in Quantum Chemistry*. Academic Press, New York, 1976.

[Dav02]   B. Davies. *Integral Transforms and their Applications*, volume 25 of *Applied Mathematical Sciences*. Springer-Verlag, Berlin, 3rd edition, 2002.

[dFK67]   D. G. de Figueiredo and L. Karlowitz. On the radial projection in normed spaces. *Bull. Amer. Math. Soc.*, 73:364–368, 1967.

[DG90]   R. M. Dreizler and E. K. U. Gross. *Density Functional Theory*. Springer-Verlag, New York Berlin Heidelberg, 1990.

[Die69]   J. A. Dieudonné. *Foundations of Modern Analysis*. Academic Press, New York, 1969.

[Don69]   W. F. Donoghue. *Distributions and Fourier Transforms*. Academic Press, New York, 1969.

[DS58]   N. Dunford and J. T. Schwartz. *Linear Operators. Part I: General Theory*. Interscience Publisher, New York, 1958.

[Edw79]   R. E. Edwards. *Fourier-Series. A Modern Introduction*, volume 1. Springer-Verlag, New York, 2 edition, 1979.

[EG73]   H. Epstein and V. Glaser. The rôle of locality in perturbation theory. *Ann. Inst. Henri Poincaré A*, 19:211, 1973.

[Esc96]   H. Eschrig. *The Fundamentals of Density Functional Theory*. Teubner Verlag, Leipzig, 1996.

[ET83]   I. Ekeland and T. Turnbull. *Infinite dimensional optimization and convexity*. Chicago lectures in mathematics. Univ. of Chicago Press, Chicago, 1983.

[Fer27] E. Fermi. Un metodo statistico per la determinazione di alcune proprieta dell'atome. *Rend. Accad. Naz. Lincei*, 6:602–607, 1927.

[GF68]   H. Grauert and W. Fischer. *Differential und Integralrechnung II*, volume 36 *Heidelberger Taschenbücher*. Springer-Verlag, Berlin, Heidelberg, New York, 1968.

[GM97]   H. Grosse and A. Martin. *Particle physics and the Schrödinger equation*, volume 6, *Cambridge monographs on particle physics, nuclear physics and cosmology*. Cambridge Univ. Press, Cambridge, 1997.

[Gol80] H. H. Goldstine. *A history of the calculus of variations from the 17th through the 19th century*, volume 5, *Studies in the History of Mathematics and Physical Sciences*. Springer-Verlag, New York, 1980.

[GP90]   A. Galindo and P. Pascual. *Quantum Mechanics I*. Texts and Monographs in Physics. Springer-Verlag, Berlin, Heidelberg, New York, 1990.

[GŠ72]   I. M. Gel'fand and G. E. Šilov. *Generalized Functions II: Spaces of Fundamental and Generalized Functions*. Academic Press, New York, 2nd edition, 1972.

[GŠ77]   I. M. Gel'fand and G. E. Šilov. *Generalized Functions I: Properties and Operations*. Academic Press, New York, 5 edition, 1977.

[Haa96] R. Haag. *Local quantum physics: Fields, particles, algebras*. Texts and Monographs in Physics. Springer-Verlag, Berlin, 1996.

[Hep69] K. Hepp. *Théorie de la renormalisation*, volume 2, *Lecture Notes in Physics*. Springer-Verlag, Berlin, Heidelberg, New York, 1969.

[HK64]   P. Hohenberg and W. Kohn. Inhomogeneous electron gas. *Phys. Rev. B*, 136:864–871, 1964.

[Hör67] L. Hörmander. *An Introduction to Complex Analysis in Several Variables*. Van Nostrand, Princeton, NJ, 1967.

[Hör83a] L. Hörmander. *The Analysis of Linear Partial Differential Operators. 1. Distribution Theory and Fourier Analysis*. Springer-Verlag, Berlin, Heidelberg, New York, 1983.

[Hör83b] L. Hörmander. *The analysis of linear partial differential operators. 2. Differential operators of constant coefficients*. Springer-Verlag, Berlin, Heidelberg, New York, 1983.

[HT96]   S. Hildebrandt and A. Tromba. *The Parsimonious Universe. Shape and Form in the Natural World.* Springer-Verlag, Berlin, 1996.

[Ish95]   C. J. Isham. *Lectures on Quantum Theory: Mathematical and Structural Foundations.* Imperial College Press, London, 1995.

[Jam74]   M. Jammer. *The Conceptual Development of Quantum Mechanics.* J. Wiley, New York, 1974.

[Jau73]   J. M. Jauch. *Foundations of Quantum Mechanics.* Addison-Wesley, Reading, Massachusetts, 1973.

[JLJ98]   J. Jost and X. Li-Jost. *Calculus of Variations*, volume 64, *Cambridge Studies in Advanced Mathematics.* Cambridge University Press, 1998.

[Jos65]   R. Jost. *The General Theory of Quantized Fields.* American Math. Soc., Providence, RI, 1965.

[Kak39]   S. Kakutani. Some Characterizations of Euclidean Spaces. *Jap, Journ. Math.*, 16:93–97, 1939.

[Kan88]   A. Kaneko. *Introduction to Hyperfunctions.* Mathematics and its Applicatons (Japanese Series). Kluwer Academic Publishers, Dordrecht Boston London, 1988.

[Kat66]   T. Kato. *Perturbation Theory for Linear Operators.* Springer-Verlag, Berlin, Heidelberg, New York, 1966.

[KG82]   A. A. Kirilov and A. D. Gvisbiani. *Theorems and Problems in Functional Analysis.* Springer-Verlag, New York, 1982.

[Kom73a] H. Komatsu, editor. *Hyperfunctions and pseudo-differential equations.* Springer Lecture Notes 287, Springer-Verlag, Berlin, Heidelberg, New York, 1973.

[Kom73b] H. Komatsu. Ultradistributions I, Structure theorems and a characterization. *J. Fac. Sci. Univ. Tokyo, Sect. IA, Math.*, 20:25–105, 1973.

[Kom77]   H. Komatsu. Ultradistributions II, The kernel theorem and ultradistributions with support in a manifold. *J. Fac. Sci. Univ. Tokyo Section IA*, 24:607–628, 1977.

[Kom82]   H. Komatsu. Ultradistributions III. Vector valued ultradistributions and the theory of kernels. *J. Fac. Sci. Univ. Tokyo Sect. IA*, 29:653–717, 1982.

[KS65]   W. Kohn and L. J. Sham. Self consistent equations including exchange and correlation effects. *Phys. Rev. A*, 140:1133–1138, 1965.

References

[Lei79] H. Leinfelder. A geometric proof of the spectral theorem for unbounded self-adjoint operators. *Math. Ann.*, 242:85–96, 1979.

[Lie83] E. Lieb. Density functionals for Coulomb systems. *Int. J. of Quantum Chemistry*, XXIV:243–277, 1983.

[Lju34] L. A. Ljusternik. On conditional extrema of functions. *Mat. Sbornik*, 41(3), 1934.

[LL01] E. Lieb and M. Loss. *Analysis*, volume 14, *Graduate Studies in Mathematics*. AMS, 2nd edition, 2001.

[LS77] E. Lieb and B. Simon. The Thomas–Fermi theory of atoms, molecules and solids. *Adv. in Math.*, 23:22–116, 1977.

[LT91] M. Ledoux and M. Talagrand. *Probability in Banach spaces: isoperimetry and processes*, volume 23, *Ergebnisse der Mathematik und ihrer Grenzgebiete; Folge 3*. Springer-Verlag, Berlin, 1991.

[Mar64] A. Martineau. Distributions et valeur au bord des fonctions holomorphes. In *Theory of distributions, Proc. Intern. Summer Inst.*, pages 193–326, Lisboa, 1964. Inst. Gulbenkian de Ciéncia.

[MR01] J. Mehra and H. Rechenberg. *The Historical Development of Quantum Theory*, volume (1 -) 6. Springer-Verlag, New York, 2001.

[MS57] J. Mikusinski and K. Sikorski. *The elementary theory of distributions*. PWN, Warsaw, 1957.

[Nag98] A. Nagy. Density functional and application to atoms and molecules. *Physics Reports*, 298:1–79, 1998.

[NB01] S. Nagamachi and E. Brüning. Hyperfunction quantum field theory: Analytic structure, modular aspects, and local observable algebras. *Journ. of Math. Phys.*, 42(1):1–31, January 2001.

[NM76] S. Nagamachi and N. Mugibayashi. Hyperfunction quantum field theory. *Commun. Math. Phys.*, 46:119–134, 1976.

[NN90] T. Nishimura and S. Nagamachi. On supports of Fourier hyperfunctions. *Math. Japonica*, 35:293–313, 1990.

[Oss78] R. Osserman. The isoperimetric inequality. *Bull. Amer. Math. Soc.*, 84:1182–1238, 1978.

[Pea88] D. B. Pearson. *Quantum Scattering and Spectral Theory*, volume 9, *Techniques of Physics*. Academic Press, London, 1988.

[PY89] R. G. Parr and W. Yang. *Density Functional Theory of Atoms and Molecules*. Oxford University Press, Oxford, 1989.

[Rem98] R. Remmert. *Theory of Complex Functions*, volume 122, *Graduate Texts in Mathematics*. Springer-Verlag, Berlin, Heidelberg, 4th edition, 1998.

[RR73] A. P. Robertson and W. J. Robertson. *Topological Vector Spaces*. Cambridge University Press, Cambridge, 1973.

[RS75] M. Reed and B. Simon. *Fourier analysis. Self-adjointness*, volume 2 of *Methods of modern mathematical physics*. Academic Press, New York, 1975.

[RS78] M. Reed and B. Simon. *Analysis of Operators*, volume 4, *Methods of Modern Mathematical Physics*. Academic Press, New York San Francisco London, 1978.

[RS79] M. Reed and B. Simon. *Scattering Theory*, volume 3, *Methods of Modern Mathematical Physics*. Academic Press, New York, 1979.

[RS80] M. Reed and B. Simon. *Functional Analysis*, volume 1, *Methods of Modern Mathematical Physics*. Academic Press, New York, 2nd edition, 1980.

[Rud73] W. Rudin. *Functional Analysis*. McGraw Hill, New York, 1973.

[Rud80] W. Rudin. *Principles of Mathematical Analysis*. Physik-Verlag, Weinheim, 1980.

[Sat58] M. Sato. On a generalization of the concept of functions. *Proc. Japan Acad.*, 34:126–130 and 604–608, 1958.

[Sat59] M. Sato. Theory of hyperfunctions I. *J. Fac. Sci., Univ. Tokyo, Sect. I*, 8:139–193, 1959.

[Sat60] M. Sato. Theory of hyperfunctions II. *J. Fac. Sci., Univ. Tokyo, Sect. I*, 8:387–437, 1960.

[Sch57] L. Schwartz. *Théorie des distributions*, volume 1. Hermann, Paris, 2nd edition, 1957.

[Sch81] M. Schechter. *Operator Methods in Quantum Mechanics*. North Holland, New York, 1981.

[Str00] M. Struwe. *Variational methods: applications to nonlinear partial differential equations and Hamiltonian systems*, volume 34, *Ergebnisse der Mathematik und ihrer Grenzgebiete, Folge 3*. Springer-Verlag, Berlin, 3rd edition, 2000.

[SW64] R. F. Streater and A. S. Wightman. *PCT, Spin and Statistics, and All That*. Benjamin, New York, 1964.

[Thi92]   W. Thirring. *A Course in Mathematical Physics: Classical Dynamical Systems and Classical Field Theory*. Springer study edition. Springer-Verlag, New York, 1992.

[Thi02]   W. Thirring. *Quantum Mathematical Physics: Atoms, Molecules and Large Systems*, volume 3. Springer-Verlag, Heidelberg, New York, 2002.

[Tho27]   L. H. Thomas. The calculation of atomic fields. *Proc. Camb. Philos. Soc.*, 23:542–548, 1927.

[Trè67]   F. Trèves. *Topological Vector Spaces, Distributions and Kernels*. Academic Press, New York, 1967.

[Vai64]   M. M. Vainberg. *Variational Methods for the Study of Nonlinear Operators*. Holden Day, London, 1964.

[Vla71]   V. S. Vladimirov. *Equations of Mathematical Physics*, volume 3, *Pure and Applied Mathematics*. Dekker, New York, 1971.

[vN67]   J. von Neumann. *Mathematical Foundations of Quantum Mechanics*, volume 2, *Investigations in Physics*. Princeton Univ. Press, Princeton, NJ, 1967.

[WG64]   A. S. Wightman and L. Gårding. Fields as operator-valued distributions in relativistic quantum theory. *Arkiv för Fysik*, 28:129–184, 1964.

[Wid71]   D.V. Widder. *An introduction to Transform Theory*, volume 42, *Pure and Applied Mathematics*. Academic Press, New York, 1971.

[YCB82]   C. Dewitt-Morette with M. Dillard-Bleick and Y. Choquet-Bruhat. *Analysis, Manifolds and Physics*. North Holland Publ. Comp., 1982.

[Zei85]   E. Zeidler. *Variational methods and optimization*, volume 3, *Nonlinear Functional Analysis and its Applications*. Springer-Verlag, New York, 1985.

[Zem87]   A. H. Zemanian. *Distribution theory and transform analysis. An introduction to generalized functions with applications*. McGraw-Hill, New York, Dover edition, 1987.

Index

o, 388
p-ball
    open, 10

Banach
    reflexive
        examples of, 381
Weierstrass theorems
    generalized, 378

adjoint, 47
adjoint operator, 251
almost periodic functions, 224
angle
    between two vectors, 189

Baire space, 448
bilinear form, 455
    continuous, 455
    separately continuous, 455
boundary value, 154
bounded
    linear function, 17
    pointwise, 241, 449
    uniformly or norm, 241

Breit–Wigner formula, 37

calculus of variations, 373
Carathéodory functions, 420
carrier, 163
Cauch–Riemann equations, 126
Cauchy estimates, 120
Cauchy sequence
    in $\mathcal{D}'(\Omega)$, 35
Cauchy's integral formula I, 118
Cauchy's integral formula II, 119
Cauchy's principal value, 33
chain rule, 55
change of variables, 54
class $\mathcal{C}^n$, 390
coercive, 383
commutation relations
    of Heisenberg, 280
compact
    locally, 237
    relatively, 381
    sequentially, 380
complete, 190, 214
    sequentially, 16
completeness relation, 214

completion
    of a normed space, 443
    of an inner product space, 443
constraint, 403
continuous
    at x_0, 17
    on X, 17
convergence
    of sequences of distributions, 35
    of series of distributions, 40
convex, 396
convex minimization, 382
convolution
    - product of functions, 83
    equation, 99
    in $S(\mathbb{R}^n)$, 85
    of distributions, 92
core
    of a linear operator, 257
    of a quadratic form, 266

delta function
    Dirac's, 3
delta sequences, 36
density matrix
    or statistical operator, 308
derivative, 388
    of a distribution, 48
    weak or distributional, 64
Dido, 374
Dido's problem, 409
differentiable
    twice, 390
direct method
    of the calculus of variations, 376
direct orthogonal sum, 201
direct sum
    of Hilbert spaces, 228
Dirichlet
    boundary conditions, 418
    Laplacian, 419
Dirichlet boundary condition, 376
Dirichlet form, 267
Dirichlet integral, 375
Dirichlet problem, 375, 417

distribution
    Dirac's delta, 33
    local order, 30
    order, 30
    periodic, 55
    regular, 5
    tempered, 42
    with support in x_0, 57
distributions, 29
    of compact support, 29, 43
    regular, 31, 32
    tempered, 29
divergence type, 423
domain
    of a linear operator, 248
dual
    algebraic, 9
    topological, 18, 27

eigenvalue problem
    nonlinear, 423
elementary solution
    advanced resp. retarded, 112
equi-continuous, 449
Euclidean spaces, 193
Euler, 373
Euler–Lagrange
    necessary condition of, 394
Euler–Lagrange equation, 376, 401
existence of a minimum, 380
extension
    of a linear operator, 249
    of a quadratic form, 266

Fock space
    Boson, 234
    Fermion, 234
form domain, 266
form norm, 266
form sum, 273
Fourier
    hyperfunction, 164
    transform, 129
    transformation, 129
        inverse, 133

on $\mathcal{S}(\mathbb{R}^n)$, 136
on $\mathcal{S}'(\mathbb{R}^n)$, 137
on $L^2(\mathbb{R}^n)$, 142
Fourier expansion, 214
Fréchet differentiable, 388
Fredholm alternative, 330
Friedrich's extension, 272
Fubini's theorem
    for distributions, 79
function
    Heaviside, 49
    strongly decreasing, 20
functional, 375
    analytic, 163
functional calculus
    of self-adjoint operators, 356
fundamental theorem of algebra, 121
fundamental theorem of calculus, 392

Gâteaux derivative, 398
Gâteaux differential, 397
generalized functions, 5
    Gelfand type $\mathcal{S}_a^b$, 162
Gram determinant, 204, 205
Gram–Schmidt orthonormalization, 213
Green's function, 106

Hadamard's principal value, 34
Hamilton operator
    free, 261
heat equation, 110, 147
Heisenberg's uncertainty relation, 311
Helmholtz differential operator, 146
Hermite
    functions, 221
    polynomials, 221
Hilbert cube, 304
Hilbert space, 191
    separable, 211, 212
Hilbert space basis, 212
Hilbert sum
    or direct sum of Hilbert spaces, 228
Hilbert transform, 96

Hilbert–Schmidt operator, 308
hlctvs
    metrizable, 445
holomorphic, 117
hyperfunctions, 164
hypo-elliptic, 107
hypo-ellipticity
    of $\bar{\partial}$, 116

inequality
    Bessel's, 188
    Cauchy–Schwarz–Bunjakowski, 189
    Schwarz', 188
    triangle, 189
infinitesimal generator, 299
inner product space, 187
integral
    with respect to a spectral family, 344
integral equation, 179
integral of functions, 391
integral operators
    of Hilbert–Schmidt, 278
isometry, 297
isoperimetric problem, 374

Kato perturbation, 314
    of free Hamiltonian, 316
Knotensatz, 220

Lagrange multiplier, 403
Lagrange multipliers
    existence of, 407
Laguerre
    functions, 222
    polynomials, 221
Laplace operator
    fundamental solution, 108
Laplace operator, 145
Laplace transform
    of distributions, 158
    of functions, 158
Laurent expansion, 123
Lebesgue space, 195

Legendre
    polynomials, 222
Leibniz, 373
Leibniz formula, 53
lemma
    of Wielandt, 281
    of Riemann–Lebesgue, 129
    of Riesz, 329
    of Weyl, 105
length, 187
level surface, 404
linear hull
    or span, 200
linear operator
    bounded, 250, 275
    closable, 255
    closed, 253
    closure of, 255
    core of, 257
    essentially self-adjoint, 256, 258
    from $\mathcal{H}$ into $\mathcal{K}$, 248
    of multiplication, 259
    positive, 288
    product or composition, 250
    self-adjoint, 256, 258
    symmetric, 256
    unbounded, 250, 276
local extrema
    necessary and sufficient conditions for, 394
locally integrable, 15
lower semi-continuous, 379

matrix spaces, 193
maximal, 214
maximal self-adjoint part
    of A, 351
Maxwell's equations
    in vacuum, 112
metric, 18
metric space, 441
    completion, 442
metrizable
    HLCTVS, 18
minimax principle, 415

minimizer
    existence, 380
    uniqueness, 380
minimizing point, 396
minimizing sequence, 376
minimum (maximum)
    global, 394
    local, 393
modulus
    of an operator, 289
momentum operator, 260
monotone, 396
muliplicator space, 53
multiplication operator
    bounded, 277
    unbounded, 278

Neumann series, 287
Newton, 373
norm, 8
    Hilbertian, 192
    induced by an inner product, 189
    of a bounded linear operator, 275
norm topology, 190

ONB
    characterization of, 213
open ball, 441
operator
    compact, 301
    completely continuous, 301
    inverse, 248
    of finite rank, 302
    trace, 305
    trace class, 304
orthogonal, 187
    complement $M^\perp$, 199
orthonormal, 187
    basis, 212
    polynomials, 219
    system, 187

parallelogram law, 192
Parseval relation, 214
partial differential operator

linear constant coefficients, 50
linear elliptic, 420
Pauli matrices, 234
Poisson equation, 109
polar decomposition, 289
polarization identity, 192
pole
    of finite order, 124
pre-Hilbert space, 187
product
    inner or scalar, 186
    of distributions and functions, 52
    rule, 52
projection theorem
    for closed subspaces, 201
    for convex sets, 413
projector
    or projection operator, 293
pseudo function, 35
Pythagoras
    theorem of, 188

quadratic form, 265
    closable, 266
    closed, 266
    continuous, 266
    densely defined, 265
    first representation theorem, 269
    lower bound, 266
    minimization, 383
    positive, 266
    second representation theorem, 271
    semi-bounded
        from below, 265
    symmetric, 265
quadratic functional, 383

Radon measure, 67
regular (critical) point, 394
regular points
    of an operator A, 318
regularization
    of distributions, 87

regularizing sequence, 88
renormalization, 58
residue, 124
resolution of the identity, 295, 341
resolvent, 318
    identity, 319
    set, 318
restriction
    of a linear operator, 249
Riemann integral, 311
Rodrigues' formula, 221

Schrödinger operator
    free, 148
Schwartz distributions, 159
self-adjoint
    geometric characterization, 339
semi-metric, 18
semi-norm, 8
    comparable, 11
    smaller than, 10
semi-norms
    system of, 11
        filtering, 11
    systems of
        equivalent, 12
sequence
    Cauchy, 15, 190
    converges, 15, 190
sequence space, 194
set
    open, 8
singularity
    essential, 124
    isolated, 124
    removable, 124
Sobolev
    constant, 417
    embeddings, 417, 424
    inequality, 417
    space, 417
Sobolev spaces, 196
Sokhotski–Plemelji formula, 37
solution
    classical or strong, 50, 105

470   Index

distributional or weak, 50, 105
fundamental, 102, 106
spectral family, 341
spectrum, 318
    continuous, 321
    discrete, 321
    essential, 362
    point, 321
square root lemma, 288
state
    bound, 365
    scattering, 365
Stone's formula, 370
Sturm–Liouville problem, 426
subset
    meager, 447
    nonmeager, 447
    nowhere dense, 447
    of first category, 447
    of second category, 447
subspace
    absolutely continuous, 359
    invariant, 351
    reducing, 351
    singular, 359
    singularly continuous, 359
support
    of a distribution, 41
    of a spectral family, 341
    of an analytic functional, 163
    of Fourier hyperfunctions, 166
    singular, 42
support condition, 91

tangent space
    existence of, 405
Taylor expansion with remainder, 392
tensor product
    for distributions, 79
    of functions, 72
    of Hilbert spaces, 231
        totally anti-symmetric, 233
        totally symmetric, 233
    of vector spaces, 230
    projective

    of E, F, 74
    of p, q, 74
test function space
    $\mathcal{D}(\Omega)$, 19
    $\mathcal{E}(\Omega)$, 21
    $\mathcal{S}(\Omega)$, 21
theorem
    Baire, version 1, 447
    Baire, version 2, 448
    Baire, version 3, 448
    Banach–Saks, 243
    Banach–Steinhaus, 241, 449
    Cauchy, 118
    closed graph, 453
    convolution, 141
    de Figueiredo–Karlovitz, 193
    extension of linear functionals, 207
    Fréchet–von Neumann–Jordan, 192
    Hörmander, 143
    Hellinger–Toeplitz, 277
    Hilbert–Schmidt, 329
    identity of holomorphic functions, 122
    inverse mapping, 453
    Kakutani, 193
    Kato–Rellich, 314
    Liouville, 121
    of F. Riesz, 236
    of residues, 125
    open mapping, 452
    Plancherel, 142
    Riesz–Fischer, 196
    Riesz–Fréchet, 206
    Riesz–Schauder, 328
    spectral, 347
    Stone, 298
    Weyl, 363
topological complement, 406
topological space, 8
    Hausdorff, 12, 13
topology, 7
    defined by semi-norms, 11
    of uniform convergence, 16

topology on $\mathfrak{B}(\mathcal{H})$
    norm or uniform, 284
    strong, 284
    weak, 284
total subset, 203
trace
    of trace class operators, 306
trace norm, 306

ultradifferentiable functions, 168
ultradifferential operator, 169
ultradistributions, 168
uniform boundedness principle, 241, 450
unitary operator, 297
unitary operators
    n-parameter group, 300
    one-parameter group, 298
upper semi-continuity, 379

variation
    nth, 400
vector space
    locally convex topological, 7
    topological, 7

wave operator, 111
weak
    Cauchy sequence, 238
    convergence, 238
    limit, 238
    topology, 237
weak topology
    $\mathcal{D}'(\Omega)$, 35
Weierstrass theorem
    Generalized I, 382
    Generalized II, 383
Weyl's criterium, 321
Wiener–Hopf operators, 280